Geometry from the Pacific Rim

Geometry from the Pacific Rim

Proceedings of the Pacific Rim Geometry Conference
held at National University of Singapore,
Republic of Singapore, December 12–17, 1994

Editors

A. J. Berrick
Bonaventure Loo
Hong-Yu Wang

Walter de Gruyter · Berlin · New York 1997

Editors

A. J. Berrick, B. Loo, H.-Y. Wang
Department of Mathematics
National University of Singapore
Kent Ridge 119260,
Republic of Singapore

1991 Mathematics Subject Classification:
53-06; 53Axx, 53Bxx, 53Cxx, 58Cxx, 58Dxx, 58Gxx

♾ Printed on acid-free paper which falls within the guidelines of the ANSI to ensure permanence and durability.

Library of Congress Cataloging-in-Publication-Data

Geometry from the Pacific Rim : proceedings of the Pacific Rim Geometry Conference held at National University of Singapore, Republic of Singapore, December 12–17, 1994 / editors, A. J. Berrick, Bonaventure Loo, Hong-Yu Wang.
p. cm.
ISBN 3-11-014792-0 (alk. paper)
1. Geometry–Congresses. 2. Topology–Congresses. I. Berrick, A. J., 1948– . II. Loo, Bonaventure, 1959– . III. Wang, Hong-Yu, 1957–
QA440.G467 1997
516–dc21 97-10529
CIP

Die Deutsche Bibliothek – Cataloging-in-Publication-Data

Geometry from the Pacific Rim : proceedings of the Pacific Rim Geometry Conference held at National University of Singapore, Republic of Singapore, December 12–17, 1994 / ed. A. J. Berrick ... – Berlin ; New York : de Gruyter, 1997
ISBN 3-11-014792-0

Printed in Germany. Typeset using the authors' TEX files: I. Zimmermann, Freiburg.
Printing: Arthur Collignon GmbH, Berlin. Binding: Lüderitz & Bauer, Berlin.
Cover design: Thomas Bonnie, Hamburg.

Preface

During the week of 12 to 17 December 1994, the National University of Singapore's Mathematics Department played host to the Pacific Rim Geometry Conference—Singapore, 1994. This event was prompted by the success of a similar meeting in Hong Kong just two years earlier; many of its participants had urged the setting up of an arithmetic progression of such gatherings.

Some 122 researchers from 15 different countries (including several remote from the shores of the Pacific) attended the conference. As indicated in the following pages, there were 11 invited, and a further 38 contributed, lectures on topics ranging from geometric analysis and differential geometry to algebraic geometry and topology.

We are pleased to present herein research and expository articles contributed by a significant proportion of the participants, including almost all of the invited speakers. All contributions have been refereed: referees' very high praise for a number of the papers is most gratifying.

As well as thanking the referees for their valuable efforts, we would also like to record our gratitude to those who were responsible for such a successful conference. Financial support came from the Department's Lee Kong Chian Centre for Mathematical Research, the Lee Foundation, the International Mathematical Union, the International Science Foundation and the Singapore Mathematical Society. Hard work was supplied by the other members of the Organizing Committee, namely Roger Chen, Cheung Leung Fu, Leonard M. C. Leung, Fred P. F. Leung, Brett McInnes, Peter Pang, Tan Ser Peow, To Wing Keung, Xu Xingwang and Zhang De-Qi. Valuable ideas and encouragement were provided by Peng Tsu Ann (Head of NUS Mathematics Department), Shing-Tung Yau and Peter Li. Finally, some splendid mathematics came from the conference participants. We feel that the following pages provide ample evidence of this.

Jon Berrick
Bonaventure Loo
Wang Hong-Yu

Table of Contents

Gauss–Manin connexions, logarithmic forms and hypergeometric functions

*A. G. Aleksandrov** *and S. Tanabé*†

Abstract. A computational approach in the study of Gauss–Manin connexions associated with simple isolated singularities of hypersurfaces or, more generally, of complete intersections is described. In the cases of A_μ and S_μ-singularities solutions of the corresponding systems of differential equations are presented in the form of generalized hypergeometric functions.

1991 Mathematics Subject Classification: Primary 32S30, 33C20, 14D05; Secondary 32S25, 14M10, 14F40

1. Introduction

In a general context the notion of local Gauss–Manin connexion appeared in works of P. Deligne and B. Malgrange. Probably one of the first who considered this notion in the framework of singularity theory was E. Brieskorn. In his famous work [9] it was proved that the connexion associated with the 1-parameter principal deformations of isolated hypersurface singularities can be presented by systems of ordinary differential equations with regular singularities. Thus, the classical Legendre equation can be considered as the Gauss–Manin connexion associated with the minimal versal deformation of simple hypersurface singularity A_2. In particular, it follows from his results that after matrix transformations with meromorphic entries such equations reduce to the ones having poles of the first order.

The next step was made by K. Saito. Calculating the connexion associated with the total 3-parameter miniversal deformation of an A_3-singularity he came to the conclusion [21] that a very convenient representation of the corresponding system may be obtained if one presents the coefficients of this system as logarithmic differential forms.

*Partially supported by Grant N3B000 from the International Science Foundation ISF and Russian Foundation of Fundamental Investigations RFFI, and by Grant 4373 from the International Association INTAS.

†Partially supported by the Japan Society for the Promotion of Sciences JSPS.

In another manner the notion of Gauss–Manin connexion was investigated by S. Ishiura and M. Noumi [15]. They described the Gauss–Manin systems for the A_μ-singularities by means of a certain Hamiltonian representation. M. Noumi also treated some particular cases of linear deformations of the Pham singularities [18]. For these cases he gave a concrete representation of the solutions to the Gauss–Manin system in terms of known types of generalized hypergeometric functions.

Gauss–Manin systems arose naturally also in works of K. Aomoto [5], [7], I. M. Gelfand [11] and their followers. They studied the integrals of type

$$J(t) = \int F_0^{\lambda_0}(z,t) F_1^{\lambda_1}(z,t) \ldots F_m^{\lambda_m}(z,t)\, dz$$

where $F_j(z,t)$, $1 \leq j \leq m$, are linear functions with respect to the variables $z = (z_1, \ldots, z_n)$ and $F_0(z,t)$ is a linear or at most a quadratic function. In this work, we consider situations where $m = 1$ and both $F_0(z,t)$ and $F_1(z,t)$ are quadratics or higher order polynomials. In fact, this point is an essential difference from earlier investigations devoted to the study of various representations of Gauss–Manin connexions for nonisolated singularities given by special arrangements of hyperplanes.

We will discuss here a computational approach in the study of Gauss–Manin connexions associated with simple isolated singularities of hypersurfaces or, more generally, of complete intersections. In the cases of A_μ and S_μ-singularities solutions of the corresponding systems of differential equations are presented in the form of generalized hypergeometric functions.

Acknowledgement. The first author takes the opportunity to thank the organizers of the conference, especially the Chairman Professor Jon Berrick, for their kind invitation and warm hospitality during the meeting. Both the authors are also deeply indebted to the referee for valuable advice and remarks and to the editors of the proceedings for their encouragement and help during the preparation of an electronic version of this manuscript.

2. Local Gauss–Manin connexion associated with an isolated singularity

Following B. Malgrange's approach, E. Brieskorn [9] calculated Gauss–Manin connexions associated with isolated hypersurface singularities. The main idea was used in the case of isolated complete intersection singularities by G.-M. Greuel (see [13]).

Let us regard an analytic mapping $f : X \to S$ of complex space germs, where S is smooth and X has isolated complete intersection singularities only. We will also suppose that f is *flat* with smooth fibres outside a certain hypersurface D in S called the *discriminant* set or, equivalently, the set of critical values of the mapping f. The coherent sheaf $\mathcal{H}^p = H^p(f_*\Omega^\bullet_{X/S}, d)$ is defined as the p-th cohomology group of the relative de Rham complex. This sheaf restricts on $S \setminus D$ to the sheaf associated to the bundle of cohomology groups $H^p(X_t, \mathbb{C})$, $t \in S \setminus D$ of the fibres X_t. The transference of the cohomology classes H^p along the tangent directions on the complement $S \setminus D$ induces a connexion

$$\nabla_{X/S} : \mathcal{H}^p(f_*\Omega^\bullet_{X/S}) \longrightarrow \mathcal{H}^p(f_*\Omega^\bullet_{X/S}) \otimes \Omega^1_S(D) \tag{1}$$

given by the rule

$$\nabla_{X/S}[\omega] = \sum_{i=1}^{k} h[\alpha_i] \otimes dt_i / h$$

where h is the defining function of the discriminant set D and $\Omega^1_S(D)$ denotes the $\mathcal{O}_S$-module of meromorphic 1-forms with polar singularities along D. Here the symbol $[\alpha]$ means the corresponding relative de Rham cohomology class of α in $f_*\Omega^p_{X/S}/d(f_*\Omega^{p-1}_{X/S})$. There is the decomposition

$$d\omega = \sum_{i=1}^{k} df_i \wedge \alpha_i, \quad \alpha_i \in f_*\Omega^p_X.$$

The formula (1) implies immediately that the Gauss–Manin connexion has poles of the first order in the case where h has no multiple factors.

Definition. It is said that a meromorphic differential 1-form $\omega \in \Omega^1_S(D)$ has *logarithmic* poles along D if ω and its differential $d\omega$ have poles at worst of the first order along D. In such a case the form ω is called the *logarithmic* differential form. Denote by $\Omega^1_S(\log D)$ the $\mathcal{O}_S$-module of *logarithmic* differential forms, that is, the module of meromorphic forms with at worst logarithmic poles along D.

In general (for example, in the case of a divisor D with *non-normal* crossings), the module of logarithmic differential forms $\Omega^1_S(\log D)$ does not coincide with the set of meromorphic differential forms having poles of the first order along D (cf. Theorem 4.6 below).

Of course one can define analogously the module of logarithmic differential p-forms denoted by $\Omega^p_S(\log D)$. Evidently, the usual differentiation d acts on these modules. Thus, we get the *logarithmic de Rham* complex $(\Omega^\bullet_S(\log D), d)$. It should be remarked that the logarithmic de Rham complex possesses various remarkable algebraic and analytic properties (see [23], [1], [2], [3]).

Let us return to the formula (1). In fact, Saito's considerations [21], [22] imply that the connexion can also be rewritten in the following way:

$$\nabla_{X/S} : \mathcal{H}^p(f_*\Omega^\bullet_{X/S}) \to \mathcal{H}^p(f_*\Omega^\bullet_{X/S}) \otimes \Omega^1_S(\log D), \tag{1'}$$

if one defines the connexion on a slightly larger module. It should be underlined that the difference between (1) and (1′) as well as the difference between the tensor factors $\Omega^1_S(D)$ and $\Omega^1_S(\log D)$, respectively, plays an important role in applications (see §5).

3. Gauss–Manin systems associated with A_μ-singularities

In this section, we shall explain how to obtain important properties on A_μ-type Gauss–Manin systems and their solutions with the aid of a result from [17], [20].

3.1. Fundamental non-trivial differential systems. Let us consider period integrals for the minimal versal deformation of the singularity A_μ, given by the equation $z^{\mu+1} + s_{\mu-1}z^{\mu-1} + \cdots + s_1 z + s_0 = 0$:

$$K_i^\lambda(s) = \int z^i (z^{\mu-1} + s_{\mu-1}z^{\mu-1} + \cdots + s_1 z + s_0)^\lambda \, dz, \quad i = 0, \dots, \mu+1. \tag{2}$$

Proposition 3.1 ([17], [20]). *The period integrals $K_0^\lambda(s), \dots, K_{\mu+1}^\lambda(s)$ satisfy the following overdetermined system of differential equations:*

$$\sum_{\ell=0}^{\mu-1} s_\ell \frac{\partial}{\partial s_0} K_{\ell+i}^\lambda + \frac{\partial}{\partial s_0} K_{\mu+1+i}^\lambda = \lambda K_i^\lambda, \quad 0 \le i \le \mu - 1, \tag*{$(3)_i$}$$

$$\sum_{\ell=1}^{\mu-1} \ell s_\ell \frac{\partial}{\partial s_0} K_{\ell+j}^\lambda + (\mu+1)\frac{\partial}{\partial s_0} K_{\mu+1+j}^\lambda = -(j+1)K_j^\lambda, \quad -1 \le j \le \mu - 1. \tag*{$(4)_j$}$$

The proof is based essentially on syzygy relations between basis elements of the cohomology group $H^1(X_s, \mathbb{C})$.

It is not difficult to transform the above system to the following form:

Proposition 3.2. *The integrals $K_0(s), \dots, K_{\mu-1}(s)$, $s = (s_0, \dots, s_{\mu-1})$, satisfy the following system of differential equations:*

$$S\frac{\partial}{\partial s_0} \boldsymbol{K}(s) = (L + V(s_2, \dots, s_{\mu-1}))\boldsymbol{K}(s), \tag{5}$$

where $\boldsymbol{K}(s) = (K_0(s), \dots, K_{\mu-1}(s))$, and $V(s_2, \dots, s_{\mu-1})$ can be presented in the following form:

$$\frac{1}{(\mu+1)^2}\begin{bmatrix} 0 & 0 & 0 & 0 & \cdots & 0 & 0 \\ 0 & 0 & 0 & 0 & \cdots & 0 & 0 \\ 2s_{\mu-1} & 0 & 0 & 0 & \cdots & 0 & 0 \\ 3s_{\mu-2} & 2\cdot 2s_{\mu-1} & 0 & 0 & \cdots & \vdots & \vdots \\ 4s_{\mu-3} & 3\cdot 2s_{\mu-2} & 2\cdot 3s_{\mu-1} & 0 & \cdots & \vdots & \vdots \\ \vdots & \vdots & \vdots & \ddots & \cdots & \vdots & \vdots \\ (\mu-1)s_2 & 2(\mu-2)s_3 & 3(\mu-3)s_4 & \cdots & (\mu-2)2s_{\mu-1} & 0 & 0 \end{bmatrix}.$$

The entries $v_{i,j}$ of the matrix $V(s_2, \dots, s_{\mu-1})$ are determined by the following rule:

$$(j+1)v_{i,j} = j v_{i+1,j+1}, \quad v_{1,j} = (j-1)s_{\mu-j+2}, \quad 1 \le i \le \mu-2, \quad 3 \le j \le \mu.$$

The matrix S admits the following representation

$$S = s_0 \cdot \mathrm{id}_\mu + C(s')$$

with a polynomial matrix $C(s')$ *and*

$$L = \mathrm{diag}\Big(\lambda + \frac{1}{\mu+1}, \ldots, \lambda + \frac{\mu}{\mu+1}\Big).$$

Remark 1. It should be remarked that the formula (5) reminds us of the expression (2.1) from [15]. It was obtained by using the flat coordinate systems for A_μ-singularities.

Corollary 3.3. *The solutions to the system* (5) *can be presented in the following expansion form:*

$$K(s_0, s') = \Big(\mathrm{id}_\mu + \sum_{k\geq 0} \frac{1}{k!} \prod_{j=0}^{k-1} D(s')\big(L - j \cdot \mathrm{id}_\mu + V(s')\big)\Big(\frac{s_0}{\Delta_\mu}\Big)^k\Big) K(0, s'), \quad (6)$$

where $\Delta_\mu(s') = \det |S(0, s')|$, *and* $D(s') = \Delta_\mu \cdot S^{-1}(0, s')$ *for* $s' = (s_1, \ldots, s_{\mu-1})$.

Proof. First of all, observe that

$$\frac{\partial}{\partial s_0} S = \mathrm{id}_\mu. \quad (7)$$

Thus,

$$S(\frac{\partial}{\partial s_0})^2 K(s) + \frac{\partial}{\partial s_0} K(s) = (L + V(s'))\frac{\partial}{\partial s_0} K(s),$$

$$\begin{aligned}\Big(\frac{\partial}{\partial s_0}\Big)^2 K(s) &= S^{-1}(L + V(s') - \mathrm{id}_\mu)\frac{\partial}{\partial s_0} K(s)\\ &= S^{-1}(L + V(s') - \mathrm{id}_\mu) S^{-1}(L + V(s')) K(s).\end{aligned}$$

In a similar way,

$$\Big(\frac{\partial}{\partial s_0}\Big)^j K(s) = \prod_{j=0}^{k-1} \big(S^{-1}(L + V(s') - j\,\mathrm{id}_\mu)\big) K(s).$$

It remains to apply Taylor's expansion theorem to $K(s)$ at the point that lies outside the discriminant set and to make an analytic continuation. □

Remark 2. In the reference [15] the main motivation for introducing the flat coordinate systems $(t_1, t') = (t_1(s), \ldots, t_\mu(s))$ is the possibility of writing down the system in a quite convenient manner:

$$\Big(t_1 \cdot \mathrm{id}_\mu - A_1(t_2, \ldots, t_\mu)\frac{\partial}{\partial t_1} K\Big) = L\,K,$$

where $A_1(t')$ is a matrix with polynomial entries and L denotes the same matrix as in Proposition 3.2. In other words, one can avoid the usage of the matrix $V(s')$ in the description of the system. The absence of this term yields several advantages. For example, the corresponding expression (6) is simplified, while $D(s')$ commutes with $(L - j \cdot \mathrm{id}_\mu)$.

3.2. Adjacencies. Let us discuss useful relations between properties of the adjacencies of A_μ-singularities and solutions of the corresponding Gauss–Manin systems. Corollary 3.3 entails several properties of $\boldsymbol{K}(s)$. First, we see hierarchy relations among period integrals that arise from the adjacency relation of A_μ-singularities. Set

$$K^j_{A_\mu}(s) = \int_{\gamma(s)} z^j (z^{\mu+1} + s_{\mu-1} z^{\mu-1} + \cdots + s_1 z + s_0)^\lambda \, dz,$$

where for simplicity we will omit the index λ in the notation $K^j_{A_\mu}(s)$. Applying the derivatives of fractional order, we get

$$K^j_{A_{\mu+1}}(0, s') = \left(\frac{\partial}{\partial s_2}\right)^\lambda \left(\frac{\partial}{\partial s_1}\right)^{-\lambda} K^j_{A_\mu}(s'), \quad 0 \le j \le \mu,$$

where $s' = (s_1, \ldots, s_\mu)$. Together with the formula (6), we see that

$$\begin{aligned} \boldsymbol{K}_{A_{\mu+1}}(s_0, s')^t &= \left(\mathrm{id}_{\mu+1} + \sum_{k\ge 0} \frac{1}{k!} \prod_{j=0}^{k-1} D(s')(L - j \cdot \mathrm{id}_{\mu+1} + V(s'))\left(\frac{s_0}{\Delta_\mu}\right)^k\right) \\ &\quad \times \left(\frac{\partial}{\partial s_2}\right)^\lambda \left(\frac{\partial}{\partial s_1}\right)^{-\lambda} \boldsymbol{K}_{A_\mu}(s'), \end{aligned}$$

where

$$\boldsymbol{K}_{A_\mu}(s') = \left(K^0_{A_\mu}(s'), \ldots, K^\mu_{A_\mu}(s')\right) \text{ and } \boldsymbol{K}_{A_{\mu+1}}(s) = \left(K^0_{A_{\mu+1}}(s), \ldots, K^\mu_{A_{\mu+1}}(s)\right).$$

Thus, the period integrals associated with an $A_{\mu+1}$-singularity can be expressed by means of those associated with an A_μ-singularity. This situation is nothing but a reflection of a phenomenon that an A_μ-singularity is adjacent to an $A_{\mu+1}$-singularity.

$$A_1 \mid_{s_{\mu-1}=0} \leftarrow A_2 \mid_{s_{\mu-2}=0} \leftarrow \cdots \leftarrow A_{\mu-1} \mid_{s_1=0} \leftarrow A_\mu \mid_{s_0=0} .$$

One can illustrate the adjacency phenomenon characterizing the solutions to the Gauss–Manin system from the following viewpoint. First recall Noumi's formula (see [19] Theorem 5 or, equivalently, [18] Theorem 3.10):

$$\begin{aligned} K^k_{A_\mu}(s_0, \ldots, s_{\mu-1}) &= \delta_{\mu+1-k,0} s_0^{-\mu/(\mu+1)} \\ &\quad + \sum_{r=1}^{\infty} (-1)^r \left(\frac{\mu-k}{\mu+1} + r - 1\right) \cdot S'_{r(\mu+1)-k} s_0^{-\frac{\mu-k}{\mu+1} - r} \end{aligned} \tag{8}$$

where $k = 0, \ldots, \mu - 1$, $S_i'(s_1, \ldots, s_{\mu-1}) = S_i(0, s_1, \ldots, s_{\mu-1})$, $i = 2, 3, \ldots$, and $S_i(s)$ will be described below. In fact, the polynomials S_i' are presumably characterized by an $A_{\mu-1}$-singularity. In this respect, we observe a similitude between the expression (6) and Noumi's formula (8). On the other hand, the formula (8), however, differs essentially from (6). Really, the former is expanded at $s_0 = +\infty$ while the latter presents an expansion around $s_0/\Delta_\mu(s') = 0$.

Since the reference [19] is less accessible for non-Japanese readers, we explain how to obtain the polynomials $S_i(s)$, where $s = (s_0, \ldots, s_{\mu-1})$. Let us consider the minimal versal deformation of an A_μ-singularity

$$F(z, s) = z^{\mu+1} + s_{\mu-1} z^{\mu-1} + \cdots + s_1 z + s_0.$$

Its fractional power admits a formal expansion in Laurent series which is denoted by $f(z)$:

$$f(z) = F(z, s)^{\frac{1}{\mu+1}} = z + \tilde{t}_2(s) z^{-1} + \tilde{t}_3(s) z^{-2} + \cdots .$$

For such a series $f(z)$ one can find another formal series $g(y)$,

$$g(y) = y - \tilde{s}_2(s) y^{-1} - \tilde{s}_3(s) y^{-2} + \cdots ,$$

so that $g(f(z)) = z$. It turns out that the two sequences $\tilde{t} = (\tilde{t}_2, \tilde{t}_3, \ldots)$ and $\tilde{s} = (\tilde{s}_2, \tilde{s}_3, \ldots)$ are related between themselves by the formulae:

$$\begin{aligned} \tilde{s}_k &= \tfrac{1}{k-1} \operatorname{Res}(f(z)^k dz), \\ \tilde{t}_k &= \tfrac{1}{k-1} \operatorname{Res}(g(y)^k dy). \end{aligned}$$

Evidently, $\tilde{s}_n = 0$. The polynomial $S_i(s)$ is defined by

$$S_i(s) = (\mu + 1)\tilde{s}_k(s) \quad \text{for } i = 2, 3, \ldots .$$

One can find further details in the paper [15].

3.3. Period integrals as products of generalized hypergeometric functions. Now we shall describe how the function in two variables

$$K^j_{A_\mu}(s_0, 0, \ldots, 0, s_i, 0, \ldots, 0)$$

can be expressed by generalized hypergeometric functions in Pochhammer's sense. For this examine in detail the case of an A_3-singularity. First one observes

$$\begin{aligned} D(0, s_1, 0) &= \Delta_3(0, s_1, 0) \cdot S^{-1} |_{s_0 = s_1 = 0} \\ &= \begin{bmatrix} 0 & 0 & 9s_1^2 \\ -9s_1^3/4 & 0 & 0 \\ 0 & -9s_1^3/4 & 0 \end{bmatrix}. \end{aligned}$$

Hence, we have that $D(0,0,s_2)(L+V(s')-j\cdot \mathrm{id}_3)$ is equal to the matrix

$$\begin{bmatrix} 0 & 0 & 9(\lambda+3/4-j)s_1^2 \\ -9/4(\lambda+1/4-j)s_1^3 & 0 & 0 \\ 0 & -9/4(\lambda+1/2-j)s_1^3 & 0 \end{bmatrix}.$$

Consequently, $K^0_{A_3}(s_0,s_1,0)$ is equal to the following expression:

$$2^{12}s_0^2\cdot {}_3F_2\Big(\tfrac{17}{12}-\tfrac{\lambda}{3},\tfrac{13}{6}-\tfrac{\lambda}{3},\tfrac{23}{12}-\tfrac{\lambda}{3};\tfrac{5}{3},\tfrac{4}{3}\mid \tfrac{2^{20}s_0^3}{3^3s_1^4}\Big)\cdot K^0_{A_3}(0,s_1,0)$$

$$-(\tfrac{5}{4}-\lambda)(\tfrac{1}{2}-\lambda)\tfrac{9s_1^5}{4}\cdot {}_3F_2\Big(\tfrac{17}{12}-\tfrac{\lambda}{3},\tfrac{7}{6}-\tfrac{\lambda}{3},\tfrac{11}{12}-\tfrac{\lambda}{3};\tfrac{1}{3},\tfrac{2}{3}\mid \tfrac{2^{20}s_0^3}{3^3s_1^4}\Big)\cdot K^1_{A_3}(0,s_1,0)$$

$$-2^8(\lambda+\tfrac{3}{4})\tfrac{s_0}{s_1^2}\cdot {}_3F_2\Big(\tfrac{1}{2}-\tfrac{\lambda}{3},\tfrac{1}{4}-\tfrac{\lambda}{3},\tfrac{3}{4}-\tfrac{\lambda}{3};\tfrac{4}{3},\tfrac{2}{3}\mid \tfrac{2^{20}s_0^3}{3^3s_1^4}\Big)\cdot K^2_{A_3}(0,s_1,0),$$

where ${}_3F_2(\alpha_1,\alpha_2,\alpha_3;\beta_1,\beta_2\mid z)$ denotes the generalized hypergeometric function of Pochhammer type:

$${}_3F_2(\alpha_1,\alpha_2,\alpha_3;\beta_1,\beta_2\mid z)=\sum_{j\geq 0}\frac{(\alpha_1;j)(\alpha_2;j)(\alpha_3;j)}{(\beta_1;j)(\beta_2;j)j!}z^j.$$

It is a well known fact (see Introduction), that the period integral of the following form can be expressed by means of Legendre's function

$$K^j_{A_3}(0,s_1,0)=\int_\gamma z^j(z^4+s_1z)^\lambda\,dz=\int_\gamma z^{\lambda+j}(z^3+s_1)^\lambda\,dz$$

where γ stands for a double loop encircling two points among the set

$$\{z: z^3+s_1=0\}.$$

3.4. Universal deformation for A_3-singularities. In concluding this section, we briefly describe how to obtain results similar to those of K. Saito's [21] concerning uniformization equations for the case of an A_3-singularity with the aid of Proposition 3.1.

Proposition 3.4. *For the period integrals $\boldsymbol{K}_{A_3}(s)=(K_0^\lambda(s),K_1^\lambda(s),K_2^\lambda(s))$, $s=(s_0,s_1,s_2)$, associated with an A_3-singularity, we have a Picard–Fuchs system of the following type:*

$$d\,\boldsymbol{K}_{A_3}(s)=\Omega_3\cdot(L+V(s_2))\cdot \boldsymbol{K}_{A_3}(s), \tag{9}$$

where

$$L=\mathrm{diag}(\lambda+1/4,\lambda+1/2,\lambda+3/4),$$

$$V(s_2)=\begin{bmatrix} 0 & 0 & 0 \\ 0 & 0 & 0 \\ \frac{s_2}{8} & 0 & 0 \end{bmatrix},$$

$$\Omega_3 = \frac{1}{\Delta_3}\begin{bmatrix} \frac{1}{3}(d\Delta_3 - s_2\eta_3) & \frac{-1}{4}\eta_2 & -\eta_3 \\ \frac{s_1}{4}\eta_3 & \frac{1}{3}(d\Delta_3 + s_2\eta_3/2) & \frac{-1}{4}\eta_2 \\ \frac{s_1}{16}\eta_2 & \frac{s_1}{4}\eta_3 & \frac{1}{3}(d\Delta_3 + s_2\eta_3/2) \end{bmatrix},$$

$$\Delta_3 = 16s_0^3 - \frac{27}{16}s_1^4 + 9s_0s_1^2s_2 - 8s_0^2s_2^2 - \frac{1}{4}s_1^2s_2^3 + s_0s_2^4,$$

$$\eta_2 = (4s_1s_2^2 + 48s_0s_1)ds_0 + (16s_0s_2^2 - 64s_0^2 - 6s_1^2s_2)ds_1 + (9s_1^3 - 32s_0s_1s_2)ds_2,$$

$$\eta_3 = -(9s_1^2 - 8s_0s_2 + 2s_2^3)ds_0 + s_1(12s_0 + s_2^2)ds_1 - (16s_0^2 + \frac{3}{2}s_1^2s_2 - 4s_0s_2^2)ds_2.$$

One sees that it is possible to make a transition from our coefficient matrix to Saito's coefficient matrix Ω by means of a linear transform induced by an invertible matrix $\Gamma \in GL(3, \mathbb{C})$:

$$\Omega = d\Gamma \cdot \Gamma^{-1} + \Gamma \cdot \Omega_3 \cdot (L + V(s_2)) \cdot \Gamma^{-1}.$$

We also remark that our system (9) has been obtained as a result of the differentiation of a set of the concrete period integrals (and of the usage of some basic properties of logarithmic forms from the paper [1]), while Saito computed the coefficients of the matrix Ω in view of the standard integrability condition $d\Omega = \Omega \wedge \Omega$.

A simple comparison between the formula (9) and K. Saito's result yields that the weight λ in the former and the parameter k in the latter are related in the following way:

$$\lambda = \frac{1}{2}(k - 1).$$

We see that the solutions to the "uniformization equation" of type I from the paper [21], §3 for every parameter value k can be obtained by the aid of our formula (6) in making use of period integrals of type A_2 (that is, the Legendre function). It should be noted that K. Saito resolved the uniformization equation only in the case when $k = 0$, (or equivalently for $\lambda = -1/2$ in our notation). In addition it follows from our calculations that the system of type II (loc. cit.), whose solutions "do not have an Euler integral representation", arises in an essentially different context than the Gauss–Manin systems associated to the A_3-singularity.

4. Gauss–Manin systems for complete intersection S_μ-singularities

Now we give an outline of the theory of Gauss–Manin systems associated with isolated complete intersection singularities (see [4] for further details). The isolated complete intersection singularity X_0 of type S_μ, $\mu \geq 5$, from the list of M. Giusti [14], is determined by the following pair of equations defined on the germ $(\mathbb{C}^3, 0)$:

$$\begin{cases} f_1(x, y, z) = x^2 + y^2 + z^\nu = 0, \\ f_2(x, y, z) = \quad yz \quad = 0, \end{cases}$$

where we set $\nu = \mu - 3$ for the sake of convenience. Let us consider the system of equations

$$\begin{cases} F_1(x, y, z, s) = f_1(x, y, z) + s_\nu z^{\nu-1} + \cdots + s_2 z + s_1 y - t_1 = 0, \\ F_2(x, y, z, s) = \qquad f_2 + s_{\nu+1} x - t_2 \qquad = 0, \end{cases}$$

that defines the total space $X \subset \mathbb{C}^3 \times S$ of the minimal versal deformation $\varphi : X \to S$ of the germ X_0. Here S is the μ-dimensional smooth germ with coordinates (s, t), $s = (s_1, s_2, \ldots, s_{\nu+1})$, $t = (t_1, t_2)$. Denote by $X_{(s,t)}$ the fiber of the miniversal deformation φ over the point $(s, t) \in S$.

We will consider integrals of the following type:

$$I_j(s, t) = \int_{\gamma(s,t)} q_j(x, y, z)\, dx \wedge dy \wedge dz / dF_1 \wedge dF_2,$$

where the integration is taken along some regularization of a real vanishing cycle $\gamma(s, t) \in H_1(X_{(s,t)}, \mathbb{C})$ (see [11]) and $q_j(x, y, z)$ are polynomials. It is well known that there is an isomorphism:

$$H^1(X_{(s,t)}, \mathbb{C}) \cong \Omega^3_{\mathbb{C}^3} / (f_1 \Omega^3_{\mathbb{C}^3} + f_2 \Omega^3_{\mathbb{C}^3} + df_1 \wedge df_2 \wedge \Omega^1_{\mathbb{C}^3}).$$

It can be realized by multiplying the 2-form $df_1 \wedge df_2$ by the elements of the cohomology group on the left side. It is not difficult to see that in the case of an $S_{\nu+3}$-singularity the above quotient space is generated by the following monomial differential forms:

$$\{1, z, z^2, \ldots, z^\nu, y, x\}\, dx \wedge dy \wedge dz.$$

Our aim is to compute the Gauss–Manin system (1)

$$\nabla_{X/T} : \mathcal{H}^1(f_* \Omega^\bullet_{X/T}) \longrightarrow \mathcal{H}^1(f_* \Omega^\bullet_{X/T}) \otimes \Omega^1_T(D)$$

associated with the so-called *principal* deformation of the singularity X_0, that is, the deformation over the (t_1, t_2)-parameter subspace $T \subset S$ in the base space of the miniversal deformation of the singularity X_0.

In effect, Leray's residue theorem yields the identity

$$\frac{d}{dt_i} \int_{\gamma(s,t)} \omega = \int_{\gamma(s,t)} \frac{d\omega}{df_i}, \quad i = 1, 2,$$

which has been proved in [9]. This implies that one can calculate the Gauss–Manin system as the relation between the integrals instead of that between cohomology classes.

Thus, the corresponding system of differential equations describes nontrivial relations between the integrals

$$I_j(s,t) = \int_{\gamma(s,t)} q_j(x,y,z)\, dx \wedge dy \wedge dz / df_1 \wedge df_2,$$

and their derivatives. Here the functions $q_j(x,y,z)$, $j = 1, \dots, \mu$, are the monomials $1, z, z^2, \dots, z^\nu, y, x$ (cf. [13], 2.3), that is,

$$\frac{z^j\, dx \wedge dy \wedge dz}{df_1 \wedge df_2} = \frac{z^{j-1}\, dz}{2x}, \quad 0 \le j \le \nu,$$
$$\frac{y\, dx \wedge dy \wedge dz}{df_1 \wedge df_2} = \frac{t_2\, dz}{z^2 x},$$
$$\frac{x\, dx \wedge dy \wedge dz}{df_1 \wedge df_2} = \frac{-dy}{2y}.$$

Evidently the last form can be easily integrated. So *nontrivial* integrals for which the differential equation shall be calculated are given by the following set of differential forms, consisting of $\mu - 1$ elements:

$$\{z^{-2}\, dz/x, z^{-1}\, dz/x, dz/x, \dots, z^{\nu-1}\, dz/x\}.$$

Suppose that $s_{\nu+1} = 0$. Then the integrals $I_j(s,t)$ can be expressed in the following manner:

$$\begin{aligned} I_j(s,t) &= \int_{\gamma(s,t)} \frac{z^{j-1}\, dz}{x} \\ &= \int \frac{z^{j-1}\, dz}{((t_2/z)^2 + z^\nu + s_\nu z^{\nu-1} + \dots + t_1 + s_1(t_2/z))^{1/2}} \\ &= \int \frac{z^j\, dz}{(z^{\nu+2} + s_\nu z^{\nu+1} + \dots + t_1 z^2 + s_1 t_2 z + t_2{}^2)^{1/2}} \end{aligned}$$

where $(x,y,z) \in \gamma(s,t)$, $-1 \le j \le \nu$. Denote

$$J_{j+2}(t) = -2\frac{\partial}{\partial t_1} I_j(0,t) = \int \frac{z^{j+2}\, dz}{(z^{\nu+2} + t_1 z^2 + t_2{}^2)^{3/2}}.$$

Now the system of equations which satisfy the integrals $J_1(t), \dots, J_{\nu+2}(t)$ shall be investigated. We remember here the notation (2) for $K_i(t)$ from the previous section. It is easy to see that the following relations between $J_i(t)$ and $K_i(t)$ hold:

$$-1/2 K_i^{-3/2}(t_2^2, 0, t_1, 0, \dots, 0) = J_i(t), \quad i = 1, \dots, \nu+2. \tag{10}$$

As remarked above, in the case of $S_{\nu+3}$-singularities, we have exactly $\nu+2$ nontrivial period integrals $J_1(t), \dots, J_{\nu+2}(t)$ which correspond to the integrals $K_1(s_0, s_2, 0, \dots, 0), \dots, K_{\nu+2}(s_0, s_2, 0, \dots, 0)$ in view of the relationship (10). In order to simplify the system that appeared in Proposition 3.1, we consider a set of μ period integrals $K_0(s', 0, \dots, 0), K_1(s', 0, \dots, 0), \dots, K_{\nu+2}(s', 0, \dots, 0)$, (the notation $s' = (s_0, s_2)$ will be used in the sequel). The superscript λ can be omitted in the case of no necessity of specification. As a matter of fact, these μ period integrals are not independent elements of a certain $\mathcal{D}$-module over $\mathbb{C}[s][\frac{\partial}{\partial s_0}]$, that is, there are relations between the integrals. The first one is the following (see $(4)_0$):

$$2s_2 \frac{\partial}{\partial s_0} K_2(s', 0) + (\nu+2) \frac{\partial}{\partial s_0} K_{\nu+2}(s', 0) = -K_0(s', 0).$$

It means that $K_{\nu+2}$ is uniquely determined by K_0 and K_2 if we take into account the homogeneity of $K_{\nu+2}$. The latter follows easily from the definition of the integral:

$$\left(s_0 \frac{\partial}{\partial s_0} + \frac{\nu}{\nu+2} s_2 \frac{\partial}{\partial s_2}\right) K^{\lambda}_{\nu+2}(s', 0) = \left(\lambda + \frac{\nu+3}{\nu+2}\right) K^{\lambda}_{\nu+2}(s', 0).$$

Notice that $K_{\nu+1}$ has no relation with the K_j's except with K_1. It gives us the second relation:

$$(2s_2, \nu+2) \frac{\partial}{\partial s_0} (K_1, K_{\nu+1})^t = 0.$$

Using homogeneity of the relation one can rewrite it in the following form

$$K_{\nu+1}(s', 0) = \frac{-2s_2}{(\nu+2)} K_1(s', 0).$$

Hence it is enough to calculate the system of equations for $\nu+1$ integrals $K_0, \dots, K_\nu$ in order to get the corresponding system for $K_1, \dots, K_{\nu+2}$.

Proposition 4.1. *The integrals $K_0(s', 0), \dots, K_\nu(s', 0)$ satisfy the following system of differential equations:*

$$(s_0 \cdot \mathrm{id}_{\nu+1} + C(s_2)) \frac{\partial}{\partial s_0} \boldsymbol{K} = (L + V(s_2)) \boldsymbol{K},$$

where the column vector $(K_0, \dots, K_\nu)$ is denoted by $\boldsymbol{K}$, and

$$C(s_2) = \begin{bmatrix} 0 & 0 & 0 & 0 & 0 & \cdots & \frac{\nu}{\nu+2}s_2 \\ 0 & 0 & 0 & \frac{\nu}{\nu+2}s_2 & 0 & \cdots & 0 \\ 0 & 0 & 0 & 0 & \frac{\nu}{\nu+2}s_2 & \cdots & \vdots \\ \vdots & \vdots & \vdots & \vdots & \cdots & \ddots & \vdots \\ 0 & 0 & 0 & 0 & \ddots & \ddots & \frac{\nu}{\nu+2}s_2 \\ 0 & \frac{-2\nu}{(\nu+2)^2}s_2^2 & 0 & 0 & \cdots & \cdots & 0 \\ 0 & 0 & \frac{-2\nu}{(\nu+2)^2}s_2^2 & 0 & \cdots & \cdots & 0 \end{bmatrix},$$

$$L = \operatorname{diag}\Big(\lambda + \frac{1}{\nu+2}, \dots, \lambda + \frac{\nu+1}{\nu+2}\Big),$$

$$V(s_2) = \begin{bmatrix} 0 & 0 & \cdots & 0 \\ \vdots & \vdots & \cdots & \vdots \\ 0 & 0 & \cdots & 0 \\ \frac{2\nu}{(\nu+2)^2} s_2 & 0 & \cdots & 0 \end{bmatrix}.$$

Corollary 4.2. *The period integrals $J_0(t), \dots, J_{\nu+2}(t)$ for the isolated complete intersection $S_{\nu+3}$-singularity satisfy the following system of differential equations:*

$$\begin{aligned} (t_2^2 \cdot \mathrm{id}_{\nu+1} + C(t_1))\frac{\partial}{\partial t_2} \boldsymbol{J}(t) &= 2t_2(L + V(t_1))\boldsymbol{J}(t), \\ J_{\nu+1}(t) &= \frac{-2t_1}{(\nu+2)} J_1(t), \\ \frac{\partial}{\partial t_2} J_{\nu+2}(t) &= \frac{-2}{(\nu+2)}\Big(t_1 \frac{\partial}{\partial t_2} J_2(t) + t_2 J_0(t)\Big), \end{aligned}$$

where $\boldsymbol{J}(t) = (J_0(t), \dots, J_\nu(t))^t$. The matrix L is the same as that in the preceding proposition and the matrices $C(t_1)$ and $V(t_1)$ can be obtained from the corresponding matrices from the preceding proposition by substituting the variable t_1 for s_2.

Taking into consideration the system obtained in Proposition 3.2, let us try to solve the system of equations to get an explicit form of the integrals $K_0(s,0), \dots, K_{\nu+2}(s,0)$:

$$s_0 \frac{\partial}{\partial s_0} K_j^\lambda + s_2 \frac{\partial}{\partial s_0} K_{j+2}^\lambda + \frac{\partial}{\partial s_0} K_{\nu+2+j} = \lambda K_j^\lambda, \quad 0 \le j \le \nu, \qquad (5)_j$$

$$2s_2 \frac{\partial}{\partial s_0} K_{j+2}^\lambda + (\nu+2)\frac{\partial}{\partial s_0} K_{\nu+2+j}^\lambda = -(j+1)K_j^\lambda, \quad -1 \le j \le \nu. \qquad (11)_j$$

Subtract the relation $(11)_j$ multiplied by $1/(\nu+2)$ from $(5)_j$. We obtain recursive relations between the period integrals $K_0(s,0), \dots, K_{\nu+2}(s,0)$:

$$\begin{aligned} \Big(s_0 \frac{\partial}{\partial s_0} - \Big(\lambda + \frac{j+1}{\nu+2}\Big)\Big) K_j^\lambda &= -\frac{\nu s_2}{(\nu+2)} \frac{\partial}{\partial s_0} K_{j+2}^\lambda, \quad 0 \le j \le \nu - 2, \\ \Big(s_0 \frac{\partial}{\partial s_0} - \Big(\lambda + \frac{\nu}{\nu+2}\Big)\Big) K_{\nu-1}^\lambda &= \frac{-\nu s_2}{(\nu+2)} \frac{\partial}{\partial s_0} K_{\nu+1}^\lambda = \frac{\nu s_2^2}{(\nu+2)^2} \frac{\partial}{\partial s_0} K_1^\lambda, \\ \Big(s_0 \frac{\partial}{\partial s_0} - \Big(\lambda + \frac{\nu+1}{\nu+2}\Big)\Big) K_\nu^\lambda &= \frac{-\nu s_2}{(\nu+2)} \frac{\partial}{\partial s_0} K_{\nu+2}^\lambda. \end{aligned}$$

By virtue of the commutation relation

$$\Big[s_0 \frac{\partial}{\partial s_0}, \frac{\partial}{\partial s_0}\Big] = -\frac{\partial}{\partial s_0},$$

that is to say

$$\Big(s_0\frac{\partial}{\partial s_0}-\alpha\Big)\frac{\partial}{\partial s_0}=\frac{\partial}{\partial s_0}\Big(s_0\frac{\partial}{\partial s_0}-\alpha-1\Big),$$

we can deduce differential equations satisfied by $K_j(s,0)$ from the above recursive relations.

Thus we obtain the following statement.

Proposition 4.3.

(1) *Assume $\nu=2m$. Then the following differential equations of order $(m+1)$ are satisfied by the integrals K_{2j}^{λ}, for all $0\le j\le m$:*

$$S_{0,j-1}(\vartheta+m-j,\lambda)S_{j,m}(\vartheta-j,\lambda)K_{2j}=\frac{-2s_2}{\nu+2}(\vartheta-(\lambda+\tfrac{1}{2})+m-j+1)\psi^m K_{2j}. \tag{12$_{2j,e}$}$$

For K_{2j+1}, where $0\le j\le m$ we have similar equations of order m:

$$T_{0,j-1}(\vartheta+m-j,\lambda)T_{j,m-1}(\vartheta-j,\lambda)K_{2j+1}=\frac{-2s_2}{\nu+2}\psi^m K_{2j+1}. \tag{12$_{2j+1,e}$}$$

(2) *Assume $\nu=2m+1$. The differential operators annihilating the period integrals K_j^{λ}, $0\le j\le\nu$, are of order $(\nu+1)$ for both even and odd cases:*

$$\begin{aligned}&S_{0,j-1}(\vartheta+\nu-j,\lambda)T_{0,m-1}(\vartheta+m-j+1,\lambda)\\&\quad\times S_{j,m}(\vartheta-j,\lambda)(\vartheta-\Big(\lambda+\frac{\nu+1}{\nu+2}\Big)+\nu-j)K_{2j}(s)\\&=\Big(\frac{2s_2}{\nu+2}\Big)^2\psi^{2m+1}(\vartheta-\Big(\lambda+\tfrac{1}{2}+j)\Big)K_{2j}(s),\end{aligned} \tag{12$_{2j,o}$}$$

$$\begin{aligned}&T_{0,j-1}(\vartheta+\nu-j,\lambda)S_{0,m-1}(\vartheta+m-j+1,\lambda)\\&\quad\times T_{j,m}(\vartheta-m,\lambda)(\vartheta-\Big(\lambda+\frac{\nu}{\nu+2}\Big)+\nu-j-1)K_{2j+1}\\&=\Big(\frac{2s_2}{\nu+2}\Big)\psi^{2m+1}(\vartheta-(\lambda+j+m))K_{2j+1},\end{aligned} \tag{12$_{2j+1,o}$}$$

where we denote

$$S_{\alpha,\beta}(X,\lambda)=\prod_{\ell=\alpha}^{\beta}\Big(X-\Big(\lambda+\frac{-\nu\ell+1}{\nu+2}\Big)\Big),$$

$$T_{\alpha,\beta}(X,\lambda)=\prod_{\ell=\alpha}^{\beta}\Big(X-\Big(\lambda+\frac{-\nu\ell+2}{\nu+2}\Big)\Big),$$

$$\psi=\Big(\frac{-\nu s_2}{\nu+2}\Big)\frac{\partial}{\partial s_0},\qquad \vartheta=s_0\frac{\partial}{\partial s_0}.$$

Corollary 4.4.

(1) *Assume $\nu = 2m$. Then we get the following differential equations of order $(m+1)$ satisfied by the period integrals $J_{2j}(t_1, t_2)$, $J_{2j+1}(t_1, t_2)$, $0 \le j \le m$, depending on parameters of the principal deformation for $S_{\nu+3}$:*

$$S_{0,j-1}\big(\theta + m - j, -\tfrac{3}{2}\big) S_{j,m}\big(\theta - j, -\tfrac{3}{2}\big) J_{2j}(t) = \tfrac{-2t_1}{\nu+2}(\theta + m - j + 2)\varphi^m J_{2j}(t), \tag{13$_{2j,e}$}$$

$$T_{0,j-1}\big(\theta + m - j, -\tfrac{3}{2}\big) T_{j,m-1}\big(\theta - j, -\tfrac{3}{2}\big) J_{2j+1} = \tfrac{-2t_1}{\nu+2}\varphi^m J_{2j+1}. \tag{13$_{2j+1,e}$}$$

(2) *Assume $\nu = 2m + 1$. Then we get the following differential equations of order $(2m + 2)$ for the period integrals:*

$$\begin{aligned} S_{0,j-1}\big(\theta &+ \nu - j, -\tfrac{3}{2}\big) T_{0,m-1}\big(\theta + m - j + 1, -\tfrac{3}{2}\big) \\ &\times S_{j,m}\big(\theta - j, -\tfrac{3}{2}\big)\big(\theta - \tfrac{\nu+3}{2(\nu+2)} + \nu - j\big) J_{2j}(t) \\ &= \Big(\frac{2t_1}{\nu+2}\Big)^2 \varphi^{2m+1}\big(\theta - (\lambda + \tfrac{1}{2} + j)\big) J_{2j}(t), \end{aligned} \tag{13$_{2j,o}$}$$

$$\begin{aligned} T_{0,j-1}\big(\theta &+ \nu - j, -\tfrac{3}{2}\big) S_{0,m-1}\big(\theta + m - j + 1, -\tfrac{3}{2}\big) \\ &\times S_{j,m}\big(\theta - m, -\tfrac{3}{2}\big)\big(\theta - \tfrac{\nu+6}{2(\nu+2)} + \nu - j - 1\big) J_{2j}(t) \\ &= \frac{2t_1}{\nu+2}\varphi^{2m-1}(\theta - j - m) J_{2j+1}(t), \end{aligned} \tag{13$_{2j+1,o}$}$$

where the differential operators $S_{\alpha,\beta}$, $T_{\alpha,\beta}$ are those defined in Proposition 4.3 *while the other notations are as follows:*

$$\varphi = \frac{-\nu t_1}{2(\nu+2)t_2}\frac{\partial}{\partial t_2}, \qquad \theta = \frac{t_2}{2}\frac{\partial}{\partial t_2}.$$

Let us try to write down the solutions of the differential equations obtained in Corollary 4.4. As they have quite similar forms, we restrict ourselves to write down the solutions for equations $(13)_{0,e}$ and $(13)_{0,o}$ only.

Theorem 4.5.

(1) *The case $\nu = 2m$. Equation $(13)_{0,e}$ has $(m + 1)$ solutions $\mathcal{U}_k(t)$, $0 \le k \le m$, that can be expressed in the following series*

$$\mathcal{U}_k(t) = t_1^{-(3\nu+4)/2\nu}\tau^{\rho_k} U_k(\tau^m, -\tfrac{3}{2}),$$

where

$$U_k(x, \lambda) = \sum_{\ell \ge 0} a_{m\ell+k}(\lambda) x^\ell,$$

$$a_{m\ell+k}(\lambda) = \prod_{j=0}^{m-1} \frac{((k-\lambda-1)/m - j/(m+1); \ell)}{((j+k)/m - 1; \ell)} \frac{(-\lambda + 1/(\nu+2); \ell)}{(m - (\lambda + \frac{1}{2}); \ell)},$$

$$\tau = (1/\nu)^{1/m}(-\nu t_1/(\nu+2))^{\frac{-m-1}{m}} t_2^2,$$

$(\alpha; \ell) = \Gamma(\alpha+\ell)/\Gamma(\alpha)$, and $0 \le k \le m$. For $k = 0, \dots, m-1$, the characteristic exponents $\rho_k = k$, while $\rho_m = \lambda + \frac{1}{2} = -1$.

(2) *The case $\nu = 2m+1$. The equation $(13)_{0,o}$ has $(2m+2)$ solutions $\mathcal{V}_k(t)$, $0 \le k \le 2m+1$, that can be expressed in the following series*

$$\mathcal{V}_k(t) = t_1^{-(3\nu+4)/2\nu} \sigma^{\rho_k} V_k(\sigma^{2m+1}, -\tfrac{3}{2}),$$

where

$$V_k(x, \lambda) = \sum_{\ell \ge 0} a_{(2m+1)\ell+k}(\lambda) x^\ell,$$

$$a_{(2m+1)\ell+k}(\lambda) = \frac{\prod_{j=1}^{m+1} \left(\frac{k-\lambda-1+(\nu j+1)/(\nu+2)}{\nu} - 1; \ell\right)\left(\frac{j}{\nu+2} + \frac{k+m-\lambda}{\nu-1}; \ell\right)}{\prod_{j=1}^{2m+1}((k+j)/\nu - 1; \ell)((k-\lambda-\frac{1}{2})/\nu; \ell)},$$

$$\sigma = \left(-4\left(\frac{t_1}{\nu+2}\right)^{\nu+2}\right)^{-1/\nu} t_2^2, \quad 0 \le k \le 2m+1.$$

For $k = 0, \dots, 2m$, the characteristic exponents $\rho_k = k$, while $\rho_{2m+1} = \lambda + \frac{1}{2} = -1$.

Proof. In view of the substitution $t_1 = s_2$, $t_2^2 = s_0$, we solve equation $(12)_{0,e}$ (respectively $(12)_{0,o}$) to get the solutions of equation $(13)_{0,e}$ (respectively $(13)_{0,o}$). In order to obtain the recursive relation between $a_{m\ell+k}$ and $a_{m(\ell+1)+k}$ (respectively, between $a_{(2m+1)\ell+k}$ and $a_{(2m+1)(\ell+1)+k}$) we consider the following trivial equality:

$$(\tau\frac{\partial}{\partial\tau} - \alpha)\tau^r = (r-\alpha)\tau^r.$$

Further calculations are achieved in an elementary manner. □

Remark 3. The functions $U_k(x, \lambda)$ (respectively $V_k(x, \lambda)$) introduced in the above theorem can be regarded as generalized hypergeometric functions

$${}_{m+1}F_m(\alpha_1^{(k)}, \dots, \alpha_{m+1}^{(k)}; \beta_1^{(k)}, \dots, \beta_m^{(k)} | x) = \sum_{\ell \ge 0} \frac{(\alpha_1^{(k)}; \ell) \dots (\alpha_{m+1}^{(k)}; \ell)}{(\beta_1^{(k)}; \ell) \dots (\beta_m^{(k)}; \ell)} \frac{x^\ell}{\ell!}$$

(respectively,

$${}_{2m+2}F_{2m+1}(\gamma_1^{(k)}, \dots, \gamma_{2m+2}^{(k)}; \delta_1^{(k)}, \dots, \delta_{2m+1}^{(k)} | x)$$

in Pochhammer's sense (see [8], [16]). This can be checked easily, as we find the factor $(0; \ell) = \ell!$ in every denominator of the expansion coefficients. The indices $\alpha_1^{(k)}, \beta_1^{(k)}, \dots$, *etc.*, are obtained from the expansion coefficients.

Regarding the relationship with the logarithmic differential forms studied by K. Saito, we obtain the following result.

Theorem 4.6. *Let $D = \{t \in \mathbb{C}^2; t_2\varphi(t_1, t_2) = 0\}$ be the discriminant set of the principal deformation for an $S_{\nu+3}$-singularity. Then the Gauss–Manin system for the period integrals $\boldsymbol{J} = (J_0, \ldots, J_\nu)^t$ associated with an $S_{\nu+3}$-singularity permits a representation as a Picard–Fuchs system (total differential system) with coefficients from $\Omega^1_S(\log D)$ as follows:*

$$dJ = L\left(A\frac{d(t_2\varphi)}{t_2\varphi} + (H_1(t_1) + H_2(t_1, t_2))\frac{\iota_\eta(\omega)}{t_2\varphi}\right)J, \tag{14}$$

where L is the diagonal matrix corresponding to the weights of the basis of the cohomology group which appeared in Proposition 3.2*,*

$$A = 2\,\mathrm{diag}(1/(2m+1), \ldots, 1/(2m+1)), \qquad \textit{if} \quad \nu = 2m,$$
$$A = 4\,\mathrm{diag}((2m+1)/(4m+3), \ldots, (2m+1)/(4m+3)), \quad \textit{if} \quad \nu = 2m+1,$$

$$H_1(t_1) = -m^{m-1}\frac{t_1^m}{(2m+1)(-m-1)^{m+1}}\begin{bmatrix} 2m & 0 & 0 & 0 & \cdots & 0 \\ 0 & 1 & 0 & 0 & \cdots & 0 \\ \vdots & \ddots & \vdots & \cdots & \ddots & \vdots \\ 0 & 0 & 0 & \ddots & \ddots & 0 \\ 0 & 0 & 0 & \cdots & 1 & 0 \\ 0 & 0 & 0 & \cdots & 0 & 1 \end{bmatrix} \quad (\nu = 2m),$$

$$H_1(t_1) = \frac{8(2m+1)^{2m+1}t_1^{2m+2}}{(4m+3)(2m+3)^{2m+3}}\begin{bmatrix} 2(2m+1) & 0 & 0 & 0 & \cdots & 0 \\ 0 & -1 & 0 & 0 & \cdots & 0 \\ \vdots & & \ddots & \vdots & \cdots \ddots & \vdots \\ 0 & & 0 & 0 & \ddots \quad \ddots & 0 \\ 0 & & 0 & 0 & \cdots \; -1 & 0 \\ 0 & & 0 & 0 & \cdots \quad 0 & -1 \end{bmatrix} \quad (\nu = 2m+1),$$

$H_2(t)$ is a matrix with polynomial entries with zero diagonal part. All the matrices given above have the size $(\nu+1)\times(\nu+1)$. The functions that appear in the definition of the divisor D have the following forms:

$$\varphi(t_1, t_2) = t_2^{2m} - 2(2m)^m(-t_1/(2m+2))^{(m+1)}, \quad \nu = 2m,$$

$$\varphi(t_1, t_2) = t_2^{4m+2} + 4(2m+1)^{2m+1}(t_1/(2m+3))^{(2m+3)}, \quad \nu = 2m+1.$$

The holomorphic 1-form $\iota_\eta(\omega)$ is defined by the interior product between the Euler vector field $\eta = \nu t_1\frac{\partial}{\partial t_1} + \frac{(\nu+2)}{2}t_2\frac{\partial}{\partial t_2}$ and the volume form $\omega = dt_1 \wedge dt_2$.

5. Further remarks and problems

Let us consider a reduced hypersurface in $\mathbb{C}^n$

$$D = \{t \in \mathbb{C}^n : h(t) = h_1(t) \cdot \ldots \cdot h_m(t) = 0\}, \quad D = \bigcup_{i=1}^{m} D_i,$$

where $D_i = \{t \in \mathbb{C}^n : h_i(t) = 0\}$ are the *irreducible* components of D. Assume that D is a divisor *with normal crossings*, that is,

1) D_i intersects D_j transversely, $i \neq j$,

2) $\dim D_i \cap D_j \cap D_k \leq n-3$ for $i \neq j, k$, $j \neq k$.

It is known (see [23], (2.9)) that $\Omega^1_{\mathbb{C}^n}(\log D)$ is generated by

$$dh_1/h_1, \ldots, dh_m/h_m \tag{15}$$

as an $\mathcal{O}_{\mathbb{C}^n}$-module. In such a case, a "Pfaff system of Fuchsian type" has been defined in quite a natural manner for a set of unknown functions $\mathbf{I} = (I_1, \ldots, I_\mu)$:

$$d\boldsymbol{I} = (\sum_{j=1}^{m} A_j \frac{dh_j}{h_j})\boldsymbol{I}, \tag{16}$$

where $A_j \in \operatorname{End}(\mathbb{C}^\mu) \otimes \mathcal{O}_{\mathbb{C}^n}$. Pfaff systems of this type were studied in [12]. It has been shown in Theorem 3.1, however, when the components of the divisor D do not intersect transversely, there appear logarithmic differential forms like $\iota_\eta(\omega)$ in (14), in addition to those of type (15). That is to say, it is natural to think of a class of systems

$$d\,\mathbf{I} = (A\frac{dh}{h} + \sum_{j=1}^{n-1} B_j \frac{\omega_j}{h})\boldsymbol{I} \tag{17}$$

with $\omega_j \in \operatorname{Tors} \Omega^1_D$ and A, $B_j \in \operatorname{End}(\mathbb{C}^\mu) \otimes \mathcal{O}_{\mathbb{C}^n}$ satisfying the integrability condition:

$$dA\frac{dh}{h} + \sum_{j=1}^{n-1} dB_j \frac{\omega_j}{h} + B_j d(\frac{\omega_j}{h}) = \sum_{j=1}^{n-1} (B_j A - AB_j)\frac{\omega_j dh}{h^2}.$$

Expression (17) above is appropriate in describing a Pfaff system with $\Omega^1_{\mathbb{C}^n}(\log D)$ coefficients, in virtue of the following exact sequence (see [3]:

$$0 \longrightarrow \frac{dh}{h}\mathcal{O}_{\mathbb{C}^n} + \Omega^1_{\mathbb{C}^n} \longrightarrow \Omega^1_{\mathbb{C}^n}(\log D) \xrightarrow{h} \operatorname{Tors} \Omega^1_D \longrightarrow 0.$$

Here $\operatorname{rank}(\operatorname{Tors} \Omega^1_D) = n-1$ and $\Omega^1_D = \Omega^1_{\mathbb{C}^n}/(h \cdot \Omega^1_{\mathbb{C}^n} + dh \cdot \mathcal{O}_{\mathbb{C}^n})$. Furthermore it shall be noted that in the case when $\Omega^1_{\mathbb{C}^n}(\log D)$ is a free $\mathcal{O}_{\mathbb{C}^n}$-module, the *integral variety* defined by the dual free module $\operatorname{Der}_{\mathbb{C}^n}(\log D)$ coincides with $D = \{t \in \mathbb{C}^n : h(t) = 0\}$ (see [23] (1.9)).

It should be noted that in the case of a divisor D with normal crossings the $\mathcal{O}_D$-module of torsion differentials Tors Ω^1_D is generated by the following $(m-1)$ differential 1-forms

$$hdh_1/h_1, \ldots, \widehat{hdh_i/h_i}, \ldots, hdh_m/h_m, \quad 1 \le i \le m,$$

where the i-th form hdh_i/h_i is omitted. Therefore, in this situation the system (16) can be considered as a special case of (17). Thus one may regard the system of type (17) as a natural generalization of "the Pfaff system of Fuchsian type" to the general case of an arbitrary *reduced reducible* hypersurface.

For this kind of systems we see readily that

Proposition 5.1. *For a polynomial $h(t) \in \mathbb{C}[t]$ with no multiple factors the system (17) appears to be a system with regular singularities along D (not only along regular part D^{reg} of D !).*

Proof. First observe that (17) has regular singularities along the regular part D^{reg}, then apply a theorem of P. Deligne ([10] Théorème 4.1) to our situation. Namely, this theorem states that a regular singularity of a system along D^{reg} entails a regular singularity along the whole D. □

The following two questions were proposed by K. Aomoto concerning systems of type (17).

Question 1. Let us consider an arbitrary representation $\rho \in \pi_1(\mathbb{C}^n \setminus D)$. Is it possible to find a system of form (17) such that its solutions induce ρ as their monodromy representation? In other words, is the class (17) wide enough for the existence of solutions to the Riemann–Hilbert problem?

K. Aomoto gave a positive answer (see [6]) to this problem in the case where ρ is contained in a unipotent subgroup of $\mathrm{GL}(\mu, \mathbb{C})$.

Question 2. Describe the cases when there exists an appropriate finite covering space X over $\mathbb{C}^n$:

$$\begin{array}{ccc} \pi : X & \to & \mathbb{C}^n \\ \cup & & \cup \\ \widetilde{D} & \to & D \end{array}$$

such that the preimage of a system of type (17) under π has the following form

$$d\tilde{I} = \left(\sum_{j=1}^{m} A_j \frac{dh_j}{h_j}\right)\tilde{I}$$

where $A_j \in \mathrm{End}(\mathbb{C}^\mu) \otimes \mathcal{O}_X$, $h_j \in \mathcal{O}_X$, $j = 1, \ldots, m$ and $\widetilde{D} = \cup_{j=1}^m \{z \in X : h_j(z) = 0\}$. It was remarked by K. Aomoto that the system (14) turns out to be the case in question because its solutions are described by Pochhammer's hypergeometric functions that are interpreted as solutions to the following system [8]:

$$\frac{d}{dz}J = (\frac{A_1}{z} + \frac{A_2}{z-1})J,$$

where $A_1, A_2 \in \mathrm{End}(\mathbb{C}^\mu)$.

References

[1] A. G. Aleksandrov, Milnor numbers of non-isolated Saito singularities, Functional. Anal. Appl. 21 (1) (1987), 1–10 (in Russian).

[2] A. G. Aleksandrov, Non-isolated Saito singularities, Mat. Sb. 21 (4) (1988), 554–567 (in Russian).

[3] A. G. Aleksandrov, Nonisolated hypersurface singularities, Adv. Soviet Math. 1, 211–246. Amer. Math. Soc., Providence, RI, 1990.

[4] A. G. Aleksandrov, S. Tanabé, Computing Gauss–Manin systems for complete intersection singularities S_μ, Georgian Math. J. 3 (5) (1996), 401–422.

[5] K. Aomoto, Les équations aux différences et les intégrales des fonctions multiformes, J. Fac. Sci. Univ. Tokyo Sect. IA Math. 22 (1975), 271–297.

[6] K. Aomoto, Fonctions hyperlogarithmiques et groupes de monodromie unipotents, J. Fac. Sci. Univ. Tokyo Sect. IA Math. 25 (1978), 149–156.

[7] K. Aomoto, Configurations and Invariant Gauss–Manin Connections of Integrals I, Tokyo J. Math. 5 (1982), 249–287.

[8] P. Appell and J. Kampé de Fériet, Fonctions hypergéométriques et hypersphériques, Paris 1926.

[9] E. Brieskorn, Die Monodromie der isolierten Singularitäten von Hyperflächen, Manuscripta Math. 2 (1970), 103–161.

[10] P. Deligne, Equations différentielles à points singuliers réguliers, Lecture Notes in Math. 163, Springer-Verlag, 1970.

[11] I. M. Gelfand, V. A. Vasiliev, A. V. Zelevinski, Generalized hypergeometric functions on complex Grassmannians, Funktsional Anal. i Prilozhen. 21 (1) (1987), 23–38 (in Russian).

[12] R. Gerard et A. H. M. Levelt, Étude d'une classe particulière de système de Pfaff du type de Fuchs sur l'espace projectif complexe, J. Math. Pures Appl. 51 (1972), 189–217.

[13] G.-M. Greuel, Der Gauß–Manin-Zusammenhang isolierter Singularitäten von vollständigen Durchschnitten, Math. Ann. 214 (3) (1975), 235–266.

[14] M. Giusti, Classification des singularités isolées simples d'intersections completès, in: Singularities, Proc. Symp. Pure Math. 40, Part 1, 457–494. Amer. Math. Soc., Providence, RI, 1983.

[15] S. Ishiura and M. Noumi, A calculus of the Gauss–Manin system of type A_ℓ, I,II, Proc. Japan Acad. Ser. A Math. Sci. 58 (1982), 13–16, 62–65.

[16] A. H. M. Levelt, Hypergeometric functions, Proc. Konink. Nederl. Akad. Wetensch. 64 (1961), 190–211.

[17] Kh. M. Malikov, Redefinition of the differential systems for versal integrals of type A, D, E, Differentsial'nye Uravneniya 18 (8) (1982), 1394–1400 (in Russian).

[18] M. Noumi, Expansion of the solutions of a Gauss–Manin system at a point of infinity, Tokyo J. Math. 7 (1984), 1–60.

[19] M. Noumi, Gauss–Manin system and the flat coordinate system (Connection with the expansion of the solutions at ∞), RIMS Kokyuroku 481, 1983, 62–72 (in Japanese).

[20] V. P. Palamodov, Deformations of complex spaces, Itogi Nauki i Tekhniki, Sovrem. Probl. Mat. 10, 123–221, VINITI, Moscow, 1986 (in Russian)

[21] K. Saito, On the uniformization of complements of discriminant loci, in: Hyperfunctions and Linear partial differential equations, RIMS Kokyuroku 287 (1977), 117–137.

[22] K. Saito, On the periods of primitive integrals I. preprint, RIMS, Kyoto Univ., 1982.

[23] K. Saito, Theory of logarithmic differential forms, J. Fac. Sci. Univ. Tokyo, Sect. IA Math. 27 (2) (1980), 265–291.

Laws of trigonometry in symmetric spaces

Helmer Aslaksen and Hsueh-Ling Huynh

Abstract. This paper consists of two parts. In the first part, we reformulate the work of E. Leuzinger on trigonometry in noncompact symmetric spaces. In the second part, we outline an alternative method using invariants of the isotropy group representation. Appropriately formulated, these methods apply to both compact and noncompact symmetric spaces. This work is contained in the Ph.D. dissertation of H.-L. Huynh.

1991 Mathematics Subject Classification: 53C35, 15A72, 15A24

1. What is trigonometry?

Trigonometry is fundamental to the study of the classical geometries. First of all, it gives congruence conditions for triangles. In Euclidean, hyperbolic and spherical spaces, two triangles are congruent if and only if they satisfy the side-angle-side (SAS) or side-side-side (SSS) condition.

Given a triangle, we also want to associate to it metric quantities that are invariant under the group of isometries. The laws of trigonometry can then be interpreted as relations between such quantities. Given two sides and their subtended angle, the law of cosines determines the third side, and given two sides and one angle adjacent to the third side, the law of sines determines the second adjacent angle.

In a general Riemannian manifold, the law of cosines can be generalized as a comparison theorem for geodesic triangles in the form of an inequality. On the other hand, it is clear that congruence conditions among configurations of points can only be found in spaces with a high degree of symmetry. Only in such spaces can we obtain exact formulas relating invariant geometric quantities.

We will hence consider a globally symmetric space X. We will also be interested in configurations of n points, where n may be greater than 3, and we will refer to this as trigonometry, too. Let X^n be the space of ordered subsets of n points. Given a group of isometries G of X, we let G act diagonally on X^n. That is, $(x_1, \dots, x_n) \to (gx_1, \dots, gx_n)$ for $g \in G$. Let U be a G-invariant open dense subset of X^n. Once U is specified, an n-tuple $(x_1, \dots, x_n) \in U$ is said to be in general position.

A trigonometric quantity with respect to G is a G-invariant function $\alpha\colon U \to \mathcal{F}$, where $\mathcal{F}$ is a set on which G acts trivially. We will also call the induced function

$\alpha: U/G \to \mathcal{F}$ a trigonometric quantity. Let R be a subset of $\mathcal{F}$. We say that α satisfies the trigonometric relation R if $\alpha(U) \subseteq R$. Finally, α is said to be a congruence condition with respect to G if it is an injective function into $\mathcal{F}$.

For example, let X be the Euclidean plane and let $U \subset X^3$ consist of the 3-tuples of distinct points. Then $\alpha: U \to \mathbb{R} \times \mathbb{R} \times \mathbb{R} \times S^1$, which maps an ordered triangle to its three sides and the angle subtended by the first two sides, is a trigonometric quantity with respect to the group of orientation-preserving isometries. It satisfies the trigonometric relation

$$\{ (a, b, c, \theta) \mid a^2 + b^2 - c^2 - 2ab\cos\theta = 0 \}.$$

It is also a congruence condition with respect to the same group.

2. The moduli space of n points in symmetric spaces

By definition, a Riemannian (globally) symmetric space X is a Riemannian manifold with central symmetry through every point $x \in X$, i.e., the diffeomorphism that reverses all the geodesics through x is a global isometry. We will make use of some basic facts about symmetric spaces. Proofs and details can be found in [He], [Mo] or [Wo].

Let $I(X)$ be the full isometry group of X, and let $I_0(X)$ be the component of the identity. Then $I_0(X)$ acts transitively on X. Let G be a Lie group such that $I_0(X) \subseteq G \subseteq I(X)$ and pick a basepoint $x_0 \in X$. Let K be the subgroup of G that fixes x_0 and let σ be the geodesic symmetry at x_0. Then $X = G/K$, and $(G_\sigma)_0 \subseteq K \subseteq G_\sigma$, where G_σ is the set of fixed points of σ and $(G_\sigma)_0$ is the identity component of G_σ. If X is of noncompact type, then K will be a maximal compact subgroup of G.

We would like to determine X^n/G or U/G, where U is a suitably chosen set of points in general position. Let $\mathfrak{p}$ be the tangent space of X at x_0. Under the exponential map $\operatorname{Exp}: \mathfrak{p} \to X$, we have a correspondence $X^n/G \leftrightarrow (\bigoplus^n \mathfrak{p})/K$ in the noncompact case, and $U/G \leftrightarrow (\bigoplus^n \mathfrak{p})/K$ in the compact case. The ring of invariant polynomial functions, $\mathbb{R}[\bigoplus^n \mathfrak{p}]^K$, is the coordinate ring of the space $(\bigoplus^n \mathfrak{p})/K$, and we wish to describe the moduli space of n points by giving a set of generators and relations for this ring. Note that the generators will then be trigonometric quantities, and the relations will be trigonometric relations.

For the classical geometries, the full isotropy group is $O(m)$. The isotropy representation is the standard action on $\mathbb{R}^m$, and the generators and relations of the ring $\mathbb{R}[\bigoplus^n \mathfrak{p}]^K$ are given by the First and Second Fundamental Theorems of invariant theory of H. Weyl ([We]). For symmetric spaces of rank greater than one, our knowledge of $\mathbb{R}[\bigoplus^n \mathfrak{p}]^K$ is less complete.

On the other hand, the geometry of two points in a symmetric space is well understood. Consider the set of flats in X. These are the maximal, flat, totally geodesic subspaces. They are all of the same dimension r, which is the rank of the symmetric space X. Each flat is isometric to a Euclidean space in case X is of noncompact type, and a flat torus in case X is compact. Every point of X is contained in a flat, G permutes the flats transitively and K permutes transitively all those that contain the point x_0. Given a flat F, there is an Abelian subgroup A of G that restricts to the full group of Euclidean translations on F. Let us choose a base flat F_0 containing x_0. Then each $x \in F_0$ is ax_0 for some $a \in A$, and each $x \in X$ is ky for some $k \in K$ and $y \in F_0$.

It follows that if we have two points in X, then we may assume that the first point is x_0 and the second point is $x_1 \in F_0$. Hence $X^2/G \cong F_0/M'$, where M' is the subgroup of K that preserves the set F_0. Under the correspondence $a \to ax_0$, the action of M' on F_0 is equivalent to the action of M' on A given by $a \to mam^{-1}$, and M' is the normalizer of A in K. Furthermore, $M = Z_K(A)$, the centralizer of A in K, is the subgroup of K that fixes F_0 pointwise. The Weyl group $W = M'/M$ decomposes F_0 into a union of Weyl chambers. For each chamber C, every W-orbit in F_0 intersect C exactly once. And W permutes the set of chambers simply and transitively. Now let us choose among the chambers a base chamber C_0, so that it contains the point x_1. Collecting all these facts together, we obtain the following description of the moduli of 2 points.

$$X^2/G \cong K\backslash G/K \cong A/W \cong C_0.$$

To specify a pair of points (y, y') up to congruence with respect to the full group of isometries $I(X)$, we first move them by an element of $g \in G$ to a pair $(gy, gy') = (x_0, x_1) = (x_0, ax_0)$, where x_0 is the base point, x_1 lies in the base chamber and $a \in A$. Define $M^* \subseteq I(X)$ to be the subgroup that fixes the point x_0 and preserves the set C_0. Then

$$X^2/I(X) \cong M^*\backslash A/W.$$

The coset M^*a is called the interval determined by the ordered pair of points (y, y').

The structure of $\mathbb{R}[\mathfrak{p}]^K$ is completely determined by a theorem of Chevalley. Let $\mathfrak{A}$ be the linear subspace of $\mathfrak{p}$ that exponentiates to F_0. The adjoint action of K on $\mathfrak{p}$ restricts to an action of M' on $\mathfrak{A}$, with kernel M. Hence the group W acts on $\mathfrak{A}$, and we can state the following theorem.

Theorem 1.

$$\mathbb{R}[\mathfrak{p}]^K \cong \mathbb{R}[\mathfrak{A}]^W \cong \mathbb{R}[\sigma_1, \dots, \sigma_r],$$

where $\mathbb{R}[\sigma_1, \dots, \sigma_r]$ *is a free polynomial algebra on the generators* $\sigma_1, \dots, \sigma_r$.

3. The laws of sines and cosines for noncompact symmetric spaces

We will now consider triangles, i.e., the case of three points. The symmetric spaces of rank one turn out to be the so-called two-point homogeneous spaces, and consist of the classical geometries, the complex and quaternionic projective and hyperbolic spaces and the Cayley projective and hyperbolic plane. The trigonometry of the classical geometries can be summed up in the well-known laws of sines and cosines. Subsequently, the trigonometry of $\mathbb{P}^n(\mathbb{C})$ was studied by W. Blaschke and H. Terheggen ([BT]) in 1939, after partial results by J. L. Coolidge ([Co]) in 1921. A different approach was taken by P. A. Sirokov ([Sir]), who obtained more complete results. Sirokov died in 1944, but the results were found among his papers, and published under his name in 1957 by A. P. Sirokov, A. Z. Petrov and B. A. Rozenfeld ([Sir]). B. A. Rozenfeld ([Ro]) later generalized Sirokov's results to the other compact two-point homogeneous spaces, i.e., quaternionic projective space and the Cayley projective plane. In 1986, W.-Y. Hsiang ([Hs]) independently developed the trigonometry of all the two-point homogeneous spaces, giving a unified and differential-geometric treatment that applies equally well to the compact and noncompact cases. In 1987, U. Brehm ([Br]) modified Hsiang's results, using an approach similar to [BT].

The first trigonometric relations for spaces of rank greater than one were given in 1988 in the Ph.D. thesis of H. Aslaksen ([As]) under the supervision of W.-Y. Hsiang. He obtained trigonometric formulas for $SU(3)$ by using Lie theory and direct manipulation of matrices. In his 1990 Ph.D. thesis, E. Leuzinger ([Le]) found the laws of sines and cosines for symmetric spaces of noncompact type. The essential idea is to consider the motions of Weyl chambers, rather than points or geodesic segments. In this section we will describe a reformulation of Leuzinger's method.

Fix a symmetric space of noncompact type X. Recall that a geodesic is said to be regular if it is contained in exactly one flat. We will say that a triple of points (A, B, C) is in general position if the geodesic segments AB, BC, CA are all regular, and they lie in three distinct flats. (Note that if A, B, C lie in the same flat, then Euclidean trigonometry applies, but the Euclidean rotations in the flat are generally not elements of $I(X)$.)

Suppose we have two chambers $\mathcal{A}$ and $\mathcal{B}$ with a common apex x but lying in two distinct flats. By an element of $g \in G$ we can move x to the base point x_0, and $\mathcal{A}$ to the base chamber $\mathcal{C}_0$. Then $g\mathcal{B} = kg\mathcal{A} = k\mathcal{C}_0$ for some $k \in K$. The element k is determined only up to right multiplication by elements of the subgroup M. Furthermore, left multiplication by M^* does not affect the congruence class of the pair of chambers with respect to $I(X)$. The double coset M^*kM is called the angle between the ordered pair of chambers $(\mathcal{A}, \mathcal{B})$ with a common apex.

Given a triple (A, B, C) in general position, we may assume that A is the base point x_0 and B lies in the base chamber $\mathcal{C}_0$. In Figure 1, the intervals BC, AC, AB are represented by the elements M^*a, M^*b, M^*c with $a, b, c \in A$.

$\mathcal{C}_0'$ is the translate of $\mathcal{C}_0$ along the supporting flat of AB to the apex B, $\mathcal{C}_2$ is the chamber with apex B containing BC, $\mathcal{C}_2'$ is the translate of $\mathcal{C}_2$ along the supporting flat of BC to the apex C, $\mathcal{C}_1$ is the chamber with apex A containing AC, and $\mathcal{C}_1'$ is the translate of $\mathcal{C}_1$ along the supporting flat of AC to the apex C.

The angle at $(\mathcal{C}_0, \mathcal{C}_1)$ is M^*k_AM, at $(\mathcal{C}_0', \mathcal{C}_2)$ is M^*k_BM, and at $(\mathcal{C}_1', \mathcal{C}_2')$ is M^*k_CM. The function

$$(A, B, C) \to (M^*a, M^*b, M^*c, M^*k_AM, M^*k_BM, M^*k_CM)$$

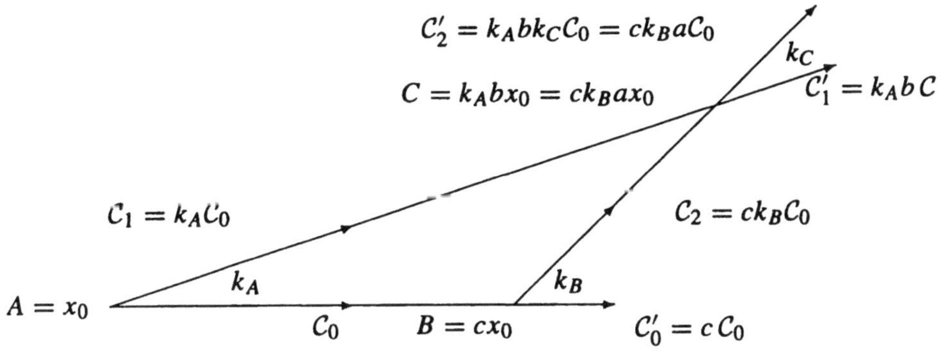

Figure 1: A triangle in general position in X

is a trigonometric quantity with respect to $I(X)$. Our aim is to find relations among its components. For suitable representatives of intervals and angles, the following equations relate the various chambers.

$$
\begin{aligned}
C_0' &= c\, C_0 \\
C_2 &= c k_B c^{-1} C_0' = c k_B C_0 \\
C_2' &= (c k_B) a (c k_B)^{-1} C_2 = c k_B a C_0 \\
C_1 &= k_A C_0 \\
C_1' &= k_A b\, C_0 \\
C_2' &= k_A b k_C (k_A b)^{-1} C_1' = k_A b k_C C_0
\end{aligned} \tag{1}
$$

Equating the two expressions for C_2', we obtain the following important relation.

$$(k_A b k_C) C_0 = (c k_B a) C_0. \tag{2}$$

This gives the following fundamental relation at group level.

Theorem 2. *It is possible to choose representatives for the cosets corresponding to the sides and angles so that*

$$(k_A b k_C) M = (c k_B a) M.$$

In general, the quantities appearing on the two sides of this relation are not determined by the intervals and angles alone. In other words, the representatives for the sides and angles cannot be chosen arbitrarily from the cosets they represent. The following theorem from [Le] clarifies this matter. It is essentially the SAS condition.

Theorem 3. *The set of triples (c, k_B, a) that satisfy Theorem 2 consists of all the representatives of a diagonal double coset $M^*(c, k_B M, a)$. The function*

$$(A, B, C) \to M^*(c, k_B M, a)$$

is a congruence condition.

We will now show how we can derive the laws of trigonometry from Theorem 2, using the KAK and KNA decompositions of G. For any group H and subgroups E, F of H, a function on H is said to be (E, F)-invariant if it is invariant under left multiplication by E and right multiplication by F. Note that M normalizes K and centralizes A, so that $KM = MK$ and $AM = MA$ are subgroups of G. The next two theorems follow easily from Theorem 2.

Theorem 4 (Law of cosines). *Suppose $f: G \to \mathbb{R}$ is a (K, KM)-invariant function. Then*

$$f(b) = f(ck_Ba).$$

If there are r such f's that are functionally independent, then M^*b can be uniquely computed from the data $M^*(c, k_B, a)$.

Theorem 5 (Law of sines). *Suppose $g: G \to \mathbb{R}$ is a (K, AM)-invariant function. Then*

$$g(bk_C) = g(ck_B).$$

Thus the problem of finding trigonometric formulas is reduced to the search for functions on the group G that are (K, K)- and (K, AM)-invariant. We will briefly indicate how that can be done.

We have $K\backslash G/K \cong A/W \cong \mathfrak{A}/W$. By Theorem 1, $\sigma_1, \dots, \sigma_r$ induce r independent (K, K)-invariant functions on G via the exponential map.

Recall the Iwasawa Decomposition $G = KNA$, where N is a unipotent subgroup and the map $K \times N \times A \to G$ is a diffeomorphism. Now M centralizes A and normalizes N. It is easy to check that a function $\bar{g}$ on N that invariant under conjugation by M can be extended to a function g on G that is (K, AM)-invariant. (This is the reason why we write the Iwasawa decomposition as KNA rather than KAN.)

Finally, we say that a set of sine laws is complete if, by using these laws, M^*k_BM can be determined uniquely from M^*k_CM, M^*c and M^*b.

4. The space $\mathrm{SL}(m, \mathbb{R})/\mathrm{SO}(m)$

Let $X = \mathcal{P}(m, \mathbb{R})$ be the space of $m \times m$ positive definite symmetric matrices with determinant 1. The group $G = \mathrm{SL}(m, \mathbb{R})$ acts on X by $(g, p) \to gpg^t$. The stabilizer K of the base point $x_0 = I_m$ is $\mathrm{SO}(m)$. X is an irreducible symmetric space of noncompact type, whose central symmetry at I is given $p \to p^{-1}$. The tangent space of $\mathcal{P}(m, \mathbb{R})$ at I is the space $\mathfrak{p}$ of symmetric matrices with trace equal to 0. The exponential map $\mathrm{Exp}: \mathfrak{p} \to \mathcal{P}(m, \mathbb{R})$ is defined by the usual matrix exponential.

The base chamber can be chosen to be A^+, the set of diagonal matrices $\mathrm{diag}(a_1, \dots, a_m)$ with $a_1 \dots a_m = 1$ and $a_1 > \dots > a_m > 0$. The rank of X is $m - 1$. The Abelian subgroup $A \subset G$ that induces Euclidean translations on this choice of base flat is the group $A(m)$ of diagonal matrices with determinant 1. $M = Z_K(A)$ is the finite group consisting of $\mathrm{diag}(\epsilon_1, \ldots, \epsilon_m)$, with $\epsilon_i = \pm 1$ and $\epsilon_1 \ldots \epsilon_m = 1$. The laws of sines and cosines can now be stated as follows ([Le]).

Theorem 6. *Consider $\mathcal{P}(m, \mathbb{R}) = \mathrm{SL}(m, \mathbb{R})/\mathrm{SO}(m)$. Define $p \colon \mathrm{SL}(m, \mathbb{R}) \to \mathcal{P}(m, \mathbb{R})$ by $p(g) = g^t g$. The functions $F_i(g) = \mathrm{tr}(p(g)^i)$, $i = 1, \ldots, m-1$, form a complete and independent set of (K, K)-invariant functions on $\mathrm{SL}(m, \mathbb{R})$. From these functions a complete set of cosine laws can be obtained by Theorem 4.*

Let $\vec{g}_i$ be the column vectors of g. The functions

$$G_{ij}(g) = (\vec{g}_i, \vec{g}_j)^2 |\vec{g}_i|^{-2} |\vec{g}_j|^{-2},$$

for $i, j = 1, \ldots, m$, form a complete set of (K, AM)-invariant functions on $\mathrm{SL}(m, \mathbb{R})$. From these functions a complete set of sine laws can be obtained by Theorem 5.

The (K, K)-invariant functions given by Chevalley's theorem can be taken to be the elementary symmetric functions in the eigenvalues of $p(g)$, which are related to $\mathrm{tr}(p(g)^i)$ by Newton's formulas. The functions G_{ij} are associated with the Gram-Schmidt process, which gives the KNA-decomposition for $\mathrm{SL}(m, \mathbb{R})$. Note that N is the subgroup of upper triangular matrices with unit diagonal.

For $m = 2$, the space X is the hyperbolic plane, and Theorem 6 reduces to the hyperbolic sine and cosine laws

$$\cosh a = \frac{\cos A + \cos B \cos C}{\sin B \sin C},$$

$$\frac{\sin A}{\sinh a} = \frac{\sin B}{\sinh b} = \frac{\sin C}{\sinh c}.$$

For these and other explicit formulas that can be derived from Theorem 6, we refer to [Le].

The symmetric space $\mathcal{P}(m, \mathbb{R})$ is universal in a certain sense. Given a symmetric space $X = G/K$ of noncompact type, we can find an irreducible, faithful representation $\rho \colon G \to \mathrm{SL}(m, \mathbb{R})$, which in addition is self-adjoint, i.e., $\rho(G)^t \subseteq \rho(G)$. The representation ρ induces an embedding of X into $\mathcal{P}(m, \mathbb{R})$ as a totally geodesic subspace, and the subgroup of $\mathrm{SL}(m)$ that preserves X acts by isometries on X. The base flat and base chamber in X are the restrictions of those in $\mathcal{P}(m, \mathbb{R})$. The subgroup $A \subset G$ that induces translations on the base flat is the restriction of the group $A(m)$. (But note that a regular geodesic in X need not be regular in $\mathcal{P}(m, \mathbb{R})$.) Furthermore, the maximal compact subgroup $K = O(m) \cap G$, after we have identified G with its image under ρ. (See [Mo].)

From these facts it follows that the restrictions of $(O(m), O(m))$-invariant functions to G are (K, K)-invariant and the restrictions of $(O(m), A(m)M(m))$-invariant functions to G are (K, AM)-invariant. In fact, complete sets of invariant functions can be obtained in this way. With Theorem 6, this gives us an alternative method for getting the sine and cosine laws for a general symmetric space of noncompact type.

We now turn to a discussion of congruence conditions for the space $\mathcal{P}(m, \mathbb{R})$.

Theorem 7. *The SSS condition fails to be a congruence condition for $\mathcal{P}(m, \mathbb{R})$ when $m \geq 3$.*

Proof. From Theorem 3 we know that the space of triangles has the form $M^*\backslash(A^+ \times K/M \times A^+)$. Since the group M^* is finite for the space $\mathcal{P}(m, \mathbb{R})$, the dimension of the space of triangles is

$$2\dim(A^+) + \dim(K) = 2(m-1) + (m-1)m/2,$$

while the dimension of the space of three pairs is $3(m-1)$. □

Theorem 8. *Consider two n-tuples of points, $x = (x_1, \ldots, x_n)$ and $y = (y_1, \ldots, y_n)$, in general position in $\mathcal{P}(m, \mathbb{R})$. If all the corresponding triples $(x_{i_1}, x_{i_2}, x_{i_3})$ and $(y_{i_1}, y_{i_2}, y_{i_3})$ are congruent with respect to* $\mathrm{SL}(m, \mathbb{R})$*, then x and y are congruent.*

Proof. We may assume that the first three points of the two triples coincide, and that $x_1 = y_1 = I$, $x_2 = y_2 = a \in A(m)$, and $x_3 = y_3 = p$. For some given k, let $x_k = q$ and $y_k = q'$. We wish to show that $q = q'$.

Now by assumption we have $(I, a, q) \cong (I, a, q') \bmod \mathrm{SL}(m)$. Hence there is an $\epsilon = \mathrm{diag}(\epsilon_1, \ldots, \epsilon_m) \in M$ such that $q' = \epsilon q \epsilon$. Since M is finite, for each ϵ there must be an open set of q for which this equation holds.

By assumption we also have $(I, p, q) \cong (I, p, q')$. Hence $\mathrm{tr}(pq) = \mathrm{tr}(pq') = \mathrm{tr}(p\epsilon q\epsilon)$ for an open set of (p, q). Rewriting this in terms of matrix entries, we have $\sum p_{ij}q_{ij} = \sum p_{ij}q_{ij}\epsilon_i\epsilon_j$. But this is only possible if $\epsilon_i = \epsilon_j$ for all i, j. In other words, $\epsilon = 1$ and $q = q'$, as desired. □

5. Compact symmetric spaces

We will now discuss how to approach the compact case. We must restrict our attention to configurations of points that are "small". For example, we may require that the whole configuration stays within the conjugate locus of each one of its points. The flats then become flat tori, and the chambers are closed linear simplexes.

We will consider $\mathrm{SU}(m)/\mathrm{SO}(m)$, the dual of $\mathrm{SL}(m, \mathbb{R})/\mathrm{SO}(m)$.

Theorem 9. *Consider* $\mathrm{SU}(m)/\mathrm{SO}(m)$. *Let A be the set of diagonal matrices in* $\mathrm{SU}(m)$, *and let M be the finite group consisting of* $\mathrm{diag}(\epsilon_1, \ldots, \epsilon_m)$, *with $\epsilon_i = \pm 1$ and $\epsilon_1 \ldots \epsilon_m = 1$. Let (u, v) be the symmetric bilinear form $u_1v_1 + \cdots + u_mv_m$ on $\mathbb{C}^m$.*

1. *The functions $f_i(u) = \mathrm{tr}(uu^t)^i$, $i = 1, \ldots, m$, form a complete and independent set of (K, K)-invariant functions on* $\mathrm{SU}(m)$. *From the real and imaginary parts of these functions we can obtain a complete set of cosine laws for* $\mathrm{SU}(m)/\mathrm{SO}(m)$ *by Theorem 4.*

2. *Let $\vec{u_i}$ be the column vectors of u. The functions*

$$g_{ij}(u) = (\vec{u_i}, \vec{u_j})^2(\vec{u_i}, \vec{u_i})^{-1}(\vec{u_i}, \vec{u_j})^{-1},$$

for $i, j = 1, \dots, m$, form a complete set of (K, AM)-invariant functions on $\mathrm{SU}(m)$. *From the real and imaginary parts of these functions we can obtain a complete set of sine laws for* $\mathrm{SU}(m)/\mathrm{SO}(m)$ *by Theorem* 5.

The symmetric spaces $\mathrm{SU}(m)/\mathrm{SO}(m)$ and $\mathrm{SL}(m, \mathbb{R})/\mathrm{SO}(m)$ are dual to one another. In particular, their Lie algebras $\mathfrak{su}(m)$ and $\mathfrak{sl}(m, \mathbb{R})$ are real forms of $\mathfrak{sl}(m, \mathbb{C})$. They also have the same isotropy group $\mathrm{SO}(m)$. Comparing Theorems 6 and 9, we see that we have found a set of functions on $\mathrm{SL}(m, \mathbb{C})$ that restrict to (K, K)- and (K, AM)-invariant functions on the groups of both symmetric spaces.

This is a general phenomenon, and gives rise to a correspondence between the trigonometric laws for dual pairs of symmetric spaces. The trigonometric laws described above are expressed in terms of the coordinates of the corners of the triangles, i.e., matrices in $\mathcal{P}(m, \mathbb{C})$ or $\mathrm{SU}(m)$. But we can also write $x_i = \exp V_i$ where V_i is a tangent vector. This will give relations between the tangent vectors describing the sides of the triangle. We will call such relations tangential trigonometric relations. Such relations have many advantages, and they will be discussed in detail later on. In particular, it is easier to describe the relationship between the laws for compact and noncompact spaces using tangential relations. Let $\mathfrak{p}(m, \mathbb{C})$ denote the set of symmetric matrices with complex entries and trace 0. Let $\mathrm{SO}(m)$ act on this space by conjugation. The restriction of this representation to the symmetric matrices with real entries gives us the isotropy representation of $\mathrm{SL}(m, \mathbb{R})/\mathrm{SO}(m)$, while its restriction to those with purely imaginary entries yields the isotropy representation of $\mathrm{SU}(m)/\mathrm{SO}(m)$. The complex exponential map restricts to the exponential maps of each of the two spaces. Thus the substitution $V \to iV$ establishes a correspondence between the tangential trigonometric relations for the two spaces. The substitutions $\lambda \to i\lambda$, $\mathrm{tr}(V_{i_1} \dots V_{i_d}) \to i^d \,\mathrm{tr}(V_{i_1} \dots V_{i_d})$ transform a tangential trigonometric formula on X to such a formula on its dual symmetric space X'.

An obvious question is now how the results of this section are related to the laws of trigonometry for $\mathrm{SU}(3)$ derived in [As]. If we consider $\mathrm{SU}(3)$ as a compact Lie group and try to study the geometry of Cartan cells, we can easily derive the cosine laws, but it is not clear how to get the sine laws. In order for this method to work, we have to view $\mathrm{SU}(3)$ as $\mathrm{SU}(3) \times \mathrm{SU}(3)/\mathrm{SU}(3)$, or imbed it in $\mathrm{SL}(6, \mathbb{R})/\mathrm{SO}(6)$. But in Section 6 we will discuss a method which is more comparable to the results in [As], and which also applies to both the compact and the noncompact case.

6. Invariants of n points and their relations

Let $X = G/K$ be a symmetric space of compact or noncompact type. Consider an n-tuple of points $x = (x_1, \dots, x_n)$. By Theorem 8, its congruence class is determined once we know the congruence class of all its subtriples. Hence all the congruence invariants of x and their relations can be expressed in terms of the invariants and relations associated to the subtriples. Nevertheless, it is of interest to discuss the invariants of n points and the trigonometric relations among such invariants directly. This will also give us an alternative method for deriving the laws of trigonometry for both compact and noncompact spaces. It turns out that this method is also more suitable for computation.

Suppose there are two observers, A and B, attached to the points x_1 and x_2, respectively. Let $g \in G$ be an isometry that takes x_1 to the base point I and $h \in G$ an isometry that takes x_2 to I. (We can assume that the basepoint is I since any noncompact space can be imbedded in $\mathcal{P}(m, \mathbb{R})$ and any compact space can be imbedded in SO(m).) Then

$$\begin{aligned}(x_1, x_2, \ldots, x_n) &\cong (gx_1, gx_2, \ldots, gx_n) = (I, p_2, p_3, \ldots, p_n)\\ &\cong (hx_1, hx_2, \ldots, hx_n) = (q_1, I, q_3, \ldots, q_n).\end{aligned} \tag{3}$$

Let $\mathcal{T} = \{\tau\}$ be a complete set of K-invariant functions in the n variables $x_1, \ldots, x_n$. (Such a set exists since K is compact.) We define the trigonometric data observed by A to be the set of values

$$\tau(gx_1, gx_2, \ldots, gx_n) = \tau(I, p_2, p_3, \ldots, p_n). \tag{4}$$

This is independent of the choice of g, and it is a trigonometric quantity associated to the n-tuple x.

Similarly, we define the trigonometric data observed by B to be the set of values

$$\tau(hx_1, hx_2, \ldots, hx_n) = \tau(q_1, I, q_3, \ldots, q_n). \tag{5}$$

If we set $g = p_2{}^{-1/2}$, then $g = g^t$ and $z \to gzg^t = gzg$ is an isometry which moves p_2 to I. Hence there is an isometry $k \in K$ such that

$$(gIg, gp_2g, gp_3g, \ldots, gp_ng) = (p_2^{-1}, I, gp_3g, \ldots, gp_ng) = k(q_1, I, q_3, \ldots, q_n).$$

(Notice that q_1 and p_2^{-1} are conjugate. This corresponds to the side joining x_1 and x_2.) But then the data observed by B is of the form

$$\tau(q_1, I, q_3, \ldots, q_n) = \tau(p_2{}^{-1}, I, p_2{}^{-1/2}p_3p_2{}^{-1/2}, \ldots, p_2{}^{-1/2}p_np_2{}^{-1/2}), \tag{6}$$

and since the right hand side only involves the p_i, we might be able to express it in terms of the trigonometric data observed by A, i.e., $\tau(I, p_2, \ldots, p_n)$. This will give us the trigonometric relations we desire. In particular, for $n = 2$ we get the laws of trigonometry for triangles.

Since any noncompact symmetric space can be imbedded in $\mathcal{P}(m, \mathbb{R})$ and any compact symmetric space can be imbedded in SO(m), we need to consider orthogonal invariants of symmetric or skewsymmetric matrices. In either case we can take $\mathcal{T}$ to be a set of functions of the form $\tau(x) = \operatorname{tr}(x_{i_1} \cdots x_{i_k})$ ([Sib]). Then the trigonometric

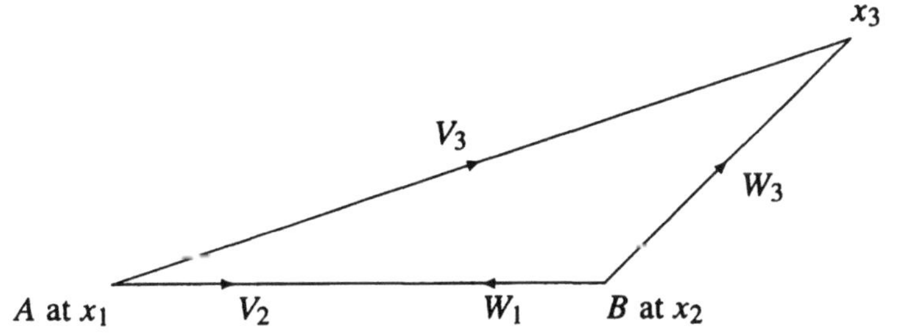

Figure 2: Relating the tangential trigonometric data of two observers

relations we require are of the form

$$\begin{aligned}\operatorname{tr}(q_{i_1}q_{i_2}\cdots q_{i_k}) &= \operatorname{tr}(p_2^{-1/2}p_{i_1}p_2^{-1}p_{i_2}\cdots p_2^{-1}p_{i_k}p_2^{1/2}) \\ &= \operatorname{tr}(p_2^{-1}p_{i_1}p_2^{-1}p_{i_2}\cdots p_2^{-1}p_{i_k}),\end{aligned} \tag{7}$$

where p_1 and q_2 are taken to be I. In other word, for $f(x_1, \ldots, x_n)$ we get

$$f(q_1, q_3, q_4 \ldots, q_n) = f(p_2^{-1}, p_2^{-1}p_3, p_2^{-1}p_4 \ldots, p_2^{-1}p_n). \tag{8}$$

See [AH] for details on how to reduce such expressions. We can also express this in terms of tangential data. We can write $p_i = e^{V_i}$ and $q_i = e^{W_i}$, where $V_i, W_i \in \mathfrak{p}$ and try to find relations between the tangent vectors V_i and W_i. Notice that W_2 and $-V_1$ are conjugate, since they represent the side joining x_1 and x_2 (see Figure 2).

Let $\mathcal{T}' = \{\tau'\}$ be a complete set of K-invariant functions in the $n-1$ variables $U_1, \ldots, U_{n-1}$. (Such a set exists since K is compact.) We define the tangential data observed by A to be the set of values

$$\tau'(V_2, \ldots, V_n), \tag{9}$$

and the tangential data observed by B to be the set of values

$$\tau'(W_1, W_3, \ldots, W_n). \tag{10}$$

Since we can imbed our space into $\mathcal{P}(m, \mathbb{R})$ or $SO(m)$, the tangential trigonometric data observed by A is of the form

$$\operatorname{tr}(V_{i_1}\cdots V_{i_k}) \tag{11}$$

and the tangential trigonometric data observed by B is of the form

$$\operatorname{tr}(W_{i_1}\cdots W_{i_k}). \tag{12}$$

These are the corner invariants of Hsiang ([Hs]) and Aslaksen ([As]).

Symmetric functions in the eigenvalues $\lambda_i(V_j)$ of V_j can easily be determined from $\operatorname{tr}(V_i^k)$, so we will also include them as part of the tangential trigonometric data.

Geometrically, the λ_i measure the interval between the observer and the observed point.

We call relations between the tangential data at the different corners for tangential trigonometric relations. Relation (8) now becomes

$$f(e^{W_1}, e^{W_3}, e^{W_4} \cdots, e^{W_n}) = f(e^{-V_2}, e^{-V_2}e^{V_3}, e^{-V_2}e^{V_4} \cdots e^{-V_2}e^{V_n}). \tag{13}$$

This will give us tangential trigonometric relations once we are able to express the trigonometric data observed at the group level,

$$\operatorname{tr}(e^{V_{i_1}} \cdots e^{V_{i_k}}), \tag{14}$$

in terms of the tangential trigonometric data

$$\operatorname{tr}(V_{i_1} \cdots V_{i_k}). \tag{15}$$

This is accomplished by the following identity.

Theorem 10. *Let M, $P \in M(m, \mathbb{C})$ and let f be an analytic function in one variable. Let $\lambda(M) = (\lambda_1(M), \ldots, \lambda_m(M))$ be the eigenvalues of M. Then*

$$\operatorname{tr}(f(M)P) = \sum_{k=0}^{m-1} h_k(\lambda(M)) \operatorname{tr}(M^k P),$$

where the h_k are of the form

$$h_k(\lambda(M)) = \sum_i f(\lambda_i(M)) R_{ik}(\lambda(M)),$$

and the R_{ik} are symmetric, rational functions in $\lambda(M)$.

Proof. We give the proof in the case where M is diagonalizable with distinct eigenvalues, and P is arbitrary. This forms an open dense subset of matrices, and the remaining cases will follow by proceeding to the limit.

We can write $M = gdg^{-1}$ where d is diagonal. Then $gf(d)g^{-1} = f(M)$ also. Furthermore, since the eigenvalues of d are distinct, the set $1, d, d^2, \ldots, d^{m-1}$ form a basis for the set of diagonal matrices, $\mathfrak{A}$.

There is another basis $e_1, e_2, \ldots, e_m$ for $\mathfrak{A}$, where e_i is the matrix with 1 at the ii-position and 0 elsewhere. These two bases are related as follows

$$(1\ d\ d^2\ \ldots\ d^{m-1}) = (e_1\ e_2\ \ldots\ e_n)V(\lambda),$$

where $V(\lambda)$ is the Vandermonde matrix

$$\begin{pmatrix} 1 & \lambda_1 & \lambda_1^2 & \cdots & \lambda_1^{m-1} \\ 1 & \lambda_2 & \lambda_2^2 & \cdots & \lambda_2^{m-1} \\ & & & \vdots & \\ 1 & \lambda_m & \lambda_m^2 & \cdots & \lambda_m^{m-1} \end{pmatrix}.$$

We let $R_{ik}(\lambda)$ be the ki-th entry of $V(\lambda)^{-1}$ and define $h_k(\lambda)$ as in the statement of the theorem. Now $e_i = \sum_k R_{ik}(\lambda)d^k$. Hence

$$\operatorname{tr}(f(M)P) = \operatorname{tr}(gf(d)g^{-1}P) = \operatorname{tr}(f(d)g^{-1}Pg)$$
$$= \sum_{i,k} f(\lambda_i)\operatorname{tr}(R_{ik}d^k g^{-1}Pg) = \sum_{i,k} f(\lambda_i)R_{ik}\operatorname{tr}(M^k P) = \sum_k h_k(\lambda)\operatorname{tr}(M^k P).$$

The theorem now follows. The functions R_{ik} can be computed explicitly by inverting V. □

As a remark, we would like to point out that by setting $f(M) = M^m$, we can derive the Cayley–Hamilton Theorem from Theorem 10.

We can now take $f = \operatorname{Exp}$ and apply Theorem 10 repeatedly to the trigonometric relations (7). The resulting trigonometric formulas are relations among the tangential data. But notice that the tangential relations are just restatements of the original relations. We must first determine the set of K-invariants of in $(x_1, \dots, x_n)$. These computations have been carried out for SU(3) and SL(3, ℂ)/ SU(3), giving an alternative to the methods used by Aslaksen ([As]) and Leuzinger ([Le]). The details will be published elsewhere ([AH]).

References

[As] Aslaksen, H., Laws of trigonometry on SU(3), Trans. Amer. Math. Soc. 317 (1990), 127–142.

[AH] Aslaksen, H., Huynh, H.-L., Laws of trigonometry of SU(3) and SL(3, ℂ)/ SU(3), Preprint, National Univ. of Singapore, 1995.

[BT] Blaschke, W., Terheggen, H., Trigonometria Hermitiana, Rend. Sem. Mat. Univ. Roma (4) 3 (1939), 153–161.

[Br] Brehm, U., The Shape invariant of triangles in two-point homogeneous spaces, Geom. Dedicata 33 (1990), 59–76.

[Co] Coolidge, J. L., Hermitian metrics, Ann. of Math. 22 (1921), 11–28.

[Du1] Dubnov, J. S., Sur une généralisation de l'équation de Hamilton-Cayley et sur les invariants simultanés de plusieurs affineurs, Trudy. Sem. Vektor. Tenzor. Anal. 2–3 (1935), 351–367.

[Du2] ——, Complete system of invariants of two affinors in centro-affine space of two or three dimensions, Trudy. Sem. Vektor. Tenzor. Anal. 5 (1941), 250–270 (in Russian), Math. Reviews 8 (1947), 95.

[DI] Dubnov, J. S., Ivanov, V., Sur l'abaissement du degré des polynomes en affineurs, C. R. (Doklady) Acad. Sci. URSS 41 (1943), 95–98.

[He] Helgason, S., Differential Geometry, Lie Groups, and Symmetric Spaces, Academic Press, New York 1978.

[Hs] Hsiang, W.-Y., On the laws of trigonometries of two-point homogeneous spaces, Ann. Global Anal. Geom. 7 (1989), 29–45.

[Hu] Huynh, H.-L., General trigonometic relations in symmetric spaces, with subsequent reflections on the nature of metrically conservative societies, Ph.D. thesis, UC Berkeley, 1993.

[Le] Leuzinger, E., On the trigonometry of symmmetric spaces, Comment. Math. Helv. 67 (1992), 252–292.

[Le1] Lew, J. S., The generalized Cayley–Hamilton Theorem in n dimensions, Z. Angew. Math. Phys. 17 (1966), 650–653.

[Le2] —, Reducibility of matrix polynomials and their traces, Z. Angew. Math. Phys. 18 (1967), 289–293.

[Mo] Mostow, G. D., Strong Rigidity of Locally Symmetric Spaces, Ann. of Math. Stud. 78, Princeton Univ. Press, Princeton 1973.

[Pr] Procesi, C., The invariant theory of $n \times n$ matrices, Adv. in Math. 19 (1976), 306–381.

[Ra] Razmyslov, Y. P., Trace identities of full matrix algebras over a field of characteristic zero, Math. USSR-Izv. 8 (1974), 727–760.

[Ri] Rivlin, R. S., Further remarks on the stress deformation relations for isotropic materials, J. Rat. Mech. Anal. 4 (1955), 681–702.

[Ro] Rozenfeld, B. A., On the theory of symmetric spaces of rank one, Mat. Sb. (N.S.) 41 (83) (1957), 373–380 (in Russian), Math. Reviews 20 (1959), 2756 a, b.

[Sib] Sibirskii, K. S., Algebraic invariants for a set of matrices, Siberian Math. J. 9 (1968), 115–124.

[Sir] Sirokov, P. A., On a certain type of symmetric space, Mat. Sb. (N.S.) 41 (83) (1957), 361–372 (in Russian), Math. Reviews 20 (1959), 2755 a, b.

[Te] Terheggen, H., Zur analytischen Geometrie auf der Geraden von Hermite als Grenzfall der Geometrie in der Hermitischen Ebene und ihr Zusammenhang mit der gewöhnlichen sphärischen Trigonometrie, Jahresber. Deutsche Math. Verein. 50 (1940), 24–35.

[We] Weyl, H., The Classical Groups, Princeton Univ. Press, Princeton 1946.

[Wo] Wolf, J. A., Spaces of Constant Curvature, McGraw-Hill, New York 1967.

Determinants and extremal metrics in conformal geometry

Sun-Yung A. Chang and Paul C. Yang***

0. Introduction

In the recent literature, there are many instances where extremal functions and optimal constants in Sobolev inequalities play important role in the solution of geometric variational problems. A well known example is the solution of the Yamabe problem provided by Aubin [Au] and Schoen [Sc-1]: On a compact closed Riemannian manifold (M, g) of dimension $n \geq 3$, among metrics $g_u = u^{\frac{4}{n-2}} g$ conformal to g, the infimum of $\int R_u \, dV_u$ is attained under the constraint $\mathrm{Vol}(g_u) = \mathrm{Vol}(g)$, where R_u is the scalar curvature of g_u; and hence the scalar curvature of the minimizing metric is a constant function. The scalar curvature R_u is given by the differential equation

$$c_n \Delta u + R_u u^{\frac{n+2}{n-2}} = Ru \quad \text{on } M, \tag{0.1}$$

where R is the scalar curvature of (M, g), $c_n = 4\frac{n-1}{n-2}$. The problem to minimize $\int_M R_u \, dV_u$ over $\mathrm{Vol}(g_u) = \mathrm{Vol}(g)$ is equivalent to the problem to find a positive function u which extremizes the following Sobolov inequality:

$$Y(M, g) \left(\int_M u^{\frac{2n}{n-2}} \, dV_g \right)^{\frac{n-2}{n}} \leq c_n \int_M |\nabla u|^2 \, dV_g + \int_M Ru^2 \, dV_g \tag{0.2}$$

where $Y(M, g)$ is the optimal constant in the inequality.

In dimension two, one may view the uniformization theorem as the solution of the corresponding Yamabe problem. On a compact surface (M, g) within the class of metrics $g_w = e^{2w} g$ conformal to g, one can find a metric g_w with constant Gaussian curvature $K[w] \equiv 1, 0$, or -1 according to the sign of $\int K \, dV_g$. The Gaussian

---

* Research supported in part by NSF Grant DMS-9401465.

** Research supported in part by NSF Grant DMS-9300881.

curvature K_w is given by the differential equation

$$\Delta w + K_w e^{2w} = K \quad \text{on } M. \tag{0.3}$$

According to the Gauss–Bonnet theorem, $\int_M K_w \, dV_{g_w} = 4\pi\chi(M)$, where $\chi(M)$ is the Euler characteristic of M, a topological constant. Thus it does not make sense to minimize $\int_M K_w \, dV_{g_w}$. The variational functional with (0.3) as Euler equation is given by

$$J[w] = \log \frac{\int_M dV_{g_w}}{\int_M dV_g} - \left(\int_M |\nabla w|^2 \, dV_g + 2 \int_M K w \, dV_g \right). \tag{0.4}$$

To prove that $J[w]$ attains its supremum, one needs to employ a sharp inequality of Moser–Trudinger ([M], [T]): there exists a constant $c = c(M, g)$ so that if $w \in C^1(M)$ then

$$\int_M e^{4\pi w^2} \, dV_g \leq c, \text{ if } \int_M |\nabla w|^2 \, dV_g \leq 1 \quad \text{and} \quad \int_M w \, dV_g = 0. \tag{0.5}$$

There is an interesting connection of the functional (0.4) to a geometric variational problem. For a Riemannian metric g, let $\{0 < \lambda_1 \leq \lambda_2 \leq \cdots\}$ be the spectrum of the Laplacian Δ_g. Let $\zeta(s) = \sum \lambda_i^{-s}$ for $\Re s > 1/2$, then through a meromorphic continuation of the zeta function using the heat kernel expansion of Δg, $-\zeta'(0)$ is well-defined, and one may define $\log \det \Delta g$ to be $-\zeta'(0)$ (cf. [R-S]). In [Po], Polyakov further computed the logarithm of the ratio of determinant of two conformally related metrics $g_w = e^{2w} g$ on a compact surface without boundary.

$$F[w] = \log \frac{\det \Delta_w}{\det \Delta} = \frac{1}{3} \int_M (|\nabla w|^2 + 2Kw) \, dV_g \tag{0.6}$$

under the normalization that $\mathrm{Vol}(g_w) = \mathrm{Vol}(g)$. Notice that $F[w]$ is essentially the same as the functional $J[w]$ in (0.4). In a series of papers, Osgood–Phillips–Sarnak ([OPS-1], [OPS-2], [OPS-3]) have further studied the functional $F[w]$, and they have shown among other things that $F[w]$ enjoys a certain compactness property on account of the Moser–Trudinger inequality, and proved that in each conformal class, the functional $F[w]$ attains its extrema at the constant curvature metrics.

To generalize formula (0.6) to manifolds of higher dimensions, we need to recall some definitions. Recall that a metrically defined operator A is said to be conformally covariant if under the conformal change of metric $g_w = e^{2w} g$, the pair of corresponding operators A_w and A are related by

$$A_w(\varphi) = e^{-bw} A(e^{aw} \varphi). \tag{0.7}$$

The basic examples of such operators are the conformal Laplacian $L = -c_n \Delta + R$, where $c_n = \frac{4(n-1)}{n-2}$ and R is the scalar curvature, then (0.7) holds with $a = \frac{n-2}{2}$ and $b = \frac{n+2}{2}$; for the Dirac operator $\not\nabla$ on spinors, (0.7) holds with $a = \frac{n-1}{2}$, $b = \frac{n+1}{2}$.

Consider the interesting fourth order operator on 4-manifolds due to Paneitz ([Pa]):

$$P\varphi = \Delta^2\varphi + \delta\left(\frac{2}{3}RI - 2\mathrm{Ric}\right)d\varphi, \tag{0.8}$$

where δ denotes the divergence, d the deRham differential, and Ric the Ricci curvature of the metric. Under the conformal change $g_w = e^{2w}g$, P undergoes the transformation $P_w = e^{-4w}P$. For general dimension n, the existence of such an operator P with $P_w = e^{-nw}P_n$ for even dimensions is verified in [GJMS]. However, it is only explicitly known for the Euclidean space R^n and hence for the spheres S^n. The explicit formula for P on S^n has appeared in Branson [B-1], see also Beckner [Be].

When the dimension of a closed manifold is odd, it was shown in Branson [B-2] that $\log\det L_g$ is a conformal invariant. Thus the next natural dimension to study the generalized Polyakov formula (0.8) is four.

In this paper, we will survey some of the results in the study of extremal functions and extremal metrics for the generalized Polyakov formula in dimension four. We organize the paper as follows. In Section 1, we study the log determinant formula for 4-manifolds without boundary, citing results mainly from [BO], [BCY] and [CY]. In Section 2, we discuss the Paneitz operator and its associated curvature Q, we also give a geometric application of the extremal metric, the main references being [CY] and [CGY]. In Section 3, we study the log-determinant formula (0.6) for compact manifolds with boundary, citing results from [Al], [BG], [CQ-1] and [CQ-2]. Finally in Section 4, we discuss some PDE problems related to the curvature operators which arise naturally in the discussion of the log-determinant formulas above, we also discuss some open questions.

1. Extremal metrics on compact closed 4-manifolds

Suppose (M, g) is a compact, closed 4-manifold, and suppose A is a conformally covariant operator satisfying (0.7) with $b - a = 2$. In [BO] Branson–Ørsted gave an explicit computation of the normalized form of $\log \frac{\det A_w}{\det A}$ which may be expressed as:

$$F[w] = \gamma_1 I[w] + \gamma_2 II[w] + \gamma_3 III[w] \tag{1.1}$$

where γ_1, γ_2 and γ_3 are constants depending only on A and

$$I[w] = 4\int |C|^2 w\, dV_g - \left(\int |C|^2\, dV_g\right)\log \fint e^{4w}\, dV_g$$

$$II[w] = \langle Pw, w\rangle + 4\int Qw\, dV_g - \left(\int Q\, dV_g\right)\log \fint e^{4w} dV_g,$$

$$III[w] = 12\left(Y(w) - \frac{1}{3}\int (\Delta R)\, w\, dV_g\right),$$

where C is the Weyl tensor, $12Q = -\Delta R + R^2 - 3|\mathrm{Ric}|^2$ and $Y(w) = \int \left(\frac{\Delta(e^w)}{e^w}\right)^2 - \frac{1}{3}\int R|\nabla w|^2$. The quantities Q and $|C|^2$ occur in the Gauss–Bonnet integrand:

$$4\pi^2\chi(M) = \int \left(\frac{|C|^2}{8} + Q\right) dV_g, \tag{1.2}$$

where $\chi(M)$ denotes the Euler characteristic of the manifold M. It is also important to remark that the functional $III[w]$ [BCY] may be written as

$$III[w] = \frac{1}{3}\left[\int R_w^2\, dV_w - \int R^2\, dV_g\right]$$

so that when the background metric is assumed to be the Yamabe metric in a positive conformal class, the functional III is non-negative.

In [BCY] we made two observations. The first is that on the standard 4-sphere (S^4, g_0) the functional $F[w]$ for the conformal Laplacian L (and Dirac square $\nabla\!\!\!/^2$) is extremized in a strong way, that each term $II[w]$ and $III[w]$ are extremized by the standard metric $g_w = g_0$. For $III[w]$ this is a consequence of Obata's result ([Ob]) that the constant scalar curvature conformal metrics on S^4 are standard. For the functional $II[w]$, this is a consequence of Beckner's [Be] inequality which is valid for all dimensions, (see discussion in Section 4). The second observation is that the functional $F[w]$ enjoys certain compactness properties for the operators L and $\nabla\!\!\!/^2$ for most compact locally symmetric Einstein 4-manifolds. The basic analytic inequality required is Adam's inequality [Ad]: on (M^4, g), there exists a constant C_M such that

$$\int e^{32\pi^2 w^2} \le C_M \quad \text{if} \int w = 0 \text{ and } \int (\Delta w)^2 \le 1. \tag{1.3}$$

In [CY], we continue the study of the log-determinant formula (1.1) on general 4-manifolds. The first result is to extend the compactness criteria to a more general class of 4-manifolds; then we isolate the functional $II[w]$ and provide certain compactness criteria. We also provide certain apriori estimates for solutions of the Euler equations for the functional II.

We define conformal invariants

$$\begin{aligned} k_d &= -\gamma_1 \int |C|^2\, dV_g - \gamma_2 \int Q\, dV_g \\ &= (-\gamma_2)4\pi^2\chi(M) + \left(\frac{\gamma_2}{8} - \gamma_1\right)\int |C|^2\, dV_g \end{aligned} \tag{1.4}$$

and

$$\begin{aligned} k_p &= \int Q\, dV_g \\ &= 4\pi^2\chi(M) - \frac{1}{8}\int |C|^2 dV_g. \end{aligned} \tag{1.4'}$$

We have the following existence results:

Theorem 1.1. *If the functional F satisfies $\gamma_2 < 0$, $\gamma_3 < 0$, and $k_d < (-\gamma_2)8\pi^2$, then* $\sup\{F[w] \mid w \in W^{2,2}\}$ *is attained by some function w_d and the metric $g_d = e^{2w_d} g_0$ satisfies the equation*

$$\gamma_1 |C_d|^2 + \gamma_2 Q_d - \gamma_3 \Delta_d R_d = -k_d \cdot \mathrm{Vol}(g_d)^{-1}. \tag{1.5}$$

Further, all functions $\varphi \in W^{2,2}$ satisfy the inequality

$$k_d \log \fint e^{4(\varphi - \tilde{\varphi})}\, dV_d \leq (-\gamma_2)\langle P\varphi, \varphi\rangle - 12\gamma_3 Y_d(\varphi). \tag{1.6}$$

where $\tilde{\varphi}$ denotes the mean value of φ with respect to the metric g_d, and $\fint$ denotes $\frac{1}{\mathrm{Vol}(M, g_d)} \int_M dV_d$.

For the functional II, we have a similar existence result:

Theorem 1.2. *If $k_p < 8\pi^2$, and assume that P is an non-negative operator with kernel of P consisting of constant functions. Then* $\inf\{II(w) \mid w \in W^{2,2}\}$ *is achieved by some function w_p and the metric $g_p = e^{2w_p} g$ satisfies the equation*

$$Q_p = k_p \,\mathrm{Vol}(g_p)^{-1}, \tag{1.7}$$

and all functions $\varphi \in W^{2,2}$ satisfy the inequality

$$k_p \log \fint e^{4(\varphi - \tilde{\varphi})}\, dV_p \leq \langle P\varphi, \varphi\rangle \tag{1.8}$$

where $\tilde{\varphi}$ is the mean value of φ with respect to the metric g_p. Further the equation (1.7) *may be expressed in terms of the conformal factor w_p:*

$$Pw_p + 2Q = 2Q_p e^{4w_p}. \tag{1.9}$$

In particular, for the operators L and ∇^2, we obtain existence results for extremal metrics of the corresponding log-determinant functional. We also obtain existence of extremals for the functional II in many cases. Thus for a large class of conformal 4-manifolds, we have the existence of several extremal metrics in addition to the Yamabe metric. It is an interesting problem to study the relation among these metrics. For example, we found in [BCY] that on S^4 all these extremal metrics coincide. In order to identify these extremal metrics in special circumstances, we provide some uniqueness results.

Theorem 1.3. *If $k_d \leq 0$, the extremal metric g_d for the functional F corresponding to the conformal Laplacian operator L is unique.*

Theorem 1.4. *If $k_p \leq 0$, P is non-negative and* $\ker P = \{constants\}$, *then the extremal metric g_p is unique.*

These uniqueness assertions are obtained as consequences of the convexity of the corresponding functionals. We were able to identify some of the extremal metrics with known metric in special circumstances. We remark that:

1) Theorem 1.1 applies to all functionals of the form $\gamma_1 I + \gamma_2 II + \gamma_3 III$. What is required in the statement of Theorem 1.1 is that $\gamma_2\gamma_3 > 0$. In case $\gamma_2 > 0, \gamma_3 > 0$ we substitute infimum for supremum of $F[w]$ in the statement of the theorem.

2) The main examples of functionals, $F[w]$, which appear in the form of Theorem 1.1 above, are quotients of two zeta function determinants $\log(\det A_w / \det A)$, where A_w is the operator A evaluated with respect to the conformal metric $g_w = e^{2w} g$. In [BO], for operators A or powers of A which satisfy the conformal covariance property:
$$A_w(\varphi) = e^{-bw} A(e^{aw}\varphi),$$
explicit formulae for quotient of function determinant are computed and is of the form $F[w]$. Also up to a multiplicative constant, the coefficients $\gamma_1, \gamma_2, \gamma_3$ in $F[w]$ are computed for $A = -\Delta + \frac{R}{6}$, the conformal Laplacian operator, for which the conformal covariance property holds with $b = 3$ and $a = 1$, and for $A = \nabla\!\!\!\!/^2$, the square of the Dirac operator, for which the conformal covariance property holds with $b = \frac{3}{2}$ and $a = \frac{5}{2}$.

In the following, we combine the computation in [BO] with the the above theorem. We remark that in [BO], the coefficients of $F[w]$ is expressed in terms of $(\beta_1, \beta_2, \beta_3)$, with $(\beta_1, \beta_2, \beta_3) = (\gamma_1, \gamma_2, 6\gamma_3)$.

Corollary 1.5. *For the conformal Laplace operator $L = -\Delta + \frac{R}{6}$ defined on a compact closed 4-manifold (M, g), we have $(\gamma_1, \gamma_2, \gamma_3) = (1, -4, -\frac{2}{3})$, $k_d = 16\pi^2\chi(M) - \frac{3}{2}\int |C|^2$ so that for the functional $F[w] = \gamma_1 I[w] + \gamma_2 II[w] + \gamma_3 III[w]$, $\sup\{F[w] \mid w \in W^{2,2}\}$ is attained with extremal metric satisfying (1.5) and (1.6) when $k_d < 32\pi^2$.*

In this case, if zero is not an eigenvalue of L, then setting
$$P_{L_w}(g_w) = \left(\frac{\mathrm{Vol}(g_w)}{\mathrm{Vol}(g)}\right)^{\frac{1}{2}\zeta_{L_w}(0)} \det L_w,$$
we find that
$$F[w] = (4\pi)^{-2}\frac{1}{90}\log\left(P_L(g)/P_{L_w}(g_w)\right).$$

Corollary 1.6. *For the square of the Dirac operator $\nabla\!\!\!\!/^2$ defined on a compact closed 4-manifold (M, g) we have $(\gamma_1, \gamma_2, \gamma_3) = (7, -88, -\frac{14}{3})$, $k_d = 352\pi^2\chi(M) - 18\int |C|^2$ so that for the functional $F[w] = \gamma_1 I[w] + \gamma_2 II[w] + \gamma_3 III[w]$, $\sup\{F[w] \mid w \in W^{2,2}\}$ is achieved by an extremal metric satisfying (1.5), (1.6) when $k_d < 704\pi^2$.*

In this case if we assume zero is not an eigenvalue of ∇^2, then setting

$$P_{\nabla^2_w}(g_w) = \left(\frac{\mathrm{Vol}(g_w)}{\mathrm{Vol}(g)}\right)^{\frac{1}{2}\zeta_{\nabla^2_w}(0)} \det \nabla^2_w$$

we find that $F[w] = (4\pi)^{-2}\frac{1}{180}\log\left(P_{\nabla^2_w}(g_w)/P_{\nabla^2}(g)\right)$.

We examine the situations in which we can identify some of the extremal metrics. Let g_d denote the extremal metric for the determinant of the conformal Laplacian, g_p denote the extremal metric for the functional II, and g_y denote the Yamabe metric. In [CY], these metrics are examined for several families of canonical metrics. Here are some examples: We denote $k'_p = 4k_p$.

1) For S^4 we have $|C|^2 = 0$, $\chi = 2$ hence $k_d(S^4) = k'_p(S^4) = 32\pi^2$. In this case all three extremal metrics exist and we have the coincidence of all three with the standard metric: $g_d = g_p = g_y = g_0$.
2) For $\mathbb{H}^2 \times \mathbb{H}^2$, we have $\chi = (4\pi^2)^{-1} \cdot$ volume, $|C|^2 = \frac{16}{3}$. For any lattice Γ, $k_d(\mathbb{H}^2 \times \mathbb{H}^2/\Gamma) = -4 \cdot \text{volume} < 0$, $k'_p = \frac{4}{3}\text{volume}$. So we have $g_0 = g_d = g_y$. When Σ is a genus 2 hyperbolic surface, we have $k'_p(\Sigma \times \Sigma) = \frac{4}{3}16\pi^2 < 32\pi^2$. In the standard metric for $\Sigma \times \Sigma$ we have
$$P = \Delta^2 + \Delta.$$
If we take Σ to have a pinching neck so that $\lambda_1(\Sigma) \ll 1$, then we see that $\lambda_1(P)$ is small negative. For such a conformal structure on $\Sigma \times \Sigma$, the functional II is not bounded from below, and the standard metric is a saddle point for II.
3) $S^1 \times S^3$ is known to be conformally flat and by varying the radius t of S^1 we have a 1-parameter family of conformally flat structures $S^1(t) \times S^3$. We have $\chi = 0$ and $|C|^2 = 0$, so that $k_d(S^1 \times S^3) = k'_p(S^1 \times S^3) = 0$. It follows that $g_0 = g_d = g_p$. The last equality holds since the Euler equation is a linear equation. We recall that Schoen [Sc-2] has shown that there exists a t_0 such that for $t \leq t_0$, $g_y = g_0$; while $g_y \neq g_0$ when $t > t_0$. This is an instance where the extremal metric g_d is different from the Yamabe metric.

2. A geometric application

From the discussion in the section above, we observe that the Paneitz operator P in (0.8) and its associated curvature Q on a 4-manifold is a natural analogue of the operator $-\Delta$ and the Gaussian curvature K on compact surface. For example (P, Q) satisfies

$$-Pw + 2Q_w e^{4w} = 2Q \tag{2.1}$$

similar to the role the pair $(-\Delta, K)$ plays in (0.3). To strengthen the analogy, we note that the curvatures K and Q occur as integrands in the Chern–Gauss–Bonnet formula. On general compact 4-manifolds, one can study apriori bounds for general solutions satisfying (2.1) using methods similar to that in the much studied problem of "prescribing Gaussian curvature" on a compact surface. Some apriori estimates for solutions of this equation is also given in [CY].

Another potential application to results in Section 1 is to use the existence of extremal metric with $Q \equiv$ constant to study 4-manifolds. We will describe here a result which points to partial success in this direction.

Suppose (M^4, g) is a compact, locally conformally flat 4-manifold with positive scalar curvature. Then by a result of Bourguignon ([Bo]), the second Betti number of M^4 satisfies $\beta_2 = 0$. Consequently if M^4 is orientable, we have

$$\chi(M^4) = \beta_0 - \beta_1 + \beta_2 - \beta_3 + \beta_4 = 2 - 2\beta_1. \tag{2.2}$$

(2.2) should be compared to the familiar formula for compact surfaces

$$\chi(M^2) = 2 - 2g, \tag{2.3}$$

where g is the genus of M^2. Thus for such 4-manifold $\chi(M^4) \leq 2$. The following classification theorem which can be derived using Schoen's result on the Yamabe problem, is given an alternative proof by Gursky in [G] without using the positive mass theorem.

Theorem 2.1 ([Gu, Theorem A]). *Let M be a compact 4-manifold. There is an $\varepsilon_0 > 0$ such that if M admits a metric g of non-negative scalar curvature whose Weyl curvature C_g satisfies*

$$\int |C_g|^2 \, dV_g < \varepsilon_0$$

then $\chi(M) \leq 2$. Furthermore, if $\chi(M) > 0$ then M also admits a metric of constant positive scalar curvature; hence M is diffeomorphic to S^4 if $\chi(M) = 2$, and M is diffeomorphic to $\mathbb{RP}^4$ if $\chi(M) = 1$.

One might hope to use a similar strategy as the proof in Theorem 2.1 above to characterize such M^4 with $\chi(M^4) = 0$. But the study of Schoen [Sc-2] of the Yamabe equation for the conformally flat manifolds $M^4(t) = S^3 \times S^1(t)$ suggests there is a subtle difficulty. He showed that depending on the parameter t, the product metric may or may not be the Yamabe metric which minimize $\int R_g \, dV_g$. The existence and uniqueness results in Section 1 indicate that the log determinant functional is a suitable variational problem for the conformal classes of $M^4(t)$. According to Corollary 1.5, if (M^4, g) is LCF with $R > 0$ and $\chi(M^4) = 0$, then $k_d = 0$, hence an extremal metric g_d for the log determinant functional $F[w] = \log\det L_w / \det L$ exists and is unique. Our next result shows that in this case, if R_d is a constant, then (M^4, g_d) is isometric to the quotient of $S^3 \times S^1$ with the standard product metric.

Theorem 2.2. *Suppose* (M^4, g) *is compact, LCF with* $R > 0$ *and* $\chi(M^4) = 0$, *then* $k_d = 0$ *hence an extremal metric* g_d *for the functional* $F[w] = \log\det L_w/\det L$ *exists and is unique. If in addition* R_d *is a constant then* $(\widetilde{M}^4, g_d)$, *the universal cover of* (M, g_d), *is isometric to* $S^3 \times \mathbb{R}^1$ *with the product metric.*

For another analogy between the curvature Q and the scalar curvature R, we adopt the point of view of Fischer and Marsden ([FM]) in their study of the scalar curvature. They considered R as a map from $\mathcal{M}(M)$, the space of Riemannian metrics on a manifold M, to $C^\infty(M)$. The derivative of R at a metric g is denoted R'_g and is a linear map from $\Sigma_2(M)$, the space of symmetric 2-tensors on M, to $C^\infty(M)$. Its L^2-adjoint $(R'_g)^* : C^\infty(m) \to \Sigma_2(M)$. If R'_g fails to be surjective then R is not a submersion at g and we say that g is R-singular. This condition is equivalent to $(R'_g)^*$ possessing a non-trivial kernel; hence there is a function $f \equiv 0$ such that

$$(R'_g)^* f = 0. \tag{2.4}$$

In order to amplify the parallels between the work of Fischer and Marsden on the scalar curvature and our work on Q, we are going to state their result in a slightly different (but in out view more natural) manner than what appears in [F-M]. To this end, recall that the Einstein gravitational tensor $T(g) = \mathrm{Ric} - \frac{1}{2}Rg$ is the gradient of the total scalar curvature

$$\mathcal{S}[g] = \int R_g dV_g.$$

We then have

Theorem 2.3 ([F-M]). *Suppose M is a compact manifold and g is R-singular. Then R is a constant and* $(\frac{n+2}{n-2})R \in \operatorname{spec}(L_g)$. *Furthermore,* (2.4) *admits constant solutions if and only if* $T(g) \equiv 0$.

Remarks.

1) If $T(s) = 0$, then by tracing we see that $R = 0$, hence g is Ricci-Flat.
2) R-singular metrics which are LCF have been classified (see [K] and [La]).

Interpreting Q as a map $Q : \mathcal{M}(M^4) \to C^\infty(M^4)$ and letting Q'_g denote its derivative at g, we say that g is Q-singular if there exists a non-trivial solution to

$$(Q'_g)^*(f) = 0. \tag{2.5}$$

Recall that the Bach tensor $B(g)$ is the gradient of the Weyl functional

$$\begin{aligned}\mathcal{W}[g] &= \int |C_g|^2\, dV_g \\ &= 32\pi^2\chi(M^4) - 8\int Q_g\, dV_g.\end{aligned}$$

Therefore, $-\frac{1}{8}B(g)$ is the gradient of the total Q curvature:

$$\mathcal{Q}[g] = \int Q_g \, dV_g.$$

Theorem 2.4. *Suppose M^4 is a compact manifold and g is Q-singular. Then Q is a constant and $8Q \in \operatorname{spec}(P_g)$. Furthermore,* (2.5) *admits constant solutions if and only if $B(g) \equiv 0$ and $Q \equiv 0$.*

Theorem 2 sheds some light on why the case where (M^4, g) is LCF and $\chi(M^4) = 0$ should be difficult. For if $g_d = e^{2w_d}g$ has constant scalar curvature then the Euler equation for Q_d becomes

$$Q_d \equiv 0,$$

and we conclude that g_d is Q-singular. This is analogous to the study of the scalar curvature map near a Ricci-flat metric, which is one of the most technically difficult parts of [F-M].

3. Zeta functional determinant on 4-manifolds with boundary

Let M be a compact Riemannian manifold with smooth boundary. In dimension two, Alvarez [Al] considered the Laplacian with the Dirichlet boundary condition on compact surfaces with boundary and derived the following Polyakov (cf. [Po], [RS]) type formula:

$$\log \frac{\det \Delta[\omega]}{\det \Delta[0]} = -\frac{1}{6\pi}\Big\{\frac{1}{2}\int_M \omega\Delta\omega + \int_M \omega K[0] + \frac{1}{2}\oint_{\partial M} \omega(-N\omega) + \oint_{\partial M} \omega k[0]\Big\}$$
$$-\frac{1}{4\pi}\Big\{\oint_{\partial M} (k\,ds)[\omega] - \oint_{\partial M} (k\,ds)[0]\Big\}, \tag{3.1}$$

where K is the Gaussian curvature of M, k is the geodesic curvature of ∂M and N is the inward unit normal. There are two types of variational problems for the Alvarez formula. We will name the conformal change of metrics preserving $\operatorname{Vol}(M)$ as the conformal variation of type I, and name the conformal change of metrics preserving $\operatorname{Vol}(\partial M)$ as the conformal variation of type II. Osgood, Phillips and Sarnak in [OPS-1], [OPS-2], [OPS-3] have proved the existence of extrema of the zeta functional determinant under conformal variations of both types (see also [On]). A crucial tool which they have used to establish their compactness result is the following sharp inequalites of Moser–Trudinger type

$$\log \int_{S^2_+} e^{2\phi}\, dv \le \frac{1}{2\pi}\int_{S^2_+} |\nabla\phi|^2\, dv + 2\int_{S^2_+} \phi \frac{dv}{2\pi}, \tag{3.2}$$

and of Lebedev–Milin type

$$\log \oint_{\partial D} e^{(u-\bar{u})} \frac{d\theta}{2\pi} \leq \frac{1}{4}\left\{ \int_D u \Delta u \frac{dx\,dy}{\pi} + 2 \oint_{\partial D} u \frac{\partial}{\partial n} u \frac{d\theta}{2\pi} \right\}, \tag{3.3}$$

where $\bar{u}$ is the average of u on the boundary ∂D of the unit disk in the plane.

We look for an Alvarez–Polyakov formula in higher dimensions for a pair of operators (A, B). We assume that A is a self-adjoint, geometric differential operator of order $2l$ with positive definite leading symbol, and B is a differential operator defined on the Cauchy data for A on ∂M. We will assume further that (A, B) satisfies the following conformal assumptions.

Conformal Assumptions. Suppose that both A and B are conformally covariant of bidegree (a_1, a_2) and (b_1, b_2) in the following sense

$$A[\omega](f) = e^{-a_1\omega} A[0](e^{a_2\omega} f),$$
$$B[\omega](g) = e^{-b_1\omega} B[0](e^{b_2\omega} g),$$

for any $f \in C^\infty(M)$, $g \in C^\infty(\partial M)$. And assume that

$$B[0](e^{a_2\omega} g) = 0 \text{ if and only if } B[\omega](g) = 0,$$

for any $\omega \in C^\infty(\bar{M})$.

Example. The typical examples of pairs (A, B) which satisfy all three assumptions above are

$$A = \mathcal{L} = \Delta + \frac{n-2}{4(n-1)} R$$

the conformal Laplacian of bidegree $(\frac{n+2}{2}, \frac{n-2}{2})$, and R is the scalar curvature, and B is either the Dirichlet boundary condition or

$$B = \mathcal{R} = -N + \frac{n-2}{2(n-1)} H$$

the Robin operator of bidegree $(\frac{n}{2}, \frac{n-2}{2})$, where H is the trace of the second fundamental form of the boundary ∂M.

Under these assumptions, Branson and Gilkey [BG-1] have established a general theory to compute the zeta functional determinant for ratio of two conformal metrics. In particular, they have computed some formula for the determinant on 4-manifolds with boundary. In [CQ-1] we rederive the determinant formula for 3- and 4-manifolds with boundary in geometric forms. One of the basic observations which we have made in [CQ-1] is a discovery of a boundary operator of bidegree (0,3) on the boundary of a 4-manifold. To explain the operator, we first observe that on compact surface, the geodesic curvature k, which has appeared in the Alvarez formula above, is associated with the the Neumann operator $P_1 = -N$ in the following way:

$$-N\omega + k = k_w e^{\omega} \text{ on } \partial M. \tag{3.4}$$

In dimension 4, we have found a boundary curvature invariant

$$T = -\frac{1}{2}NJ + JH - R_{aNbN}L_{ab} + \frac{1}{9}H^3 - \frac{1}{3}\operatorname{Tr} L^3 + \frac{1}{3}\tilde{\Delta}H,$$

which is associated to the operator

$$P_3 = -\frac{1}{2}N\Delta - \tilde{\Delta}N + \frac{2}{3}H\tilde{\Delta} + L_{ab}\tilde{\nabla}_a\tilde{\nabla}_b + \frac{1}{3}\tilde{\nabla}H \cdot \tilde{\nabla} + (F - 2J)N,$$

where $J = R/6$, L is the second fundamental form of ∂M, $F = R_{aNaN}$, $\tilde{\Delta}$ denotes the Laplacian on the boundary and $\tilde{\nabla}$ denotes the gradient on the boundary. The curvature T transforms under conformal change of metrics as

$$P_3\omega + T = T_w e^{3\omega}, \tag{3.5}$$

like (2.1) and (3.4). In this case the Chern–Gauss–Bonnet formula for compact 4-manifolds with boundary may be written as

$$\chi(M) = (32\pi^2)^{-1}\int_M (|C|^2 + 4Q)\,dx + (4\pi^2)^{-1}\oint_{\partial M}(T - \mathcal{L}_4 - \mathcal{L}_5)\,dy, \tag{3.6}$$

where $\mathcal{L}_4$, $\mathcal{L}_5$ are some pointwise conformally invariant boundary curvatures. Actually, the original term which has appeared in the Chern–Gauss–Bonnet formula was the term $\mathcal{S}$ in the place of T, where $T = \mathcal{S} + \frac{1}{3}\tilde{\Delta}H$. It turns out that the discovery of the pair (P_3, T) is the crucial step in [CQ-1],[CQ-2] to establish some sharp Sobolev Trace inequality of Lebedev–Milin type, which in turn lead to the identification of some extremal metrics of the log-determinant functional for some model cases.

Having defined P_3 and T, we may rewrite the formula of the heat coefficient a_4 given by Branson and Gilkey (cf. [BG-1]) and after some lengthy computations, in [CQ-1] the following formula was derived.

Theorem 3.1. *Let M be a 4-manifold with smooth boundary. Suppose that (A, B) satisfies the assumptions as above and that (A, B) has no zero eigenvalues. Then*

$$\begin{aligned} -\frac{1}{2l}\log\frac{\det(A,B)[\omega]}{\det(A,B)[0]} &= \beta_1\int_M \omega(|C|^2\,dv)[0] + \beta_2\int_M D[\omega]\,dx[0] \\ &+ \frac{1}{2}\beta_3\Big\{\int_M (J^2\,dv)[\omega] - \int_M (J^2\,dv)[0]\Big\} + \beta_4\oint_{\partial M}\omega(\mathcal{L}_4\,ds)[0] \\ &+ \beta_5\oint_{\partial M}\omega(\mathcal{L}_5\,ds)[0] + \mu_1\oint_{\partial M}\big((T\,ds)[\omega] - (T\,ds)[0]\big) \end{aligned} \tag{3.7}$$

$$+\mu_2 \oint_{\partial M} \big((\tilde{R}H\,ds)[w] - (\tilde{R}H\,ds)[0]\big) + \mu_5 \oint_{\partial M} \big((H^3\,ds)[\omega] - (H^3\,ds)[0]\big)$$
$$+\mu_3 \oint_{\partial M} \big(((RH - 3FH)\,ds)[\omega] - ((RH - FH)\,ds)[0]\big)$$
$$+\mu_4 \oint_{\partial M} \big(((FH - 3R_{aNbN}L_{ab})\,ds)[\omega] - ((FH - 3R_{aNbN}L_{ab})\,ds)[0]\big)$$
$$+\mu_6 \oint_{\partial M} \big(((H^3 - 9\operatorname{Tr} L^3)\,ds)[\omega] - ((H^3 - 9\operatorname{Tr} L^3)\,ds)[0]\big),$$

where

$$D[\omega] = \frac{1}{4}\int_M \omega P_4\omega\,dv + \frac{1}{2}\int_M \omega Q\,dv + \frac{1}{2}\oint_{\partial M} \omega P_3\omega\,ds + \oint_{\partial M} \omega T\,ds$$
$$+\frac{1}{12}\Big\{\oint_{\partial M}(-R - \frac{1}{3}H^2 + 3F)\omega_N\,ds - \oint_{\partial M} H\omega_{;NN}\,ds - \oint_{\partial M} H\tilde{\Delta}\omega\,ds\Big\}, \tag{3.8}$$

for 11 *constants* $\{\beta_\nu\}_{\nu=1}^5$ *and* $\{\mu_i\}_{i=1}^6$ *depending only on* (A, B).

Example. Suppose that $A = \mathcal{L}$ the conformal Laplacian and $B = \mathcal{R}$ the Robin operator. Then, we have, up to a common scale, (cf. [BG-2]):

$\beta_1 = 2$	$\beta_2 = -8$	$\beta_3 = -8$	$\beta_4 = -4$
$\beta_5 = -8$	$\mu_1 = 60$	$\mu_2 = -\frac{4}{3}$	$\mu_3 = -\frac{50}{9}$
$\mu_4 = -20$	$\mu_5 = -\frac{22}{9}$	$\mu_6 = -\frac{44}{27}$.	

Table 3.1

For the purpose of stating our results on 4-manifolds more clearly, we shall modify the notions of two types of conformal variations which are more restrictive than the original definitions.

Definition 3.2. Suppose that (M, g) is a 4-manifold with boundary ∂M and that $H = 0$. Then a metric $g_\omega = e^{2\omega}g$ is called a *conformal variation of type I* if

1. $\mathrm{Vol}(M)[\omega] = \mathrm{Vol}(M)[0]$
2. $H[\omega] = H[0] = 0$
3. $\oint_{\partial M}(T\,ds)[\omega] = \oint_{\partial M}(T\,ds)[0]$.

Definition 3.3. Suppose that (M, g) is a 4-manifold with boundary ∂M and that $H = 3$. Then a metric $g_\omega = e^{2\omega}g$ is called a *conformal variation of type II* if

1. $\mathrm{Vol}(\partial M)[\omega] = \mathrm{Vol}(\partial M)[0]$
2. $H[\omega] = H[0] = 3$
3. $\oint_{\partial M}(T\,ds)[\omega] = \oint_{\partial M}(T\,ds)[0]$.

Conformal Variations of Type I. We may obtain $W^{2,2}$-compactness of isospectral sets within conformal variations of type I on general 4-manifolds by applying the following sharp inequality of Adams type for the Neumann problem.

Lemma 3.4. *For any function ϕ with Neumann boundary condition $N\phi = 0$ on a given 4-manifold M with smooth boundary ∂M, one has*

$$\int_M e^{\alpha|\phi-\bar{\phi}|^2}\,dv \le C(M,\alpha) \quad \text{if} \int_M |\Delta\phi|^2\,dv \le 1, \tag{3.9}$$

for any $\alpha < 16\pi^2$. Consequently one has

$$\int_M e^{4(\phi-\bar{\phi})}\,dv \le \frac{4}{\alpha}\int_M |\Delta\phi|^2\,dv + \log C(M,\alpha). \tag{3.10}$$

In the following discussion, we shall name

$$a_4(A,B) = \beta_1 \int |C|^2 + \beta_2\Big\{\tfrac{1}{2}\int Q + \oint T\Big\} + \beta_4 \oint \mathcal{L}_4 + \beta_5 \oint \mathcal{L}_5$$

the conformal index of the operator (A, B). Notice that $a_4(A, B)$ does not depend on the choice of the background metric for metrics in the same conformal class. As we shall see in the following results, $a_4(A, B)$ plays the same role as the corresponding index $a_4(A)$ (which was denoted by k_d in the previous section and in [CY]) does for compactness results on zeta functional determinants. As a consequence of the sharp inequality (3.10) we have the following compactness result.

Theorem 3.5. *Suppose that (A, B) satisfies the assumptions as above, and assume that $\beta_2, \beta_3 < 0$, and $\frac{1}{\beta_2}a_4(A, B) < 4\pi^2$. Then within conformal variations of type I, $F[\omega]$ is bounded from above and*

$$\|\omega - \omega_b\|^2_{W^{2,2}} + \|\omega - \omega_b\|^4_{W^{1,4}} \le C\,|F[\omega]| + C, \tag{3.11}$$

where ω_b is the average of ω over ∂M and C does not depend on ω.

Since we can control ω_b through the spectral information of the metric g_w, we have as a direct consequence of Theorem (3.5) the following result.

Corollary 3.6. *Suppose that (A, B) satisfies the three assumptions. And suppose that $-a_4(A, B) < (-\beta_2)4\pi^2$ and $\beta_2, \beta_3 < 0$. Then within conformal variations of type I, the metrics g_ω with the same spectrum as g_0 satisfy the estimate*

$$\|\omega\|_{W^{2,2}} \le C, \tag{3.12}$$

where C is a constant which does not depend on ω.

We also established the uniqueness of the extremal metric of functional determinant if it exists, as a consequence of the convexity property of functional determinant (cf. [CY]).

Theorem 3.7. *Suppose that (A, B) satisfies the assumptions as above, and suppose that $2\beta_3 \le \beta_2 \le \beta_3 < 0$, and $a_4(A, B) \ge 0$. In addition assume that $(\beta_2 + 6\mu_4)L[0]$ is nonnegative definite. Then the maximum of $F[\omega]$ under conformal variations of type I is unique when it exists.*

Example 3.8. We have some simple examples where all the assumptions in Theorem 3.7 above are satisfied. We will consider the boundary value problems where $A = \mathcal{L}$ and B is either the Dirichlet or Robin boundary operator. In this case, $\beta_1 = 2$ and $\beta_2 = \beta_3 = -8$ (see Table 3.1). Applying the Chern–Gauss–Bonnet formula, we obtain:

1) For $S^2 \times S^2_+$, we have Vol $= 8\pi^2$, $\chi = 2$, $|C|^2 = \frac{16}{3}$, and $L = 0$. Therefore we have $a_4 = \frac{128}{3}\pi^2 > 0$
2) For $\Sigma_g \times S^2_+$, where Σ_g is a hyperbolic surface of genus $g > 1$, we know $\chi = 2-2g$, $|C| = 0$, and $L = 0$. Therefore $a_4 = 64(g-1)\pi^2 > 0$.

Conformal Variation of Type II. The problems involved in the study of conformal variations of type II is in general more difficult than that of the study of conformal variations of type I. We will demonstrate, in the special case when $M = B^4$, some geometric and analytic aspects to study the zeta functional determinant under conformal variations of this type. We hope that the discussion of this special case will shed light for the study of the problem for general manifolds. We now fix the standard metric of B^4 as the background metric g_0 and write the Alvarez–Polyakov formula on B^4.

Lemma 3.9. *Suppose that (A, B) satisfy the assumptions as above. Let $(M, \partial M)$ be (B^4, S^3) with the standard metric g_0. If g_ω is a conformal variation of type II, then the Alvarez–Polyakov formula is*

$$F[\omega] = -\frac{1}{2l}\log\frac{\det(A,B)[\omega]}{\det(A,B)[0]} = \beta_2 D[\omega] + \frac{1}{2}\beta_3\int_{B^4} J^2[\omega]e^{4\omega} + 3\mu_3\oint_{S^3}(R-3F)[\omega]e^{3\omega}\,dy + 3\mu_2\Big\{\oint_{S^3}\tilde{R}[\omega]e^{3\omega}\,dy - \oint_{S^3}6\,dy\Big\} \tag{3.13}$$

where

$$D[\omega] = \tfrac{1}{4}\int_{B^4}\omega(P_4\omega)\,dx + \tfrac{1}{2}\oint_{S^3}\omega(P_3\omega)\,dy + 2\oint_{S^3}\omega\,dy - \frac{1}{4}\oint_{S^3}(\omega_N + \omega_{;NN})\,dy \tag{3.14}$$

and

$$P_3 = -\frac{1}{2}N\Delta - \tilde{\Delta}N + \tilde{\Delta} \text{ and } P_4 = \Delta^2.$$

In the discussion below, the following sharp inequality of Lebedev–Milin type plays the same role for the compactness results as the classical Lebedev–Milin inequality does for compact surface with boundary.

Lemma 3.10. *Suppose $\omega \in C^\infty(\overline{B^4})$, and assume ω satisfies*
(1) $H[\omega] = H[0] = 3$ *($N\omega|_{S^3} = 1 - e^\omega$),*
(2) $\oint_{S^3}(R\,dy)[\omega] = \oint_{S^3}(R\,dy)[0] = 0$.
Then

$$\log\left\{\frac{1}{2\pi^2}\oint_{S^3} e^{3\omega}\,dy\right\} \le \frac{3}{4\pi^2}D[\omega], \tag{3.15}$$

or equivalently,

$$\log\left\{\frac{1}{2\pi^2}\oint_{S^3} e^{3(\omega-\bar\omega)}\,dy\right\} \le \frac{3}{4\pi^2}\left\{\frac{1}{4}\int_{B^4}\omega P_4\omega\,dx + \frac{1}{2}\oint_{S^3}\omega P_3\omega\,dy - \frac{1}{4}\oint_{S^3}(\omega_N + \omega_{;NN})\,dy\right\}, \tag{3.15'}$$

where $\bar\omega = \frac{1}{2\pi^2}\oint_{S^3}\omega\,dy$. Moreover equality holds if and only if $e^{2\omega}g_0$ is isometric to g_0.

As a consequence of (3.15′) and the sharp Sobolev inequality on S^3, we can establish the following theorem.

Theorem 3.11. *Suppose that (A, B) satisfies the three assumptions and that (A, B) has no zero eigenvalue. Let $(M, \partial M)$ be (B^4, S^3) and let g_ω be a conformal variation of type II defined on B^4. Assume further that g_ω satisfies $\oint_{S^3}\{(R\,dy)[\omega]-(R\,dy)[0]\} = 0$, and that β_2, β_3 and $\mu_2+\frac{3}{2}\mu_3$ have the same sign. Then the zeta functional determinant stays in the same sign and attains zero at and only at the metrics isometric to the standard metric.*

Example 3.12. Recall from Table (3.1) in the first part of this article that $\beta_2 = -8$, $\beta_3 = -8$, and $3\mu_2+\frac{9}{2}\mu_3 = -29$ when $A = \mathcal{L}$ the conformal laplacian and $B = \mathcal{R}$ the Robin boundary operator. Therefore $(\mathcal{L}, \mathcal{R})$ satisfies all the assumptions in Theorem 3.10.

We discuss the zeta functional determinants within another natural subclass of conformal variations of type II.

Definition 3.13. We call g_ω a *restricted conformal variation of type II* from g_0 if it is a conformal variation of type II and $P_4[0](\omega) = 0$ in M.

For example, when $(M, g_0) = (B^4, g_0)$, $P_4[0](\omega) = (\Delta)^2[0](\omega)$. Metrics g_ω conformal variation of type II are in the restricted class if and only if w is biharmonic on B^4.

Theorem 3.14. *Suppose that (A, B) satisfies the three assumptions. Assume further that (A, B) has no zero eigenvalue. Let $(M, \partial M)$ be (B^4, S^3). Then within the set of restricted conformal variations of type II, the zeta functional determinant stays in the same sign and attains zero at and only at metrics which are isometric to the standard metric, provided that $\beta_2, \beta_3, \mu_2, \mu_3$ have the same sign.*

This result is established as a consequence of three sharp inequalities: the sharp inequality of Beckner [Be], the sharp Sobolev inequality on S^3, and the sharp Sobolev trace inequality due to Escobar and Beckner [E-1], [Be]. It is very interesting to see that all these three sharp inequalities on S^3 appear simultaneously as parts of the log-determinant formula for (B^4, S^3).

Example 3.15. β_2, β_3, μ_2 and μ_3 are all negative when $(A, B) = (\mathcal{L}, \mathcal{R})$. See Table 3.1 in the first part of this article. Meanwhile they are not of the same sign when $(A, B) = (\mathcal{L}, \mathcal{D})$.

4. Higher dimensional generalizations

In this section, we will discuss some higher dimensional analogues of the Paneitz operator, and some partial results for the zeta functional determinant in higher dimensions.

For a manifold M of dimension not equal to four, in general the existence and precise form of the Paneitz operator with conformal covariant properties (0.7) with $(a, b) = (0, n)$ and (2.1) is unknown. But as we have mentioned before in the article [GJML], it proved that when n is even, an operator of order n with conformal covariant property of index $(0, n)$ exists; also when $M = R^n$ with the Euclidean metric, the operator $P_n = (-\Delta)^{\frac{n}{2}}$. Using stereographical projection pulling S^n to R^n, one can then compute the form of P_n on S^n. It turns out that this is precisely the operator which has appeared in Beckner's [B] inequality. And the precise form of P_n on S^n is

$$P_n = \prod_{k=0}^{\frac{n}{2}} (-\Delta + k(n-k-1)) \quad \text{when } n \text{ is even}$$

$$= \left(-\Delta + \left(\frac{n-1}{2}\right)^2\right)^{\frac{1}{2}} \prod_{k=0}^{\frac{n-3}{2}} (-\Delta + k(n-k-1)) \quad \text{when } n \text{ is odd.}$$

In view of (0.1), (0.3) and (2.1), an analogous question to ask is that of prescribing Q in the following equation

$$-P_n w + Q e^{nw} = (n-1)! \quad \text{on } S^n. \tag{4.1}$$

It turns out that equation (4.1) is a more natural extension of equation (0.3) than that of equation (0.1) of the problem of prescribing scalar curvature. This point of view is explained in [CY].

We remark that for $n = 2$, $P_2 = -\Delta$; for $n = 4$; $P_4 = -\Delta(-\Delta + 2) = (-\Delta)(-\Delta + \frac{R_0}{6})$ is the Paneitz operator which we have discussed in the previous sections of this paper.

In [BCY] and [CQ-2], extremal metrics for the zeta functional determinant on (S^4, g_0), resp. (B^4, S^3, g_0) where g_0 = surface measure on S^4, B^4 resp. are identified as metrics isometric to g_0. One crucial inequality which we use repeatedly is the following sharp inequality of Beckner:

$$\log \int_{S^4} e^{4w} \leq 4 \int_{S^4} w + \frac{1}{3} \int_{S^4} w P_4(w) \tag{4.2}$$

where $\int$ denotes the integration $\frac{1}{\text{Vol}(S^4)} \int_{S^4} dV g_0$. It turns out in [Be-1], (4.2) is established as a special case of a general sharp inequality on S^n:

$$\log \int_{S^n} e^{nw} \leq n \int_{S^n} w + \frac{n}{2(n-1)!} \int_{S^n} w P_n(w) \tag{4.3}$$

where P_n is as in (4.1). Beckner's original proof of (4.3) depends on a sharp inequality of Lieb–Young. In [CY, section 4] we gave an alternative proof of (4.3) by exploring the conformal covariant property of P_n. In [B-2], Branson computed the zeta functional determinant formula on (S^6, g_0) and again identified the extremal metrics for the formula as metrics isometric to g_0. It is perhaps not surprising that the inequality (4.3) in the special case $n = 6$ again plays a role in Branson's work; but the formula also involves other sharp inequalities of tensor forms.

We would like to point out that there is a natural definition of a fourth order Paneitz operator S_n in all dimensions $n \geq 5$ which enjoys a conformal covariance property with respect to conformal changes in metrics. The relation of this operator to the Paneitz operator in dimension four is completely analogous to the relation of the conformal Laplacian to the Laplacian in two dimensions. $S_n = (-\Delta)^2 + \delta(a_n R + b_n R_{i,j})d + \frac{n-4}{2} Q_n$ for n-dimensional manifold for $n \geq 5$; where

$$Q_n = c_n |\rho|^2 + d_n R^2 - \frac{1}{2(n-1)} \Delta R,$$

and $a_n = \frac{(n-2)^2+4}{4(n-1)(n-2)}$, $b_n = -\frac{4}{n-2}$, $c_n = -\frac{2}{(n-2)^2}$, $d_n = \frac{n^3-4n^2+16n-16}{8(n-1)^2(n-2)^2}$ are dimensional constants. ($P = S_4$, $Q_4 = Q$). The following table explains the analogy among the pairs (Δ, K), (P, Q), (L, R) and (S_n, Q_n).

$n=2 \quad \Delta w + K_w e^{2w} = K$	$n=4 \quad -Pw + 2Q_w e^{4w} = 2Q$
Yamabe equation for $g_u = u^{\frac{4}{n-2}} g$	Q-equation for $g_u = u^{\frac{4}{n-4}} g$
$n \geq 3 \quad Lu = Ru^{\frac{n+2}{n-2}}$	$n \geq 5 \quad S_n u = Q_n u^{\frac{n+4}{n-4}}$
Covariance property $g_u = u^{\frac{4}{n-2}} g$,	Covariance property $g_u = u^{\frac{4}{n-4}} g$
$n \geq 3 \quad L_u(\psi) = u^{-\frac{n+2}{n-2}} L_g(u\phi)$	$n \geq 5 \quad (S_n)_u(\phi) = u^{-\frac{n+4}{n-4}} S_n(u\phi)$

In a current joint work of J. Qing, W. S. Wang and P. Yang, a systematic study of the generalized prescribing curvature equation for (S_n, Q_n) is made, and some existence and regularity results of the Yamabe equation type have been obtained for the Q-equations.

References

[Ad] D. Adams, A sharp inequality of J. Moser for higher order derivatives, Ann. of Math. 128 (1988), 385–398.

[Al] O. Alvarez, Theory of strings with boundary, Nuclear Phys. B 216 (1983), 125–184.

[Au] T. Aubin, Nonlinear Analysis on Manifolds, Monge–Ampere equations, Grundlehren Math. Wiss. 252, Springer-Verlag, 1982.

[Be] W. Beckner, Sharp Sobolev inequalities on the sphere and the Moser–Trudinger inequality, Ann. of Math. 138 (1993), 213–242.

[Bo] J. P. Bourguignon, Les varietes de dimension 4 à signature non nulle dont la courbure est harmonique sont d'Einstein, Invent. Math. 68 (1981), 263–286.

[B-1] T. Branson, Group representations arising from Lorentz conformal geometry, J. Funct. Anal. 74 (1987), 199–293.

[B-2] —–, Sharp Inequality, the Functional Determinant and the Complementary Series, preprint, 1992.

[BCY] T. Branson, S.-Y. A. Chang and P. C. Yang, Estimates and extremal problems for the log-determinant on 4-manifolds, Comm. Math. Phys. 149 (1992), 241–262.

[BG-1] T. Branson and P. Gilkey, The functional determinant of a 4-dimensional boundary value problem, preprint 1992, Trans. Amer. Math. Soc. 344 (1994), 479–532.

[BG-2] —–, The Asymptotics of the Laplacian on a Manifold with Boundary, Comm. Partial Differential Equations 15 (1990), 245–272.

[BO] T. Branson and B. Ørsted, Explicit functional determinants in four dimensions, Proc. Amer. Math. Soc. 113 (1991), 669–682.

[CGY] S.-Y. A. Chang, M. Gursky and P. Yang, On a fourth order invariant in conformal geometry, preprint.

[CQ-1] S.-Y. A. Chang and J. Qing, The zeta functional determinants on 4-manifolds with boundary I — the formula, preprint, 1995.

[CQ-2] —, The zeta functional determinants on 4-manifolds with boundary II — extremal metrics and and compactness of isospectral set, preprint, 1995.

[CY] S.-Y. A. Chang and P. C. Yang, Extremal Metrics of Zeta Functional Determinants on 4-Manifolds, Ann. of Math. 142 (1995), 171–212.

[E-1] J. Escobar, Sharp constant in a Sobolev trace inequality, Indiana Univ. Math. J. 37 (1988), 687–698.

[E-2] —, The Yamabe problem on manifolds with boundary, J. Differential Geom. 35 (1992), 21–84.

[FM] A. Fisher and J. Marsden, Deformations of the scalar curvature, Duke Math. J. 42 (1975), 519–547.

[GJML] C. R. Graham, R. Jenne, L. Mason, and G. Sparling, Conformally invariant powers of the Laplacian, I: existence, J. London. Math. Soc. 46 (1992), 557–565.

[Gu] M. Gursky, Locally conformally flat four- and six-manifolds of positive scalar curvature and positive Euler characteristic, Indiana Univ. Math. J. 43 (1994), 474–774.

[K] O. Kobayashi, A differential equation arising from the scalar curvature, J. Math. Soc. Japan 34 (1982), 665–675.

[L] J. Lafontaine, Sur la geometrie d'une generalisation de l'equation differentielle d'Obata, J. Math. Pures App. 62 (1983), 63–72.

[M] J. Moser, A sharp form of an inequality by N. Trudinger, Indian Math. J. 20 (1971), 1077–1091.

[Ob] M. Obata, Certain conditions for a Riemannian manifold to be isometric with a sphere, J. Math. Soc. Japan 14 (1962), 333–340.

[On] E. Onofri, On the positivity of the effective action in a theory of random surfaces, Commun. Math. Phys. 86 (1982), 321–326.

[OPS-1] B. Osgood, R. Phillips, and P. Sarnak, Extremals of determinants of Laplacians, J. Funct. Anal. 80 (1988), 148–211.

[OPS-2] —, Compact isospectral sets of surfaces, J. Funct. Anal. 80 (1988), 212–234.

[OPS-3] —, Moduli space, heights and isospectral sets of plane domians, Ann. of Math. 129 (1989), 293–362.

[Pa] S. Paneitz, A quartic conformally covariant differential operator for arbitrary pseudo-Riemannian manifolds, preprint 1993.

[Po] A. Polyakov, Quantum geometry of Bosonic strings, Phys. Lett. B 103 (1981), 207–210.

[RS] D. B. Ray and I. M. Singer, R-torsion and the Laplacian on Riemannian manifolds, Adv. Math. 7 (1971), 145–210.

[Sc-1] R. Schoen, Conformal deformation of a Riemannian metric to constant scalar curvature, J. Differential Geom. 20 (1984), 479–495.

[Sc-2] ——, Variational theory for the total scalar curvature functional for Riemannian metrics and related topics, in: Topics in Calculus of Variations (M. Giaquinta, ed.), Lecture Notes in Math. 1365, Springer-Verlag, Berlin 1989.

[T]] N. Trudinger, Remarks concerning the conformal deformation of Riemannian structure on compact manifolds, Ann. Scuola Norm. Sup. Pisa Cl. Sci. 3 (1968), 265–274.

The infinity behavior of the solutions to the harmonic map heat equation

Hyeong In Choi

Abstract. We discuss the continuity at infinity of the solutions of the harmonic map heat equation. Our result is a pointwise version which says that both the ball convergence condition and continuity at infinity at one point is sufficient for the continuity at that infinity point. We also discuss the role that the ball convergence condition plays. In particular, we construct an example where the ball convergence condition is satisfied and the initial data is of sublinear growth, but with bad modulus of continuity. In this case, the continuity at infinity of the solution of the linear heat equation fails to be continuous. We also discuss a slightly stronger ball convergence condition which is valid for most known examples that guarantees the continuity at infinity.

1991 Mathematics Subject Classification: 58E20, 53C21, 58G32.

0. Introduction

One of the interesting questions in the study of non-compact complete manifolds is that of infinity behavior of geometric objects. In this paper, we look at the infinity behavior of the solution of the harmonic map heat equation into the target space which has "good" convexity property. We will also look at the more refined questions of the continuity at infinity of the solution of the linear heat equation, which was motivated by our study of harmonic maps.

First, we need to clarify our terms. Since the concept of infinity is rather hazy, we envision it as approaching ideal points at infinity. As a non-compact complete manifold has no boundary at infinity, one somehow artificially attaches an ideal boundary point at infinity, and studies continuity at that point.

The simplest one is certainly the one-point compactification, and the pertinent result in this case is the Liouville type theorem. But we are concerned with more general ideal boundary points. One well known example is the so-called Eberlein–O'Neill boundary of a complete simply connected manifold of nonpositive sectional

---

Supported in part by BSRI-95-1416 and by GARC

curvature [7]. There are many other constructions of ideal boundary points, and the geometric study of such ideal boundary points is an interesting subject in itself. However, certain analytic properties of geometric objects depend only on some very general structures of the ideal boundary point, regardless of its specific geometric construction. One such example is the *ball convergence condition* introduced by Donnelly and Li [6]. It is a very weak condition, but still, it is enough to guarantee the continuity at infinity of the solution of the linear heat equation. In Section 1, we generalize the Donnelly–Li result to the continuity at infinity *at one point* of the solution of the harmonic map heat equation. In doing so, we need to overcome the presence of topology of the target manifold, which naturally leads us into the study of a pointwise version of the linear heat equation problem.

The results in Section 2 stem from our attempt to understand how the ball convergence condition plays a role in the continuity at infinity of the solution of the linear heat equation. We give a counter-example in which continuity fails, and then show how to modify the ball convergence condition to fix the result.

1. Harmonic map heat equation

In this section, we summarize our results in [2]. Let M be a complete, non-compact manifold with Ricci curvature bounded below by a nonpositive constant. (This is a standing assumption on M.)

We first give our basic set up: Let $M' = M \cup \{\overline{x}\}$ be a topological space with one ideal point $\overline{x}$ attached to M. M' is assumed to be topologized in such a way that M is a proper dense open subset of M'. For lack of terminology, we call such M' a *one-point extension* of M. It should be emphasized that M' is *not* assumed to be compact. For example, suppose $\overline{M} = M \cup M(\infty)$ is a compactification of M, and let $\overline{x} \in M(\infty)$. Then $M' = M \cup \{\overline{x}\}$ with the induced topology is a one-point extension of M. We say that a one-point extension M' satisfies the *ball convergence condition* if for any $r > 0$ and for any sequence $\{x_n\}$ in M converging to $\overline{x}$, the geodesic ball $B_r(x_n)$ centered at x_n with radius r converges to $\overline{x}$ as $n \to \infty$.

Donnelly and Li first gave the above definition, and proved the following global result: Suppose M is a complete manifold without boundary, and let $\overline{M}$ be a compactification of M with the ideal boundary $M(\infty)$ such that every point of $M(\infty)$ satisfies the ball convergence condition. And let $\phi : \overline{M} \to \mathbb{R}$ be a continuous function. Suppose $h(x,t)$ is the solution of the following initial value problem:

$$\Big(\Delta - \frac{\partial}{\partial t}\Big)h(x,t) = 0 \tag{1.1}$$
$$h(x,0) = \phi(x).$$

Then for any $t > 0$, $h(x,t)$ converges to $\phi(\overline{x})$ as $x \to \overline{x}$.

Our first main result is the following one-point generalization of the Donnelly–Li result for Lipschitz data.

Theorem 1.1. *Suppose M is a complete Riemannian manifold with Ricci curvature bounded below by $-K \leq 0$. Let $M' = M \cup \{\overline{x}\}$ be a one-point extension of M satisfying the ball convergence condition. Suppose $\phi : M' \to \mathbb{R}$ is a continuous function such that $\phi|_M$ is Lipschitz. Let $f(x,t)$ be the solution of the linear heat equation* (1.1) *with initial data ϕ. Then, for each fixed t, $\lim_{x\to\overline{x}} f(x,t) = \phi(\overline{x})$.*

The basic philosophy behind of the proof of Theorem 1.1 is that first, the neighborhood of $\overline{x}$ contains a ball of arbitrarily large radius when we take its center to converge to $\overline{x}$, and second, one can control the probability the Brownian motion exiting a ball of fixed size in terms of its radius and other geometric data. The proof follows from the standard probabilistic method by carefully estimating the exit probability. One difficulty here is that the initial data is unbounded, so the exit probability estimate has to be sharper. The next theorem is our nonlinear generalization of Theorem 1.1.

Theorem 1.2. *Suppose that M is a complete Riemannian manifold with Ricci curvature bounded below by a constant, and N is a complete Riemannian manifold such that its sectional curvature K_N satisfies $-\kappa \leq K_N \leq 0$ and the norm of the covariant derivative of the curvature tensor is bounded by a constant. Let $M' = M \cup \{\overline{x}\}$ be a one-point extension of M satisfying the ball convergence condition. Assume further that M' is connected, locally path connected, and semi-locally 1-connected. Let u be the unique solution of the harmonic map heat equation:*

$$\left(\Delta - \frac{\partial}{\partial t}\right)u^i(x,t) = g^{\alpha\beta}(x)\Gamma^i_{jk}(u(x,t))\frac{\partial u^j}{\partial x^\alpha}(x,t)\frac{\partial u^k}{\partial x^\beta}(x,t) \qquad (1.2)$$
$$u(x,0) = f(x),$$

with initial data $f : M \to N$ which is a C^1 map with bounded energy density. Assume that $f(x)$ can be continuously extended to $f : M' \to N$. Then, for any t, $u(x,t)$ converges to $f(\overline{x})$, as x goes to $\overline{x}$.

There are many interesting points to be discussed. First, we give an

Outline of the proof of Theorem 1.2. Let $\widetilde{M}'$ and $\widetilde{N}$ be the universal coverings of M' and N with covering maps p_M and p_N respectively [11]. Let $M'' = \widetilde{M}' - p_M^{-1}\{\overline{x}\}$. Note that M'' is a complete covering manifold of M. Each element of $\pi_1(M')$ can be identified with a deck transformation of $\widetilde{M}'$ which is also an isometry of M''. One can view $\pi_1(N)$ similarly. Let $\tilde{f} : \widetilde{M}' \to \widetilde{N}$ be a lifting of f. Then for each $\sigma \in \pi_1(M')$, there exists $\tau \in \pi_1(N)$ such that $\tilde{f}(\sigma x) = \tau\tilde{f}(x)$, for all $x \in M''$. By [11], the initial value problem (1.1) has a unique solution $\tilde{u} : M'' \times [0,\infty) \to \widetilde{N}$ such that $\tilde{u}(x,0) = \tilde{f}(x)$. By an easy application of the maximum principle, $\tilde{u}$ descends to a solution u of the harmonic map heat equation with the initial map f. Fix $\overline{z} \in p_M^{-1}(\overline{x})$. To complete the proof, we have only to show that $\tilde{u}(x,t) \to f(\overline{z})$ as $x \to \overline{z}$, for each fixed $t \geq 0$. Since $\widetilde{N}$ is a Cartan–Hadamard manifold, we can choose a global normal coordinate $(y^1, \ldots, y^n)$ on $\widetilde{N}$ with respect to which $\tilde{f}$ can be represented by $(\tilde{f}^1, \ldots, \tilde{f}^n)$ where $\tilde{f}^i$ is a function on $\widetilde{M}'$. Define a map $h : M'' \times [0, \infty) \to \widetilde{N}$ by simply declaring $h = (h^1, \ldots, h^n)$ with respect to the global normal coordinates $(y^1, \ldots, y^n)$ of $\widetilde{N}$. Consider the following initial value problem for the linear heat equation

$$\left(\Delta - \frac{\partial}{\partial t}\right) h^i(x, t) = 0$$
$$h^i(x, 0) = \tilde{f}^i(x),$$

for $i = 1, \ldots, n$ and $x \in M''$. Since $h^i(x, t)$ is the solution of the above initial value problem, Theorem 1.1 implies that $h^i(x, t)$ converges to $\tilde{f}^i(\overline{z})$ as x goes to $\overline{z}$, which means that $h(x, t) \to \tilde{f}(\overline{z})$ as $x \to \overline{z}$. One trouble in our proof is that h, unlike $\tilde{u}$, has no equivariance property, which makes it impossible to push h down to a map from $M \times [0, \infty)$ to N. This forces us to work in the universal cover in which many nice properties downstairs break down. Define $\psi : \widetilde{N} \times \widetilde{N} \longrightarrow \mathbb{R}$ by $\psi(y_1, y_2) = \phi(\rho(y_1, y_2))$ for $(y_1, y_2) \in \widetilde{N} \times \widetilde{N}$, where $\phi : \mathbb{R} \to \mathbb{R}$ is a C^2 function such that for some constant $C' > 0$, $\phi(0) = \phi'(0) = \phi''(0) = 0$, $0 \leq \phi'(t) \leq C'$, and $\phi''(t) \geq 0$ for all $t \in [0, \infty)$. Note that

$$D^2\psi = \phi' D^2 \rho + \phi'' d\rho^2,$$

where d and D^2 are taken over $\widetilde{N} \times \widetilde{N}$. Since $D^2\rho$ blows up like $1/\rho$ as $\rho \to 0$ and $\phi'(0) = \phi''(0) = 0$, ψ is C^2 even along the diagonal. One can then check that

$$\left(\Delta - \frac{\partial}{\partial t}\right) \psi(\tilde{u}(x, t), h(x, t)) \geq -C_1 e^{C_2|h|} |\nabla h|^2,$$

where C_1 and C_2 are constants depending only on C' and the geometry of M and N. In particular, it is independent of $\tilde{u}$ or h. Here we used the following Lemma 1.3 on the exponential growth property of the Christoffel symbol of $\widetilde{N}$. Now

$$\left(\Delta - \frac{\partial}{\partial t}\right) \cosh C_3|h| \geq C_3' e^{\frac{C_3}{2}|h|} |\nabla h|^2,$$

for suitable constant C_3'. We get from these two inequalities, by suitably choosing constants C_3 and C_4, that

$$\left(\Delta - \frac{\partial}{\partial t}\right) \{\psi(\tilde{u}, h) + C_4 \cosh C_3|h|\} \geq 0.$$

The function enclosed by the braces in the above inequality has initial value $C_4 \cosh C_3|\tilde{f}|$. Now let v be the solution of the linear heat equation (1.1) with the initial data $v(x, 0) = \cosh C_3|\tilde{f}(x)|$. Then the maximum principle implies that

$$\psi(\tilde{u}(x, t), h(x, t)) \leq C_4 \{v(x, t) - \cosh C_3|h(x, t)|\}. \tag{1.3}$$

Now by Theorem 1.1, $v(x,t)$ converges to $\cosh C_3|\tilde{f}(\overline{z})|$ as x goes to $\overline{z}$. Therefore, $\psi(\tilde{u}(x,t), h(x,t))$ converges to 0 as x goes to $\overline{z}$. Since we have proved above that $h(x,t)$ converges to $\tilde{f}(\overline{z})$ as x goes to $\overline{z}$, the proof is now complete.

Lemma 1.3. *Suppose N is a simply connected complete Riemannian manifold such that its sectional curvature K_N satisfies $-\kappa \leq K_N \leq 0$ and the norm of the covariant derivative of the curvature tensor is bounded by a constant. Let $(x^1, \dots, x^n)$ be a normal coordinate centered at a fixed point $p \in N$. Let $ds^2 = h_{\alpha\beta}dx^\alpha dx^\beta$ be a Riemannian metric. Let $\Gamma^\alpha_{\beta\gamma}$ be the Christoffel symbol. Then $\Gamma^\alpha_{\beta\gamma} = O(e^{C|x|})$ on N for some constant C depending only on the dimension, the bounds of the sectional curvature and the covariant derivative of the curvature tensor of N.*

Remarks. The basic philosophy of our approach in Theorem 1.2 is that if the image lies in a convex ball, the solution of the harmonic map heat equation, which is a solution of the nonlinear problem (1.2), is well approximated in the C^0 sense by a map which can be constructed by solving a corresponding linear problem. In other words, the nonlinear terms in (1.2) can be handled very well by simply dropping them. In a sense, our method is a variant of that used in [1]. However, unlike in [1], the target manifold N now has topology, which prevents us from utilizing the normal co-ordinates. Therefore the method of [1] cannot be applied. It is important to note that our generalization of the Donnelly–Li result is necessary for our program even when we assume that $\overline{M}$ is a compactification of M such that every point of $M(\infty)$ satisfies the ball convergence condition and the initial map f is continuous up to $M(\infty)$. The Donnelly–Li result cannot be used as linear ones approximating nonlinear ones for our purpose. The reason for this difficulty is again the presence of topology in N. Our basic set up is to represent the map u as an n-tuple of functions, and to do so, we need global co-ordinates for the target. For this reason, we are forced to go to the universal covers of M and N and lift all the data to the universal covers. In doing so, the universal cover of $\overline{M}$ is no longer compact, and the lifted initial map is no longer bounded, therefore we can assume neither the ball convergence condition being valid at every point at infinity nor the continuity of the initial map up to every point at infinity. Namely, we are forced to deal with the generalization of the Donnelly–Li result, i.e., Theorem 1.1, out of this necessity.

Another interesting point is inequality (1.3). In our proof, virtually no knowledge is available on $\tilde{u}$ which is the lift of u. Unlike in [1], $\tilde{u}$ is now unbounded, and the distance between $\tilde{u}$ and the approximating map h, which comes from the linear problem, is not controllable. We were able to overcome this difficulty by carefully choosing the function ψ which replaces the role of the naively defined distance function. By doing so, the terms involving $\tilde{u}$ miraculously disappear in our calculation and we are left with terms expressible solely in terms of h. The upshot is that the discrepancy between $\tilde{u}$ and h is controlled by a quantity expressed in terms of the solutions of the *linear* heat equation. For this reason, we would like to call this a new form of C^0 estimate.

When the target has nonpositive sectional curvature, the existence of the solution of (1.2) was proved by Li and Tam. However, if the target manifold has positive curvature, it is no longer true that (1.2) has a solution for all time. The solution may blow up in finite time, and it may develop singularities. However, when it is confined to a convex geodesic ball $B_\tau(p)$ of radius τ and center at p in N, the following long time existence holds.

Theorem 1.4. *Suppose that M is a complete Riemannian manifold with Ricci curvature bounded below by a constant and N is a complete Riemannian manifold with sectional curvature bounded above by a constant $\mu > 0$. Assume that $\tau < \min\{\frac{\pi}{2\sqrt{\mu}}$, injectivity radius of N at $p\}$. Let $f \in C^1(M, N)$ be such that $f(M) \subset B_\tau(p)$ and $\sup_M e(f) < \infty$. Then (1.2) has a unique solution u defined on $M \times [0, \infty)$ with the initial map f such that u has the following properties:*

(a) $u(M \times [0, T]) \in B_\tau(p)$ *for all* $T > 0$,

(b) $\sup_{M\times[0,T]} e(u) < \infty$ *for all* $T > 0$.

With this long time existence theorem, the maximum principle technique which is similar to, but harder than, those used in the proof of Theorem 1.2 gives the following result:

Theorem 1.5. *Suppose N is a complete Riemannian manifold whose sectional curvature is bounded above by a constant $\mu > 0$. Let $p \in M$. Assume that $\tau < \min\{\frac{\pi}{2\sqrt{\mu}}$, injectivity radius of N at $p\}$. Let $f \in C^1(M, B_\tau(p)) \cap C(\overline{M}, B_\tau(p))$. Let u be the unique solution of (1.2) given by Theorem 1.4 with the initial map f. If $\overline{M}$ satisfies the ball convergence criterion, then $u(\bar{x}, t) = f(\bar{x})$ for all $(\bar{x}, t) \in M(\infty) \times [0, \infty)$.*

2. Linear heat equation

In this section, we study more refined questions on the infinity behavior of the solution of the linear heat equation

$$\begin{aligned}\Big(\frac{1}{2}\Delta - \frac{\partial}{\partial t}\Big)u(x,t) &= 0, \qquad (x,t) \in M \times [0,\infty), \\ u(x,0) &= f(x), \qquad x \in M.\end{aligned} \tag{2.1}$$

Let X_t be a Brownian motion on M. Then $u(x,t)$ can be written in terms of probability by

$$u(x,t) = E_x[f(X_t)].$$

In Theorem 1.1, the ball convergence condition and the Lipschitzness of the initial data were important. This raises general question on the role that the ball convergence condition plays in relation with the growth rate of the initial data and its modulus of continuity. The purpose of this section is to clarify how these conditions lead into the continuity at infinity of solution of the heat equation.

First, we give an easy generalization, Theorem 2.1, which says that as long as the initial data has the exponential modulus of continuity, the ball convergence condition gives the continuity at infinity. It is interesting to note that our proof also shows that the condition on modulus of continuity can be dropped altogether, as is the case in [6], when the initial data is continuous and bounded.

Theorem 2.1. *Suppose M is a complete m-dimensional Riemannian manifold with Ricci curvature bounded below by $-(m-1)k^2$. Let $M' = M \cup \{\overline{x}\}$ be a one-point extension of M satisfying the ball convergence condition. Let $f : M' \to \mathbb{R}$ be a continuous function. Suppose f satisfies the following modulus of continuity: i.e., $|f(x) - f(y)| \le \phi(\gamma(x, y))$ and $\phi(r) \le c_1 e^{c_2 r}$, for some continuous function $\phi : \mathbb{R}^+ \to \mathbb{R}$ and positive constants c_1, and c_2. Let $u(x,t)$ be the solution of the initial value problem* (2.1). *Then, for each fixed t, $\lim_{n\to\infty} u(x_n, t) = f(\overline{x})$, for any sequence $\{x_n\}$ converging to $\overline{x}$.*

Next, we construct an example of a one-point extension of $\mathbb{R}^2$ such that the ball convergence condition is satisfied at the point $\overline{0} \in M(\infty)$ and a sublinearly growing initial data which is continuous on $M' = M \cup \{\overline{0}\}$, but whose modulus of continuity grows far faster than the exponential function. In this case, we show that $u(x,t)$ does not converge to $f(\overline{0})$ as $x \to \overline{0}$. This example indicates that the condition on the modulus of continuity is more important than the actual growth rate.

Example. We now construct an example that shows Theorem 1.1 or 2.1 fails even with the ball convergence condition when the exponential modulus of continuity condition is not valid. We first construct an appropriate initial data f, and then a one-point extension of $\mathbb{R}^2$.

(E.1) Initial data. Define $f(x,y)$ in $\mathbb{R}^2$ as follows: Let $s_k = (k^3, 0)$ for sufficiently large positive integer k. We use the notation $B(p, R)$ to denote the ball of radius R with the center $p \in \mathbb{R}^2$. Define A_k by

$$A_k = \left[B(s_k, k) - B(s_k, \sqrt{\log\log k})\right] \cap \left\{(x,y) : |y| \ge \frac{1}{2}\sqrt{\log\log k}\right\}.$$

Now define $f(x)$ in $B(x_k, k)$ by $f(x) = \sqrt{k}$ in A_k, and extend f to a continuous function on $\mathbb{R}^2$ in such a way that $f = 0$ outside an ϵ-neighborhood of A_k.

Note that f grows sublinearly. In fact, its growth is $O(|x|^{\frac{1}{6}})$. But the modulus of continuity grows far faster than the exponential growth.

(E.2) One-point extension. Consider the curve given as the graph of $y = a(\log\log x)^{\frac{1}{4}}$ for $x > 100$ and $a \in \mathbb{R}$. When $a = 0$, this curve corresponds to the positive x-axis, which we call $\bar{0}$. Attaching $\bar{0}$ to $M = \mathbb{R}^2$, we obtain a one point extension $M' = M \cup \{\bar{0}\}$ of $M = \mathbb{R}^2$. M' is topologized as follows: First, $\mathbb{R}^2$ is given the usual topology. For $a > 0$ and $\alpha > 100$, define

$$U_{a,\alpha} = \left\{(x, y) \in \mathbb{R}^2 \; : \; x > \alpha, \, |y| < a\left(\log\log x\right)^{\frac{1}{4}}\right\}.$$

The family of such $U_{a,\alpha}$ defines a sub-basis of topology at $\bar{0}$.

One can then easily see that M' satisfies the ball convergence condition.

(E.3) Value at $\bar{0}$. Obviously, f is continuous on M', and $f(p) \to 0$ as $p \to \bar{0}$. But $e^{t\Delta/2} f(s_k)$ does not converge to 0, even if $s_k \to \bar{0}$.

Proof. Check

$$\begin{aligned} E_{s_k}[f(X_t)] &\geq \frac{2}{\sqrt{2\pi t}} \int_{\frac{\pi}{6}}^{\frac{5\pi}{6}} \int_{\sqrt{\log\log k}}^{k} \sqrt{k} \exp(-\frac{\rho^2}{2t})\, \rho \, d\rho \, d\phi \\ &= \sqrt{\frac{2tk}{\pi}} \frac{2\pi}{3} \left[\frac{1}{(\log k)^{\frac{1}{2t}}} - \exp(-\frac{k^2}{2t}) \right]. \end{aligned}$$

Thus $E_{s_k}[f(X_t)] \to \infty$ as $k \to \infty$. This example shows that the most important condition for Theorem 1.3 is the exponential modulus of continuity, not the actual growth rate of f, which in this case is sublinear. □

The rest of this section grew out of our attempt to understand how the example in this section fails to be continuous. In most compactifications such as attaching the Eberlein–O'Neill boundary to the complete, simply-connected, nonpositively curved manifolds, something stronger than mere ball convergence condition holds, which we define as ψ-strong ball convergence condition. One interesting consequence of our strong ball convergence condition is that we no longer need the condition on modulus of continuity of initial data. We only need the growth rate of the initial data in relation with the growth of ψ. For example, we have the continuity at infinity if the initial data is exponentially growing, and $\psi(r) = r^\beta$ for some $\beta > \frac{1}{2}$.

We now define a stronger version of the ball convergence condition which is satisfied by many important examples such as attaching the Eberlein–O'Neill boundary to complete, simply-connected, nonpositively curved manifolds.

Definition 2.1. Let $\psi : \mathbb{R}^+ \to \mathbb{R}^+$ be an increasing function. A point $\bar{x} \in M'$ *satisfies the ψ-strong ball convergence condition*, if for any neighborhood U of $\bar{x}$, there exists $\alpha \in \mathbb{R}^+$ such that for any sequence $\{x_n\}$ in M converging to $\bar{x}$, the geodesic ball $B_{\alpha\psi(a_n)}(x_n)$ centered at x_n with radius $\alpha\psi(a_n)$ is contained in U for sufficiently large n, where $a_n = \rho(x_n)$ is the distance from the fixed point y_0 in M to x_n.

Remark 2.1. It may be possible to generalize our definition by considering more general functions of $\psi(a_n)$ instead of linear functions $\alpha\psi(a_n)$.

We now illustrate below a use of our ψ-strong ball convergence condition. Notice the sharpness of our result in relation to the growth rate (*) beyond which the result obviously fails.

We say a function f on M satisfies the growth condition (*), if

$$\lim_{r\to\infty} \sup_{x\in B_r(y_0)} \frac{\log|f|(x)}{r^2} = 0. \tag{*}$$

Theorem 2.2. *Let M be a complete m-dimensional Riemannian manifold with Ricci curvature bounded below by a nonpositive constant. Let $M' = M \cup \{\overline{x}\}$ be a one-point extension of M satisfying the ψ-strong ball convergence condition at $\overline{x}$ where $\psi(r) = r$. Let $X_t : M \to M$ be the Brownian motion. Let U be a neighborhood of $\overline{x}$ in M'. Let $f : M' \to \mathbb{R}$ be a function which is continuous on M', and satisfies the condition (*). Let $u(x,t)$ be the solution of* (2.1) *with the initial data f. Then, fixing t, $\lim_{n\to\infty} u(x_n,t) = f(\overline{x})$, for any sequence $\{x_n\}$ converging to $\overline{x}$.*

Remark 2.2. The same proof can be employed to prove the continuity at infinity when either

(i) $f(r)$ and $\psi(r) \le Cr^\alpha$, for some $\alpha > 0$ and $C > 0$, or

(ii) $f(r) \le Ce^{\alpha r}$, for some $\alpha > 0$ and $C > 0$ and $\psi(r) \le Cr^\beta$ for some $\beta > \frac{1}{2}$.

One may also find similar conditions.

Remark 2.3. It is an interesting question if Theorem 2.2 still holds when the Ricci curvature bound is relaxed by an integral condition as in [8] in connection with the condition of the stochastic completeness of a Riemannian manifold.

Remark 2.4. The details of the proof of the results in Section 2 will appear elsewhere.

References

[1] Aviles, P., Choi, H. I., and Micallef, M., Boundary Behavior of Harmonic Maps on Non-smooth Domains and Complete Negatively Curved Manifolds, J. Funct. Anal. 99 (1991), 293–331.

[2] Chi, D. P., Choi, H. I., and Kim, H. J., Heat Equation for Harmonic Maps of the Compactification of Complete Manifolds, to appear in J. Geom. Analysis.

[3] Choi, H. I., Asymptotic Dirichlet Problems for Harmonic Functions on Riemannian Manifolds, Trans. Amer. Math. Soc. 28 (1984), 691–716.

[4] Choi, H. I., On the Liouville Theorem for Harmonic Maps, Proc. Amer. Math. Soc. 85 (1982), 91–94.

[5] Choi, H. I., Kim, H. J., and Lee, D. S., Continuity at Infinity of Solutions of Heat Equation on Complete Manifolds, to appear.

[6] Donnelly, H. and Li, P., Heat Equation and Compactifications of Complete Riemannian Manifolds, Duke Math. J. 51 (1984), 667–673.

[7] Eberlein, P. and O'Neill, B., Visibility Manifolds, Pacific J. Math. 46 (1973), 45–109.

[8] Hsu, P., Heat Semigroup on a Complete Riemannian Manifold, Ann. Probab. 17 (1989), 1248–1254.

[9] Karp, L. and Li, P., The heat equation on complete manifolds, preprint.

[10] Kendall, W., The radial part of Brownian motion of a manifold: a semimartingale property, Ann. Prob. 15-4 (1987), 1491–1500.

[11] Li, P. and Tam, L. F., The Heat Equation and Harmonic Maps of Complete Manifolds, Invent. Math. 105 (1991), 1–46.

[12] Li, P. and Yau, S.-T., On the parabolic kernel of Schrödinger operator, Acta Math. 156 (1986), 153–201.

[13] Liao, G. G. and Tam, L. F., On the Heat Equation for Harmonic Maps from Non-compact Manifolds, Pacific J. Math. 153-1 (1992), 129–145.

[14] Prat, J., Étude asymptotique du mouvement brownien sur une variété à courbure négative, C.R. Acad. Sci. Paris Ser. A 280 (1975), 1539–1542.

[15] Saloff-Coste, L. and Strook, D. W., Opérateurs uniformement sous-elliptiques sur les groupes de Lie, J. Funct. Anal. 98 (1991), 97–121.

On the absence of critical dimensions for the subelliptic Laplacian on the Heisenberg group

K. S. Chou and D. Geng *†

Abstract. Existence and nonexistence of positive solutions of the problem for the subelliptic Laplacian on the Heisenberg group involving critical exponent is studied for all dimensions. It is shown that there is no critical dimension for the problem.

1991 Mathematics Subject Classification: 35J70, 35D05.

1. Introduction

The concept of critical dimensions for elliptic problems with critical nonlinearity was first conceived in the celebrated paper of Brezis and Nirenberg [1]. They studied the problem

$$\begin{cases} -\Delta u = u^p + \lambda u & u > 0 \quad \text{in } \Omega \\ u = 0 & \text{on } \partial\Omega, \end{cases} \tag{1}$$

where $p = (n+2)/(n-2)$, $n \geq 3$, and Ω is a bounded domain in $\mathbb{R}^n$ $(n \geq 3)$ with smooth boundary. When $n \geq 4$, they proved that (1) is solvable if $\lambda \in (0, \lambda_1)$, where λ_1 is the first Dirichlet eigenvalue for the Laplacian. However, when $n = 3$, the infimum of $\{\lambda : \lambda > 0, (1) \text{ is solvable}\}$ is positive. Subsequent studies by Pucci–Serrin [8] and Edmunds–Fortunato–Jannelli [4] for a similar problem for the biharmonic operator demonstrate the same property, namely, when $n = 5, 6, 7$, the infima of $\{\lambda : \lambda > 0,$ the problem is solvable$\}$ are positive. They call 5, 6, and 7 the critical dimensions for the biharmonic operator. Now, critical dimensions for various elliptic problems have been found; see, for instance, Egnell [5] for the p-Laplacian; Chou–Geng [2] for weighted Laplacians and Chou–Geng–Yan [3] for the Hessian operators on balls.

In this note we show that the analogous problem for the subelliptic Laplacian Δ_{H^n} on the Heisenberg group H^n has no critical dimension.

*Department of Mathematics, the Chinese University of Hong Kong, Shatin, Hong Kong.
†Supported by Driect Grant, CUHK.

2. Main result

We recall that $\boldsymbol{H}^n$ is a Lie group with underlying manifold $\mathbb{C}^n \times \mathbb{R}$ ($n \geq 1$), endowed with the composition law:

$$(z, t) \circ (z', t') = (z + z', t + t' + 2\,\mathrm{Im}\, z \cdot z')$$

where $z \cdot z' = \sum_{j=1}^{n} z_j \cdot z_j'$ for z, $z' \in \mathbb{C}^n$. Setting $z_j = x_j + iy_j$, $(x_1, \ldots, x_n, y_1, \ldots, y_n, t)$ forms a real coordinate system for $\boldsymbol{H}^n$. In this coordinate system the vector fields

$$\begin{cases} X_j = \dfrac{\partial}{\partial x_j} + 2y_j \dfrac{\partial}{\partial t}, \\ Y_j = \dfrac{\partial}{\partial y_j} - 2x_j \dfrac{\partial}{\partial t}, \end{cases} \qquad j = 1, \ldots, n$$

generate the real Lie algebra of all left-invariant vector fields on $\boldsymbol{H}^n$. Now the subelliptic Laplacian of the Heisenberg group is defined by

$$\Delta_{\boldsymbol{H}^n} = \sum_{j=1}^{n} (X_j^2 + Y_j^2).$$

In $\boldsymbol{H}^n$ the distance is defined as follows:

$$d((z, t), (z', t')) = d((z', t')^{-1} \circ (z, t))$$

where

$$d(z, t) = |t + i|z|^2|^{\frac{1}{2}}.$$

Let D be a bounded smooth domain in $\boldsymbol{H}^n$. Consider

$$\begin{cases} -\Delta_{\boldsymbol{H}^n} u = u^p + \lambda u & \text{in } D, \\ u > 0 & \text{in } D, \\ u = 0 & \text{on } \partial D, \end{cases} \tag{2}$$

where $p = (Q + 2)/(Q - 2)$ and $Q = 2n + 2$ is the homogeneous dimension of $\boldsymbol{H}^n$.
We shall prove

Theorem 1.
(i) *There exists a solution for* (2) *if* $\lambda \in (0, \lambda_1)$, *where* λ_1 *is the first Dirichlet eigenvalue of* $\Delta_{\boldsymbol{H}^n}$.

(ii) *Suppose D is δ-starshaped* [6] *with respect to some* $(z_0, t_0) \in D$. *Then there is no solution for* (2) *if* $\lambda \notin (0, \lambda_1)$.

Denote the Hilbert space of all functions $u \in L^2(D)$ whose distributional derivatives $X_j u$, $Y_j u \in L^2(D)$ $(j = i, \ldots, n)$ by $S_1^2(D)$. The norm in $S_1^2(D)$ is given by

$$\|u\|^2 = \int_D \left(|\nabla_{\boldsymbol{H}^n}|^2 + |u|^2\right) dz\, dt$$

where

$$|\nabla_{H^n} u|^2 = \sum_{j=1}^{n} \left[(X_j u)^2 + (Y_j u)^2 \right];$$

also let $\overset{\circ}{S}{}_1^2(D)$ be the completion of $C_0^\infty(D)$ under the above norm. We have the continuous embedding of Sobolev-type [6]

$$\overset{\circ}{S}{}_1^2(D) \hookrightarrow L^{\frac{2Q}{Q-2}}(D).$$

By a weak solution u of (2) we mean that $u \in \overset{\circ}{S}{}_1^2(D)$ and it satisfies

$$\int_D \nabla_{H^n} u \cdot \nabla_{H^n} v \, dz \, dt = \int_D f(u) v \, dz \, dt$$

for all $v \in \overset{\circ}{S}{}_1^2(D)$ where $f(u) = \lambda u + u^p$. If $u \in \overset{\circ}{S}{}_1^2(D)$ is a weak solution of (2), from the results in [6]:

$$u \in L^p_{\mathrm{loc}}(\bar{D}) \quad \text{for every } p \in [2, \infty),$$

and $u \in \Gamma^2_{\mathrm{loc}}(\bar{D})$, where $\Gamma^2(\bar{D})$ is the space of all continuous functions on $\bar{D}$ such that $X_j u$, $X_j^2 u$, $Y_j u$ and $Y_j^2 u$ are continuous in D and can be continuously extended to $\bar{D}$.

3. Proof of Theorem 1

With the aid of the inequality of Poincaré-type for the Dirichlet integral on H^n one can show that the first Dirichlet eigenvalue of $(-\Delta_{H^n})$ satisfies

$$\lambda_1 = \inf \left\{ \frac{\|u\|^2}{\int_D u^2 \, dz \, dt} \ : \ u \in \overset{\circ}{S}{}_1^2(D) \setminus \{0\} \right\} > 0$$

and there exists a positive function $\varphi \in \overset{\circ}{S}{}_1^2(D)$ such that

$$\begin{cases} -\Delta_{H^n} \varphi = \lambda_1 \varphi & \text{in } D \\ \varphi = 0 & \text{on } \partial D \end{cases} \tag{3}$$

If u is a solution of (2), it follows from (3) that

$$\begin{aligned} \lambda_1 \int_D u\varphi \, dz \, dt &= -\int_D \varphi \Delta_{H^n} u \\ &= \int_D u^p \varphi + \lambda \int_D u\varphi \\ &> \lambda \int_D u\varphi \, dz \, dt. \end{aligned}$$

Hence we have $\lambda_1 > \lambda$.

The second part of Theorem 1 is essentially contained in the following.

Lemma 1 ([6], Theorem 2.2). *Suppose D is star-shaped with respect to some $(z_0, t_0) \in D$. Then the problem,*

$$\begin{cases} -\Delta_{H^n} u = f(u) & \text{in } D, \\ u = 0 & \text{on } \partial D, \end{cases} \tag{4}$$

has no non-trivial non-negative solution $u \in \Gamma^2(\overline{D})$ provided f is locally Lipschitz, $f(0) = 0$ and

$$2QF(u) - (Q-2)uf(u) \le 0, \quad u \ge 0,$$

where $F(u) = \int_0^u f(t)dt$.

Now, letting $f(u) = u^p + \lambda u$, $p = (Q+2)/(Q-2)$, we have

$$F(u) = \tfrac{Q-2}{2Q} u^{2Q/(Q-2)} + \tfrac{1}{2}\lambda u^2$$

and

$$2QF(u) - (Q-2)uf(u) = 2\lambda u^2 \le 0$$

if $\lambda \le 0$. This completes the proof of Theorem 1(ii).

Next we prove the first part of the theorem. We shall follow [1]. For $\lambda \in (0, \lambda_1)$, let

$$S_\lambda = \inf\left\{\int_D |\nabla_{H^n} u|^2 - \lambda \int_D u^2 \;:\; u \in \overset{\circ}{S}{}^2_1(D),\ \int_D |u|^{p+1} = 1\right\}.$$

If S_λ is achieved by $u_0 \in \overset{\circ}{S}{}^2_1(D)$, then there exists μ such that

$$-\Delta_{H^n} u_0 - \lambda u_0 = \mu u_0^p.$$

Without loss of generality we may suppose $u_0 \ge 0$ in D, and it is clear that $\mu = S_\lambda$. Now if $\lambda < \lambda_1$, then $\mu > 0$ and $S_\lambda^{1/(p-1)} u_0$ is a solution of (2). Note that u is positive by the strong maximum principle in [6]. In the following we let $S = S_0$.

Lemma 2. *Suppose $S_\lambda < S$. Then S_λ is attained.*

The proof of the lemma is almost identical to Lemma 1.2. in [1].

Lemma 3. *If $\lambda > 0$, then $S_\lambda < S$.*

The proof of this lemma relies on the best constant and minimizers for the Sobolev inequality

$$\left(\int_{H^n} |u|^{2Q/(Q-2)}\, dz\, dt\right)^{(Q-2)/Q} \le S^{-1/2} \left(\int_{H^n} |\nabla_{H^n} u|^2\, dz\, dt\right)^{\frac{1}{2}}, \tag{5}$$

for all $u \in \overset{\circ}{S}{}^2_1(D)$.

In [7] Jerison and Lee found that the best constant S is attained only at the function $u = K|w+i|^n$, $K \in \mathbb{C}$, as well as those obtained from u by left translation and dilations on the Heisenberg group. Substituting u into (5) we find the best constant

$$S = \pi n^2 \left(\frac{4}{n!}\right)^{\frac{1}{n+1}}.$$

Consider the quotient

$$J_\lambda(w) = \frac{\int_D |\nabla_{H^n} w|^2\, dz\, dt - \lambda \int_D |w|^2\, dz\, dt}{\left[\int_D |w|^{2Q/(Q-2)}\, dz\, dt\right]^{(Q-2)/Q}};$$

we shall show that $J_\lambda(w) < S$ for some w provided $\lambda > 0$. Without loss of generality we assume $0 \in U \subset\subset D$, where U is an open set in H^n. Let

$$w = \varphi u_\varepsilon(z,t) = \frac{\varphi}{[t^2 + (r^2+\varepsilon)^2]^{n/2}}$$

where $r^2 = |x|^2 + |y|^2$ and u_ε is an extremal function in the Sobolev inequality (after some dilation); $\varphi \in C_0^\infty(D)$, $\operatorname{supp}\varphi \subset\subset D$, $\varphi(U) \equiv 1$, $0 \le \varphi \le 1$.

We estimate each term in $J_\lambda(w)$ as $\varepsilon \to 0$.

$$\begin{aligned}
\frac{\partial w}{\partial x_j} &= \varphi_j u_\varepsilon - \frac{2n(r^2+\varepsilon)\varphi x_j}{[t^2+(r^2+\varepsilon)^2]^{n/2+1}},\\
\frac{\partial w}{\partial y_j} &= \varphi_{n+j} u_\varepsilon - \frac{2n(r^2+\varepsilon)\varphi y_j}{[t^2+(r^2+\varepsilon)^2]^{n/2+1}},\\
\frac{\partial w}{\partial t} &= \varphi_t u_\varepsilon - \frac{n\varphi t}{[t^2+(r^2+\varepsilon)^2]^{n/2+1}},
\end{aligned}$$

where $\varphi_j = \frac{\partial \varphi}{\partial x_j}$, $\varphi_{n+j} = \frac{\partial \varphi}{\partial y_j}$ and $\varphi_t = \frac{\partial \varphi}{\partial t}$.

$$\begin{aligned}
|\nabla_{H^n} w|^2 &= \sum_{j=1}^n \left[\left(\frac{\partial w}{\partial x_j} + 2y_j \frac{\partial w}{\partial t}\right)^2 + \left(\frac{\partial w}{\partial y_j} - 2x_j \frac{\partial w}{\partial t}\right)^2\right]\\
&= \frac{4n^2 r^2 \varphi^2}{[t^2+(r^2+\varepsilon)^2]^{n+1}} + 4r^2 u^2 \varphi_t^2 - \frac{8nr^2\varphi t}{[t^2+(r^2+\varepsilon)^2]^{n/2+1}}\\
&\quad + u^2 \sum_{j=1}^n [\varphi_j^2 + \varphi_j^2] - \frac{4n(r^2+\varepsilon)u\varphi}{[t^2+(r^2+\varepsilon)^2]^{n/2+1}} \sum_{j=1}^n [x_j\varphi_j + y_j\varphi_{n+j}]\\
&\quad - 4n\left(u\varphi_t - \frac{nt\varphi}{[t^2+(r^2+\varepsilon)^2]^{n/2+1}}\right) \sum_{j=1}^n [y_j\varphi_j - x_j\varphi_{n+j}].
\end{aligned}$$

Thus as $\varepsilon \to 0^+$, we have

$$\begin{aligned}\int_{H^n} |\nabla_{H^n} w|^2\, dz\, dt &= \int_{\mathbb{R}^{2n+1}} \sum_{j=1}^{n} \left[(X_j w)^2 + (Y_j w)^2\right] dx\, dy\, dt \\ &= \omega \int_{\mathbb{R}^{2n+1}} \frac{4n^2 r^2 \varphi^2}{[t^2 + (r^2+\varepsilon)^2]^{n+1}}\, dx\, dy\, dt + O(1) \\ &= 8n^2 \omega \int_0^\infty \int_0^\infty \frac{r^{2n+1}}{[t^2 + (r^2+\varepsilon)^2]^{n+1}}\, dr\, dt + O(1),\end{aligned}$$

where ω is the area of the unit sphere in $\mathbb{R}^{2n}$. Reversing the order of the integration, we have

$$\begin{aligned}\int_{H^n} |\nabla_{H^n} w|^2\, dz\, dt &= 8n^2 \omega \int_0^\infty \frac{r^{2n+1} dr}{(r^2+\varepsilon)^{2n+1}} \int_0^\infty \frac{dt}{(1+t^2)^{n+1}} + O(1) \\ &= 8n^2 \varepsilon^{-n} \left(\frac{1}{2}\frac{\Gamma(n+1)\Gamma(n)}{\Gamma(2n+1)}\right)\left(\frac{1}{2}\frac{\Gamma(\frac{1}{2})\Gamma(n+\frac{1}{2})}{\Gamma(n+1)}\right) + O(1) \\ &\equiv C_1 \varepsilon^{-n} + O(1)\end{aligned}$$

as $\varepsilon \to 0^+$;

$$\begin{aligned}\int_D w^{2Q/(Q-2)} dz dt &= 2\omega \int_0^\infty \frac{r^{2n-1} dr}{(r^2+\varepsilon)^{2n+1}} \int_0^\infty \frac{dt}{(1+t^2)^{n+1}} + O(1) \\ &= 2\omega \varepsilon^{-n-1} \left(\frac{1}{2}\frac{\Gamma(n+1)\Gamma(n)}{\Gamma(2n+1)}\right)\left(\frac{1}{2}\frac{\Gamma(\frac{1}{2})\Gamma(n+\frac{1}{2})}{\Gamma(n+1)}\right) + O(1) \\ &\equiv C_2 \varepsilon^{-n-1} + O(1)\end{aligned}$$

as $\varepsilon \to 0^+$. Finally we estimate $\int w^2\, dz\, dt$. First, for $n > 1$:

$$\begin{aligned}\int_D w^2\, dz\, dt &= 2\omega \int_0^\infty \frac{r^{2n-1} dr}{(r^2+\varepsilon)^{2n-1}} \int_0^\infty \frac{dt}{(1+t^2)^{n}} + O(1) \\ &= 2\omega \varepsilon^{-n+1} \left(\frac{1}{2}\frac{\Gamma(n-1)\Gamma(n)}{\Gamma(2n-1)}\right)\left(\frac{1}{2}\frac{\Gamma(\frac{1}{2})\Gamma(n-\frac{1}{2})}{\Gamma(n)}\right) + O(1) \\ &\equiv C_3 \varepsilon^{-n+1} + O(1)\end{aligned}$$

as $\varepsilon \to 0^+$. If $n = 1$, there exist R_1 and R_2: $0 < R_1 \le R_2$ and

$$B_{R_1} \times (-R_1, R_1) \subset D \subset B_{R_2} \times (-R_2, R_2)$$

where

$$B_R = \{(x, y) \in \mathbb{R}^2 \; : \; r^2 = x^2 + y^2 < R\}.$$

Hence

$$\int_{B_{R_1}\times(-R_1,R_1)} w^2\,dz\,dt \le \int_D w^2\,dz\,dt \le \int_{B_{R_2}\times(-R_2,R_2)} w^2\,dz\,dt.$$

Now we compute

$$\begin{aligned}\int_{B_R\times(-R,R)} w^2\,dz\,dt &= 2\omega\int_0^R\int_0^R \frac{r\,dr\,dt}{t^2+(r^2+\varepsilon)^2} + O(1)\\ &= 2\omega\int_0^R r\frac{\arctan\frac{R}{r^2+\varepsilon}dr}{r^2+\epsilon}dt + O(1)\\ &\equiv C_3'|\log\varepsilon| + O(1), \qquad C_3'>0,\end{aligned}$$

as $\varepsilon\to 0^+$. Now we compute the quotient $J_\lambda(w)$. If $n>1$,

$$\begin{aligned}J_\lambda(w) &= \frac{C_1\varepsilon^{-n}-\lambda C_3\varepsilon^{-n+1}+O(1)}{(C_2\varepsilon^{-n-1}+O(1))^{n/(n+1)}}\\ &= S-\lambda C_4\varepsilon + O(1)\varepsilon^n, \qquad C_4>0,\end{aligned}$$

as $\varepsilon\to 0^+$. Hence, for $\lambda>0$ and ε sufficiently small, we have

$$S_\lambda \le J_\lambda(w) < S.$$

When $n=1$, we have

$$\begin{aligned}J_\lambda(w) &= \frac{C_1\varepsilon^{-1}-\lambda C_3'|\log\varepsilon|+O(1)}{(C_2\varepsilon^{-2}+O(1))^{1/2}}\\ &= S-C_4'\varepsilon|\log\varepsilon| + O(1)\varepsilon, \qquad C_4'>0.\end{aligned}$$

So the same conclusion holds.

The proof of Theorem 1(i) is completed. □

References

[1] H. Brezis , L. Nirenberg, Positive solutions of nonlinear elliptic equations involving critical Sobolev exponents, Comm. Pure Appl. Math. 36 (1983), 437–477.

[2] K. S. Chou, D. Geng, On the critical dimension of a semilinear degenerate elliptic equation involving critical Sobolev–Hardy exponent, Nonlinear Anal. 26 (1996), 1965–1984.

[3] K. S. Chou, D. Geng, S. S. Yan, Critical dimensions of a Hessian equation involving critical exponent and a related asymptotic results, J. Differential Equations 129 (1996), 79–110.

[4] D. E. Edmunds, D. Fortunato, E. Jannelli, Critical exponents, critical dimensions and the biharmonic operator, Arch. Rational Mech. Anal. 112 (1990), 269–289.

[5] H. Egnell, Elliptic boundary value problems with singular coefficient and critical nonlinearities, Indiana Univ. Math. J. 38 (1989), 235–251.

[6] N. Garofalo, E. Lanconell, Existence and nonexistence results for semilinear equations on the Heisenberg group, Indiana Univ. Math. J. 41 (1992), 71–98.

[7] D. Jerison, J. M. Lee, Extremals for the Sobolev inequality on the Heisenberg group and the CR Yamabe problem, J. Amer. Math. Soc. 1 (1988), 1–13.

[8] P. Pucci, J. Serrin, Critical exponents and critical dimensions for polyharmonic operators, J. Math. Pure Appl. 69 (1990), 55–83.

Minimal triangulation, complementarity and projective planes

Basudeb Datta

Abstract. Brehm and Kühnel proved that if M^d is a combinatorial d-manifold with $3d/2+3$ vertices and $|M^d|$ is not homeomorphic to S^d then the combinatorial Morse number of M^d is three and hence $d \in \{0, 2, 4, 8, 16\}$ and $|M^d|$ is a manifold like a projective plane in the sense of Eells and Kuiper.

We discuss the existence and uniqueness of such combinatorial manifolds. We also present the following result: "Let M_n^d be a combinatorial d-manifold with n vertices. M_n^d satisfies complementarity if and only if $d \in \{0, 2, 4, 8, 16\}$ with $n = 3d/2 + 3$ and $|M_n^d|$ is a manifold like a projective plane".

1991 Mathematics Subject Classification: 57Q15

1. Introduction

Recall that a *simplicial complex* is a set of non-empty finite sets such that every non-empty subset of an element is also an element. For $i \geq 0$, the elements of size $i+1$ are called the i-dimensional simplices (in short, i-simplices) of the complex. For $i = 1, 2, 3$, the i-simplices are also called the edges, triangles and tetrahedra of the complex, respectively. For a simplicial complex X, the maximum of k such that X has a k-simplex is called the *dimension* of X.

The union of all the simplices in a simplicial complex X is called the *vertex-set* of X and is denoted by $V(X)$. Elements of $V(X)$ are called vertices of X. If U is a non-empty subset of $V(X)$ then the induced subcomplex of X, denoted by $X(U)$, on U is the simplicial complex whose simplices are those simplices of X which are contained in U. If X_1, X_2 are two simplicial complexes, then a *simplicial isomorphism* from X_1 to X_2 is a bijection $\pi : V(X_1) \to V(X_2)$ such that for $\sigma \subseteq V(X_1)$, σ is a simplex of X_1 if and only if $\pi(\sigma)$ is a simplex of X_2. The complexes X_1, X_2 are called (simplicially) *isomorphic* when such an isomorphism exists.

A simplicial complex is usually thought of as a prescription for constructing a topological space (called the *geometric realization* of X and denoted by $|X|$) by pasting together geometric simplices. Formally, $|X|$ is the subspace of $[0, 1]^{V(X)}$ consisting of the functions $f : V(X) \to [0, 1]$ such that the support $\{v \in V(X) : f(v) \neq 0\}$ is a simplex of X and $\sum_{v \in V(X)} f(v) = 1$. If σ is a simplex then $|\sigma| := \{f \in |X| : \sum_{v \in \sigma} f(v) = 1\}$ is called the *geometric carrier* of σ or the *geometric simplex* corresponding to σ.

For any set V with $\#(V) = d + 2$ ($d \geq 0$), let K be the simplicial complex whose simplices are all the non-empty proper subsets of V. Then $|K|$ is homeomorphic to the d-dimensional sphere S^d. This simplicial complex is called the *standard d-sphere* and is denoted by S^d_{d+2}.

By a *subdivision* of a simplicial complex K we mean a simplicial complex K' together with a homeomorphism s from $|K'|$ onto $|K|$ which is simplexwise linear, i.e., maps the geometric carrier of each simplex of K' linearly into the geometric carrier of a simplex of K.

Two simplicial complexes K and L are called *combinatorially equivalent* (denoted by $K \approx L$) if they have isomorphic subdivisions. Clearly, if K and L are combinatorially equivalent then $|K|$ and $|L|$ are homeomorphic. In fact, in this case, $|K|$ is PL-homeomorphic to $|L|$ and, conversely, if $|K|$ and $|L|$ are PL-homeomorphic with their induced PL-structure then $K \approx L$ (see [9]).

If σ is a simplex of a simplicial complex X then the *link* of σ in X, denoted by $\mathrm{Lk}_X(\sigma)$, is by definition the simplicial complex whose simplices are the simplices τ of X such that τ is disjoint from σ and $\sigma \cup \tau$ is a simplex of X.

A simplicial complex X is called a *combinatorial d-sphere* if it is combinatorially equivalent to the standard d-sphere. A d-dimensional simplicial complex X is called a *combinatorial d-manifold* if the link of each 0-simplex is a combinatorial $(d - 1)$-sphere. We know ([9]) that, for a simplicial complex X, $|X|$ with the induced piecewise linear structure is a PL-manifold if and only if X is a combinatorial manifold.

Let X_1, X_2 be two simplicial complexes with disjoint vertex sets. Then their *join* $X_1 * X_2$ is the simplicial complex whose simplices are those of X_1 and of X_2, and the (disjoint) unions of simplices of X_1 with simplices of X_2.

A simplicial complex K is said to satisfy *complementarity* if exactly one of each complementary pair of non-empty vertex-sets is a simplex of the complex. If $v \notin V(S^d_{d+2})$ then $K_d := S^d_{d+2} \cup \{v\}$ satisfies complementarity. If L satisfies complementarity and A is a non-empty set disjoint from $V(L)$ then $L * \mathcal{P}(A)$ satisfies complementarity, where $\mathcal{P}(A)$ denotes the power set of A.

Let K be a combinatorial d-manifold. A map $g : U$ ($\subseteq |K|$) $\to \mathbb{R}^n$ is called a *combinatorial map* if there exists a subdivision (K', s) of K and a homeomorphism h (not necessarily s, see [6]) from $|K'|$ to $|K|$ such that $g_i \circ h : h^{-1}(U) \to \mathbb{R}$ is simplexwise linear for all $i \in \{1, \ldots, n\}$. A map $f : |K| \to \mathbb{R}$ is called *combinatorial-non-degenerate* (*cnd*) if for each $a \in |K|$ there exists a chart (U, φ) with $\varphi : U \to \mathbb{R}^d$ combinatorial and there exists $\lambda_a \in \mathbb{R}^+$ such that

either $\varphi_d(x) = \lambda_a(f(x) - f(a))$ for all $x \in U$ (such an $a \in |K|$ is said to be *combinatorial-ordinary*), or there exists a $k \in \{0, \ldots, d\}$ with

$$-\max\{|\varphi_1(x)|, \ldots, |\varphi_k(x)|\} + \max\{|\varphi_{k+1}(x)|, \ldots, |\varphi_d(x)|\} = \lambda_a(f(x) - f(a))$$

for all $x \in U$ (such an a is called a *combinatorial critical point of index k*).

Let K be a combinatorial manifold. A function $f : |K| \to \mathbb{R}$ is called *regular simplexwise linear* (*rsl*) if it is simplexwise linear and $f(u) \neq f(v)$ for any two vertices $u \neq v$. If f is rsl then f is cnd (see [6], [10]).

Let $\mu(f) :=$ number of combinatorial critical points of f and $\mu(K) := \min\{\mu(f) : f \text{ is cnd}\}$. $\mu(K)$ is called the *combinatorial Morse number* of K.

In 1962, Eells and Kuiper [6] proved that a combinatorial d-manifold M^d with combinatorial Morse number $\mu(M^d) = 3$ has dimension $d = 0, 2, 4, 8$ or 16. If $d = 0$, $|M^d|$ consists of three points. If $d = 2$ then $|M^d|$ is the real projective plane. For $d =$ 4, 8 or 16, $|M^d|$ is a simply connected cohomology projective plane over the complex numbers, quaternions or Cayley numbers respectively. Each of the manifolds of the above type is called a *manifold like a projective plane*. This classification turned up in the 1987 paper [3] of Brehm and Kühnel on combinatorial manifolds with few vertices. Specifically, they proved the following.

Let M_n^d be a combinatorial d-manifold with n vertices:

(BK1) if $n < 3\lceil d/2 \rceil + 3$, then $|M_n^d|$ is homeomorphic to S^d;

(BK2) if $n = 3(d/2) + 3$ and $|M_n^d|$ is not homeomorphic to S^d, then $d = 0, 2, 4, 8$ or 16 and $|M_n^d|$ must be a 'manifold like a projective plane'.

It is classically known that there exists a unique (up to simplicial isomorphism) 6-vertex triangulation (denoted by $\mathbb{R}P_6^2$) of the real projective plane $\mathbb{R}P^2$.

In the case $d = 4$, a PL simply connected cohomology $\mathbb{C}P^2$, being a PL 4-manifold, is uniquely smoothable by a result of Hirsch and Mazur [8], and by Freedman's classification [7] of 4-manifolds, a smooth simply connected cohomology $\mathbb{C}P^2$ is actually homeomorphic to $\mathbb{C}P^2$. Thus, a non-sphere 9-vertex combinatorial 4-manifold triangulates $\mathbb{C}P^2$, and may be denoted by $\mathbb{C}P_9^2$.

Consider a 9-vertex simplicial complex $\mathcal{X}$, whose vertices are the points of the affine plane $\mathbb{P}$ over the 3-element field. The 36 maximal faces (4-simplices) of $\mathcal{X}$ are as follows. Fix a set Π of three mutually parallel lines of $\mathbb{P}$ (i.e., Π is a parallel class of lines in $\mathbb{P}$). Let the lines in Π be l_0, l_1, l_2, in a fixed cyclic orientation. For $0 \leq i \leq 2$ and for any point x on l_i, the set $(l_i \cup l_{i+1}) \setminus \{x\}$ is a face of $\mathcal{X}$. (Addition in the subscript is modulo 3). This gives $3 \times 3 = 9$ faces of $\mathcal{X}$. Also, for any two (distinct) intersecting lines m_1, m_2 of $\mathbb{P}$ outside Π, $m_1 \cup m_2$ is a face of $\mathcal{X}$. This gives $(9 \times 6)/2 = 27$ faces of $\mathcal{X}$. The remaining simplices are, of course, the non-empty subsets of these 36 maximal faces. Then (see [2]) $\mathcal{X}$, thus constructed, triangulates $\mathbb{C}P^2$. In [2] we proved: "If $\mathcal{Y}$ is an arbitrary 9-vertex complex projective plane, then $\mathcal{Y}$ is isomorphic to the combinatorial manifold $\mathcal{X}$ constructed above. Thus the 9-vertex $\mathbb{C}P^2$ is unique up to simplicial isomorphism."

In [4] Brehm and Kühnel constructed three (non-isomorphic) combinatorial 8-manifolds with 15 vertices, each of them triangulates the same topological manifold which is not S^8.

In [1] Arnoux and Marin proved that $\mathbb{C}P^2_9$ satisfies complementarity and they remarked that a similar proof works for other cases of the equality in (BK2). In this article we present

Theorem. *Let M^d_n be a combinatorial d-manifold with n vertices. M^d_n satisfies complementarity if and only if $d \in \{0, 2, 4, 8, 16\}$ with $n = 3(d/2) + 3$ and $|M^d_n|$ is a 'manifold like a projective plane'.*

Conjecture. *If a pseudomanifold K satisfies complementarity then K is a combinatorial manifold.*

2. Preliminaries

For a simplicial complex K with n vertices let $f_i(K) :=$ number of i-simplices in K. Thus, $f_0(K) = n$ and $f_i(K) \leq \binom{n}{i+1}$ for all i. Let $\nu(K) := \max\{i : f_{i-1}(K) = \binom{n}{i}\}$.

For $n > p$, let V be a set of n distinct points on the curve $\{(t, t^2, \ldots, t^p) \in \mathbb{R}^p : t \in \mathbb{R}\}$. Let $C(n, p)$ be convex hull of V. Then it is known (see [12]) that: (i) $C(n, p)$ is a simplicial (i.e., each proper face is a geometric simplex) polytope. So the boundary $\partial C(n, p)$ defines a simplicial complex C^{p-1}_n (namely, a set σ is a simplex of C^{p-1}_n if and only if σ is a proper non-empty subset of V such that the affine hull of σ is disjoint from the convex hull of $V \setminus \sigma$) such that $|C^{p-1}_n|$ homeomorphic to S^{p-1}, and (ii) $f_{i-1}(C^{p-1}_n) = \binom{n}{i}$ for $0 < i \leq \lfloor \frac{p}{2} \rfloor$ and $\nu(C^{p-1}_n) = \lfloor \frac{p}{2} \rfloor$ for $n > p + 1$. The Upper Bound Conjecture (established by Stanley in [12]) says:

(UBC) If K is a triangulation of S^{p-1} with n vertices, then $f_j(K) \leq f_j(C^{p-1}_n)$, for $j = 0, 1, \ldots, p - 1$.

Lemma 1. *If M is a combinatorial d-manifold with n vertices and $n > d + 2$ then $\nu(M) \leq \lfloor \frac{d}{2} \rfloor + 1$.*

Proof. We may assume $\nu(M) \geq 2$ (since if $\nu(M) = 1$ then there is nothing to prove). By (UBC) we have $f_i(\mathrm{Lk}_M(\{v\})) \leq f_i(C^{d-1}_{n-1})$ for each $v \in V(M)$. Therefore $\nu(\mathrm{Lk}_M(\{v\})) \leq \nu(C^{d-1}_{n-1}) = \lfloor \frac{d}{2} \rfloor$ for each $v \in V(M)$ and hence $\nu(M) \leq \lfloor \frac{d}{2} \rfloor + 1$. □

Remark. Lemma 1 also follows (without using (UBC)) from Dehn–Sommerville equations (which hold for any triangulation of S^{d-1}).

Let K be a simplicial complex with vertex set V. If $V_1 \subseteq V$ then $|K(V_1)|$ is a deformation retract of the open set $A = |K| \setminus |K(V \setminus V_1)|$, $|K(V \setminus V_1)|$ is a deformation retract of the open set $B = |K| \setminus |K(V_1)|$ and $|K| = A \cup B$.

For a space X, if Y is a subset of X and α is a cohomology class in X then $\alpha|_Y := i^*(\alpha)$ is called the restriction of α to Y, where $i : Y \hookrightarrow X$ is the inclusion map. (Here the coefficient ring is $\mathbb{Z}$ or $\mathbb{Z}/2\mathbb{Z}$.) We know (e.g., see [1]) that if X is the union of two open sets A and B and two cohomology classes α and β of X are such that $\alpha|_A = 0$ and $\beta|_B = 0$ then $\alpha \cup \beta = 0$ in X. From these we get

Lemma 2 (Compare [1, Lemma 1, p. 189]). *Let K be a simplicial complex with vertex-set V. Let α and β be two cohomology classes of $|K|$ and $V_1 \subseteq V$. If $\alpha|_{|K(V_1)|} = 0$ and $\beta|_{|K(V \setminus V_1)|} = 0$ then $\alpha \cup \beta = 0$ in $|K|$.*

Let L be a subcomplex of a simplicial complex K. If $L \cong S^c_{c+2}$ then the sum of all the c-simplices in L is a c-cycle (with coefficients in $\mathbb{Z}/2\mathbb{Z}$). This c-cycle is called a *simple c-sphere* and is also denoted by L. Then we have

Lemma 3. *Let K be a simplicial complex of dimension d and for $0 < c \leq d$ let K contain all the c-simplices. If $z \in Z_c(K, \mathbb{Z}/2\mathbb{Z})$ then z is a sum of simple spheres.*

Proof. If $\#(V(z)) = c + 2$ then z is itself a simple c-sphere. So, we may assume $\#(V(z)) = k + 2$ with $k > c$. Let $S := S^k_{k+2}(V(z))$. Then $z \in Z_c(S, \mathbb{Z}/2\mathbb{Z})$, and hence $z \in B_c(S, \mathbb{Z}/2\mathbb{Z})$. Let $z = \partial y$, where $y \in C_{c+1}(S, \mathbb{Z}/2\mathbb{Z})$. Now, if $y = \sigma_1^{c+1} + \cdots + \sigma_m^{c+1}$, then $z = \partial\sigma_1^{c+1} + \cdots + \partial\sigma_m^{c+1} \pmod 2$. Clearly each $\partial\sigma_j^{c+1}$, $j = 1, \ldots, m$, is a simple c-sphere in K. This proves the lemma. □

Let K be a d-dimensional simplicial complex with $d + c + 3$ vertices $(1 \leq c \leq d)$ and having the property that every $(d-1)$-simplex is a face of exactly two d-simplices. For a c-simplex τ in K, let $b(\tau) :=$ number (mod 2) of d-simplices which are disjoint from τ, and for a c-chain $\sum_i \tau_i$ define

$$b\,(\sum_i \tau_i) := \sum_i b(\tau_i) \pmod 2. \tag{2.1}$$

For a simple c-sphere Σ in K, let $S_\Sigma := \{\tau \in K : \tau$ is a $(d-1)$-simplex disjoint from Σ and $\{v\} \cup \tau \in K$ for some vertex v in $\Sigma\}$ and let $T_\Sigma := \{\sigma \in K : \sigma$ is a d-simplex and disjoint from $\Sigma\}$. Clearly $\#(T_\Sigma) \in \{0, 1\}$. Then

$$\#(T_\Sigma) = 0 \Rightarrow b(\Sigma) = 2\#(S_\Sigma) = 0 \pmod 2. \tag{2.2}$$

$$\#(T_\Sigma) = 1 \Rightarrow b(\Sigma) = \#(S_\Sigma) + (c+2)\#(T_\Sigma) = d + c + 3 \pmod 2. \tag{2.3}$$

3. Proof of the theorem

Let M be a combinatorial $2c$-manifold having vertex-set V with $\#(V) = 3c + 3$ and $|M|$ be a manifold like a projective plane. Since $\mathbb{R}P_6^2$ satisfies complementarity, we may assume that $c \in \{2, 4, 8\}$. Then there exists $\alpha \in H^c(|M|, \mathbb{Z}/2\mathbb{Z})$ with $\alpha \cup \alpha \neq 0$. Let $V_1 \subseteq V$ with $\#(V_1) = 2c + 1$. Then V_1 must satisfy

(C1) either V_1 is a $2c$-simplex and $M(V \setminus V_1) \cong S^c_{c+2}$, or $V \setminus V_1$ is a $(c+1)$-simplex.

If V_1 is a $2c$-simplex then, by Lemma 2, $\alpha|_{|M(V \setminus V_1)|} \neq 0$ and hence $M(V \setminus V_1) \cong S^c_{c+2}$.

If V_1 is not a $2c$-simplex, we want to show that $V \setminus V_1$ is a $(c + 1)$-simplex. Let $K := M \cup \{\tau \subseteq V \;:\; 1 \leq \#(\tau) \leq c\}$. The obvious extension of a cocycle a representing α by setting it to be zero on the c-simplices added, is seen to be a cocycle on K. (Infact, any extension of a will be a cocycle on K.) The corresponding cohomology class (also denoted by α) clearly restrict to α on M. Define b as in (2.1). Then we have:

(C2) if Σ is a simple c-sphere in K, then $b(\Sigma) \neq 0$ if and only if there is a (unique) $2c$-simplex σ^{2c} disjoint from Σ,

(C3) b is a cocycle,

(C4) if β is the cohomology class of b then $\beta \cup \beta \neq 0$.

Here c is even and d (the dimension of K) is $2c$, therefore (C2) follows from (2.2) and (2.3).

To prove (C3) we have to show that if Σ is a boundary of a $(c + 1)$-simplex then $b(\Sigma) = 0$. Since Σ is a boundary of a $(c + 1)$-simplex, $\alpha|_{|M(U)|} = 0$, where U is the set of vertices in Σ, and this implies, by Lemma 2, that $\alpha|_{|M(V \setminus U)|} \neq 0$ and hence $V \setminus U$ is not a $2c$-simplex. Thus there does not exist any $2c$-simplex disjoint from Σ and hence by (C2) $b(\Sigma) = 0$. This proves (C3).

Let σ^{2c} be a $2c$-simplex in K. Let U be the set of $c + 2$ vertices not in σ^{2c}. Then by Lemma 2 $\alpha|_{|K(U)|} \neq 0$ and hence $K(U) \cong S^c_{c+2}$ and $\alpha(K(U)) = 1$. Also by (C2) $\beta(K(U)) = 1$. Therefore, $(\alpha + \beta)(K(U)) = 0$. Hence by Lemma 2 $(\alpha + \beta) \cup (\alpha + \beta) = 0$. Thus $\beta \cup \beta = \alpha \cup \alpha \neq 0$. This proves (C4).

Let Σ_0 be the c-skeleton of $K(V \setminus V_1)$. Then $\Sigma_0 \cong S^c_{c+2}$ and by (C2) $\beta(\Sigma_0) = 0$ and hence $\beta|_{|K(V \setminus V_1)|} = 0$. Therefore, by Lemma 2, $\beta|_{|K(V_1)|} \neq 0$. Hence, by Lemma 3, there exists a simple c-sphere Σ' in $K(V_1)$ with $b(\Sigma') = 1$ and therefore by (C2) there exists a $2c$-simplex σ^{2c} disjoint from Σ'. Then $\sigma^{2c} \supseteq V \setminus V_1$ and therefore $V \setminus V_1$ is a $(c + 1)$-simplex in K and hence in M. This proves (C1).

It follows from (C1) that all the c-simplices are in M (and hence $K = M$). So, to prove complementarity it is enough to show that:

(C5) if $U \subseteq V$ with $\#(U) = c + 2 + i$ $(0 \leq i < c - 1)$ then exactly one of U and $V \setminus U$ is a simplex.

We will prove (C5) by induction on i. From (C1) it is true when $i = 0$. So, assume $0 < i < c - 1$ (and hence $c = 4$ or 8) and (C5) is true for all U with $\#(U) = c + 2 + (i - 1)$. Let $\#(U) = c + 2 + i$. By Lemma 2 both of U and $V \setminus U$ cannot be simplices. If $V \setminus U$ is not a simplex, then $V \setminus (U \setminus \{v\})$ is not a simplex for any vertex v in U and $\#(U \setminus \{v\}) = c + 2 + (i - 1)$. Hence by the induction hypothesis $U \setminus \{v\}$ is a simplex. Thus $M(U)$ contains all the $(c + i)$-simplices and $i > 0$. Therefore $\alpha|_{|M(U)|} = 0$ and hence, by Lemma 2, $\alpha|_{|M(V \setminus U)|} \neq 0$. Then, by Lemma 3, $M(V \setminus U)$ contains a simple c-sphere Σ' which is not the boundary of a $(c + 1)$-simplex. Therefore, by (C1), U' $(= V \setminus V(\Sigma'))$ is a $2c$-simplex and clearly $U \subseteq U'$. So, U is a simplex. Thus, (C5) is true for i. Hence by induction (C5) is true. This proves the "if part" of the Theorem.

Conversely, let M be an n-vertex combinatorial d-manifold satisfying complementarity. It is trivial from the definition that, for $d = 0$, $|M|$ consists of three points, and since there is clearly no 1-manifold satisfying complementarity, we may take $d \geq 2$.

Since no set containing more than $d + 1$ vertices is a simplex, $n \leq 2d + 3$ and each set containing at most $n - d - 2$ vertices is a simplex. That is, for $i \leq n - d - 3$, all i-simplices occur in M. More generally $f_i(M) + f_{n-i-2}(M) = \binom{n}{i+1}$. As each vertex forms a 0-simplex, therefore $n > d + 2$. Thus, $d + 2 < n \leq 2d + 3$. Therefore, by Lemma 1, $n - d - 2 \leq \nu(M) \leq \lfloor \frac{d}{2} \rfloor + 1$. This implies that $n \leq d + \lfloor \frac{d}{2} \rfloor + 3$ and hence by (BK1) and (BK2) $|M|$ is either a sphere or a 'manifold like a projective plane'. Since M satisfies complementarity, exactly half of the $2^n - 2$ non-empty proper vertex sets are simplices of M. So $\sum_{i \geq 0} f_i(M) = 2^{n-1} - 1$, which is odd. Therefore the Euler characteristic of M is odd, and hence $|M|$ cannot be a sphere. This completes the proof. □

Acknowledgement. The author is thankful to Dr. Bhaskar Bagchi for suggesting this problem and for numerous useful conversations. The financial support received from the conference organisers is gratefully acknowledged.

References

[1] P. Arnoux and A. Marin, The Kühnel triangulation of complex projective plane from the view-point of complex crystallography (Part II), Mem. Fac. Sci. Kyushu Univ. Ser. A 45 (1991), 167–244.

[2] B. Bagchi and B. Datta, On Kühnel's 9-vertex complex projective plane, Geom. Dedicata 50 (1994), 1–13.

[3] U. Brehm and W. Kühnel, Combinatorial manifolds with few vertices, Topology 26 (1987), 465–473.

[4] U. Brehm and W. Kühnel, 15-vertex triangulation of an 8-manifold, Math. Ann. 294 (1992), 167–193.

[5] B. Datta, Combinatorial manifolds with complementarity, Proc. Indian Acad. Sci. (Math. Sci.) 104 (1994), 385–388.

[6] J. Eells, jr., and N. H. Kuiper, Manifolds which are like projective planes, Inst. Hautes Études Sci. Publ. Math. 14 (1962), 181–222.

[7] M. Freedman, The topology of four dimensional manifolds, J. Differential Geom. 17 (1983), 357–454.

[8] M. Hirsch and B. Mazur, Smoothing of piecewise linear manifolds, Ann. of Math. Stud. 80, Princeton Univ. Press, Princeton 1974.

[9] J. F. P. Hudson, Piecewise linear topology, W. A. Benjamin Inc., New York 1969.

[10] W. Kühnel, Triangulations of manifolds with few vertices, in: F. Tricerri (ed.)., Advances in Differential Geometry and Topology, World Scientific, Singapore 1990, 59–110.

[11] W. Kühnel and T. F. Banchoff, The 9-vertex complex projective plane, Math. Intelligencer 5 (3) (1983), 11–22.

[12] R. P. Stanley, The Upper Bound Conjecture and Cohen-Macaulay Rings, Stud. Appl. Math. LIV (1975), 135–142.

Compact infrasolvmanifolds are smoothly rigid

*F. T. Farrell and L. E. Jones**

1. Statement of results

Denote the connected component containing the identity element of a topological group G by G^0. We will use the letter S throughout this paper to denote a Lie group which is both solvable and simply connected. Recall that S is diffeomorphic to $\mathbb{R}^n$ for some integer n. Let $\mathrm{Aff}(S)$ denote the group of all affine motions of S; i.e., $\mathrm{Aff}(S)$ is the semi-direct product $S \rtimes \mathrm{Aut}(S)$ where $\mathrm{Aut}(S)$ is the group of all automorphisms of S.

Definition 1.1. A compact infrasolvmanifold M is a compact orbit space of the form $M = \Gamma \backslash S$ where S is a solvable and simply connected Lie group acted upon by a closed torsion-free subgroup $\Gamma \subset \mathrm{Aff}(S)$ satisfying

1. Γ^0 is contained in the nil radical of S and
2. the closure of the image of Γ in $\mathrm{Aut}(S)$ is compact.

The orbit space $\Gamma \backslash S$ is a closed manifold with a natural smooth structure and its universal cover is the homogeneous space $\Gamma^0 \backslash S$ which is diffeomorphic to $\mathbb{R}^m$ where $m = \dim S - \dim \Gamma^0$. Also Γ / Γ^0 is the group of all deck transformations; hence $\pi_1(M) = \Gamma/\Gamma^0$. Since we do not discuss noncompact infrasolvmanifolds in this paper, we will sometimes refer to a *compact infrasolvmanifold* by the shorter name *infrasolvmanifold*.

Note that Definition 1.1 generalizes the notion of compact Riemannian flat manifold and compact infranilmanifold which are its special cases when S is respectively abelian and nilpotent. It also generalizes the notion of compact solvmanifold which is the special case when $\Gamma \subseteq S$. The following rigidity statement is the main result of this paper.

---

* Both authors were supported in part by the National Science Foundation.

Theorem 1.2. *Any two compact infrasolvmanifolds, of dimension different from* 4, *whose fundamental groups are isomorphic are smoothly diffeomorphic. In fact, any given isomorphism of fundamental groups is induced by a smooth diffeomorphism.*

In the cases of compact Riemannian flat manifolds and compact infranilmanifolds, Theorem 1.2 was previously proven by Bieberbach [3] and Lee–Raymond [11], respectively. The isomorphism in these cases is in fact induced by an affine diffeomorphism [3] and [9]. And Mostow [13] previously showed that Theorem 1.2 is also true for solvmanifolds; however an affine diffeomorphism is not always possible in this case. One difficulty preventing an affine diffeomorphism is that a group Γ can be a lattice in two different simply connected solvable Lie groups; but in at most one simply connected nilpotent Lie group because of Malcev's rigidity theorem [12]. Our proof of Theorem 1.2 relies heavily on Mostow's paper [13]. But it also uses smoothing theory [10] and the topological rigidity result [4]. It is in fact an immediate consequence of [4] (and [16] in dimension 3) that in all dimensions (including dimension 4) Theorem 1.2 is true if diffeomorphism is weakened to homeomorphism in its conclusion. The results needed from [4] and [10] in our argument depend on a large quantity of topological work which Mostow of course does not need in [13]. It is perhaps possible that a proof of Theorem 1.2 could be constructed which is independent of this topological work.

The pair (S, Γ) occurring in Definition 1.1 is called a *presentation* for M. The image of Γ in $\mathrm{Aut}(S)$ is called the *holonomy group* of this presentation. The presentation is said to be discrete if Γ is a discrete subgroup of $\mathrm{Aff}(S)$. Compact Riemannian flat manifolds and compact infranilmanifolds M both have discrete presentations (Γ, S) with finite holonomy groups and with S abelian when M is flat Riemannian and nilpotent when M is an infranilmanifold. Also every flat Riemannian torus and every compact nilmanifold has a discrete presentation with trivial holonomy. However there are compact solvmanifolds which do *not* admit a discrete presentation with trivial holonomy. For example, the Klein bottle (which is a solvmanifold) does not have such a presentation since it is not orientable. We prove, apropos this discussion, the following fact in the appendix to this paper.

Lemma 1.3. *Every compact infrasolvmanifold has a presentation with finite holonomy group.*

The proof of Lemma 1.3 is independent of the rest of this paper. This is necessary since Lemma 1.3 is used in proving Theorem 1.2.

2. Proof of the main result

This entire section is devoted to proving Theorem 1.2. We start with some definitions and two preliminary lemmas.

An (N, T)-bundle is a smooth fiber bundle $p : E \to T$ with fiber a closed nilmanifold N, base space a closed torus T, and whose structure group is $\mathrm{Aff}(N)$; i.e., the group of all affine notions of N. We furthermore require that $\pi_1(N)$ is a characteristic subgroup of $\pi_1(E)$. Let $\mathcal{G}(E)$ denote the group consisting of all smooth bundle maps $f : E \to E$ which cover affine motions on T. Give $\mathcal{G}(E)$ the compact-open topology. Let $\mathcal{E}(E)$ denote the space of all self homotopy equivalences of E; $\mathcal{E}(E)$ is also given the compact-open topology.

Lemma 2.1. *The natural inclusion map $\sigma : \mathcal{G}(E) \to \mathcal{E}(E)$ is a (weak) homotopy equivalence.*

The proof of Lemma 2.1 is an exercise in obstruction theory. It makes use of the facts that the inclusions

$$\mathrm{Aff}(N) \subset \mathcal{E}(N) \quad \text{and} \quad \mathrm{Aff}(T) \subset \mathcal{E}(T)$$

are both weak homotopy equivalences together with the well known calculation that

$$\pi_i \mathcal{E}(M) = \begin{cases} \mathrm{Out}(\pi_1 M) & \text{if } i = 0 \\ \mathrm{Center}(\pi_1 M) & \text{if } i = 1 \\ 0 & \text{if } i > 1 \end{cases}$$

where M denotes any aspherical manifold. We will only show why the inclusion map $\mathrm{Aff}(N) \subset \mathcal{E}(N)$ is a weak homotopy equivalence and allow the reader to flesh out the remainder of the argument. Present N by $N = \mathcal{N}/\pi$ where $\mathcal{N}$ is a simply connected nilpotent Lie group and π is a discrete cocompact subgroup isomorphic to $\pi_1(N)$. Recall that the affine group of $\mathcal{N}$ is

$$\mathrm{Aff}(\mathcal{N}) = \mathrm{Aut}(\mathcal{N}) \ltimes \mathcal{N}.$$

Because of Malcev's Theorem, $\mathrm{Aut}(\pi)$ is identified to the subgroup of $\mathrm{Aut}(\mathcal{N})$ consisting of all $\alpha \in \mathrm{Aut}(\mathcal{N})$ such that $\alpha(\pi) = \pi$. Let G be the closed subgroup of $\mathrm{Aff}(\mathcal{N})$ given by the semi-direct product

$$G = \mathrm{Aut}(\pi) \ltimes \mathcal{N}.$$

Then G maps naturally onto $\mathrm{Aff}(N)$. Denoting the kernel of this epimorphism by $\hat{\pi}$, we have that

$$\hat{\pi} = \{(C_a, a^{-1}) \mid a \in \pi\}$$

where C_a denotes the inner automorphism $C_a(x) = axa^{-1}$. Note that π is isomorphic to $\hat{\pi}$ via the map $a \mapsto (C_a, a^{-1})$. Also notice that $\mathcal{N}$ is the connected component of G containing its identity element and that

$$\mathcal{N} \cap \hat{\pi} = \mathrm{center}(\pi).$$

We can thus identify the quotient group $\mathcal{N}/\mathrm{center}(\pi)$ with $\mathrm{Aff}_0(N)$ which is the connected component of $\mathrm{Aff}(N)$ containing its identity element. This identification together with the fact that $\mathcal{N}$ is contractible yields the following calculation:

$$\pi_i \operatorname{Aff}(N) = \begin{cases} \text{center}(\pi) & \text{if } i = 1. \\ 0 & \text{if } i > 1. \end{cases}$$

Recall next that $\pi_0 \operatorname{Aff}(N)$ can be identified as the quotient group

$$\operatorname{Aff}(N)/\operatorname{Aff}_0(N).$$

This group equals $G/(\hat{\pi}\mathcal{N})$ which in turn equals $\operatorname{Aut}(\pi)/\operatorname{Int}(\pi)$ since the image of $\hat{\pi}$ under the canonical homomorphism

$$G = \operatorname{Aut}(\pi) \ltimes \mathcal{N} \to \operatorname{Aut}(\pi)$$

is the group $\operatorname{Int}(\pi)$ of all inner automorphisms of N, hence

$$\pi_0 \operatorname{Aff}(N) = \operatorname{Out}(\pi).$$

We have thus shown that $\operatorname{Aff}(N)$ and $\mathcal{E}(N)$ have isomorphic homotopy group; in fact, it is easily seen that the inclusion map induces these isomorphisms. This completes our dimension of Lemma 2.1.

We now state our second preliminary result.

Lemma 2.2. *Let M_i $(i = 1, 2)$ be a pair of compact infrasolvmanifolds and $f : \pi_1(M_1) \to \pi_1(M_2)$ be an isomorphism. Then there exists an (N, T)-bundle $p : E \to T$ and a pair of (finite sheeted) regular covering spaces $p_i : E \to M_i$ $(i = 1, 2)$ such that*

1. *$f \circ \pi_1(p_1) = \pi_1(p_2)$ and*
2. *the deck transformations for both covering spaces are elements of $\mathcal{G}(E)$.*

Remark. The homomorphism induced by p_i between fundamental groups is denoted by $\pi_1(p_i) : \pi_1(E) \to \pi_1(M_i)$.

This lemma is a consequence of work of Mostow [13] together with later work of Auslander [1], [2]. It's proof will be given after first using the result to prove Theorem 1.2 which we rephrase as follows.

Theorem 1.2 (rephrased). *Let M_i $(i = 1, 2)$ be a pair of compact infrasolvmanifolds and $\alpha : \pi_1(M_1) \to \pi_1(M_2)$ be an isomorphism. Then α is induced by a diffeomorphism provided* $\dim M_i \neq 4$. *When* $\dim M_i = 4$, *it is at least induced by a homeomorphism.*

Proof. This result is obviously true when $\dim M_i \leq 2$. And both manifolds are Haken when $\dim M_i = 3$ (cf. [15]); hence the result follows from Waldhausen [16] in this case.

We proved earlier in [4] that α is induced by a homeomorphism $f : M_1 \to M_2$ when $\dim M_i \geq 4$. We now proceed to use the smoothing theory of Kirby–Siebenmann

[10] to show that f is homotopic to a diffeomorphism when dim $M_i \geq 5$. (We are unable to show this when dim $M_i = 4$ since [10] is not applicable in that dimension.) We assume from now on that dim $M_i \geq 5$. Let $\bar{f} : \tilde{M}_1 \to \tilde{M}_2$ be a lift of f to the universal covering spaces $\tilde{M}_i \to M_i$ $(i = 1, 2)$ whose groups of deck transformations are denoted by Γ_i and are identified with $\pi_1(M_i)$. Note that $\bar{f}$ is an α-equivariant homeomorphism and that the tangent bundles TM_i of M_i can be identified with

$$\tilde{M}_i \times_{\Gamma_i} \tilde{M}_i \underset{P_i}{\longrightarrow} M_i$$

where the map P_i is induced by projection onto the first factor of $\tilde{M}_i \times \tilde{M}_i$. Hence $\bar{f} \times \bar{f}$ induces a topological bundle map, denoted df, where $df : TM \to TM$ and covers f. (A topological bundle map is a fibre preserving map which induces homeomorphisms between fibres.)

Smoothing theory [10] yields that f is homotopic to a diffeomorphism provided df is homotopic through topological bundle maps covering f to a smooth bundle map ϕ; i.e., to a Diff$(\tilde{M}_i)$-bundle map. It actually suffices to construct something weaker; namely, a blocked bundle concordance between df and a smooth bundle map ϕ covering f. Such a concordance is a homeomorphism

$$F : TM_1 \times [0, 1] \to TM_2 \times [0, 1]$$

with $F(v, 0) = (df(v), 0)$ and $F(v, 1) = (\phi(v), 1)$, for all $v \in TM_1$, and satisfying the following condition $*$ which is a weakening of fibre preserving.

There is a triangulation $\mathcal{T}$ of M_1 such that
$$F(p_1^{-1}(\Delta) \times [0, 1]) = p_2^{-1}(f(\Delta) \times [0, 1]) \qquad (*)$$
for each simplex Δ of $\mathcal{T}$.

This weaker condition is sufficient because smoothing theory uses only the stable tangent bundles and that stably Top$(\mathbb{R}^n)$-bundles and $\overline{\text{Top}}(\mathbb{R}^n)$-bundles are equivalent theories. Here $\overline{\text{Top}}(N)$ denotes the semi-simplicial group where an n-simplex of $\overline{\text{Top}}(N)$ is a self-homeomorphism

$$F : \Delta_n \times N \to \Delta_n \times N$$

such that $F(\sigma \times N) = \sigma \times N$ for each face σ of the standard n-simplex Δ_n; its classifying space $B\,\overline{\text{Top}}(N)$ classifies blocked bundles with fibre N.

Let $p_i = E \to M_i$ be the finite sheeted regular covering spaces given by Lemma 2.2 where E is the total space of an (N, T)-bundle; let G_i denote their deck transformation groups, and let $\hat{f} : E \to E$ be a lift of f which exists because of property 1 of Lemma 2.2. Note that $\hat{f}$ is an α-equivariant homeomorphism where $\pi_1(M_i)$ acts on E via the deck transformation groups G_i. We can assume (and do) that $\bar{f}$ has been chosen to be a lift of $\hat{f}$. Consider the two bundles

$$q_i : E \times_{G_i} E \to M_i \qquad (2.2.1)$$

where q_i is induced by projection onto the first factor. The product map $\hat{f} \times \hat{f}$ induces a Top(E)-bundle map

$$Df : E \times_{G_1} E \to E \times_{G_2} E$$

which covers f. Notice that df covers Df; in particular, that

$$\tilde{M} \times_{\Gamma_i} \tilde{M} \to E \times_{G_i} E$$

are (usually) irregular covering spaces for $i = 1, 2$. We propose next to construct a blocked bundle concordance

$$K : (E \times_{G_1} E) \times [0, 1] \to (E \times_{G_2} E) \times [0, 1]$$

between Df and a Diff(E)-bundle map ψ covering f. Then a lift of K will give the desired blocked bundle concordance between df and a smooth bundle map ϕ covering f; thus proving Theorem 1.2. (The map ϕ will be a lift of ψ.)

Our first step is to construct the Diff(E)-bundle map ψ. Apropos this, note that both bundles in (2.2.1) have the subgroup $\mathcal{G}(E)$ of Diff(E) for structure group. This is because of property 2 of Lemma 2.2. By forgetting some structure, we can regard Df as merely a $\mathcal{E}(E)$-bundle map; i.e., a fiber preserving map inducing homotopy equivalences between fibres. Then we can construct a homotopy h_t, through $\mathcal{E}(E)$-bundle maps covering f, between Df and a $\mathcal{G}(E)$-bundle map ψ. This is done by a standard obstruction theory argument using Lemma 2.1; i.e., the fact that the inclusion of $\mathcal{G}(E)$ into $\mathcal{E}(E)$ is a weak homotopy equivalence.

The final step in our proof of Theorem 1.2 starts by recalling that the inclusion map

$$\tau : \overline{\mathrm{Top}}(E) \to \mathcal{E}(E)$$

is also a weak homotopy equivalence. This result is just a reformulation of Wall's topological rigidity theorem [17, Th. 15 B.1] for closed aspherical manifolds with a poly-$\mathbb{Z}$ fundamental group. (The connection to the above assertion about τ is made more transparent by [4, Th. 4.1] and [6, p. 267].) We can now construct the blocked bundle concordance K between Df and ψ posited above. This is again done by applying a standard obstruction theory argument to modify the homotopy h_t, using that τ is a weak homotopy equivalence, to change it to K. □

Remark 2.2.2. We note that the inclusion map

$$\omega : \mathrm{Top}(E) \to \mathcal{E}(E)$$

is definitely *not* a weak homotopy equivalence when dim $E > 10$; in fact, $\ker(\pi_0\omega)$ is then *not* finitely generated. (See [8, p. 9–10] for an explicit construction of a non-zero element in $\ker(\pi_0\omega)$ for the case where E is an n-dimensional torus, $n > 10$.) This assertion follows by modifying the argument given in [6, pp. 267–268] using [17, Th. 15 B. 1] in place of [6, Addendum 2.4] and [7, Th. 2.1] in place of [5, Cor. 0.6]. (We also wish to correct a misprint in [6, p. 268, line 21]; namely, the symbol '≡'

should be replaced by '$\subseteq$'.) This fact forced us to use blocked concordances instead of the more familiar bundle map homotopies in proving Theorem 1.2.

We now prove Lemma 2.2 under the extra assumption that both M_1 and M_2 have discrete presentations with finite holonomy groups. It is shown in the appendix that every infrasolvmanifold has a presentation with finite holonomy group. Hence the only additional assumption made is that there are presentations which are also discrete. This extra assumption is made so as not to obscure our argument which is easily adapted to the general situation. Hence there are simply connected solvable Lie groups G_i ($i = 1, 2$) and discrete (torsion-free) subgroups $\hat{\Gamma}_i$ of $\text{Aff}(G_i)$ such that the image of $\hat{\Gamma}_i$ in $\text{Aut}(G_i)$ is a finite group and M_i is the orbit space $G_i/\hat{\Gamma}_i$ under the affine action. After identifying $\pi_1(M_i)$ with $\hat{\Gamma}_i$, an elementary argument shows that there exist finite index, normal subgroups Γ_i of $\hat{\Gamma}_i$ such that $\Gamma_i \subset G_i$ and $f(\Gamma_1) = \Gamma_2$; hence $\phi = f|_{\Gamma_1}$ is an isomorphism between lattices Γ_i in G_i. Let F_i denote the finite quotient group $\hat{\Gamma}_i/\Gamma_i$. We now refer to the proof of Mostow's rigidity theorem for solvmanifolds as reworked by Raghunathan in [14, Th. 3.6, pp. 50–52]. Using Raghunathan's notation, there exists a diffeomorphism $\tilde{\phi} : G_1 \to G_2$ satisfying

(i) $\tilde{\phi}|_{\Gamma_1} = \phi$

(ii) $\tilde{\phi}(x\gamma) = \tilde{\phi}(x)\tilde{\phi}(\gamma)$ for all $\gamma \in \Gamma_1 \cdot G_1'$ and $x \in G$

(iii) $\tilde{\phi}|_{\Gamma_1 \cdot G_1'} : \Gamma_1 \cdot G_1' \to \Gamma_2 \cdot G_2'$ is an isomorphism between closed Lie subgroups of G_i where G_i' is the smallest connected subgroup of the nilradical N_i of G_i containing $[\Gamma_i, \Gamma_i]$.

Note that

$$\pi_i : G_i \to G_i/(\Gamma_i \cdot G_i')$$

is a principal $\Gamma_i \cdot G_i'$-bundle, denoted by ξ_i, and $\tilde{\phi}$ is a bundle map which induces a diffeomorphism $\psi : B_1 \to B_2$ where B_i denotes the base space of ξ_i; i.e., $B_i = G_i/(\Gamma_i \cdot G_i')$. Furthermore, given any diffeomorphism $\Psi : B_1 \to B_2$ in the same homotopy class as ψ, we can find a diffeomorphism $\Phi : G_1 \to G_2$ with the same properties as $\tilde{\phi}$ except that Φ induces Ψ instead of ψ. Let E_i denote the homogeneous space G_i/Γ_i and $q_i : E_i \to B_i$ be the map induced by π_i. Note that these are smooth bundles whose fibers are the closed nilmanifolds $G_i'/(\Gamma_i \cap G_i)$, denoted N_i, and with structure groups $\text{Aff}(N_i)$. Also note that Φ induces a bundle map $\Phi' : E_1 \to E_2$ which is a diffeomorphism covering Ψ. Now consider the regular covering space $E_i \to M_i$ whose group of deck transformations is identified with F_i. And note that F_i also acts on $E_i \to B_i$ via bundle maps. Consider the following statement.

Lemma 2.3. *Each B_i supports the structure of a flat Riemannian torus in such a way that F_i acts on B_i via affine maps.*

Before proving Lemma 2.3, we use it to complete the proof of Lemma 2.2. Let $\Psi : B_1 \to B_2$ be an affine diffeomorphism in the homotopy class of ψ; it exists because of both Bieberbach's and Malcev's rigidity theorems. Now identify E_1 to E_2 via the diffeomorphism Φ' covering Ψ which we discussed above. Call this common space E; it is the total space of an (N, T)-bundle which clearly satisfies the properties posited in Lemma 2.2. □

Proof of Lemma 2.3. Our proof will depend on the semisimple splitting of G as described by Auslander in [1, pp. 236–238] together with some basic results about it from his papers [1], [2]. (We abbreviate our previous notation throughout this proof by dropping the subscript i; e.g., G denotes either G_1 or G_2, B denotes B_i, etc.) Note that B is a compact solvmanifold with a presentation given by $B = G/(\Gamma \cdot G')$ and that $\pi_1(B)$ is free abelian. Let H be the maximal connected normal subgroup of G which is contained in G'. Since H is a closed subgroup, G/H is also a simply connected solvable Lie group. Denote this new group by $\mathcal{G}$ and let D denote its closed subgroup $(\Gamma \cdot G')/H$. Then $B = \mathcal{G}/D$ is another presentation of B having the additional property that $\mathcal{G}$ acts locally effectively on B. Hence $\mathcal{G}$ is of class R because of [2, Th. 3.6] and the fact that $\pi_1(B)$ is abelian. (Class R solvable Lie groups are defined in [2, p. 271].) Recall that the semisimple splitting of $\mathcal{G}$ is the Lie group

$$\mathcal{G}_s = T_{\mathcal{G}} \ltimes \mathcal{G} = T_{\mathcal{G}} \ltimes M_{\mathcal{G}}$$

where $M_{\mathcal{G}}$ is the nil-shadow of $\mathcal{G}$. And $T_{\mathcal{G}}$ is a compact (connected) abelian Lie subgroup of Aut($\mathcal{G}$) because of [2, Th. 3.5] which also yields that the intersection $\mathcal{D} = M_{\mathcal{G}} \cap D$ is a finite index, normal subgroup of D. Fix a subgroup T of $\mathcal{G}_s$ which is conjugate to $T_{\mathcal{G}}$ and note that

$$\mathcal{G}_s = T \ltimes \mathcal{G} = T \ltimes M_{\mathcal{G}}.$$

Then define a $M_{\mathcal{G}} \cap \mathcal{G}$-equivariant diffeomorphism η by

$$\mathcal{G} \underset{\sigma}{\longrightarrow} T \ltimes \mathcal{G} = T \ltimes M_{\mathcal{G}} \underset{\omega}{\longrightarrow} M_{\mathcal{G}}$$

i.e., η is the composite of inclusion σ and projection ω; it induces a diffeomorphism from $\mathcal{B} = \mathcal{G}/\mathcal{D}$ to the compact nilmanifold $M_{\mathcal{G}}/\mathcal{D}$. But $M_{\mathcal{G}}/\mathcal{D}$ is affinely equivalent to a flat torus because of Malcev's rigidity theorem and the fact that $\pi_1(\mathcal{B})$ is free abelian since $\mathcal{B}$ is a finite sheeted cover of B. We put a flat torus structure on $\mathcal{B}$ via this identification.

Let $\hat{F}$ be the finite group consisting of all self diffeomorphisms of $\mathcal{B}$ which cover elements of F (acting on B) via the covering projection $p : \mathcal{B} \to B$. To complete the proof of Lemma 2.3, it suffices to show that $\hat{F}$ acts via affine maps on $\mathcal{B}$; since if so, then $\mathcal{B}$ induces a flat affine torus structure on B and F acts on B via affine maps of this induced structure.

It is immediate that right multiplication on $\mathcal{G}$ by any fixed element of $\mathcal{G}$ is identified with an affine diffeomorphism of $M_{\mathcal{G}}$ via η. Consequently, the subgroup of $\hat{F}$ consisting of the deck transformations of p acts affinely on $\mathcal{B}$ since each deck transformation is induced by right multiplication on $\mathcal{G}$ by an element of D. And therefore B does inherit a flat torus structure from $\mathcal{B}$.

Notice next that $\hat{\Gamma}$ normalizes both G' and H; hence H is a normal (closed) subgroup of $\hat{\Gamma} \cdot G'$. Let $\hat{D}$ denote the quotient Lie group; i.e.,

$$\hat{D} = (\hat{\Gamma} \cdot G')/H$$

and note that both D and $\mathcal{D}$ are normal subgroups of $\hat{D}$ with quotient groups F and $\hat{F}$, respectively. Let C denote the (finite) holonomy group of $\hat{\Gamma}$; i.e., its image in $\mathrm{Aut}(G)$. Then $\hat{\Gamma}$ is contained in the Lie subgroup $C \ltimes G$ of $\mathrm{Aff}(G) = \mathrm{Aut}(G) \ltimes G$. Since H is a normal (closed) subgroup of $C \ltimes G$, we can form its quotient Lie group $C \ltimes \mathcal{G}$ which contains $\hat{D}$. We now recall from [1, p. 238] that we can identify C with a subgroup of $\mathrm{Aut}(\mathcal{G}_s)$ extending its action on $\mathcal{G}$; in this way $C \ltimes \mathcal{G}$ imbeds in $C \ltimes \mathcal{G}_s$. Let K be a maximal compact subgroup of $C \ltimes \mathcal{G}_s$ containing C and let $T = K^0$. Then T is a subgroup of $\mathcal{G}_s$ and is conjugate to $T_{\mathcal{G}}$ in $\mathcal{G}_s$; it is the subgroup fixed above to define the diffeomorphism $\eta : \mathcal{G} \to M_{\mathcal{G}}$.

Note that $\eta(g)$, $g \in \mathcal{G}$, is characterized as the unique solution to the following equation in $C \ltimes \mathcal{G}_s = K \ltimes \mathcal{G}$:

$$g = k\eta(g) \tag{2.3.1}$$

where $k \in K$ and $\eta(q) \in M_{\mathcal{G}}$. Note next that the action of $\hat{F}$ on $\mathcal{B}$ is determined by the subgroup $\hat{D}$ of $C \ltimes \mathcal{G}$ acting naturally on $\mathcal{G}$. But one easily sees, using (2.3.1), that each element of this action is identified with an affine diffeomorphism of $M_{\mathcal{G}}$ via η, completing the proof of Lemma 2.3. □

A. Appendix

This appendix is devoted to proving Lemma 1.3. Let (Γ, S) be a presentation for an infrasolvmanifold M. Let G denote the (compact) closure of the image of Γ in $\mathrm{Aut}(S)$. Hence Γ is a closed cocompact subgroup of the semi-direct product $S \rtimes G$. Observe that G^0 is a torus due to a result of L. Auslander; cf. [14, Th. 8.24]. The plan is to put a second semi-direct product structure on $S \rtimes G$ by constructing another closed normal subgroup $\mathfrak{S}$ in $S \rtimes G$ such that

(A.1)
1. $\mathfrak{S}$ is both solvable and simply connected;
2. Γ^0 is contained in the nil radical of $\mathfrak{S}$;
3. $\Gamma \cap \mathfrak{S}$ has finite index in Γ;
4. $S \rtimes G = \mathfrak{S} \rtimes G$ and G acts faithfully on $\mathfrak{S}$.

It is clear that, once we succeed in this plan, $(\Gamma, \mathfrak{S})$ is a new presentation for M which has finite holonomy group thus proving Lemma 1.3.

Let N denote the nilradical of S and $\mathcal{N}$ be the nilradical of $S \rtimes G$. We clearly have that N is a closed subgroup of $\mathcal{N}$. Note that $S \rtimes G^0$ is a connected solvable Lie group since G^0 is abelian by a result due to Auslander; cf [14, Th. 8.24]. We can hence apply Mostow's result [13] (cf. [14, Th. 3.3]) to the effect that $\Gamma \cap \mathcal{N}$ is cocompact in $\mathcal{N}$. Recall that a compact subgroup of a connected nilpotent Lie group must be central. Hence the intersection $G \cap \mathcal{N}$ is in the center of $\mathcal{N}$. Using the additional fact that

$$[\mathcal{N}, S] \subset (\mathcal{N} \cap S)^0 = N,$$

we see similarly that $G \cap \mathcal{N}$ acts trivially on S via conjugation; hence

$$G \cap \mathcal{N} = \{1\}.$$

And $\mathcal{N}$ is consequently simply connected.

Let L denote the quotient group of $S \rtimes G$ by its closed normal subgroup $\mathcal{N}$; i.e.

$$L = S \rtimes G/\mathcal{N}.$$

Note that L is also a quotient group of $S/N \rtimes G$. Using this observation together with the fact that S/N is abelian and G is compact, it can be shown that the nil radical of L is abelian and contains $\eta(S)$ where $\eta : S \rtimes G \to L$ is the canonical epimorphism. Arguing further, it can be shown that L is a semi-direct product

$$L = \mathbb{R}^n \rtimes K$$

where K is the compact subgroup $\eta(G)$ and η maps G isomorphically to K. We then construct a homomorphism

$$\phi : \mathbb{R}^n \rtimes K \to \mathbb{R}^n \rtimes O(n)$$

by putting a K-invariant inner product on $\mathbb{R}^n$. The kernel of this homomorphism is the subgroup $\mathcal{K}$ of K consisting of all those elements which centralize $\mathbb{R}^n$. Notice that the nil radical of L is $\mathbb{R}^n \times \mathcal{K}^0$.

Let Γ_1 denote the image of Γ in L. It is a discrete and cocompact subgroup of L since $\Gamma \cap \mathcal{N}$ is cocompact in $\mathcal{N}$. Let $\Gamma_2 = \phi(\Gamma_1)$, then Γ_2 is also discrete and cocompact in $\mathbb{R}^n \rtimes O(n)$ since $\mathcal{K}$ is compact. We have therefore that $\Gamma_2 \cap \mathbb{R}^n$ is a cocompact lattice in $\mathbb{R}^n$ by Bieberbach's Theorem since $\mathbb{R}^n \rtimes O(n)$ is the group of all rigid motions of Euclidean n-space; cf. [14, Cor. 8.26]. Note that $\phi^{-1}(\mathbb{R}^n) = \mathbb{R}^n \times \mathcal{K}$ and $\mathcal{K}^0$ is a torus with finite index in $\mathcal{K}$. We can hence find a subgroup A of $\phi^{-1}(\Gamma_2 \cap \mathbb{R}^n) \cap \Gamma_1$ satisfying

(A.2) 1. A is free abelian of rank n;

2. A has finite index in Γ_1;

3. $A \subseteq \mathbb{R}^n \times \mathcal{K}^0$.

Note that $\phi(\eta(S)) \subset \mathbb{R}^n$ since $\eta(S) \subseteq \mathbb{R}^n \times \mathcal{K}^0$. Hence $\phi \circ \eta$ induces a map $G \to O(n)$ making the following diagram commute:

$$\begin{array}{ccc} S \rtimes G & \longrightarrow & G \\ \downarrow \phi\circ\eta & & \downarrow \\ \mathbb{R}^n \rtimes O(n) & \longrightarrow & O(n). \end{array}$$

This diagram then shows that $\phi(K) = \phi(\eta(G))$ is a finite group since Γ_2 has finite holonomy group by Bieberbach's Theorem while the holonomy group of Γ is dense in G. Hence $K^0 = \mathcal{K}^0$ and $\Gamma_1(\mathbb{R}^n \times K^0) = L$. We will use this fact in verifying the following assertion.

Assertion A.3. There exists a closed normal subgroup $\mathfrak{S}_1$ of L such that

1. $\mathfrak{S}_1$ is isomorphic to $\mathbb{R}^n$;
2. $A \cap \mathfrak{S}_1$ is a (uniform) lattice in $\mathfrak{S}_1$;
3. $L = \mathfrak{S}_1 \rtimes K$.

Before verifying Assertion A.3, we use it to complete the proof of Lemma 1.3 by implementing the plan outlined at the start of this appendix.

Set $\mathfrak{S} = \eta^{-1}(\mathfrak{S}_1)$, then $\mathfrak{S}$ is a closed normal subgroup of $S \rtimes G$. Since $\mathcal{N}$ is a closed, normal, simply connected and nilpotent subgroup of $\mathfrak{S}$ with $\mathfrak{S}/\mathcal{N} \cong \mathbb{R}^n$, $\mathfrak{S}$ satisfies condition 1 of (A.1). Condition 4 is also satisfied because of property 3 of Assertion A.3 and the fact that η maps G isomorphically onto K. Note that $A \cap \mathfrak{S}_1$ has finite index in Γ_1 because of (A.2) and Assertion A.3; hence $\Gamma \cap \eta^{-1}(A \cap \mathfrak{S}_1)$ has finite index in Γ. Condition 3 of (A.1) is therefore also satisfied since

$$\Gamma \cap \eta^{-1}(A \cap \mathfrak{S}_1) \subseteq \Gamma \cap \mathfrak{S}.$$

The remaining condition 2 is also satisfied since $\Gamma^0 \subseteq N$, $N \subseteq \mathcal{N}$ and $\mathcal{N}$ is a connected, normal, nilpotent subgroup of $\mathfrak{S}$.

We now complete the discussion in this appendix by verifying Assertion A.3. Since A has finite index in Γ_1, it contains a subgroup A_1 of finite index which is normal in Γ_1. This group maps monomorphically onto a subgroup A_2 of $\mathbb{R}^n$ via projection onto the first factor of $\mathbb{R}^n \times \mathcal{K}^0 = \mathbb{R}^n \times K^0$ since K^0 is compact and A_1 is torsion-free. There is hence a homomorphism $\psi : A_2 \to K^0$ whose graph is A_1. Note that both $\mathbb{R}^n$ and K^0 are Γ_1-invariant subgroups of $\mathbb{R}^n \times K^0$ under the conjugation action since $\mathbb{R}^n$ and $\mathbb{R}^n \times K^0$ are normal in L and K^0 is *the* maximal compact subgroup of $\mathbb{R}^n \times K^0$. Therefore ψ is a Γ_1-equivariant homomorphism in terms of the induced conjugation actions on $\mathbb{R}^n$ and K^0 since A_1 is normal in Γ_1.

Since K^0 is a torus, its universal covering group is $\mathbb{R}^s$ where $s = \dim K^0$. Let $\hat{\psi} : A_2 \to \mathbb{R}^s$ be a lift of ψ; i.e., a homomorphism such that $p \circ \hat{\psi} = \psi$ where $p : \mathbb{R}^s \to K^0$ is the canonical projection. Note that $\hat{\psi}$ exists since A_2 is free abelian. The conjugation action of Γ_1 on K^0 lifts uniquely to an action of Γ_1 on $\mathbb{R}^s$ and A_1 acts trivially on $\mathbb{R}^s$ under this action. Denote the product of $\gamma \in \Gamma_1$ with $x \in \mathbb{R}^s$ by $\gamma \cdot x$.

Now $\hat{\psi}$ extends uniquely to a homomorphism $\Psi : \mathbb{R}^n \to \mathbb{R}^s$ since A_2 is a (uniform) lattice in $\mathbb{R}^n$. We next average Ψ relative to the Γ_1-actions to get a Γ_1-equivariant homomorphism $\Psi_1 : \mathbb{R}^n \to \mathbb{R}^s$ given explicitly by the formula

$$\Psi_1(x) = \frac{1}{r} \sum_{\gamma A_1 \in \Gamma/A_1} \gamma^{-1} \cdot \Psi(\gamma x \gamma^{-1})$$

where $r = [\Gamma_1 : A_1]$ and $x \in \mathbb{R}^n$.

Let $\mathfrak{S}_1$ be the graph of the composite $p \circ \Psi_1 : \mathbb{R}^n \to K^0$. It is clearly a closed subgroup of L satisfying property 1 of Assertion A.3 and is a normal subgroup because $p \circ \Psi_1$ is Γ_1-equivariant and $\Gamma_1(\mathbb{R}^n \times K^0) = L$. Property 3 is immediate. We need finally to discuss property 2. Let

$$rA_1 = \{ra \mid a \in A_1\}$$

and notice that

$$p \circ \Psi_1(ra) = \psi(ra)$$

for all $a \in A_2$. We consequently have that $rA_1 \subseteq \mathfrak{S}_1$ which implies that property 2 of Assertion A.3 is also satisfied since rA_1 has finite index in A. □

References

[1] Auslander, L., An exposition of the structure of solvmanifolds, Part I: Algebraic Theory, Bull. Amer. Math. Soc. 79 (1973), 227–261.

[2] — An exposition of the structure of solvmanifolds, Part II: G-induced flows, Bull. A.M.S., 79 (1973), 262–285.

[3] Bieberbach, L., Über die Bewegungsgruppen der Euklidischen Räume. II, Die Gruppen mit einen endlichen Fundamentalbereich, Math. Ann. 72 (1912), 400–412.

[4] Farrell, F. T., Jones, L. E., The surgery L-groups of poly-(finite or cyclic) groups, Invent. Math. 91 (1988), 559–586.

[5] —, Stable pseudoisotopy spaces of non-positively curved manifolds, J. Differential Geom. 34 (1991), 769–834.

[6] —, Topological rigidity for compact nonpositively curved manifolds, Proc. Sympos. Pure Math. 54, Part 3, Amer. Math. Soc., Providence, R. I., 1993, 229–274.

[7] —, Isomorphism conjectures in algebraic K-theory, J. Amer. Math. Soc. 6 (1993), 249–297.

[8] Hatcher, A. E., Concordance spaces, higher simple homotopy theory, and applications, Proc. Sympos. Pure Math. 32, Amer. Math. Soc., Providence, R.I., 1978, pp. 3–21.

[9] Kamishima, Y., Lee, K.-B., Raymond, F., The Seifert construction and its applications to infranilmanifolds, Quart. J. Math.Oxford Ser. (2) 34 (1983), 433–452.

[10] Kirby, R. C., Siebenmann, L. C., Foundational essays on topological manifolds, smoothings, and triangulations, Ann. of Math. Stud. 88, Princeton Univ. Press, Princeton, N.J., 1977.

[11] Lee, K.-B., Raymond, F., Rigidity of almost crystallographic groups, Contemp. Math. 44, Amer. Math. Soc., Providence, R.I., 1985, 73–78.

[12] Malcev, A. I., On a class of homogeneous spaces, Amer. Math. Soc. Transl. 39 (1951).

[13] Mostow, G. D., Factor spaces of solvable groups, Ann. of Math. 60 (1954), 1–27.

[14] Raghunathan, M. S., Discrete subgroups of Lie groups, Springer-Verlag, Berlin–Heidelberg–New York 1972.

[15] Scott, P., The geometries of three-manifolds, Bull. London Math. Soc. 15 (1983), 401–487.

[16] Waldhausen, F., On irreducible 3-manifolds which are sufficiently large, Ann. of Math. 87 (1968), 56–88.

[17] Wall, C. T. C., Surgery on compact manifolds, Academic Press, London, 1970.

Informal note on topology, geometry and topological field theory

Kenji Fukaya

0. Introduction

This note is an informal discussion on topology-geometry and topological field theory. The author would like to emphasise that this is an informal note. It means that several parts of the contents are far from being rigorous mathematics. (Of course results quoted in this note as theorems are all rigorous.) Also the author does not yet have a systematic study on some of the topics mentioned here.

The author would like to explain what the target of his recent (and coming) research is, or what he has dreamt of recently.

Our discussion is related to the following topics: "quantization of the concept of spaces", "duality between analysis and geometry", "homological algebra and homotopical algebra". Some of them do not yet have a well defined mathematical sense. We only have some phenomena or evidence which suggests their possibility.

1. Duality between geometry and analysis

Let us start from the following "definition".

Definition 1.1.[1] Let M be a space. A *geometry* on M is the study of maps from some space Y to M. An *analysis* on M is the study of maps from M to some space Y.

Example 1.2. Let $M = \mathbb{R}^2$ be the Euclidean plane. The usual Euclidean geometry is the study of lines, triangles, circles *etc*. They are maps from S^1 or $\mathbb{R}$ to $M = \mathbb{R}^2$.

The analysis on $M = \mathbb{R}^2$ is the study of functions of two variables $f(x, y)$, and so is the study of maps from $M = \mathbb{R}^2$ to $\mathbb{R}$.

1 M. Gromov mentioned this kind of "definition" in his talk at the meeting of the Japan Mathematical Society. He is not responsible for any nonsense in this note.

From this "definition" it is quite clear what the author means by "geometry is dual to analysis". Of course everything is well known in Example 1.2. Let us mention some less obvious examples.

The first example is the well known de Rham theorem. Here:

[Geometry] The singular or simplicial homology theory $H_k(M)$ is defined by considering cycles, that is studying maps $\Delta \to M$ from a simplex Δ to M.

[Analysis] De Rham cohomology is described by $\bigwedge^k(M)$, the set of all k forms and the homomorphism $d^k : \bigwedge^k \to \bigwedge^{k+1}$. Namely $H^k_{dR}(M) = \ker d^k / \operatorname{Im} d^{k-1}$.

De Rham's theorem implies that $\operatorname{Hom}(H_k(M), \mathbb{R}) \cong H^k_{dR}(M)$, the duality between geometry and analysis. We remark that this duality is basic in electromagnetism, that is, abelian gauge theory. We will return to this example in Section 3.

Let me mention some recently discovered examples that are related to the scalar curvature. There are basically two methods known which are used to study manifolds of positive scalar cruvature. One is minimal surface theory and the other is the Dirac operator. I shall only mention small parts of them.

[Geometry] Minimal surface theory. Let M be a 3-manifold of positive scalar curvature. A theorem by D. Fisher-Colbrie, R. Schoen and S. T. Yau says that if $\Sigma \to M$ is a (stable) minimal surface, then Σ is diffeomorphic to S^2. Using existence results of a minimal surface in a given homotopy class, we obtain a topological constraint on M.

[Analysis] Let M be an even-dimensional spin manifold. We then have bundles of positive and negative spinors S^+, S^-. Using the Riemannian metric we have an elliptic operator $D : S^+ \to S^-$, the Dirac operator. A theorem of Lichnerowicz implies that if the scalar curvature of M is positive then the kernel and cokernel of the Dirac operator vanish. On the other hand, the Atiyah-Singer index theorem implies that $\dim\ker D - \dim\operatorname{coker} D$ is a topological invariant of M described explicitly by the characteristic classes of M. Combining them we obtain a topological constraint on manifolds of positive scalar curvature.

Thus minimal surfaces and Dirac operators are the two basic methods used to study manifolds of positive curvature. In fact, they imply the same result sometimes, but by a different proof. It seems that there isn't a good way of explaining the relationship between these two methods. They are dual in our sense.

Problem 1.3. Find a more mathematical way to describe the duality between minimal surfaces and the Dirac operator.

There is a similar phenomenon related to the positive mass conjecture.

The next example is mirror symmetry. I cannot describe it precisely, partly because I do not have enough knowledge of what is going on in the study of mirror symmetry, and partly because the problem is still wide open.

[Geometry] Let M be a 3-dimensional Kähler manifold. Assume that the Ricci curvature of M vanishes. Consider the space

$$\mathcal{M}_k(M) = \{\varphi : S^2 \to M \mid J_M\varphi = \varphi J,\ \varphi^*[w] \cap [S^2] = k\},$$

where ω is a Kähler form on M, and J_M and J are complex structures of M and S^2 respectively. By the assumption on Ricci curvature, the virtual dimension of this space is independent of k and is equal to 0. By counting its order we obtain some numbers. This construction can be generalized to give a quantum deformation of cup product.

[Analysis] Let M be as above. Consider the sheaf cohomology $H^1(M; \mathcal{O}(T))$, where $\mathcal{O}(T)$ is the sheaf of holomorphic sections of the tangent bundle. This cohomology is given by harmonic theory. There is a cubic map called the Yukawa coupling:

$$H^1(M; \mathcal{O}(T)) \otimes H^1(M; \mathcal{O}(T)) \otimes H^1(M; \mathcal{O}(T)) \to \mathbb{C}.$$

The Mirror Conjecture asserts that the geometry of M in the above sense is equivalent to analysis of $\widehat{M}$ where $\widehat{M}$ is a Ricci flat Kähler 3-fold called the *mirror of* M.

Our fourth example is a theorem proved recently by C. Taubes [T]. Let (M, ω) be a symplectic 4 manifold.

[Geometry] (Gromov–Witten invariant) Fix an almost complex structure J on M compatible with the symplectic structure on M. Namely, assume that $g(X, Y) = \omega(JX, Y)$ is a Riemannian metric. Let $x \in H^2(M; \mathbb{Z})$. Let Σ^g be a closed surface of genus g. Set

$$\mathcal{M}_g(M, x) = \{\varphi : (\Sigma^g, J_\Sigma) \to M \mid J\varphi = \varphi J_\Sigma,\ \varphi^*[\omega] = PD(x)\}/\sim,$$

where $PD(x)$ is the Poincaré dual of x and $(\varphi, J_\Sigma) \sim (\varphi', J'_\Sigma)$ if and only if there exists a diffeomorphism $\psi : \Sigma^g \to \Sigma^g$ such that $(\varphi, J_\Sigma) = (\varphi'\psi, \psi^* J'_\Sigma)$. By the Riemann-Roch theorem, one can find the virtual dimension of $\mathcal{M}_g(M, x)$. We consider the order of this set $\mathcal{M}_g(M, x)$ counted with sign when the dimension is 0. This is, by definition, the Gromov–Witten invariant $Z_{GW}(M, x)$. (g is automatically determined.)

[Analysis] (Seiberg–Witten invariant) Let $x \in H^2(M; \mathbb{Z})$ with mod 2 reduction $x(\text{mod } 2) \equiv w_2(M)$ and let L be a complex line bundle over M with $c_1(L) = x$. Consider a $\text{Spin}_\mathbb{C}$ structure P on M (*i.e.* a principal $\text{Spin}_\mathbb{C}(4)$-bundle over M) such that $L = P \times_{\text{Spin}_\mathbb{C}(4)} U(1)$. The group $\text{Spin}_\mathbb{C}(4)$ has two (complex 2-dimensional) representations $\Delta^\pm$. Put $S^\pm = P \times_{\text{Spin}_\mathbb{C}(4)} \Delta^\pm$, hence $\det S^\pm = L$. Given a connection A on L there is a Dirac operator $D_A : \Gamma(S^+) \to \Gamma(S^-)$ induced by A and the Levi-Civita connection on TM, the tangent bundle of M. Also there is a quadratic map $\Phi : \Gamma(S^+) \to \Gamma(\bigwedge_2^+(M))$, where $\bigwedge_2^+(M)$ is the bundle of self-dual 2-forms. The Monopole moduli space is defined by

$$\{(A, u) \mid D_A u = 0,\ \Phi(u) = dA + *dA\}/\mathcal{G},$$

where $\mathcal{G}$ is the Gauge group $\mathrm{Aut}(L) = \mathrm{Map}(M, U(1))$. Counting the order of this moduli space (with sign) when its virtual dimension is 0 we get the Seiberg–Witten invariant, $Z_{SW}(M, X)$.

Taubes' theorem asserts that $Z_{GW}(M, x) = Z_{SW}(M, X)$.

Remark 1.4. The above theorem by Taubes seems to be closely related to Problem 1.3. Namely, the Seiberg–Witten invariant is defined by using the monopole equation which is a nonlinear version of the Dirac operator. Also the Seiberg–Witten invariant is closely related to scalar curvature. On the other hand, the Gromov–Witten invariant is defined by using the pseudo-holomorphic curve equation. A holomorphic curve is a minimal surface. And Witten [W2] says that the Gromov–Witten invariant is a twisted and topological version of minimal surface theory (or harmonic maps).

Now let M be a manifold. Consider $C(M)$, the space of all functions on M. The basic idea of noncommutative geometry by A. Connes and others is to consider a noncommutative ring instead of the commutative ring $C(M)$. In our terminology, this can be regarded as an "analytic approach" to generalizing the concept of spaces.

Problem 1.5. Is there a "geometric approach" which is dual to noncommutative geometry?

There is probably an answer already, that is, *string theory*. Let me remark that a generalization of the concepts of spaces is the basic problem in the quantization of gravity, and string theory is regarded as a good candidate for quantum gravity.

2. Topological field theory and homological-homotopical algebra

It seems to me that the idea of string theory to generalize the concept of space is as follows. Let M be a space. We consider the space of maps $\mathrm{Map}(\Sigma, M)$, from a surface Σ to M. We consider an "integral"

$$\int_f e^{L(f)} \mathcal{D}f, \tag{2.1}$$

where $\mathcal{D}f$ is the "Feynmann measure"; and $L(f)$ is a Lagrangian, a function on $\mathrm{Map}(\Sigma, M)$, which we do not specify. The integral is taken over the space $\mathrm{Map}(\Sigma, M)$ or a submanifold of it.

$L(f)$ may also depend on additional structures on M and Σ, for example, a vector bundle on M, a submanifold of M or Σ, *etc.* By changing such additional structures on M or Σ and also the topological type of Σ (the genus of Σ), we obtain a system of numbers. (We fix the topology and geometry of M.) We regard them as "correlation functions of 2 dimensional field theory". This field theory is two-dimensional since (in our terminology) we are discussing analysis on Σ.

Thus if we could justify (2.1) for an appropriate $L(f)$, we would get a "functor"

$$\{\text{Spaces}\} \Longrightarrow \{\text{2-dimensional field theories}\}. \tag{2.2}$$

Here the author does not know the definition of "the category of 2-dimensional field theory" because that is exactly the target of his research, which seems very difficult and time consuming.

Conjecture 2.1. The functor is locally injective. Namely there is no deformation of spaces which gives the same field theory.

If this conjecture is true, then one may say that "2-dimensional field theory" is a generalization of the concept of spaces.

However it still seems too difficult at this point to attack this kind of problems and to make these constructions mathematiclly rigorous. So instead of studying geometry we start with topology, which is somewhat easier. The schema is

$$\begin{array}{lcl} \text{Geometry} & \Longleftrightarrow & \text{(Actual) Field theory} \\ \text{Topology} & \Longleftrightarrow & \text{Topological Field theory.} \end{array}$$

More precisely, the functor {metric spaces} $\Longrightarrow$ {2-dimensional field theories}, is an (actual) field theory, and the functor {topological spaces} $\Longrightarrow$ {2-dimensional field theories} is a topological field theory.

Now we consider the problem: what is the definition of "2-dimensional field theory" that we need for this purpose? The author believes that answering this question is exactly algebraic topology. To explain why, let us recall what the basic idea of algebraic topology is.

For a moment, let us give up attempting to replace all information of topological spaces by algebras and instead, let us try to replace spaces as much as possible by algebras. Namely, our purpose here is to approximate spaces by algebras. The first step is:

Example 2.2. The aim of homology theory is to approximate the space X by the chain complex $C_*(X)$.

Thus the strategy to generalizing the concept of the space is as follows. First replace the space (manifold) by algebraic objects as much as possible. Then generalize these algebraic objects.

Remark 2.3. This approach is, in fact, not very satisfactory since what we are looking for is a *geometric* (not algebraic) notion generalizing the concept of spaces. But this is the only approach I can try for the moment.

Remark 2.4. We again recall that there is an established way to replace a space by an algebra. That is, to replace the space X by $C(X)$, the ring of functions. But as we mentioned, this is the analytic approach. What we are looking for is an approach dual to this one.

The next step to approximating a space by its (co)homology is to include cup products and similar product structures. In the analytic way, this was achieved by de Rham homotopy theory or rational homotopy theory by D. Quillen–D. Sullivan. We recall

Definition 2.5. Two simply connected spaces M and M' are said to be $\mathbb{Q}$ *homotopy equivalent* if there exist N and maps $f : N \to M$, $f' : N \to M'$ such that f and f' induce isomorphisms on homotopy groups $\otimes\mathbb{Q}$.

Roughly (and imprecisely) speaking, two spaces are $\mathbb{Q}$ homotopy equivalent if and only if "they are homotopy equivalent up to finite ambiguity".

The idea of rational homotopy theory is to approximate a space by a differential graded algebra (DGA), the cochain complex with a product structure. This is performed by using differential forms. We recall

Definition 2.6. $(\bigwedge^k, d, \cdot)$ is said to be a *differential graded algebra (DGA)* if $(\bigwedge^k, \cdot)$ is a graded ring satisfying graded commutativity. Namely, $\bigwedge^k \cdot \bigwedge^{k'} \subseteq \bigwedge^{k+k'}$, $a \cdot b = -(-1)^{\deg a \deg b} b \cdot a$. Here $\deg a = k$ if $a \in \bigwedge^k$. $(\bigwedge^k, d)$ is a chain complex (*i.e.* $d : \bigwedge^k \to \bigwedge^{k+1}$ with $dd = 0$). Finally d is a differential, that is, $d(a \cdot b) = da \cdot b + (-1)^{\deg a} a \cdot db$.

Let M be a manifold. Then the ring of differential forms $(\bigwedge^k(M), d, \wedge)$ is a DGA. Thus we have a functor

$$\{\text{manifolds}\} \to \{\text{DGA}\}. \tag{2.7}$$

Remark 2.8. In fact, the DGA $(\bigwedge^k(M), d, \wedge)$ itself does not work well for this construction since it is not finitely generated and it is too big. Sullivan's idea in [Su] is to triangulate the manifold and use piecewise linear forms with rational coefficients. We shall not discuss this point here.

Then in an imprecise way, the result by D. Quillen and D. Sullivan is described as follows:

"Theorem" 2.9. *The functor induces an isomorphism*

$$\frac{\{\text{spaces}\}}{\text{homotopy equivalence}} \cong \frac{\{\text{DGA}\}}{\text{homotopy equivalence}}.$$

Remark 2.10. I put quotation marks in the above theorem only because it is written here in an imprecise manner. The theorem is rigorously established in their papers.

Again this is an analytic approach since it uses differential forms. What is the dual and geometric approach to it? It is to consider a tree and maps from it. Here we recall:

Definition 2.11. A simplicial complex T is a *tree* if it is one-dimensional and simply connected.

(See Table 1 for example.) Then the dual in "geometry" to the construction of DGA is as follows:

dim $\Sigma = 1$: Consider the space of maps $\mathrm{Map}(T, M)$ from a tree to M.

dim $\Sigma = 2$: Consider the space of maps $\mathrm{Map}(S^2, M)$ from a Riemann sphere with marked points to M.

These ideas are discussed in more detail in [Fu1] and [Fu2] so we shall not repeat them here. Here are tables describing the main ideas of this note.

Topology	Homology	(Q)Homotopy	Differential Topology	Generalization of the notion of spaces
Geometry dim Σ=1	line	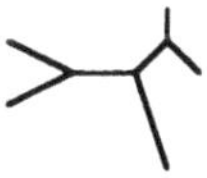 tree	 general graph	
Geometry dim Σ=2	 (S^2, 2 pts)	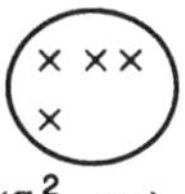 (S^2, pts)	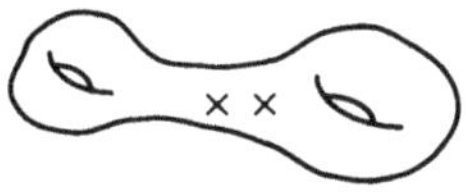 Riemann surface	
Analysis	de Rham cohomology	de Rham homotopy cyclic cohomology	non-Abelian gauge theory + gravity + ?	
	Abelian gauge theory			

Table 1

3. Degree, linking number and Vassiliev invariant

To mention something more down to earth, we recall here the work by Gauss, Vassiliev, Bar-natan, Kontsevitch, Bott–Taubes, D. Thurston *etc.* In this section, we follow Bott–Taubes' paper [BT] and D. Thurston's lecture [Th] in many points.

Let us first consider the mapping degree of a map $f : M^n \to N^n$ between two closed oriented manifolds of the same dimension. There are two simple ways to define it:

Definition 3.1.1. Let $p \in N$ be a regular value of f. Then $f^{-1}(p)$ consists of finitely many points. For each $q \in f^{-1}(p)$, we consider the differential $Df_q : T_qM \to T_pN$ of f. We put $\varepsilon_q = 1$ if Df_q preserves orientation and $\varepsilon_q = -1$ otherwise. The *mapping degree* of f is then $\deg f = \sum_{q \in f^{-1}(p)} \varepsilon_q$.

Definition 3.1.2. We choose any n-form ω on N such that $\int_N \omega \neq 0$. Then the *mapping degree* of f is equal to $\deg f = (\int_M f^*\omega)/(\int_N \omega)$.

In our terminology (3.1.1) is a geometric way and (3.1.2) is an analytic way to define the mapping degree. Their coincidence is a consequence of de Rham's theorem.

Algebra	Graph	Topology	Physics
(co)homology	line	homology	free field
cup product	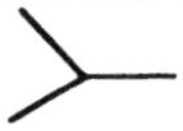	Q homotopy type	interaction tree (0 loop) amplitude
A$^\infty$ structure Massey product (related to associativity)	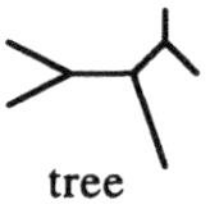 tree		
Steenrod square (related to commutativity)	automorphism of trees	homotopy type	
?	general graph	differential topology ?	higher loop amplitude

Table 2

To unify the two definitions one can use (as de Rham did) the notion of a current. We recall that a k-current T on an n-dimensional manifold M is a map $T : \bigwedge^{n-k}(M) \to \mathbb{R}$ which is $\mathbb{R}$-linear and is continuous with respect to the C^∞-topology. A k-form ω is regarded as a k-current T_ω by $T_\omega(\nu) = \int_M \omega \wedge \nu$. An oriented $(n-k)$-dimensional submanifold N is regarded as a k-current T_N by $T_N(\nu) = \int_N \nu$.

Using currents we can rewrite (3.1.1) as $\deg f = (\int_M f^*T_p)/(\int_N T_p)$. In this way the description is completely parallel to (3.1.2). We remark that the pull-back of the current f^*T is not always well-defined. In fact, in the case $T = T_N$ for a submanifold N, the pull back f^*T is well defined if and only if f is transversal to N. This is the condition that we assumed in (3.1.1).

We now use this exercise to study Gauss' formula for the linking number. Let ω_{S^2} be the volume element of S^2 such that $\int_{S^2} \omega_{S^2} = 1$. We consider a map $G : (\mathbb{R}^3 \times \mathbb{R}^3) \setminus \Delta \to S^2$ by

$$G(x, y) = \frac{x - y}{\|x - y\|},$$

where Δ is the diagonal.

Lemma 3.2. $d(G^*\omega_{S^2}) = T_\Delta$.

Proof. Of course this is well known and is a very old lemma (which probably goes back to Gauss). But the author wants to prove it since the proof suggests the relation of the lemma to the compactification of configuration space due to Fulton–Macpherson [FM] and Kontsevitch [Ko1].

First we "compactify" $(\mathbb{R}^3 \times \mathbb{R}^3) \setminus \Delta$ as follows. We take the direct product $\big((\mathbb{R}^3 \times \mathbb{R}^3) \setminus \Delta\big) \times S^2$ and a map $(\mathbb{R}^3 \times \mathbb{R}^3) \setminus \Delta \to \big((\mathbb{R}^3 \times \mathbb{R}^3) \setminus \Delta\big) \times S^2$, $(x, y) \mapsto ((x, y), G(x, y))$. The closure of the image of this map is a manifold with boundary which we denote by $\overline{(\mathbb{R}^3 \times \mathbb{R}^3) \setminus \Delta}$. Obviously $G : (\mathbb{R}^3 \times \mathbb{R}^3) \setminus \Delta \to S^2$ extends to $\overline{(\mathbb{R}^3 \times \mathbb{R}^3) \setminus \Delta}$. The boundary, $\partial\overline{(\mathbb{R}^3 \times \mathbb{R}^3) \setminus \Delta}$, of $\overline{(\mathbb{R}^3 \times \mathbb{R}^3) \setminus \Delta}$ is identified with $S(\mathbb{R}^3)$, the unit sphere bundle of the tangent bundle of $\mathbb{R}^3$.

Now let $\nu \in \bigwedge^3(\mathbb{R}^3 \times \mathbb{R}^3)$ be an arbitrary test function. Using Stokes' theorem, we calculate

$$\begin{aligned}
\big(dG^*T_{\omega_{S^2}}\big)(\nu) &= \int_{(\mathbb{R}^3\times\mathbb{R}^3)\setminus\Delta} d\nu \wedge G^*\omega_{S^2} \\
&= \int_{\overline{(\mathbb{R}^3\times\mathbb{R}^3)\setminus\Delta}} d\nu \wedge G^*\omega_{S^2} \\
&= \int_{\overline{(\mathbb{R}^3\times\mathbb{R}^3)\setminus\Delta}} d\big(\nu \wedge G^*\omega_{S^2}\big) - \int_{\overline{(\mathbb{R}^3\times\mathbb{R}^3)\setminus\Delta}} \nu \wedge dG^*\omega_{S^2} \\
&= \int_{\partial\overline{(\mathbb{R}^3\times\mathbb{R}^3)\setminus\Delta}} \nu \wedge G^*\omega_{S^2} \\
&= \int_\Delta \nu.
\end{aligned}$$

The lemma follows. □

Now let $\ell_i : S^1 \to \mathbb{R}^3$ $(i = 1, 2)$, be the link, that is, we assume that they are embeddings and that the images are disjoint from each other. We then obtain a map $\Phi : S^1 \times S^1 \to S^2$ by $\Phi(s, t) = G(\ell_1(s), \ell_2(t))$. The linking number of this link is, by definition, the degree of this map. They are calculated in two different ways:

(3.3.1) We consider a generic point $p \in S^2$ and count the order of the set $\Phi^{-1}(p)$ with sign. (See (3.1.1) for the way to define sign.)

(3.3.2) We calculate the integral $\int_{S^2} \Phi^*\omega_{S^2}$.

They give us the same number. It implies the famous formula by Gauss:

$$LK(\ell_1,\ell_2) = \int_0^1 \int_0^1 \frac{[(\ell_1(s)-\ell_2(s)) \times \frac{d\ell_1}{ds}] \cdot \frac{d\ell_2}{dt}}{\|\ell_1(s)-\ell_2(s)\|^3}\, ds\, dt. \tag{3.4}$$

It is easy to see that (3.3.2) gives the right hand side of (3.4). To see that 3.3.1 gives the left hand side, we remark that:

(a) If $(\ell_1(s), \ell_2(t))$ is unlink, we can put one of them in the region where $x \gg 0$, $|y|, |z| < 1$, and the other in the region where $x \ll 0$, $|y|, |z| < 1$. It is then easy to see that the number in (3.3.1) is zero in that case.

(b) We study how the number in (3.3.1) changes when ℓ_1 crosses ℓ_2 and goes to the other side. It is easy to see by geometric observation that it changes by ± 1.

Combining these two observations, we are done.

Remark 3.5. There is third choice for the current representing the fundamental class of S^2. Namely we take $S^1 \subseteq S^2$, and consider the current $d\alpha \wedge T_{S^1}$, where $\alpha : S^1 \to \mathbb{R}/\mathbb{Z}$ is the diffeomorphism. Using this current we obtain a formula similar to the Kontsevitch integral (see [Ba2]) to obtain the linking number. (This was pointed out by D. Thurston and D. Bar-natan in [Th].)

So far everything is classical and well-known. But this formalism can be generalized directly to study the Vassiliev invariant. Let $\ell : S^1 \to \mathbb{R}^3$ be a knot. Following Bott–Taubes we put

$$C_4 = \{(t_1,t_2,t_3,t_4) \in (S^1)^4 \mid t_1,t_2,t_3,t_4 \text{ respects the cyclic order}\}.$$

We have a map $\Phi_4 : C_4 \to (S^2)^2$, $(t_1,t_2,t_3,t_4) \mapsto [G(\ell(t_1),\ell(t_3)),\ G(\ell(t_2),\ell(t_4))]$. Unfortunately the space C_4 has a boundary. Hence the mapping degree of Φ_4 is not well-defined. One takes the following

$$C_{3,1} = \left\{ (t_1,t_2,t_3,x) \in (S^1)^3 \times \mathbb{R}^3 \,\middle|\, \begin{array}{l} t_1,t_2,t_3 \text{ respects the cyclic order} \\ x \neq \ell(t_1),\ell(t_2),\ell(t_3) \end{array} \right\}.$$

We then have a map

$$\Phi_{3,1} : C_{3,1} \to (S^2)^3$$

defined by $(t_1,t_2,t_3,x) \mapsto [G(\ell(t_1),x), G(\ell(t_2),x), G(\ell(t_3),x)]$. Bott–Taubes proved that by combining them one gets a well-defined degree. (Their approach is based on de Rham theory. D. Thurston gave a version of the counting number of the inverse of one point.)

They use a compactification of the configuration spaces C_4, $C_{3,1}$. That compactification is a generalization of $\overline{(\mathbb{R}^3 \times \mathbb{R}^3) \setminus \Delta}$ discussed in the proof of Lemma 3.2.

We mention the relation of this construction to graphs. The configuration spaces C_4 and $C_{3,1}$ are related to the graphs in Figure 1 and Figure 2 respectively.

Figure 1

Figure 2

4. Invariant of 3-manifolds and Morse homotopy

We want to generalize the construction of the last section to general 3-manifolds. Maybe the most direct generalization is as follows. Assume that our 3-manifold M is a homology 3-sphere. Fix a frame of M, that is, a trivialization of the tangent bundle TM. Fix a point p of M and put $M^\circ = M \setminus \{p\}$.

Lemma 4.1. *There exists a map $G : M^\circ \times M^\circ \setminus \Delta \to S^2$ which extends to a map from $\overline{M^\circ \times M^\circ \setminus \Delta}$ to S^2 such that its restriction to $\partial \overline{M^\circ \times M^\circ \setminus \Delta} = SM^\circ$ is obtained by the given framing of TM.*

Here $\overline{M^\circ \times M^\circ \setminus \Delta}$ is defined in a way similar to that of $\overline{(\mathbb{R}^3 \times \mathbb{R}^3) \setminus \Delta}$. The proof of Lemma 4.1 is by easy obstruction theory. However, we need some more conditions on the behavior of the map $G : M^\circ \times M^\circ \setminus \Delta \to S^2$ in the neighborhood of infinity of M°, namely, in the neighborhood of the point p of M.

Consider the 2 form $G^*\omega_{S^2}$ on $M^\circ \times M^\circ \setminus \Delta$. In a way similar to that of Lemma 3.2 we can prove that $d(G^*\omega_{S^2}) = T_\Delta$. It seems likely that one can obtain an invariant of knots using this form in a way similar to Bott–Taubes.

Remark 4.2. M. Kontsevitch [Ko1] discussed 3-manifold invariants using a 2-form ω on $M^\circ \times M^\circ \setminus \Delta$ such that $d\omega = T_\Delta$. Probably the construction he had in mind of such a form is the one discussed above. The way taken by Axelrod–Singer [AS] is to use the Riemannian metric and Green's function to find such a form ω on $M^\circ \times M^\circ \setminus \Delta$ such that $d\omega = T_\Delta$. It is easy to see that the form obtained above is directly related to the framing of M.

To study a 3-manifold invariant (instead of a knot invariant) using this 2-form, we consider the diagram in Figure 3. It means that we consider the following integral

$$\int_{((x_1,y_1),(x_2,y_2),(x_3,y_3))\in(M^\circ\times M^\circ\setminus\Delta)^3} \omega(x_1, y_1) \wedge \omega(x_2, y_2) \wedge \omega(x_3, y_3). \tag{4.3}$$

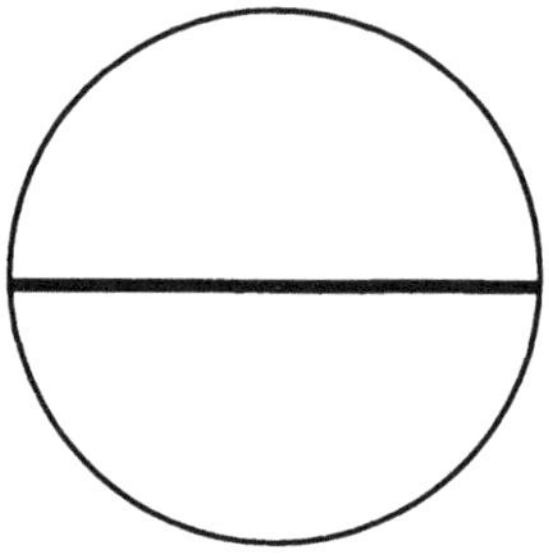

Figure 3

One needs correction terms to get a well-defined invariant. See Axelrod–Singer [AS], for details.

Problem 4.4. Find another version of the way to define the same invariant by a counting argument using the map $G : M^\circ \times M^\circ \setminus \Delta \to S^2$.

The author [Fu4], found a counting argument to define a similar (and most likely the same) invariant, in a slightly different way from that of using the map $G : M^\circ \times M^\circ \setminus \Delta \to S^2$. This approach is based on Morse theory and is roughly described as follows. Choose 3 generic functions f_1, f_2, f_3 on M. Fix a Riemannian metric on M. Let $\varphi_t^{f_i}$ be the one-parameter group of transformations associated to the gradient vector field of f_i. We use the moduli space

$$\mathcal{M}_{MS}(f_1, f_2, f_3) = \{(t_1, t_2, t_3, x, y) \in \mathbb{R}_+^3 \times M^3 \mid \varphi_{t_i}^{f_i}(x) = y,\ i = 1, 2, 3\}. \tag{4.5}$$

Counting its order we obtain a number. We also need correction terms to obtain a well-defined invariant. We omit the details which are discussed in [Fu3], [Fu4].

Remark 4.6. It is sometimes simpler to work with local coefficients. Namely we consider a flat Lie algebra bundle ν on M such that $H^*(M; \nu) = 0$. This implies that T_Δ, the current represented by the diagonal in $M \times M$ is a boundary in the local coefficients. So we do not need to remove one point p to find a current ω such that $d\omega = T_\Delta$.

Remark 4.7. The graph in Figure 3 is regarded as a Feynman diagram. Then, the invariant we have been discussing is a 2-loop amplitude. There is an interaction in this case. In fact, the interaction is related to the product structure. In Formula 4.3 it is related to the wedge product of the forms. And in Formula 4.5 it is related to intersection theory.

Originally E. Guadagnini–M. Martinelli–M. Mintchev, D. Bar-natan, Axelrod–Singer, Kontsevitch discovered this kind of construction as a perturbation theory for Chern-Simons Gauge theory or Witten's invariant. It is Kontsevitch who observed its relation to homotopical algebra and to compactification of configuration spaces.

Finally we remark that there is probably a family version of this construction, as pointed out by [AS], [Ko]. Namely, we can construct invariants of a fibre bundle whose fibre is M.

5. Relation to string theory

We next describe the relation of the construction in the last section to string theory. We have discussed this topic in the last section of [Fu4] and in [FO]. So we shall not go into details. Let us consider the cotangent bundle T^*M of M. It has a canonical symplectic form ω. We choose and fix an almost complex structure J of T^*M such that $\omega(JV, W) = g(V, W)$ is a Riemannian metric on T^*M. To each function f on M we associate Λ_f, the graph of its differential df. Λ_f is a Lagrangian submanifold of T^*M, namely $\omega|_{\Lambda_f} = 0$. We consider a surface with boundary $\Sigma_{0,3} = D^2 - \bigcup_{i=1}^2 D_i^2$ where $D_i^2 \subset D^2$ (see Figure 4).

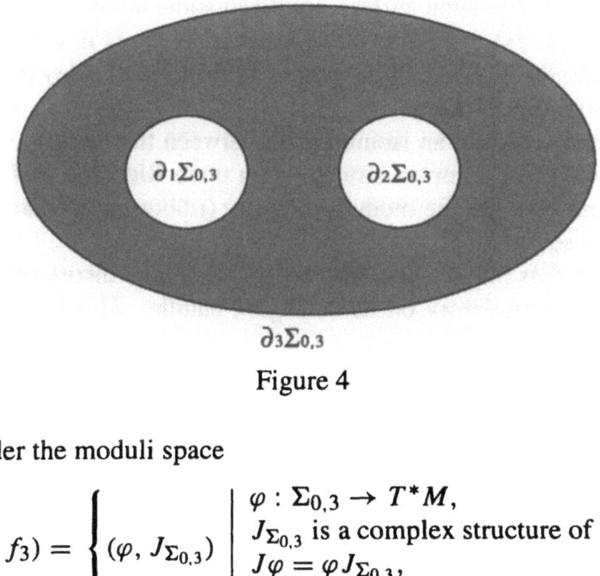

Figure 4

Then we consider the moduli space

$$\mathcal{M}_{SY}(f_1, f_2, f_3) = \left\{ (\varphi, J_{\Sigma_{0,3}}) \;\middle|\; \begin{array}{l} \varphi : \Sigma_{0,3} \to T^*M, \\ J_{\Sigma_{0,3}} \text{ is a complex structure of } \Sigma_{0,3}, \\ J\varphi = \varphi J_{\Sigma_{0,3}}, \\ \varphi(\partial_i \Sigma_{0,3}) \subseteq \Lambda_{f_i}\ i = 1, 2, 3 \end{array} \right\} \Big/ \sim$$

where $\partial_i \Sigma_{0,3}$ $(i = 1, 2, 3)$ are components of the boundary of $\Sigma_{0,3}$. The equivalence relation $\sim$ is defined as follows.

$$(\varphi, J_{\Sigma_{0,3}}) \sim (\varphi', J'_{\Sigma_{0,3}}) \iff \exists \psi \text{ such that } \varphi' = \varphi\psi,\ J'_{\Sigma_{0,3}} = \psi^* J_{\Sigma_{0,3}}.$$

We consider the order (counted with sign) of this moduli space. This is one of the correlation functions of the open string theory of T^*M. We add various correction terms to obtain a (topological) invariant of M.

This invariant is related to the one in the last section as follows:

Theorem 5.1 (Fukaya-Oh). $\mathcal{M}_{SY}(\varepsilon f_1, \varepsilon f_2, \varepsilon f_3) = \mathcal{M}_{MS}(f_1, f_2, f_3)$ *for sufficiently small* ε.

There is a similar result which asserts the relationship between the rational homotopy type and the 0-loop amplitude of the open string on the cotangent bundle [FO]. In that case, we replace $\mathcal{M}_{SY}(\varepsilon f_1, \varepsilon f_2, \varepsilon f_3)$ by the moduli space of the maps from D^2 to T^*M with an appropriate boundary condition and $\mathcal{M}_{MS}(f_1, f_2, f_3)$ with the moduli space of maps from various trees to M. (See [FO] for a precise statement.)

In the simplest case, namely in the case when the graph is a line and the Riemann surface is D^2 with two marked points on the boundary, this result was proved by Floer and used in his calculation of Floer homology of the Lagrangian intersection.

The author would like to mention only one part of the proof. Theorem 5.1 gives a relationship between two moduli spaces. The space $\mathcal{M}_{MS}(f_1, f_2, f_3)$ is the space of maps from the graph (of Figure 3) to M. The space $\mathcal{M}_{SY}(\varepsilon f_1, \varepsilon f_2, \varepsilon f_3)$ is the moduli space of maps from the open Riemann surface (of Figure 4) to T^*M.

The graph and the Riemann surface are given some additional structures. In the case of $\mathcal{M}_{MS}(f_1, f_2, f_3)$, we put positive numbers t_1, t_2, t_3 in Definition 4.5. They are regarded as lengths of edges of the graph. In the case of $\mathcal{M}_{SY}(\varepsilon f_1, \varepsilon f_2, \varepsilon f_3)$, we take a complex structure of $\Sigma_{0,3}$.

So the first step is to find an isomorphism between the moduli of metrics on the graph and the moduli of complex structures on open Riemann surfaces. A similar relation was found between the moduli of metric (ribbon) graphs and the moduli of (closed) Riemann surfaces.

Our discussion thus implies that Chern–Simons Gauge theory on a 3-manifold is equal to the open string theory on its cotangent bundle. This is the observation of Witten [W5].

This kind of equivalence is an example of the phenomenon that string theory is sometimes equivalent to the field theory of the target space. (The latter is called, in that case, the string field theory.) The duality between geometry and analysis we mentioned in this kind of cases is string theory = string field theory, where the former is geometry and the latter is analysis.

There is probably another important case for this equivalence due to [BCOV]. In their case they study the closed string theory on a Calabi–Yau 3-fold, namely a complex 3-dimensional manifold with Ricci flat curvature. Closed string theory, in this context, means counting the holomorphic maps from a closed Riemann surface. It seems to the author that their conclusion is that the "dual" to this construction is something related to the $\overline{\partial}$ operator. In the case discussed in this section, the dual to the open string is something related to the d operator. It is interesting to find some algebraic machinery which is common in four theories: $\overline{\partial}$, d, closed string and open string.

We remark that the relation of the homotopical algebra to the closed string (the operad structure in conformal field theory) is also discussed by many authors [St2], [HL], [Ge] *etc.*

6. Illusion

The author wants to state the "topological version" of Conjecture 2.1. As we sketched very briefly, we can construct a system of numbers by studying the maps from a graph to a given manifold M. Namely, we choose a Morse function and count the number of the maps from a graph to M such that each edge is a gradient line of one of the functions. There are many variants of this kind of numbers. Namely, we can involve the homology class of the moduli space of the metric structure of the graphs, we can also use various flat bundles on the target space M, and we can also involve the cohomology class of M.

Moreover, there is a family version. Namely, if there is a fibre bundle with M as a fibre, then we have again many numbers. In that case we can also use the cohomology class of the base space.

Thus, there are quite a lot of numbers obtained in that way. The author does not know what kind of algebraic structure is enjoyed by this system of numbers. He would like to call the theory which describes these numbers Morse homotopy theory (of higher genus). They are topological field theories. And in the sense of section 2, they consist of a category of field theories on one-dimensional spaces (graphs).

Conjecture 6.1. The functor

$$\{\text{smooth manifold}\} \Longrightarrow \{\text{one dimensional field theory}\}$$

is injective. In other words, Morse homotopy gives a complete invariant for differential topology. The family version is also true.

Maybe it is much too ambitious to write 6.1 as a conjecture. In fact there are only a few reasons one can believe it. Let me give some evidence.

1. Let M be simply connected, $\dim M \geq 5$. Then Sullivan's theorem says that M is determined up to finite ambiguity by its rational homotopy type and Pontrjagin class. We can find the rational homotopy type from Morse homotopy. M. Betz and R. Cohen [BC] assert that one can also find the Pontrjagin class from Morse homotopy.

2. Let $\dim M = 3$. First there is some reason to believe that Vassiliev's invariant is a complete invariant of knots. When we take its analogy we migh believe that the analogue of Vassiliev's invariant (the invariant of Chern–Simons perturbation

theory) is a complete invariant of 3-manifolds. It is almost certain that all Chern–Simons perturbation theory invariants come from Morse homotopy.

3. By simple dimension counting, it seems that the number of invariants we obtain from Morse homotopy rapidly decreases when the dimension becomes higher. We know from differential topology that in higher dimensions we do not have as many invariants as that in the case of dimension 3 or 4.

4: Let $\dim M = 2$. Then of course there are not so many invariants for 2-manifolds. So, to get something nontrivial, we need to consider the invariants of families. Namely, the invariants of fibre bundles whose fibre is a 2-manifold. Such an invariant is of course the cohomology of the mapping class group or the cohomology of the moduli space of the Riemannn surface. On the other hand, our Morse homotopy invariants are supposed to be parametrized by the cohomology class of the moduli space of graphs. But one knows that the moduli space of metric ribbon graphs is the same as the moduli space of a Riemann surface. So if we can find the correct algebraic machinery, Conjecture 6.1 in this case might be a trivial isomorphism. Namely, Morse homotopy only gives the identity map.

In fact the author does not even have very weak evidence in the case when the dimension is 4. In that case Taubes' result mentioned in Section 1 shows that the Gromov–Witten invariant is a topological invariant in many cases. (Namely, it is independent of the symplectic structure.) It is not likely that this invariant comes from Morse homotopy. So in dimension 4 one may also need 2-dimensional field theory to describe the smooth topology.

References

[At] M. Atiyah, The geometry and physics of knots, Cambridge University Press, Cambridge 1990.

[AS] S. Axelrod, I. Singer, Chern–Simons perturbation theory I, in: Proceedings of XXth DGM Conf. (S. Catto and A. Rocha eds.), World Scientific, Singapore 1992, II, Journal of Diff. Geom. 39 (1994), 173–213.

[BCOV] M. Bershadsky, S. Cecotti, H. Ooguri, C. Vafa, Kodaira–Spencer Theory of Gravity and Exact Results for Quantum String Amplitudes, Comm. Math. Phys. 165 (1994), 311–427.

[BC] M. Betz, R. Cohen, Graph moduli spaces and Cohomology Operations, Turkish J. Math. 18 (1995), 23–41.

[Ba1] D. Bar-natan, Perturbative aspects of the Chern–Simons topological field theory, PhD thesis, Princeton University 1991.

[Ba2] D. Bar-natan, Vassiliev's knot invariant, Topology (1995), 423–472.

[BT] R. Bott, C. Taubes, On the self-linking of knots, J. Math. Phys. 35 (1994), 5247–5287.

[CGPO] P. Candelas, P. Green, L. Parke, dela Ossa, A pair of Calabi–Yau manifolds as an exactly soluble superconformal field theory, Nuclear Phys. B 359 (1991), 21.

[CJS] R. Cohen, J. Jones, G. Segal, Floer's infinite dimensional Morse theory and homotopy theory, in: The Floer Memorial Volume (ed. H. Hofer, C. Taubes, A. Weinstein, E. Zehnder), Progr. Math. 133, Birkhäuser, Basel 1995 297–326.

[C] A. Connes, Noncommutative Geometry, Academic Press, London 1994.

[Fl] A. Floer, Morse theory for Lagrangian intersections, J. Differential Geom. 28 (1988), 513–547.

[Fu1] K. Fukaya, Morse homotopy, A^{∞} category and Floer homologies, in: Proceedings of Garc Workshop on Geometry and Topology (ed. H. J. Kim), Seoul National University 1994.

[Fu2] —, Morse homotopy and its quantization, to appear in the proceedings of Georgia International Conference of Topology.

[Fu3] —, Morse theory and topological field theory, to appear in Sukugaku Expositions.

[Fu4] —, Morse theory and Chern–Simons perturbation theory, Comm. Math. Phys. 181 (1996), 37–90.

[FO] K. Fukaya, Y. Oh, Zero loop open string in the cotangent bundle and Morse homotopy, preprint.

[FM] W. Fulton, R. Macpherson, A compactification of configuration space, Ann. of Math. 139 (1994), 183–225.

[Ge] E. Getzler, Operad and moduli space of genus 0 Riemann surface, in: The moduli space of curves, Progr. Math. 129, Birkhäuser, Boston 1995, 199–230.

[GMM] E. Guadagnini, M. Martinelli, M. Mintchev, Perturbative aspects of Chern–Simons field theory, Phys. Lett. B 227 (1989), 111.

[H] H. Harer, The cohomology of the moduli space of curves, Lecture Notes in Math. 1337, Springer-Verlag, 138–221.

[HL] Y. Huang, J. Lepowski, Vertex operator algebra and operads, in: The Gelfand Mathematical Seminars 1990–1992 (ed. L. Corwin, I. Gelfand, J. Lepowsky), Birkhäuser, Boston 1993, 145–161.

[Ko1] M. Kontsevitch, Feynman diagram and low dimensional topology, in: First European Congress of Mathematics, Vol. II, Progr. Math. 120, Birkhäuser, Basel 1994, 97–121.

[Ko2] —, Formal (non)-commutative Symplectic- Geometry, in: The Gelfand Mathematical Seminars 1990–1992, (ed. L. Corwin, I. Gelfand, J. Lepowsky), Birkhäuser, Boston 1993, 173–188.

[Ko3] —, A^{∞}-algebras in Mirror symmetry, preprint.

[Ko4] —, Intersection theory on the Moduli space of curves and the Matrix Airy functions, Comm. Math. Phys. 147 (1992) 1–23.

[KM] M. Kontsevitch, Y. Manin, Gromov–Witten classes, quantum cohomology, and enumerative geometry, Comm. Math. Phys. 164 (1994), 525–562.

[Pe] R. Penner, The decorated Teichmüller space of punctured surface, Comm. Math. Phys. 113 (1997), 299–339.

[R] Y. Ruan, Topological sigma model and Donaldson type invariant in Gromov theory, Duke Math. J. 83 (1996), 461–500.

[RuT] Y. Ruan, G. Tian, Mathematical theory of quantum cohomology, J. Differential Geom. 42 (1995), 259–367.

[Oh] Y. Oh, Floer cohomology of Lagrangian intersections and Pseudo-holomorphic disks, Comm. Pure Appl. Math. 46 (1993), 949–994.

[Q] D. Quillen, Rational homotopy theory, Ann. of Math. 90 (1969), 205–295.

[Sc] M. Schwartz, Morse homology, Progr. Math. 111, Birkhäuser, Basel 1993.

[St1] J. Stasheff, Homotopy associativity of H-Spaces I, Trans. Amer. Math. Soc. 108 (1963), 275–292, II, ibid. 293–312.

[St2] —–, Towards a closed string field theory: Topological and convolution algebra, Topology Proc. 17 (1992), 371–386.

[Str] K. Strebel, Quadratic differentials, Springer-Verlag, 1985.

[Su] D. Sullivan, Infinitesimal Calculations in Topology, Inst. Hautes Études Sci. Publ. Math. 74 (1978), 269–331.

[T] C. Taubes, From Seiberg–Witten equation to pseudo holomorphic curves, J. Amer. Math. Soc. 9 (1996), 845–918.

[Th] D. Thurston, Lecture at Conference "Geometry and Physics", Aarhus Denmark.

[V] V. Vassiliev, Complements of Discriminants of Smooth maps: Topology and applications, Translations of Mathematical Monographs 98, Amer. Math. Soc., Providence, R. I.,1992.

[W1] E. Witten, Supersymmetry and Morse theory, J. Differential Geom. 17 (1982), 661–692.

[W2] —–,Topological sigma model, Comm. Math. Phys. 118 (1988), 441.

[W3] —–, Quantum field theory and Jones polynomial, Comm. Math. Phys. 121 (1989), 351.

[W4] —–, Mirror manifolds and topological field theory, in: Essays on Mirror manifolds (ed. S. T. Yau), 120–158.

[W5] —–, Chern–Simons Gauge theory as a string theory, in: The Floer Memorial Volume (ed. H. Hofer, C. Taubes, A. Weinstein, E. Zehnder), Progr. Math. 133, Birkhäuser, Basel 1995, 637–678.

Singularity theory for Riemannian distance functions on non-positively curved surfaces

V. Gershkovich

Abstract. We construct Morse theory and singularity theory for Riemannian distance functions d_p on a compact non-positively curved surface M_2. Stable critical points and their indexes are described due to the connection between Min-type functions and Riemannian distance functions on non-positively curved manifolds; Min-type functions are minima of a finite number of smooth functions. We prove that the set of non-positively curved metrics such that (for a given $p \in M_2$) the distance function d_p is Morse, is open and everywhere dense in the set of non-positively curved Riemannian metrics. For a generic metric on M_2 distance functions d_q are not Morse for some points $q \in M_2$; the description of the singularities of d_q is based on Morse theory for non-positively curved Riemannian metrics $\rho(x, y)$ which are Min-type functions on $M_2 \times M_2$. We prove that a generic negatively curved metric is Morse.

I. Distance functions

0. Introduction

Several generalizations of Morse theory to different classes of non-smooth functions are due to the fact that one does not need the Hessian, but only the index of the Hessian to construct a Morse theory. This index can be interpreted as the dimension of the maximal submanifold where the function has a strict maximum.

In this way Morse theory can be extended to several classes of non-smooth functions. Stable singularities of Min-Max functions (which can be obtained by a superposition of several operations of minimum and maximum from continuous families of smooth functions) were studied in Arnold's school; results, obtained in [Br1], [Br2], [Ma1], [Ma2], show that Morse theory can be constructed for this class of functions (though it has not been done yet).

One of the purposes of such generalizations is an approach to optimization, (including non-smooth optimization) via singularity theory, see [A], [CGR1], [CGR2], [CGR3].

Another application is in geometry, where non-smooth functions arise in a natural way in "completely smooth" problems of Riemannian geometry.

In this paper we develop a Morse and singularity theory for two closely connected classes of non-smooth functions: Min-type functions, which are minima of a finite number of smooth functions on a manifold, and Riemannian distance functions on non-positively curved surfaces.

To make this paper independent we present principal facts related to Min-type functions on manifolds and to their connection with Riemannian distance functions on non-positively curved manifolds. In the first section we recall some necessary facts related to Morse theory of Min-type functions obtained in [GR]; the second section contains results on approximation of Morse Min-type functions by Morse smooth functions from [Ge]. In Section 3 we reproduce results related to relations between Min-type functions and Riemannian distance functions on negatively curved manifolds from [GR]. These results are presented without proofs.

The rest of the first chapter contains new results related to series of distance functions, interpretation of metrics on a non-positively curved manifold M_n as a distance function on $M_n \times M_n$, and initial results related to singularity theory for Riemannian distance functions.

The second chapter is devoted to Morse theory for Riemannian distance functions on negatively curved surfaces; we prove that for a given point p, the subset of non-positively curved metrics on M_2 such that d_p is a Morse distance function is open and everywhere dense.

The study of singularity theory (in Chapter III) is based on the fact that any non-positively curved Riemannian metric is a Min-type function. This allows us to describe degenerations of distance functions d_q for all $q \in M_2$ for a generic metric on M_2.

We do not discuss geometric applications of Morse theory for Riemannian distance functions. The reader can find them in the initial seminal papers [GS] and [Gr] and in recent reviews [Gro] and [Ch].

The author is grateful to Yu. Burago and M. Gromov for discussions, to J. Dodziuk for useful remarks, and to IHES for invitation and hospitality.

We use the following notations in this paper.

Notations

$r_{inj}(x)$ is the injectivity radius at x and $r_{inj} = \min_{x \in M_n} r_{inj}(x)$.

$\operatorname{diam}(M_n)$ is the diameter of M_n.

$\operatorname{Sect}_{x,\alpha}$ is the sectional curvature on a 2-plane $\alpha \in T_x M_n$, $x \in M_n$, and $\operatorname{Sect}(M_n) = \min_{x \in M_n, \alpha \subset T_x M_n} \operatorname{Sect}_{x,\alpha}$.

g is a Riemannian metric tensor.

$\rho : M_n \times M_n \to \mathbb{R}$ is the Riemannian metric ($\rho(x, y)$ is the length of a shortest curve connecting x and y).

$D_r(x)$ is the ball of radius r centered at x.

D_r^C is the ball of radius r in the simply connected Riemannian manifold $\mathcal{H}_C$ of constant curvature $-C$ (in particular, D_r^0 is the euclidean ball of radius r).

μ is the Riemannian measure on M_n.

μ_C is the Riemannian measure on $\mathcal{H}_C$.

$S_r(x)$ is the sphere of radius r centered at x.

$\pi : (\tilde{M}_n, \tilde{\rho}) \to (M_n, \rho)$ is the universal covering.

We denote all objects (points, balls, geodesics, etc.) related to the universal covering $\tilde{M}_n$ by the same letters as the corresponding objects for M_n but with tildes.

$\mathrm{Conv}\{\xi_1, \dots, \xi_n\}$ denotes the convex hull of the vectors $\xi_1, \dots, \xi_n$.

We use the letter γ with different indices to denote geodesics; we use the notation $\dot{\gamma}(p)$ to denote $\frac{d\gamma}{dt}(t)$ for t such that $\gamma(t) = p$; of course, we use such notation only when p is not a point of self-intersection for γ.

1. Morse Min-type functions

Definition 1.1 (germs of Min-type functions). Let M_n be a C^∞-smooth n-manifold. A germ of a function $f : M_n \to \mathbb{R}^1$ at $x \in M_n$ is said to be a *germ of a Min-type function* if there exist germs of smooth functions $\alpha_1, \dots, \alpha_l$ at x such that

$$f = \min\{\alpha_1, \dots, \alpha_l\} \tag{1}$$

in a neighbourhood $U(x)$.

Definition 1.2 (minimal representations). A representation (1) of a germ of a Min-type function f is said to be *minimal* if it includes the minimal possible number of smooth functions.

Definition 1.3 (global Min-type functions). A function $f : M_n \to \mathbb{R}^1$ is said to be a *Min-type function* if each germ of f is a germ of a Min-type function.

Proposition 1.1. *Let M_n be a compact manifold and $f : M_n \to \mathbb{R}^1$ be a function such that each germ of f is a germ of a Min-type function. Then there exist smooth functions $\alpha_1, \dots, \alpha_N : M_n \to \mathbb{R}^1$ such that*

$$f = \min\{\alpha_1, \dots, \alpha_N\} \tag{2}$$

on M_n.

Definition 1.4 (C^k-Min-type functions). A Min-type function f is said to be C^k-*smooth* if there exists a minimal representation (1) with C^k-smooth functions α_i.

Definition 1.5 (C^k-topology). We define the C^k*-topology* on the space of C^k-Min-type functions in such a way that a small perturbation of f is a result of independent C^k-small perturbations of α_i, $i = 1, \dots, N$ in (2).

Remark. The C^k-topology does not depend on the choice of a representation (2).

Definition 1.6 (non-degenerate points). Let $f : M_n \to \mathbb{R}^1$ be a Min-type function, let $x \in M_n$ and let $f = \min\{\alpha_1, \dots, \alpha_m\}$ be a minimal representation of f at x; x is said to be a *non-degenerate point* if the germ of f at x satisfies the following two conditions:

1) Any $m - 1$ gradients among $\operatorname{grad} \alpha_1(x), \dots, \operatorname{grad} \alpha_m(x)$ are linearly independent;

2) $\operatorname{grad}(\alpha_1 - \alpha_2), \dots, \operatorname{grad}(\alpha_{m-1} - \alpha_m)$ are linearly independent at x.

A non-degenerate point x is said to be *regular* if $\operatorname{Conv}\{\operatorname{grad} \alpha_i(x)\}$ does not include the origin, and *critical* otherwise.

Remark. The three previous definitions do not depend on the choice of a minimal representation (1).

Definition 1.7 (Morse Min-type functions). A Min-type function $f : M_n \to \mathbb{R}^1$ is said to be a *Morse* Min-type function if f is non-degenerate at all points.

Lemma 1.1. *Let $f : M_n \to \mathbb{R}^1$ be a Morse Min-type function, let x be a critical point of f and let $f = \min\{\alpha_1, \dots, \alpha_m\}$ be a minimal representation of a germ of f at x. Then the equations*

$$\alpha_1(y) = \cdots = \alpha_m(y)$$

define a germ for a smooth submanifold of dimension $(n - m + 1)$ at x.

Denote this submanifold by $\mathcal{N}_x^f$. The restriction of f on $\mathcal{N}_x^f$ is smooth (f coincides with α_1 on this submanifold, and hence coincides with each of the α_i's); in fact $\mathcal{N}_x^f$ is the germ of the maximal submanifold where f is smooth.

Lemma 1.2. *The restriction of f to $\mathcal{N}_x^f$ is smooth and x is a non-degenerate critical point for this restriction.*

Let us define the index for Min-type functions at non-degenerate critical points; we denote the index of f at x by $\operatorname{ind}_x f$.

Definition 1.8 (index). Let $f : M_n \to \mathbb{R}^1$ be a Min-type function and let x be a non-degenerate critical point for f. We define the *index of f at x* as follows

$$\operatorname{ind}_x f = (m - 1) + \operatorname{ind}_x f_{|\mathcal{N}_x^f}.$$

Theorem 1.1. *The set of Morse Min-type functions is open and everywhere dense in the space of C^2-Min-type functions.*

As is well known, see [AGV], smooth functions have simple normal forms at regular and at non-degenerate critical points. All functions on $\mathbb{R}^n$ which are regular at the origin O and which vanish at O are on the same orbit of the action of the group of local diffeomorphisms $\mathrm{Diff}^n_{\mathrm{loc}}$ of $\mathbb{R}^n$. Therefore the normal form can be chosen to be $f \sim \mathrm{const} + x_1$ at a regular point.

All smooth functions vanishing at O that have a non-degenerate critical point at O are on one of the $n+1$ orbits of the action of $\mathrm{Diff}^n_{\mathrm{loc}}$; these orbits are numbered (from 0 to n) by the index of the Hessian of f. The normal form of a germ of f at x is

$$f = \mathrm{const} - \sum_{i=1}^{k} x_i^2 + \sum_{i=k+1}^{n} x_i^2,$$

where $\{x_i\}_{i=1}^n$ is a local coordinate system around x and k is the index of f. To obtain a sensible orbit structure for Min-type functions one has to increase the transformation group (compare with [Ma1]).

Definition 1.9 (group of almost smooth homeomorphisms). A homeomorphism $\phi : (\mathbb{R}^n, 0) \to (\mathbb{R}^n, 0)$ is said to be *almost smooth* if ϕ and ϕ^{-1} are smooth outside a submanifold of positive codimension. Denote the group of almost smooth local homeomorphisms of $\mathbb{R}^n$ by Hom^n_{as}.

Denote by $\mathcal{M}_r$ the set of germs of C^2 Min-type functions which vanish at the origin and such that the origin is a non-degenerate regular point.

Theorem 1.2 (non-degenerate regular points).

1) *$\mathcal{M}_r$ is an orbit of the action of* Hom^n_{as}.
2) *The normal form of a Min-type function at a non-degenerate regular point is* $f \sim \mathrm{const} + x_1$.

Denote by $\mathcal{M}_c$ the set of germs of C^2 Min-type functions which vanish at the origin and such that the origin is a non-degenerate critical point.

Theorem 1.3 (non-degenerate critical points).

1) *$\mathcal{M}_c$ is a union of $n+1$ orbits of the action of* Hom^n_{as}.
2) *The normal form of a Min-type function at such point is*

$$f \sim \mathrm{const} - \sum_{i=1}^{k} x_i^2 + \sum_{i=k+1}^{n} x_i^2.$$

Remarks.

1) The number k coincides with the index of the Min-type function defined above, see Definition 1.9; k is the dimension of the maximal submanifold where f has a strict local maximum.

2) Certain germs of Min-type functions at degenerate critical points are equivalent to germs of Min-type functions at non-degenerate critical points under the action of Hom^n_{as}. The simplest examples are birth-death singularities defined below.

Let $f : \mathbb{R}^n \to \mathbb{R}^1$ have a birth-death singularity at the origin; then the canonical normal form of f is

$$f = \mathrm{const} + x_n^3 - \sum_{i=1}^{k} x_i^2 + \sum_{k+1}^{n-1} x_i^2,$$

see [AGV]. This function is equivalent to

$$f_1 = \mathrm{const} + x_n - \sum_{i=1}^{k} x_i^2 + \sum_{k+1}^{n-1} x_i^2$$

under the action of Hom^n_{as}; (the transformation is $x_n \to x_n^{1/3}$); in its turn, f_1 is smoothly equivalent to const $+ x_n$.

A critical point where f has a birth-death singularity can be considered as a "degenerate regular" point in the following sense: the topological type of $f^{-1}(-\infty, c)$ does not change when c goes through zero. The "classical" explanation of this effect is that the decomposition of the birth-death singularity at the origin gives two non-degenerate critical points with indices differing by one. Such points (when taken together) do not contribute to the change of homotopy type of $f^{-1}(-\infty, c)$ and can be eliminated, see [Mi2].

Proposition 1.2. *Let a degenerate critical point y of f be equivalent to a non-degenerate critical point of index k. Then the decomposition of y (by a small perturbation of f) gives the following series of points:*

1) *a non-degenerate critical point $\tilde{y}$ of index k,*
2) *a number of pairs of non-degenerate critical points with indices differing by one,*
3) *a number of (degenerate) regular points.*

This statement follows from the theorem on the approximation of a Morse Min-type function by a smooth Morse function.

2. Approximation by smooth Morse functions

Definition 2.1 (non-smooth set). Let $f : M_n \to \mathbb{R}^1$ be a Min-type function; denote by $U_f \subset M_n$ the set of points where f is not smooth.

Lemma 2.1. *Let $f : M_n \to \mathbb{R}^1$ be a Morse Min-type function. Then U_f has positive codimension in M_n.*

Denote by U_f^ϵ the ϵ-neighbourhood of U_f.

Theorem 2.1 (on an approximation). *Let M_n be a compact manifold and let $f : M_n \to \mathbb{R}^1$ be a Morse Min-type function. Let $x_1, \dots, x_l$ be critical points of f and let c_i be the index f at x_i, $i = 1, \dots, l$. Then for any positive ϵ there exists a smooth Morse function h such that*

1) *h has the same set of critical points (satisfying $\operatorname{ind}_{x_i} h = c_i$, $i = 1, \dots, l$),*
2) *h is ϵ-close to f in the C^∞-topology on $M_n \setminus U_f^\epsilon$,*
3) *h is ϵ-close to f on U_f^ϵ in the C^0-topology.*

The approximation by a smooth Morse function can be constructed easily under the assumption that the Riemannian metric on M_n is generic.

Define $h(x, t)$, $x \in M_n$, $t \in \mathbb{R}^1_+$ by the heat equation

$$\left(\frac{\partial}{\partial t} - \Delta\right) h(x, t) = 0$$

$$h(x, 0) = f_0(x),$$

and define $f_t(x)$ as $h(x, t)$.

Theorem 2.2 (approximation using heat flow). *Let $f_0 : M_n \to \mathbb{R}^1$ be a Morse Min-type function. Then f_ϵ is a C^∞-smooth Morse function for all sufficiently small positive ϵ.*

For sufficiently small ϵ, the function f_ϵ satisfies all but one of the conditions of Theorem 1.4. Namely f_ϵ has critical points $\{x_1^\epsilon, \dots, x_l^\epsilon\}$ such that x_i^ϵ is ϵ-close to x_i, $i = 1, \dots, l$ (and the index of f_ϵ at x_i^ϵ is c_i), but x_i and x_i^ϵ do not coincide. One can easily correct this using an isotopy of M_n close to the identity.

We present one more result related to heat flows on manifolds with a generic Riemannian metric. K. Uhlenbeck proved, see [U1], [U2], that a residual set of Riemannian metrics on a compact Riemannian manifold satisfies the following properties:

a) The spectrum of the corresponding Laplace operator Δ is simple.
b) All eigenfunctions of Δ are Morse.

Definition 2.2 (Uhlenbeck metrics). Let (M_n, ρ) be a compact Riemannian manifold. We call ρ an *Uhlenbeck metric* if the spectrum and the eigenfunctions of the Laplacian Δ on (M_n, ρ) satisfy conditions a and b above.

Theorem 2.3 (evolution of Morse functions). *Let M_n be a compact manifold. Let ρ be an Uhlenbeck metric on M_n. Then almost all (open and everywhere dense subset of) Morse functions $f : M_n \to \mathbb{R}^1$ satisfy the following conditions:*

1) *There exists a finite set $T = \{t_1, \dots, t_l\} \in \mathbb{R}^1$ such that f_t is Morse for $t \in \mathbb{R}^1 \setminus T$.*

2) *For every* $i = 1, \ldots, l$ *there exists a point* $x_i \in M_n$ *such that the restriction* $f_{t_i|M_n \setminus \{x_i\}}$ *is Morse.*
3) f_{t_i} *has a birth-death singularity at* x_i.

3. Distance functions on M_n

One can extend Morse inequalities, handlebody decomposition of the manifold corresponding to a Morse function, etc, to the class of Morse Min-type functions. A connection with Riemannian distance functions on non-positively curved manifolds is clarified by the following theorem.

Theorem 3.1. *Let* (M_n, ρ) *be a non-positively curved Riemannian manifold. Then every distance function* d_p, *for* $p \in M_n$ *is a Min-type function.*

To present a germ of a distance function d_p as a germ of a Min-type function, consider the universal covering

$$\pi : (\tilde{M}_n, \tilde{\rho}) \to (M_n, \rho),$$

choose $\tilde{q} \in \pi^{-1}(q)$ and let $\pi^{-1}p = \{\tilde{p}_1, \ldots, \tilde{p}_m, \ldots\}$. Hereafter we order the points $\tilde{p}_i$ in such a way that

$$\tilde{\rho}(\tilde{p}_i, \tilde{q}) \leq \tilde{\rho}(\tilde{p}_{i+1}, \tilde{q}).$$

Lemma 3.1. *Let* $p, q \in M_n$. *There exists only a finite number of points* $\tilde{p}_i$ *such that* $\tilde{\rho}(\tilde{p}_i, \tilde{q}) = \rho(p, q)$.

Denote by m the number of such points. We obtain the following (minimal) representation for a germ of a distance function d_p at q

$$d_p(y) = \min\{\tilde{d}_{\tilde{p}_1}(\tilde{y}), \ldots, \tilde{d}_{\tilde{p}_m}(\tilde{y})\}$$

for all $\tilde{y}$ in a small neighbourhood of $\tilde{q}$ and $y = \pi(\tilde{y})$ (where $\tilde{d}_{\tilde{p}_i}$ is a distance function for $\tilde{p}_i \in \tilde{M}_n$).

Let us reformulate the definition of non-degenerate points in terms of shortest geodesics $\gamma_1, \ldots, \gamma_m$ connecting p and q. (Hereafter we assume that the geodesics are parametrized by arclength.)

Proposition 3.1. *A point* q *is a non-degenerate point for* d_p *iff it satisfies the following two conditions:*

1) *For any* $1 \leq i_1 < \cdots < i_{m-1} \leq m$ *the vectors* $\dot{\gamma}_{i_1}(q), \ldots, \dot{\gamma}_{i_{m-1}}(q)$ *are linearly independent.*
2) *The vectors* $\dot{\gamma}_1(q), \ldots, \dot{\gamma}_m(q)$ *are not in an* $(m-2)$*-dimensional affine plane (that is* $\{\dot{\gamma}_i(q) - \dot{\gamma}_{i+1}(q)\}$ *are linearly independent).*

This allows us to use the concept of index, defined for non-degenerate critical points of Min-type functions, for Riemannian distance functions on non-positively curved manifolds.

Definition 3.1. (Morse distance functions). Let M_n be a non-positively curved manifold and let $p \in M_n$. The distance function d_p is said to be a *Morse distance function* if d_p is a Morse Min-type function.

Let us compare Gromov's definition and the "classical" definition of regular and critical points. The definition in the "classical style" was given above. We recall Gromov's definition now.

Definition 3.2 (Gromov's definition). Let (M_n, ρ) be a Riemannian manifold and let $p, q \in M_n$. The point q is said to be *regular* for d_p if all shortest geodesics $\gamma_1^{p,q}, \dots, \gamma_m^{p,q}$ connecting p and q satisfy the following property. There exists an open halfspace $T_q^+ M_n \subset T_q M_n$ such that $\dot{\gamma}_1^{p,q}(q), \dots, \dot{\gamma}_m^{p,q}(q) \in T_q^+ M_n$. A point q is said to be critical for d_p if q is not regular.

Definition 3.3 (multiplicity). The number $m_p(q)$ of shortest geodesics connecting p and q is said to be the *p-multiplicity of q*.

Theorem 3.2. *Let (M_n, ρ) be a Riemannian manifold and let d_p be a Morse distance function. Then Gromov's definition of critical points is equivalent to the classical one. The index of a critical point q is $m_p(q) - 1$.*

4. Distance functions on $M_n \times M_n$

Consider a Riemannian metric $\rho : M_n \times M_n \to \mathbb{R}^1$.

Proposition 4.1. *Let M_n be a non-positively curved manifold. Then ρ is a Min-type function on $M_n \times M_n$.*

Definition 4.1 (Morse metric). A non-positively curved metric ρ is said to be *Morse* if ρ is a Morse Min-type function on $M_n \times M_n$.

Let us discuss some properties of a Morse metric $\rho : M_n \times M_n \to \mathbb{R}^1$. Let $T_{p,q}(M_n \times M_n)$ denote the tangent space to $M_n \times M_n$ at $(p, q) \in M_n \times M_n$.

Proposition 4.2. *A Riemannian metric $\rho : M_n \times M_n \to \mathbb{R}^1$ is Morse iff for any points $p, q \in M_n$ the shortest geodesics $\gamma_1^{p,q}, \dots, \gamma_m^{p,q}$, connecting p and q satisfy the following conditions.*

1) *For any* $(m-1)$ *shortest geodesics* $\gamma_{i_1}, \ldots, \gamma_{i_{m-1}}$, *the tangent vectors*

$$\{\dot{\gamma}_{i_j}(p), \dot{\gamma}_{i_j}(q)\}_{j=1}^{m-1} \subset T_{p,q}(M_n \times M_n)$$

are linearly independent.

2) *The tangent vectors* $\{\dot{\gamma}_i(p), \dot{\gamma}(q)\}_{i=1}^{m}$ *do not lie in an* $(m-2)$*-dimensional affine plane in* $T_{p,q}(M_n \times M_n)$.

Let us present several examples related to Morse metrics on surfaces.

Example1. Let κ be a periodic geodesic on M_2 such that

$$\kappa(0) = \kappa(2d) = p, \kappa(d) = q, |\dot{\kappa}| \equiv 1, \rho(p,q) = d. \tag{3}$$

Then (p,q) is a critical point of index one for ρ.

Example 2. Let κ be a geodesic loop with a corner at p, satisfying (3). Then (p,q) is a regular point of ρ.

Example 3. Suppose that $p, q \in M_2$ are connected by three shortest geodesics such that p is a non-degenerate critical point for d_q (that is p is a local maximum), and q is a regular point for d_p. Then (p,q) is a regular point for ρ.

Example 4. Let $(p,q) \in M_2 \times M_2$ be a non-degenerate critical point of index three for ρ. Then p and q are connected by four shortest geodesics. Denote these geodesics as $\gamma_1, \ldots, \gamma_4$.

Proposition 4.3. *Let* ρ *be a negatively curved Morse metric on* M_2. *Let* $(p,q) \in M_2 \times M_2$ *be a non-degenerate critical point of index three. Then*

a) d_p *has a non-degenerate local maximum at* q;

b) d_q *has a non-degenerate local maximum at* p;

c) (p,q) *is not a local maximum for* ρ.

This proposition admits the following "more geometric" formulation.

Proposition 4.4. *Let* (M_2, ρ) *be a negatively curved surface. Suppose* $p, q \in M_2$ *are connected by four shortest geodesics which are in generic position. Then*

a) *there exists tangent vectors* $v_p \in T_pM_2$ *and* $v_q \in T_qM_2$ *such that*

$$\rho\big(\exp(\epsilon v_p), \exp(\epsilon v_q)\big) > \rho(p,q)$$

for small positive ϵ,

b) *there are no tangent vectors* $w_p \in T_pM_2$ *and* $w_q \in T_qM_2$ *such that*

$$\rho\big(\exp(\epsilon w_p), q\big) > \rho(p,q) \quad or \quad \rho\big(p, \exp(\epsilon w_q)\big) > \rho(p,q)$$

for a small positive ϵ.

Example 5. Let (p, q) be a non-degenerate critical point of index four for ρ. Then p and q are connected by five shortest geodesics; denote these geodesics by $\gamma_1, \ldots, \gamma_5$. Let $\alpha_i = \dot\gamma_i(p)$, $\beta_i = \dot\gamma_i(q)$, $\xi_i = \alpha_i \oplus \beta_i$. Then

a) any four vectors among $\{\xi_i\}_{i=1}^5$ are linearly independent;

b) $\{\xi_{i+1} - \xi_i\}_{i=1}^4$ are linearly independent.

c) $\mathrm{Conv}\,\{\gamma_i(p), \gamma_i(q)\}_{i=1}^5 \ni 0$.

Condition c) means that for any $v_p \in T_pM_2$, $v_q \in T_qM_2$ there exists a shortest geodesic γ connecting p and q such that

$$\langle\, \dot\gamma(p), v_p \,\rangle + \langle\, \dot\gamma(q), v_q \,\rangle > 0. \tag{4}$$

5. l-th distance function

Definition 5.1 (geodesic spectrum). Let M_n be a compact negatively curved manifold and let $p, q \in M_n$. Let $\gamma_1^{p,q}, \gamma_2^{p,q}, \ldots,$ be the sequence of geodesics connecting p, q and ordered in accordance with their lengths $l_i^{p,q} = \mathrm{length}\, \gamma_i^{p,q}$. We call

$$l_1^{p,q} \le l_2^{p,q} \le \cdots \le l_k^{p,q} \le \cdots$$

the *geodesic spectrum* of M_n at p, q.

Remark 5.1. The traditional notion of the length spectrum relates to the case $p = q$; the difference between the definitions in this case is that our definition takes into account all geodesics loops at p, unlike the classical definition which considers only periodic geodesics.

Let us define a sequence of functions $d_p^1, \ldots, d_p^j, \ldots$ on a negatively curved manifold such that $d_p^1 = d_p$.

Definition 5.2 (d_p^j). We define the *j-th distance function* as $d_p^j(q) = l_j^{p,q}$ for $j = 1, \ldots, \infty$.

The distance functions d_p^j satisfy the following properties.

Proposition 5.1. *Let M_n be a negatively curved manifold and let $p \in M_n$. Then*

1) *$d_p^j(q)$ is a continuous function of q for any j;*

2) *d_p^j depends continuously on p and on the Riemannian metric ρ on M_n.*

Corollary 5.1 (semi-continuity). *Let (M_n, ρ) be a negatively curved Riemannian manifold and let $p, q \in M_n$. Then*

$$m_{\bar p}^{\bar\rho}(\bar q) \le m_p^\rho(q)$$

for all metrics $\bar{\rho}$ sufficiently close to ρ, and all $(\bar{p}, \bar{q}) \in M_2 \times M_2$ sufficiently close to (p, q).

Proof. We have $d_p^{m_p(q)+1}(q) - d_p^1(q) > 0$. Therefore the same inequality holds for all metrics sufficiently close to ρ and for all $(\bar{p}, \bar{q})$ sufficiently close to (p, q). □

A distance functions d_p^i is not a Min-type function for $i > 1$. However they also belong to a nice class of non-smooth functions.

Definition 5.3 (Min-Max functions). Let $\{f_{i,j} : M_n \to \mathbb{R}^1\}$, $i = 1, \ldots, l$, $j = 1, \ldots, m$ be a two-parameter family of smooth functions. Then

$$f = \max_{1 \leq j \leq m} \min_{1 \leq i \leq l} f_{i,j}$$

is said to be a *Min-Max function.*

We begin with a local result.

Lemma 5.1. *Let (M_n, ρ) be a non-positively curved manifold. Then for any $p, q \in M_n$ and any integer $i > 0$ the germ of d_p^i at q is a germ of a Min-Max function.*

Proof.

1) Let $\pi : (\tilde{M}_n, \tilde{\rho}) \to (M_n, \rho)$ be the universal covering, choose $\tilde{q} \in \pi^{-1}(q)$ and let $\{\tilde{p}_1, \ldots, \tilde{p}_m, \ldots\} = \pi^{-1}(p)$.
2) Then for any $y \in M_n$ there exists positive constants K, ϵ such that d_p satisfies the following property:
$$d_p^i(y) = \max_{P \in \mathcal{T}_{i-1}} \min_{j \notin P} \tilde{d}_j(\tilde{y}),$$
where $y \in D_\epsilon(q)$, $\mathcal{T}_{i-1}$ is the set of all subsets of $\{1, \ldots, K\}$, consisting of $i - 1$ elements. □

Such a representation (for any germ of d_p^i) also gives the global result.

Proposition 5.2. *Let (M_n, ρ) be a non-positively curved compact manifold. Then d_p^i is a Max-Min function on M_n.*

One can also define i-th distance function, $\rho^i : M_n \times M_n \to \mathbb{R}^1$.

Definition 5.4 (i-th metric). We define the *i-th metric* as

$$\rho^i(p, q) = d_p^i(q).$$

Remarks.

1) Evidently, the function ρ^i is symmetric, that is, $\rho^i(x, y) = \rho^i(y, x)$.

2) $\rho^i(x,x)$ is the sequence of lengths of geodesic loops at x.

Proposition 5.3. *Let M_n be a non-positively curved compact manifold. Then ρ^i is a Min-Max function on $M_n \times M_n$.*

6. Degenerations (estimates from above)

Let M_n be an n-manifold; degenerations of a distance function d_p at q are connected with a large number (more than $n+1$) of shortest geodesics connecting pairs $p, q \in M_n$; the codimension of degeneration grows when the number of "additional" shortest geodesics increases. The upper bound for $m_p(q)$ is determined by the geometric characteristics of M_n. Hence only finitely many codimensions of degeneration are admitted for Riemannian distance functions, in contrast with classical Morse theory where there exists an infinite series of degenerations ordered by the codimension of degeneration.

Proposition 6.1. *Let (M_n, ρ) be a compact negatively curved manifold such that* $\mathrm{Sect}(M_n) \geq -C$, $r_{inj} \geq r$, *and* $\mathrm{diam}(M_n) \leq d$. *Then*

$$m_p(q) \leq \frac{\mu_C D^C_{2d}}{\mu_0 D^0_{\frac{r}{2}}}.$$

Proof.

1) Let $p, q \in M_n$,

$$\pi : (\tilde{M}_n, \tilde{\rho}) \to (M_n, \rho)$$

be the universal covering, let $\tilde{q} \in \pi^{-1}(q)$ and let $\tilde{p}_1, \ldots, \tilde{p}_{m_p} \in \pi^{-1}(p)$ be the points such that $\tilde{\rho}(\tilde{p}_i, \tilde{q}) = \rho(p,q)$, $i = 1, \ldots, m$.

2) Note that $\rho(p,q) \leq d = \mathrm{diam}(M_n)$ so that $\tilde{p}_1, \ldots, \tilde{p}_m \in D_d(\tilde{q})$.

3) We have $\mu D_d(p) \leq \mu_C D^C_d$ and $\mu_0 D^0_r \leq \mu D_r(\tilde{p}_i)$.

4) The balls $D_{\frac{r}{2}}(\tilde{p}_i)$, $i = 1, \ldots, m$ are disjoint and

$$\bigcup_{i=1}^{m_p(q)} D_{\frac{r}{2}}(\tilde{p}_i) \subset D_{d+\frac{r}{2}}(p) \subset D_{2d}(p). \tag{5}$$

5) A comparison of volumes of balls appearing on both sides of (5) gives

$$m_p(q) \leq \frac{\mu_C D^C_{2d}}{\mu_0 D^0_{\frac{r}{2}}}.$$

□

7. Degenerations (estimates from below)

Lower bounds for the number of shortest geodesics were obtained in [H].

Theorem 7.1 ([H]). *Let (M_n, ρ) be a non-positively curved manifold, $p \in M_n$. Then there exists $q \in M_n$ such that $m_p(q) \geq n+1$.*

Theorem 7.2 ([H]). *Let (M_n, ρ) be a strictly negatively curved manifold. Then there exist $p, q \in M_n$ such that $m_p^\rho(q) \geq 2n+1$.*

These theorems are very natural in the context of the theory which we develop in this paper. The first theorem simply states that there exist $n+1$ shortest geodesics, connecting p and q, for a point q where the distance function d_p has a local maximum. This estimate is sharp when d_p is Morse, in which case the index d_p at q is n and there exist precisely $n+1$ shortest geodesics connecting p and q. When q is a degenerate critical point (a degenerate local maximum) the number of shortest geodesics is greater than $(n+1)$.

The second theorem can be discussed from two different (but related) points of view.

Remark 7.1. The first point of view is that one has to consider distance functions d_p for all $p \in M_n$. The number of "additional" shortest geodesics connecting p and q is the codimension of degeneration at a critical point q. One can perturb the Riemannian metric slightly to obtain Morse distance function d_p for a given p. For some Riemannian manifolds (such as the torus T^2 with the flat metric) one can perturb the metric $\rho \to \tilde{\rho}$ in such a way that the distance functions $d_p^{\tilde{\rho}}$ are Morse for all $p \in M_2$.

For certain Riemannian manifolds this is impossible. One can obtain a Morse distance function d_p by a small perturbation of a Riemannian metric, for a given p, but not for all $p \in M_2$ simultaneously. When one considers the distance functions d_x for all $x \in M_n$ one can expect a degeneration of codimension n at a finite number of points (under the condition of jet transversality). However, some manifolds could admit smaller degenerations. Theorem 7.2 states that for any *strictly* negatively curved manifold there exists at least one point with a degeneration of codimension n. Horja, see [H], proved that for any negatively curved n-manifold there exists a pair of points connected by $2n+1$ shortest geodesics. Such points are local maxima of the distance function $\rho(x, y)$ on $M_n \times M_n$; in particular, this is always the case for the global maximum (p, q) for the metric $\rho : M_n \times M_n$, (where $\rho(p, q) = \operatorname{diam} M_n$).

Remark 7.2. The second point of view is that we have to consider a Morse metric $\rho : M_n \times M_n \to \mathbb{R}^1$. Then ρ has index $2n$ at a local maximum (p, q) and there are $2n+1$ shortest geodesics connecting p and q.

From this point of view, the explanation of the phenomenon for the torus is that the metric ρ such that the distance functions d_p are Morse for all $p \in M_2$ is not a

Morse metric. The maximum of $\rho : M_2 \times M_2 \to \mathbb{R}^1$ is attained on a two-dimensional submanifold of $T^2 \times T^2$ and this case can be analyzed in the style of Bott's generalization of Morse theory, see [Bo]). (This is a usual phenomenon for smooth functions: consider the function $f = (x+y)^2$ on $\mathbb{R}^2$ which is not Morse, but the restriction of f to any straight line parallel to one of the coordinate axes is a Morse function.)

II. Distance functions on surfaces

It is not easy to describe the locus of distance functions in the space of Min-type functions. In particular, the following statement cannot be deduced from the corresponding result for Min-type functions.

Theorem II.1. *For a generic non-positively curved metric on a surface M_2 the distance functions d_p are Morse for almost all $p \in M_2$.*

We prove this result for non-positively curved surfaces. We study the two dimensional case since ideas here can be presented in a very clear form. The multidimensional case will be discussed in forthcoming papers.

8. Degenerate points on surfaces

Definition 1.8 of non-degenerate points contains two conditions. The following lemma shows that the second condition holds for all distance functions on non-positively curved surfaces and all points with multiplicity ≤ 3.

Lemma 8.1 ([El1]). *Let M_2 be a surface equipped with a Riemannian metric ρ and let $q, x, y, z \in M_2$ be such that $\rho(q,x) = \rho(q,y) = \rho(q,z)$, and $x \neq y$, $y \neq z$, $x \neq z$. Then $(\operatorname{grad} d_x - \operatorname{grad} d_y)$ and $(\operatorname{grad} d_y - \operatorname{grad} d_z)$ are linearly independent at every point $q \in M_2$.*

Proof ([El1].) The three tangent vectors $\operatorname{grad} d_x(q)$, $\operatorname{grad} d_y(q)$ and $\operatorname{grad} d_z(q)$ have unit norm and hence cannot lie on an affine line. □

This gives the following description of degenerate regular points of the distance functions on non-positively curved surfaces.

Proposition 8.1 (degenerate regular points). *Let (M_2, ρ) be a non-positively curved compact surface and let $p, q \in M_2$. A point q is a degenerate regular point of d_p iff $m_p(q) \geq 4$ and the shortest geodesics $\gamma_1, \ldots, \gamma_{m_p(q)}$ connecting p and q satisfy the following property: There exists an open half-space $T_q^+ M_2 \subset T_q M_2$ such that $\dot\gamma_1(q), \ldots, \dot\gamma_{m_p(q)}(q) \in T_q^+ M_2$.*

There are the following two types of degenerate critical points.

Definition 8.1 (degenerate critical points with high multiplicity). Let (M_2, ρ) be a non-positively curved compact surface and let $p, q \in M_2$. A point q is a *degenerate critical point of the first type* for d_p if $m_p(q) \geq 4$ and the shortest geodesics $\gamma_1, \dots, \gamma_{m_p(q)}$ connecting p and q satisfy the following property: $\mathrm{Conv}\{\dot{\gamma}_1(q), \dots, \dot{\gamma}_{m_p(q)}(q)\} \ni O$.

Degenerate critical points of the second type are critical points with multiplicity three. A point q with $m_p(q) = 3$ is a degenerate critical point if (after a suitable renumbering of the shortest geodesics $\gamma = \gamma_1, \gamma_2, \gamma_3$ connecting p and q) $\dot{\gamma}_2(q)$ and $\dot{\gamma}_3(q)$ are linearly dependent, i.e. $-\dot{\gamma}_2(q) = \dot{\gamma}_3(q)$. Then γ_2 and γ_3 are two segments of a geodesic κ which is either self-intersecting at p or periodic. Denote $d = \rho(p, q)$. Therefore (we omit the subscript for the first geodesic)

$$\kappa(0) = \kappa(2d) = \gamma(0) = p \text{ and } \kappa(d) = \gamma(d) = q, \tag{6}$$

and we obtain the following description.

Definition 8.2 (degenerate critical points with multiplicity three). Let (M_2, ρ) be a non-positively curved compact surface and let $p, q \in M_2$. A point q is said to be a *degenerate critical point of the second type* for d_p if $m_p(q) = 3$ and there exist geodesics γ, κ satisfying (6), and such that $\dot{\kappa}(d)$ and $\dot{\gamma}(d)$ are not perpendicular.

Remark. We include an additional "generic" condition that $\dot{\kappa}(d)$ and $\dot{\gamma}(d)$ are not perpendicular. The set of geodesics which have degenerate critical point of index 3, such that κ and γ are perpendicular at q has codimension 2. The decomposition of such a singularity is more complicated than for degenerate points of the second type.

9. Generic distance functions on surfaces (multiplicity)

Fix a point $p \in M_2$. We demonstrate that for a generic metric on M_2, $m_p(q) \leq 3$ for all $q \in M_2$.

Definition 9.1 (multiple points). Let (M_2, ρ) be a non-positively curved surface, $p \in M_2$. A point q is said to be *p-multiple* if $m_p(q) \geq 3$.

Proposition 9.1. *Let (M_2, ρ) be a surface with a non-positively curved metric. Then there exists only a finite number of p-multiple points.*

Proof. A point q is p-multiple iff

$$d_p^1(q) = d_p^2(q) = d_p^3(q), \tag{7}$$

that is, $\tilde{d}_{\tilde{p}_1}(\tilde{q}) = \tilde{d}_{\tilde{p}_2}(q) = \tilde{d}_{\tilde{p}_3}(\tilde{q})$. Equations (7) define a finite set of vertices of the Dirichlet domain, compare [E]; (these vertices are p-multiple points on M_2). □

Remark. Eberlein's result holds for any metric ρ without conjugate points on M_2. However it may happen that $m_p^\rho(q) > 3$ for some points. The rest of this section is devoted to a construction of a small perturbation $\bar{\rho}$ of the metric ρ to obtain $m_p^{\bar{\rho}}(q) \leq 3$ for all $q \in M_2$. Each point q with $m_p(q) \geq 4$ can be decomposed into several multiple points $q_1, \dots, q_L$ with $m_p(q_i) = 3$, $i = 1, \dots, L$. Additional multiple points arise only due to this splitting.

Proposition 9.2. *Let M_2 be a compact surface, let $p \in M_2$ and let ρ be a non-positively curved generic metric. Then for all $q \in M_n$ we have $m_p(q) \leq 3$.*

The proposition follows from the following three lemmas.

Denote by $Q^p = \{q_1, \dots, q_l\}$ the set of multiple points for p and let Q_ϵ^p be the ϵ-neighbourhood of Q^p.

Lemma 9.1. *Let (M_2, ρ) be a non-positively curved compact surface and let $p \in M_2$. Then every sufficiently small perturbation $\bar{\rho}$ of ρ does not create p-multiple points outside Q_ϵ^p.*

Proof. We have $d_p^3 - d_p^1 > \delta(\epsilon) > 0$ on $M_2 \setminus Q_\epsilon$, and then the same inequality holds for any metric $\bar{\rho}$ sufficiently close to ρ.

Let $\gamma_1^{p,q_i}, \dots, \gamma_{m_p(q_i)}^{p,q_i}$ be the shortest geodesics connecting p and q_i, $i = 1, \dots, l$. Let $\mathcal{G}$ be the union of all shortest geodesics connecting p and all q_i, $i = 1, \dots, l$, except γ_1^{p,q_1}; and let $\mathcal{G}_\epsilon$ be the ϵ-neighbourhood of $\mathcal{G}$.

Then $\gamma_1^{p,q_1} \not\subset \mathcal{G}_\epsilon$, for sufficiently small ϵ and one can choose $x \in \gamma_1^{p,q_1} \setminus \mathcal{G}_\epsilon$. Let $\bar{g}$ be a small perturbation of the metric tensor localized in $D_\epsilon(x)$. More precisely we choose $\bar{g}$ such that the following four conditions are satisfied:

1) $\operatorname{supp}(g - \bar{g}) \subset D_{\frac{\epsilon}{2}}(x)$; (we denote by $\operatorname{supp}(g - \bar{g})$ the support of the tensor $(g - \bar{g})$);
2) C^∞-norm of $(g - \bar{g})$ is less than ϵ;
3) $(g - \bar{g})$ is positive semi-definite at each point;
4) $\bar{g}(x) = \left(1 - \frac{\epsilon}{2}\right) g(x)$.

□

Lemma 9.2. *The perturbation $g \to \bar{g}$ defined above does not create new multiple critical points outside a small ball $D_\delta(q_1)$.*

Proof.
1) New multiple points do not arise outside of Q_ϵ (see Lemma 9.1).
2) Any shortest ρ-geodesic connecting p with a point $y \in D_\epsilon(q_j)$, $j \geq 2$, does not intersect $D_\epsilon(x)$ for sufficiently small positive ϵ.

3) $\bar{\rho}$-shortest geodesics connecting p and q_i, $i = 2, \ldots, l$, do not intersect the ball $D_{\frac{\varepsilon}{2}}(x)$ for sufficiently small perturbation $g \to \bar{g}$. □

In fact we also proved the following result.

Lemma 9.3. *The p-multiple points $q_2, \ldots, q_l$ do not move, disappear, or change their multiplicities under the perturbation $\rho \to \bar{\rho}$.*

Lemma 9.4. *The p-multiplicity $m_p^{\bar{\rho}}(y)$ of any point $y \in D_\delta(q_1)$ is less than or equal to $m_p^{\rho}(q_1) - 1$.*

Proof.

1) For the metric ρ and sufficiently small δ we have

$$d_p^{m_p(q_1)+1}(y) - d_p^1(y) > \delta > 0$$

for all $y \in U_\delta(q)$. Thus the $\bar{\rho}$-multiplicity of any point $y \in D_\delta(q)$ cannot be greater than $m_p^{\rho}(q_1)$.

2) We have $d_{p_1}^{\bar{\rho}}(q_1) > d_{p_2}^{\bar{\rho}}(q_1) = \cdots = d_{p_{m_p^{\rho}(q_1)}}^{\bar{\rho}}$, hence $m_p^{\bar{\rho}}(q_1) = m_p^{\rho}(q_1) - 1$.

3) The p-multiple points are isolated according to [E]. Therefore for all q near q_1 we have $m_p^{\rho}(q) \leq 2$. For any point $q \in D_\delta(q_1) \setminus q_1$ we have $m_p^{\bar{\rho}}(q) \leq m_p^{\rho}(q) + 1$ since we have changed only one shortest geodesic. Hence $m^{\bar{\rho}}(q) \leq 3$ for $q \in D_\delta(q_1) \setminus q_1$. □

Remarks.

1) The number of non-degenerate critical points, obtained as a result of the splitting above, can be estimated roughly; the process of splitting can be presented in the following form: a point q_1 with multiplicity $m_p(q_1)$ can be split into not more than $m_p(q_1)$ points of multiplicity $(m_p(q_1) - 1)$. Some of these points are split into not more than $m_p(q_1) - 1$ points with multiplicities $m_p(q_1) - 2$, *etc.*

2) Precise estimates must take into account indices of non-degenerate multiple points ultimately obtained as a result of the splitting.

3) When we start with a degenerate regular point we obtain a collection of non-degenerate regular points, compare [GR]. Generally when we start from a degenerate critical point we obtain a number of non-degenerate regular points and non-degenerate critical points. Some of the non-degenerate critical points form pairs of non-degenerate critical points with indices differing by one. Such pairs make no contribution in changing the topological type of $f^{-1}(-\infty, c)$. This leads to the concept of "generalized" index for degenerate critical points and clarifes relations between the classical definition of degenerate points and Gromov's definition.

10. Generic distance functions (degenerate points of multiplicity three)

The results of the previous section allow us to construct a small perturbation of the Riemannian metric such that for a given p all points $q \in M_2$ have p-multiplicities ≤ 3. To construct a small perturbation of the metric which eliminates degenerate critical points with multiplicity three, we need two simple statements related to self-intersecting and periodic geodesics for generic metrics on non-positively curved manifolds.

Proposition 10.1. *Let M_2 be a surface of genus $g \geq 2$ and let $p \in M_2$. For a non-positively curved metric ρ on M_2 the set of geodesics self-intersecting at p is discrete.*

Proof. Let $\{\kappa_i\}_{i=1}^{\infty}$ is an infinite series of geodesics on M_2 self-intersecting at p and converging to κ_∞. We have $\dot\kappa_i \to \dot\kappa_\infty$ and length $\kappa_i \to$ length κ_∞. Choose their parametrization in such a way that $\kappa_i(0) = p$ and $| \dot\kappa_i | \equiv 1$. Denote by d_∞ the length of κ_∞ Then the point p is self-conjugate along κ_∞, (that is there exists a non-zero Jacobi field $J(t)$ along $\kappa_\infty(t)$ vanishing for $t = 0$ and for $t = d_\infty$. Since the surface is non-positively curved this is impossible. □

Corollary 10.1. *Let M_2 be a surface of genus $g \geq 2$ and let $p \in M_2$. The set of geodesics self-intersecting at p with length bounded from above is finite.*

In a similar way one can prove the following statement.

Proposition 10.2. *For a non-positively curved metric on M_2 the set of periodic geodesics γ such that $\gamma(0) = p$ is discrete.*

Lemma 10.1. *Let $p, q \in M_2$, let $m_p(q) = 3$ and let q be a degenerate critical point for d_p^ρ. Then there exists a small perturbation $\rho \to \bar\rho$ of the Riemannian metric such that this degenerate point disappears (and this perturbation does not create new p-multiple points).*

Proof.

1) Since q is a degenerate critical point, there exists a geodesic κ, such that $\kappa(0) = \kappa(2d) = p$, $\kappa(d) = q$ and $|\dot\kappa| \equiv 1$; κ is either self-intersecting at p or periodic; in addition there exists a shortest geodesic γ connecting p and q, such that $\gamma(0) = p$, $\gamma(d) = q$ and $|\dot\gamma| \equiv 1$.
2) Let κ_ϵ be an ϵ-neighbourhood of κ. One can choose a point $x \in \gamma \setminus \kappa_\epsilon$.
3) There are no other geodesics $\bar\kappa$, $\bar\kappa(0) = p$, self-intersecting at p (or periodic) in a certain neighbourhood of κ.

4) Let us choose a point $x \in \gamma \setminus \kappa_\epsilon$ and take a C^∞-small perturbation of the metric tensor $g \to \bar{g}$ with a support of $(g - \bar{g})$ inside an ϵ-ball $D_\epsilon(x)$; we assume that the perturbation of the metric tensor satisfies all conditions stated in lemma 9.2.
5) New multiple points cannot arise on any other geodesic with a self-intersection at p. □

When κ and γ are not orthogonal at q we obtain the following decomposition of a degenerate critical point.

Lemma 10.2. *Suppose d_p has a non-degenerate critical point of the second type at q. Then there exists a small perturbation of the metric tensor $g \to \bar{g}$, localized in a small neighbourghood of a point $x \in \gamma$, such that one obtains a non-degenerate critical point q of index one for $d_p^{\bar{\rho}}$ and a new regular point q_ϵ with $m_p^{\bar{\rho}}(q_\epsilon) = 2$; ($q_\epsilon$ is the intersection of the geodesic κ and the perturbed geodesic $\bar{\gamma}$).*

11. Proof of Theorem II.1

A simple version of the theorem on generic negatively curved metrics is the following.

Theorem 11.1. *Let M_2 be a surface with a metric ρ of non-positive curvature. Then there exists a small perturbation $\rho \to \bar{\rho}$ such the the set $\mathcal{P} = \{p \in M_2 \mid d_p \text{ is Morse}\}$ is open and everywhere dense in M_2.*

Remark. When we considered perturbations of the metric tensor $g \to \bar{g}$ (to make a distance function d_p Morse), these perturbations were concentrated around certain points on geodesics connecting p with p-multiple points. It is easy to check that if we fix a small positive δ, we can choose all perturbations of the metric tensor in such a way that $\operatorname{supp}(\bar{g} - g) \subset D_\delta(p)$.

Proof. Denote the initial metric tensor by g_0.
1) Choose a countable, everywhere dense subset $\{x_1, \ldots, x_m, \ldots\} \subset M_2$ and a small positive ϵ.
2) There exists a C^∞-small perturbation of the metric tensor $g_0 \to g_1$, such that $(g_0 - g_1)$ is supported inside $D_\epsilon(x_1)$, and makes d_{x_1} Morse.
3) Replace $\epsilon \to \frac{\epsilon}{2}$ and continue this process in such a way that we make d_{x_2} Morse using $\epsilon/2$-small C^∞-perturbation $g_1 \to g_2$ such that $\operatorname{supp}(g_2 - g_1) \subset D_{\frac{\epsilon}{2}}(x_2)$ and we preserve the Morse property of d_{x_1} (to conserve Morse properties of d_{x_1} we decrease ϵ if necessary).
4) Continue this process to obtain in the limit a C^∞-metric tensor g_∞ such that all the d_{x_i}'s are Morse. □

A more interesting version of this theorem is the following.

Theorem 11.2. *Let M_n be a surface, and let ρ be a non-positively curved metric on M_2. Then there exists a small perturbation $\rho \to \tilde{\rho}$ such the the set $\mathcal{Q} = \{p \in M_2 \mid d_p$ is not Morse $\}$ has codimension one in M_2.*

A proof of this theorem is based on the technique developed in the next chapter.

III. Singularity theory for distance functions on surfaces

In this chapter we discuss two related problems:

1) Metrics ρ with minimal possible degenerations of distance functions d_p at all points $p \in M_2$.
2) Morse metrics on $M_2 \times M_2$.

12. Morse-homogeneous metrics on T^2

Definition 12.1 (Morse-homogeneous metrics). A metric on M_n is said to be *Morse homogeneous* if distance functions d_p^ρ are Morse for all $p \in M_n$.

Remark. When a metric ρ is Morse-homogeneous, the distance functions d_p have the same number of critical points for all $p \in M_n$; denote by $\{x_1^p, \dots, x_l^p\}$ the set of critical points for d_p. The numbering of these points can be chosen in such a way that all maps $p \to x_i^p$ are continuous and $\operatorname{ind}_{x_i^p} d_p = \operatorname{ind}_{x_i^q} d_q$ for all $p, q \in M_n$.

It appears that such metrics can be constructed on the 2-torus. Let us present T^2 as a rectangle $[0, 1] \times [0, 1]$ with opposite sides identified; choose $p = (0, 0)$ and $q = (1/2, 1/2)$. For the standard Riemannian metric on T^2 we have $m_p(q) = 4$. Note that the distance function d_p is Morse on $T^2 \setminus \{q\}$ for the standard flat metric on T^2. We show that there exists a small perturbation $\tilde{\rho}$ of ρ such that $\tilde{\rho}$ is Morse homogeneous and $\max_{p,q \in T^2} m_p(q) = 3$.

Consider the universal covering $(\mathbb{R}^2, \tilde{\rho}) \to (T^2, \rho)$, where $\tilde{\rho}$ is the Euclidean metric on $\mathbb{R}^2$. Choose an orthogonal basis $\xi_1 = (1, 1)$ and $\xi_2 = (-1, 1)$ of $\mathbb{R}^2$. Choose a small positive δ and perturb the metric $\tilde{\rho} \to \tilde{\rho}^\delta$ on $\mathbb{R}^2$ in such a way that (ξ_1^δ, ξ_2), where $\xi_1^\delta = (1 - \delta)\xi_1$ is an orthonormal basis on $\mathbb{R}^2$.

Proposition 12.1. *The metric ρ^δ constructed above is Morse homogeneous on T^2 and* $\max_{p,q \in T^2} m_p(q) = 3$.

Proof. Let $o = (0, 0)$. The function d_o on T^2 has six critical points. The point o has index 0. Points $a = (0, 1/2), b = (1/2, 0), c = (1/2, 1/2)$ are non-degenerate critical points with index one. There are two non-degenerate critical points with index two $d = ((1-\delta)/2, (1-\delta)/2)$ and $e = ((1+\delta)/2, (1+\delta)/2)$.

For each $q \in T^2$, the distance function d_q has six non-degenerate critical points (obtained by a shift by q from the corresponding critical points of the function d_o). □

13. Generic metrics on negatively curved surfaces

Properties of generic strictly negatively curved metrics on surfaces are described by the following statements.

Proposition 13.1. *Let M_2 be a surface of genus ≥ 2. Then $\left(\max_{p,q\in M_2} m_p(q)\right) = 5$ for a generic strictly negatively curved metric on M_2.*

Comments. We demonstrate in following sections that a generic strictly negatively curved metric is Morse (as a function on $M_2 \times M_2$). Points of local maxima of this function have index 4, hence they are connected by five shortest geodesics.

Proof of the proposition. The proof is similar to the proof of corresponding results in Chapter II. We present an outline of the proof.

1) Points (p, q) with $m_p(q) \geq 5$ are isolated in $M_2 \times M_2$, see [H].
2) Some of these points can have multiplicity ≥ 6. Let $m_p(q_1) \geq 6$ and let $\gamma_1^1, \ldots, \gamma_{m_p(q_1)}^1$ be the shortest geodesics connecting p and q_1.
3) Consider all other points $q_2, \ldots, q_l$ with $m_p(q_j) \geq 5$ and all shortest geodesics $\gamma_1^j, \ldots, \gamma_{m_p(q_j)}^j$, $j = 2, \ldots, l$ connecting points q_j with p. Take the union of all these geodesics except γ_1^1 as $\mathcal{G}$ and denote the ϵ-neighbourhood of $\mathcal{G}$ by $\mathcal{G}_\epsilon$.
4) Choose a point $x \in \gamma_1^1 \setminus \mathcal{G}_\epsilon$. Then a small perturbation of the metric tensor inside $D_\epsilon(x)$ decreases the p-multiplicity at q_1. □

Horja ([H]) proved that points $p, q \in M_2$ such that (p, q) is a local maximum of the metric $\rho : M_2 \times M_2 \to \mathbb{R}^1$ are connected by at least 5 shortest geodesics. The previous proposition shows that for a generic metric such points are connected by exactly five shortest geodesics.

For saddle points of ρ we obtain the following statements, see Section 4.

Proposition 13.2 (index three). *Let M_2 be a compact surface and let ρ be a Morse negatively curved metric on M_2. Then there exist points $p, q \in M_2$ connected by four shortest geodesics and such that p is a local maximum for d_q and q is a local maximum for d_p (but (p, q) is not a local maximum for ρ).*

In the same way one can prove the following result.

Proposition 13.3 (index two). *Let M_2 be a compact surface and let ρ be a Morse negatively curved metric on M_2. Then there exist points $p, q \in M_2$ connected by three shortest geodesics and such that p is a local maximum for d_q and q is a local maximum for d_p (but (p, q) is not a local maximum for ρ).*

These propositions are based on the fact that a generic negatively curved metric is Morse.

Theorem 13.1. *The set of Morse negatively curved metrics on M_2 is open and everywhere dense in the space of all negatively curved metrics on M_2.*

Definition 13.1. We call a metric *almost non-positively curved* if $\operatorname{Sect} M_n < \epsilon \operatorname{diam}(M_n)^2$.

For almost non-positively curved manifolds, there are conjugate points along some geodesics. However the distance up to the nearest conjugate point is far bigger than the diameter of the manifold. Results listed in Chapter I for negatively curved manifolds can be generalized to almost non-positively curved manifolds, compare [GR].

Conjecture. The statement of the preceding theorem holds for any almost non-positively curved metric (in particular for small perturbations of the flat metric on T^2).

Remark. The author believes that the theorem can be generalized to an arbitrary dimension (without any restrictions on the sectional curvature).

Proof of Proposition 13.2.

1) There exists a critical point (p, q) of index three for any Morse function on $M_2 \times M_2$.
2) There exists four geodesics $\gamma_1, \ldots, \gamma_4$ connecting p and q.
3) We have

$$\operatorname{Conv}\{\dot\gamma_1(p), \ldots, \dot\gamma_4(p)\} = T_p M_2;$$

and

$$\operatorname{Conv}\{\dot\gamma_1(q), \ldots, \dot\gamma_4(q)\} = T_q M_2.$$

Then p is a local maximum for d_q, and q is a local maximum for d_p.

4) (p, q) is not a local maximum for ρ since the index of (p, q) is three. □

Theorem 13.1 states that there are no points of multiplicity ≥ 6 for generic metrics. We have prove that there are no degenerate points of multiplicity two. To obtain the necessary result we have to prove following three statements.

Lemma 13.1. *There are no degenerate points of multiplicity three for generic metrics.*

Lemma 13.2. *There are no degenerate points of multiplicity four for generic metrics.*

Lemma 13.3. *There are no degenerate points of multiplicity five for generic metrics.*

The remainder of the paper is devoted to a proof of these statements.

14. Points of multiplicity three

Proof of Lemma 13.1. Let $\gamma_1, \gamma_2, \gamma_3$ be the shortest geodesics connecting p and q.

1) The only possible degeneration for a point of multiplicity three is linear dependence of tangent vectors $(\dot{\gamma}_1(p), \dot{\gamma}_1(q))$ and $(\dot{\gamma}_2(p), \dot{\gamma}_2(q))$ in $T_{p,q}$ $(M_2 \times M_2)$ (after a suitable renumbering of geodesics). The second type of degeneration is impossible due to Lemma 8.1.
2) Let q be a degenerate critical point of multiplicity three for d_p. Then there exists a periodic geodesic κ (consisting on two segments γ_1 and γ_2) such that

$$\kappa(0) = \kappa(2d) = p, \quad \kappa(d) = q, \quad |\dot{\kappa}| = 1$$

and the shortest geodesic γ connecting p and q

$$\gamma(0) = p, \quad \gamma(d) = q, \quad |\dot{\gamma}| = 1.$$

3) The periodic geodesic κ is isolated (there exists only a finite number of periodic geodesics on a negatively curved manifold whose lengths are bounded from above by a given constant (in particular, with the length smaller than or equal to the diameter of M_2) is finite. Then a C^∞-small perturbation of the metric tensor inside a small ball centered at a point $x \in \gamma \setminus \kappa$ eliminates this degenerate critical point and does not create new degenerate points of multiplicity three. □

We prove several statements related to points of multiplicity three which are interesting in themselves and useful in our struggle with degenerate critical points of multiplicity four in the next section.

Definition 14.1 ($\mathcal{P}_3$). Define the set $\mathcal{P}_3$ of points $(p, q) \in M_2 \times M_2$ such that p and q are connected by three shortest geodesics $\gamma_1^{p,q}, \gamma_2^{p,q}, \gamma_3^{p,q}$, which satisfy the following conditions:

1) any two vectors among $\dot{\gamma}_1^{p,q}(p), \dot{\gamma}_2^{p,q}(p)\dot{\gamma}_3^{p,q}(p)$ are linearly independent;
2) any two vectors among $\dot{\gamma}_1^{p,q}(q), \dot{\gamma}_2^{p,q}(q)\dot{\gamma}_3^{p,q}(q)$ are linearly independent.

Proposition 14.1. *$\mathcal{P}_3$ is a two dimensional submanifold of $M_2 \times M_2$.*

Proof.

1) Let $(p, q) \in \mathcal{P}_3$; we show that for any x sufficiently close to p there exists y close to q such that x and y are connected by three shortest geodesics.

2) Choose $\tilde{q} \in \pi^{-1}(q)$; let $\pi^{-1}p = \{\tilde{p}_1, \tilde{p}_2, \ldots, \}$; perturb slightly $p \to o$. For r sufficiently close to q choose $\tilde{r} \in \pi^{-1}r$ close to $\tilde{q}$, then we have $\pi^{-1}(o) = \{\tilde{o}_1, \ldots, \tilde{o}_m, \ldots\}$, where $\tilde{o}_i$ is close to $\tilde{p}_i$ for $i = 1, 2, 3$.
3) Denote by $\Omega(\tilde{p}_1, \tilde{p}_2) \subset \tilde{M}_2$ the set of points equidistant from both $\tilde{p}_1$ and $\tilde{p}_2$.
4) $\Omega(\tilde{p}_1, \tilde{p}_2)$ and $\Omega(\tilde{p}_2, \tilde{p}_3)$ have a transversal intersection at $\tilde{q}$, hence the sets $\Omega(\tilde{o}_1, \tilde{o}_2)$ and $\Omega(\tilde{o}_2, \tilde{o}_3)$ has a transversal intersection at a unique point $\tilde{r}$ near $\tilde{q}$.
5) This gives a map $U_\epsilon(p) \to U_\epsilon(q)$; the graph of this map gives a germ of $\mathcal{P}_3$ at (p, q). □

Definition 14.2. We distinguish four types of points in $\mathcal{P}_3$:

1) $\mathcal{P}_3^{cc} = \{(p, q) \in \mathcal{P}_3 \mid p \text{ is critical for } q \text{ and } q \text{ is critical for } p\}$;
2) $\mathcal{P}_3^{cr} = \{(p, q) \in \mathcal{P}_3 \mid p \text{ is critical for } q \text{ and } q \text{ is regular for } p\}$;
3) $\mathcal{P}_3^{rc} = \{(p, q) \in \mathcal{P}_3 \mid p \text{ is regular for } q \text{ and } q \text{ is critical for } p\}$;
4) $\mathcal{P}_3^{rr} = \{(p, q) \in \mathcal{P}_3 \mid p \text{ is regular for } q \text{ and } q \text{ is regular for } p\}$.

Proposition 14.2. *Each of the four sets $\mathcal{P}_3^{cc}, \mathcal{P}_3^{cr}, \mathcal{P}_3^{rc}, \mathcal{P}_3^{rr}$ consists of several connected components of $\mathcal{P}_3$.*

Proof. We prove this only for $\mathcal{P}_3^{cc}$ since the proof is similar for the other sets. Let $(p, q) \in \mathcal{P}_3^{cc}$. Then q is a non-degenerate local maximum for d_p and p is a non-degenerate local maximum for d_q; hence for $\bar{p}$ sufficiently close to p, the function $d_{\bar{p}}$ has a unique non-degenerate local maximum $\bar{q}$ in a small neighbourhood of q, and visa versa. □

For every Morse-homogeneous metric we obtain the following result.

Proposition 14.3. *Let ρ be a Morse-homogeneous metric on M_2. Then $\mathcal{P}_3^{cc} \subset M_2 \times M_2$ is a closed submanifold, and each of the projections $(p, q) \mapsto p$ and $(p, q) \mapsto q$ defines a diffeomorphism of a connected component of $\mathcal{P}_3^{cc}$ onto M_2.*

Critical points of ρ of multiplicity three are in $\mathcal{P}_3^{cc}$; let us describe such points. Let $(p, q) \in \mathcal{P}_3^{cc}$; denote by $\gamma_1^{p,q}, \gamma_2^{p,q}, \gamma_3^{p,q}$ the shortest geodesics connecting p and q. Consider a map $\Phi : \mathcal{P}_3^{cc} \to T^2$ defined as follows. Denote the angle coordinates on T^2 by $\phi_{1,2}$ and $\phi_{2,3}$; define

$$\phi_{i,i+1}(p) = \text{angle}\big(\dot{\gamma}_i(p), \dot{\gamma}_{i+1}(p)\big), \quad \phi_{i,i+1}(q) = \text{angle}\big(\dot{\gamma}_i(q), \dot{\gamma}_{i+1}(q)\big),$$
$$\phi_{i,i+1} = \phi_{i,i+1}(p) - \phi_{i,i+1}(q)$$

for $i = 1, 2$. (The angles are assumed to be oriented and their orientation is defined by the orientation of the surface M_2.) Finally, define $\Phi(p, q) = (\phi_{1,2}, \phi_{2,3})$. Let

$O = (0, 0)$. A comparison with the definition of critical points gives the following result.

Lemma 14.1. *The points $(p, q) \in \mathcal{P}_3^{cc}$ with $\Phi(p, q) = O$ are non-degenerate critical points of index two for the metric ρ.*

15. Technical lemmas

We need several technical lemmas which we use in our study of points of multiplicity four and five in the next two sections.

Lemma 15.1. *Let M_n be a negatively curved manifold. Suppose there exist two curves $\alpha, \beta : (-\delta, \delta) \to M_n$ such that for each τ the points $\alpha(\tau), \beta(\tau)$ are connected by exactly m shortest geodesics. Then such geodesics form m continuous families, that is, there exist m continuous maps $\Gamma_i : (-\delta, \delta) \times [0, d] \to M_2$, $i = 1, \dots, m$, such that for each $\tau \in (-\delta, \delta)$ the curve $\gamma_i(t) = \Gamma_i(\tau, t)$ is a shortest geodesic connecting $\alpha(\tau)$ and $\beta(\tau)$.*

Proof. Consider the universal covering $\pi : (\tilde{M}_n, \tilde{\rho}_n) \to (M_n, \rho_n)$. Choose a family of points $\tilde{q}^\tau \in \pi^{-1}(\beta(\tau))$ to be continuously dependent on τ. Let$\{\tilde{p}_1^\tau, \dots, \tilde{p}_k^\tau, \dots\} = \pi^{-1}(\alpha(\tau))$. Any pair of points $\alpha(\tau), \beta(\tau)$ are connected by exactly m shortest geodesics, hence the maps $\tau \to \tilde{p}_i^\tau$ can be chosen in such a way that they define continuous curves for $i = 1, \dots, m$. (This defines continuous families of geodesics Γ_i, for $i = 1, \dots, m$). □

We recall the Carathéodory theorem.

Theorem 15.1 (Carathéodory). *Let $a_1, \dots, a_m \in \mathbb{R}^n$, $n < m$, are such that* $\operatorname{Conv}\{a_1, \dots a_m\} \ni O$. *Then there exist $a_{i_1}, \dots, a_{i_{n+1}}$ such that*

$$\operatorname{Conv}\{a_{i_1}, \dots, a_{i_{n+1}}\} \ni O.$$

We need two similar lemmas:

Lemma 15.2. *Let L be a linear space and suppose $v_1, \dots, v_m \in L$ satisfy the following two conditions:*

a) *$v_1, \dots, v_{m-1}$ are linearly dependent;*

b) $\operatorname{Conv}(v_1, \dots, v_m) \ni O$.

Then there exist $(m - 1)$ vectors $v_{i_1}, \dots, v_{i_{m-1}}$ such that

$$\operatorname{Conv}\{v_{i_1}, \dots, v_{i_{m-1}}\} \ni O.$$

Lemma 15.3. *Let L be a linear space; suppose $\nu_1, \dots, \nu_m \in L$ are on an affine hyperplane and*

$$\mathrm{Conv}\{\nu_1, \dots, \nu_m\} \ni O.$$

Then there exist $\nu_{i_1}, \dots, \nu_{i_{m-1}}$ which are linearly dependent.

16. Points of multiplicity four

Denote by $\mathcal{P}_4$ the set of points $(p, q) \in M_2 \times M_2$ such that p and q are connected by (at least) four shortest geodesics.

There are two types of degeneration for critical points of multiplicity four, see Section 4.

a) A point with the first type of degeneration is a point $(p, q) \in M_2 \times M_2$ where for three of the four shortest geodesics connecting p and q (denote these geodesics by $\gamma_1, \gamma_2, \gamma_3$), the vectors

$$\left(\dot\gamma_1(p),\ \dot\gamma_1(q)\right),\ \left(\dot\gamma_2(p),\ \dot\gamma_2(q)\right),\ \left(\dot\gamma_3(p),\ \dot\gamma_3(q)\right) \in T_{p,q}(M_2 \times M_2)$$

are linearly dependent.

b) A point with the second type of degeneration is a point (p, q) where vectors

$$\left(\dot\gamma_1(p),\ \dot\gamma_1(q)\right), \left(\dot\gamma_2(p),\ \dot\gamma_2(q)\right), \left(\dot\gamma_3(p),\ \dot\gamma_3(q)\right), \left(\dot\gamma_4(p),\ \dot\gamma_4(q)\right)$$

are in a two dimensional affine plane in $T_{p,q}(M_2 \times M_2)$.

Similar to Section 14 one can prove the following result.

Lemma 16.1. *Points $(p, q) \in \mathcal{P}_4$ satisfying the following two properties are isolated:*

a) *any two vectors $(\dot\gamma_i(p), \dot\gamma_i(q))$, $(\dot\gamma_j(p), \dot\gamma_j(q)) \in T_{p,q}(M_2 \times M_2)$ are linearly independent.*

b) $\mathrm{Conv}\,\{(\dot\gamma_i(p), \dot\gamma_i(q))\}_{i=1}^{3} \ni O.$

Such a degenerate critical point can be eliminated by a small perturbation of a riemannian metric tensor inside a small ball centered at a point of γ_4. Simple exercises in linear algebra, presented in the previous section, allows us to obtain the necessary result for all degenerate critical points of multiplicity four. Lemma 15.2 gives the following result.

Lemma 16.2. *Let (p, q) be a degenerate critical point of the first type of multiplicity four. Then either one can enumerate shortest geodesics connecting (p, q) in such a way that they satisfy the condition of Lemma 16.1 or two of these geodesics are segments of a periodic geodesic.*

Proposition 16.1. *Let (M_2, ρ) be a negatively curved surface such that there are no degenerate critical points for ρ with multiplicity three. Then there exists a small perturbation $\rho \to \bar{\rho}$ which eliminates all degenerate critical points of the first type with multiplicity four.*

Proof.
1) If two of the shortest geodesics are segments of a periodic geodesic, one can decrease the number of shortest geodesics connecting p and q as in Section 14.
2) If the shortest geodesics satisfy the condition of Lemma 16.1, it is sufficient to take a small perturbation of the metric tensor at a point of γ_4. □

Remark. A perturbation in the previous statement can be chosen so small that it does not create new degenerate critical points with multiplicities three and four.

The second type of degenerations consists of critical points where

$$\left\{\dot{\gamma}_i^{p,q}(p),\ \dot{\gamma}_i^{p,q}(q)\right\} - \left\{\dot{\gamma}_4^{p,q}(p),\ \dot{\gamma}_4^{p,q}(q)\right\}, \quad i = 1, 2, 3$$

are linearly dependent.

The lemma (which follows from Lemma 15.3) shows that this case can be reduced, in fact, to the previous one.

Lemma 16.3. *Let (p, q) be a degenerate critical point of the second type of multiplicity four. Then one can enumerate shortest geodesics connecting (p, q) in such a way that they satisfy the condition of Lemma* 16.1.

17. Points of multiplicity five

Proposition 17.1. *Let (M_2, ρ) be a negatively curved surface. Suppose $\rho : M_2 \times M_2$ has no degenerate critical points with multiplicity $\neq 5$. Then there exists a small perturbation of the metric tensor (localized around a finite number of points of M_2) which eliminates all degenerate critical points of multiplicity five (and does not create new multiple points).*

The scheme of the proof is the same as in the previous section.

Remark. The proof for degenerate critical points of multiplicity five can be simplified if one uses results of [H], which state that points connected by five shortest geodesics are isolated. The argument presented here works in fact in a more general situation.

18. Concluding remarks

We discuss possible generalizations of this theory and its applications. More precisely one can discuss generalizations and applications of two connected theories presented in this paper.

I.1 Morse theory and singularity theory for Min-type functions. Natural classes of non-smooth functions arise in "smooth" problems of optimization, calculus of variations, mathematical economics and so on. Some (but not all) classes are connected with operations of Maximum and Minimum (and their superpositions). The Morse theory for Min-type functions that can be obtained as minima of finite families of smooth functions was constructed in [GR]. Stable singularities for functions obtained by Max and Min operations from continuous families of smooth functions were classified in [Br1], [Br2], [Ma1], [Ma2]. (The Morse theory was not constructed for this class, but it can be done easily.)

Singularity theory for the class of Max-Min functions functions (which arose in this paper) is far more difficult, see a discussion in [CGR1], but it seems possible that it can be constructed.

These results can be applied in non-smooth optimization, [CGR1], mathematical economics in the style of S. Smale, and variational problems, see [A].

I.2 Morse theory for Riemannian distance function. This theory must be (and it probably can be) generalized to general Riemannian manifolds; it would be interesting to extend it also for some other categories of manifolds, in particular, to sub-Riemannian manifolds.

The most natural and clear applications relate to classification of stable singularities of spheres (and geodesic spheres) in Riemannian manifolds (when the the radius of a sphere is greater than the injectivity radius at the center) and singularities of sub-Riemannian spheres (with all positive radii, the injectivity radius here is zero at all points, see [Ge1]). This theory can also be useful in some open problems such as the problem of contractibility of small balls in sub-Riemannian manifolds.

More problematic, but potentially very interesting, is to use this theory for a deeper understanding of the difference between positively and non-negatively curved manifolds (this includes a number of problems such as Hopf's problem of the existence of a positively curved metric on $S^2 \times S^2$, and Gromov's conjecture that the maximum of sums of Betti numbers for non-negatively curved n-manifolds is attained on the flat n-torus (the last problem is pointed out by M. Gromov)).

The following conjecture can clarify the connection of these problems to Morse theory for Riemannian distance functions.

Gromov proved, [Gr], that there exists a constant c_n depending only on dimension, such that the sum of Betti numbers for any non-negatively curved n-manifold M_n is bounded from above by c_n. For such a manifold denote by $C(M_n)$ the minimal possible number of critical points of a Morse distance function on M_n for a non-negatively curved Riemannian metric.

The problem is: *Does there exist C_n such that $C(M_n) \leq C_n$* (for all M_n admitting a non-negatively curved metric)?

References

[A] V. I. Arnold, Catastrophe theory, Springer-Verlag, 1986.

[AGV] V. Arnold, A. Gussein-Zade, and A. Varchenko, Singularities of differentiable maps, vv. 1, 2, Nauka, Moscow, 1982.

[Bo] R. Bott, Morse theory old and new, Bull Amer. Math. Soc. 7 (1982), 331–358.

[Br1] L. N. Bryzgalova, Singularities of maximum of parametrically dependent functions, Functional Anal. Appl. 11 (1977), 49–51.

[Br2] L. N. Bryzgalova, On maximum of parametrically dependent functions, Functional Analysis and Applications, 12(1977), 66–67.

[B] Bucher, Stability of the cut locus in dimension less than or equal to 6, Invent. Math. 43 (19), 199–231.

[Ch] J. Cheeger, Critical points of distance functions and application to geometry, in: Geometric Topology (DeBartolomeis, P., Tricerri, P., ed.), Lecture Notes in Math. 1504, Springer-Verlag, 1990, 1–38.

[CGR1] B. Craven, V. Gershkovich, and D. Ralph, Morse Theory and Relations between Smooth and Non-smooth Optimization, Proc. of Optimization Day, Univ. of Ballarat, 1994.

[CGR2] B. Craven, V. Gershkovich, and D. Ralph, Morse theory and invariant concepts of convexity, Optimization 35 (1995), 99–112.

[CGR3] B. Craven, V. Gershkovich, and D. Ralph, On invariant concept of convexity for vector functions, Res. Report, Melbourne Univ, 1995.

[E] P. Eberlein The Cut Locus of Non-compact Finitely Connected Surface without Conjugate Points, Comment. Math. Helv. 51 (1976), 23–44.

[EI1] P.E. Erlich and H. C. Im Hof, Dirichlet regions without conjugate points, Comment. Math. Helv. 54 (1979), 642–658.

[EI2] P.E. Erlich and H. C. Im Hof, Metric Circles and Bisectors, Math. Z. 159 (1978), 101–105.

[Ge] V. Gershkovich, Generic heat evolution, Res. Rep. 2 (1997), Dept. of Math., Melbourne University, to appear in Russian J. Math. Phys.

[Ge1] V. Gershkovich, Sub-Riemannian manifolds as limits of Riemannian Manifolds, I. Curvature, diameter, and volume, to appear in Russian J. Math. Physics.

[GR] V. Gershkovich and H. Rubinstein, Morse theory for distance functions on non-positively curved manifolds. I. Distance functions as Min-type functions, Res. Rep. 12 (1985), Dept. of Math., Melbourne University.

[GG] M. Golubitski and V. Guillemin, Stable mappings and their singularities, Springer-Verlag, 1973.

[Gr] M. Gromov, Curvature, diameter, and Betti numbers, Comment. Math. Helv. 56 (1989), 179–195.

[Gro] K. Grove, Critical Point theory for Distance functions, Proc. Sympos. Pure Math. 54:3 (1993), 357–385.

[GS] K. Grove and K. Shiohama, A generalized sphere theorem, Ann. of Math. (2) 106, (1977), 206–211.

[H] P. Horja, On the number of geodesic segments connecting two points on manifold of non-positive curvature, preprint.

[Ma1] V. Matov, Singularities of functions of maximum on manifolds with boundaries, Trudy Sem. Petrovsk. 6 (1981), 195–221.

[Ma2] V. Matov, Topological classification of germs of maxima and minima of generic families of functions, Uspekhi Mat. Nauk 37:4 (1982), 129–130 (in Russian).

[Mi1] J. Milnor, Morse theory, Princeton Univ. Press, 1963.

[Mi2] J. Milnor, Lectures on h-cobordism theorem, Princeton Univ. Press, 1965.

[U1] K. Uhlenbeck, Generic properties of eigenfunctions, Amer. J. Math. 98 (1976), 1059–1076.

[U2] K. Uhlenbeck, Eigenfunctions of Laplace Operators, Bull. Amer. Math. Soc. 78 (1972), 1073–1076.

[W1] A. Weinstein, Generic conjugate loci, Notes Summer AMS Inst., 1968.

[W2] A. Weinstein, Generic cut locus and conjugate locus of a Riemannian manifold, Ann. of Math. 87 (1968), 29–41.

The Gromov–Lawson–Rosenberg conjecture: when do manifolds admit metrics of positive scalar curvature

Peter B. Gilkey

1991 Mathematics Subject Classification: 53C21, 58G25.

1. Introduction

In this expository note, we report on joint work with B. Botvinnik and S. Stolz [5, 6] which uses algebraic topology and the theory of elliptic partial differential equations to prove a theorem in differential geometry. Let M be a closed manifold of dimension $m \geq 5$ with finite cyclic fundamental group $\mathbb{Z}_\ell$ whose universal cover is spin. If M is non orientable, we also assume m is odd. We use the eta invariant (which is an analytic invariant) and equivariant twisted spin bordism and connective K-theory (which are topological invariants) to show that M admits a metric of positive scalar curvature (which is a question in differential geometry) if and only if a generalized index of the Dirac operator $\alpha(M)$ vanishes. This settles the Gromov–Lawson–Rosenberg conjecture in this special case.

A manifold M admits a spin structure if and only if the second Stiefel–Whitney class $w_2(M) = 0$. If $w_2(M) = 0$ and if M admits a metric of positive scalar curvature, then the Bochner–Lichnerowicz [11] formula shows that there are no harmonic spinors on M. If m is divisible by 4, this implies that the index of the Dirac operator D on M vanishes and consequently the A-roof genus $\hat{A}(M) = 0$. This integer valued invariant is computable in terms of characteristic classes and is independent of the metric g on M and of the spin structure s on M.

Let D be the Dirac operator. If $m \equiv 1 \bmod 8$, let $\hat{A}(M, s) \in \mathbb{Z}_2$ be the mod 2 reduction of $\dim(\ker(D))$. If $m \equiv 2 \bmod 8$, then $\dim\ker(D)$ is even and we let $\hat{A}(M, s) \in \mathbb{Z}_2$ be the mod 2 reduction of $\frac{1}{2}\dim\ker(D)$. Note that $\hat{A}(M, s)$ depends on the spin structure s but not on the Riemannian metric if $m \equiv 1, 2 \bmod 8$. We set $\hat{A}(M) = 0$ if m is congruent to 3, 5, 6, or 7 mod 8. If M is simply connected, the spin structure is unique and we drop the dependence upon s; in general, however, there can be inequivalent spin structures.

Since $\hat{A}(M, s) = 0$ if M admits a metric of positive scalar curvature, the A-roof genus is an obstruction to the existence of a metric of positive scalar curvature on M. Stolz [16] proved the converse holds if M is simply connected and if $m \geq 5$.

Theorem 1.1. *Let M be a closed simply connected spin manifold of dimension $m \geq 5$. Then M admits a metric of positive scalar curvature if and only if $\hat{A}(M) = 0$.*

If the fundamental group π of a spin manifold M is non-trivial, Rosenberg [13] has defined an element $\alpha(M)$ generalizing the A-roof genus which takes values in the K-theory of the reduced group C^* algebra $C_r^*(\pi)$. If M is not spin but the universal cover of M is spin, then the alpha invariant can be extended suitably. If M admits a metric of positive scalar curvature, then $\alpha(M) = 0$. The Gromov–Lawson–Rosenberg conjecture is that this is the only obstruction to the existence of a metric of positive scalar curvature in dimensions $m \geq 5$. We refer to Rosenberg and Stolz [14] for a general discussion of this conjecture.

Conjecture 1.2. *Let M be a closed connected manifold of dimension $m \geq 5$ with fundamental group π whose universal cover $\tilde{M}$ is a spin manifold. Then M admits a metric of positive scalar curvature if and only if $\alpha(M)$ vanishes.*

In this note, we report on recent joint work with Botvinnik and Stolz [5, 6] on Conjecture 1.2 in the case where $\pi = \mathbb{Z}_\ell$ is a finite cyclic group. Our results for m even and M not orientable are incomplete; this case is currently under further investigation.

Theorem 1.3. *Let M be a closed connected manifold of dimension $m \geq 5$ with fundamental group $\mathbb{Z}_\ell$ whose universal cover $\tilde{M}$ is a spin manifold.*

(a) *If $w_1(M) = 0$ and if $w_2(M) = 0$, then M admits a metric of positive scalar curvature if and only if $\hat{A}(M, s) = 0$ for every spin structure s on M.*

(b) *If $w_1(M) = 0$ and if $w_2(M) \neq 0$, then M admits a metric of positive scalar curvature if and only if $\hat{A}(\tilde{M}) = 0$.*

(c) *If $w_1(M) \neq 0$, if $w_2(M) = 0$, and if m is odd, then M admits a metric of positive scalar curvature if and only if $\alpha(M) = 0$.*

(d) *If $w_1(M) \neq 0$, if $w_2(M) \neq 0$, and if m is odd, then M always admits a metric of positive scalar curvature.*

We can work one prime at a time to prove Theorem 1.3. The case ℓ odd follows from results of Kwasik and Schultz [10] and Stolz [16]; we also refer to Botvinnik et al. [4]. The case $\ell = 2$ was handled by Rosenberg and Stolz [14]. We therefore assume $\ell = 2^q$ for $q > 1$ henceforth.

The proof of Theorem 1.3 uses a surprising amount of algebraic topology (equivariant bordism and connective K-theory) and elliptic operator theory (the eta invariant).

In §2, we shall give an equivalent formulation of Theorem 1.3 in terms of equivariant bordism and connective K-theory and we shall complete the proof of Theorem 1.3 if m is even and if $w_1(M) = 0$ or if m is odd and if $w_1(M) \neq 0$. In §3, we will introduce the eta invariant and show that this analytic invariant extends to bordism and to connective K-theory. We will compute the eta invariant of spherical space forms and of spherical space form bundles. We will complete the proof of Theorem 1.3 in the remaining case that m is odd and that M is orientable.

2. Equivariant bordism and connective K-theory

Let $B\mathbb{Z}_\ell$ be the classifying space for the cyclic group $\mathbb{Z}_\ell := \{\lambda \in \mathbb{C} : \lambda^\ell = 1\}$ and let $f : M \to B\mathbb{Z}_\ell$ define a $\mathbb{Z}_\ell$ structure on M. If M is connected, f is equivalent to giving a representation of the fundamental group of M to $\mathbb{Z}_\ell$. If $\pi_1(M) = \mathbb{Z}_\ell$, take the identity representation. Let ξ be a real vector bundle over the classifying space $B\mathbb{Z}_\ell$. We consider a triple (M, s, f) where $f : M \to B\mathbb{Z}_\ell$ and where s is a spin structure on $T(M) \oplus f^*\xi$. The twisted bordism group $\Omega(B\mathbb{Z}_\ell, \xi)$ is defined by introducing the equivalence relation $(M_1, s_1, f_1) \sim (M_2, s_2, f_2)$ if there exists a compact Riemannian manifold N with boundary $M_1 \dot{\sqcup} -M_2$ so that the structures extend to N. Since only the values of $u_i := w_i(\xi) \in H^i(B\mathbb{Z}_\ell, \mathbb{Z}_2) = \mathbb{Z}_2$ for $i = 1, 2$ are relevant, we will also denote these groups by $\Omega_m(B\mathbb{Z}_\ell, u_1 = \cdot, u_2 = \cdot)$.

The following theorem provides the crucial link between differential geometry and algebraic topology. It reduces the question of constructing a metric of positive scalar curvature to a question in equivariant bordism. We refer to [9, 15] for the proof of (a) and to [12, 13] for the application of (a) to bordism which is contained in (b).

Theorem 2.1.

(a) *Let M be a compact manifold, which is not necessarily connected, which admits a Riemannian metric of positive scalar curvature. Then any manifold which can be obtained from M by performing surgeries in codimension at least* 3 *also admits a metric with positive scalar curvature.*

(b) *Let M be a connected manifold of dimension $m \geq 5$ with cyclic fundamental group $\mathbb{Z}_\ell$. Let $[(M, s, f)] = [(M_1, s_1, f_1)]$ in $\Omega_m(B\mathbb{Z}_\ell, \xi)$. If M_1 admits a metric of positive scalar curvature, then so does M.*

Let $\mathbb{H}P^2$ denote quaternion projective space with the usual homogeneous metric. Let $\mathbb{H}P^2 \to E \to B$ be a fiber bundle where the transition functions are the group of isometries $PSp(3)$ of $\mathbb{H}P^2$. Since $\mathbb{H}P^2$ is simply connected, the projection $p : E \to B$ induces an isomorphism of fundamental groups; any $\mathbb{Z}_\ell$ structure on E arises from a $\mathbb{Z}_\ell$ structure on B. Let $T_m(B\mathbb{Z}_\ell, \xi)$ be the subgroup of $\Omega(B\mathbb{Z}_\ell, \xi)$ generated by manifolds E arising in this fashion. Let $\mathrm{ko}_m(B\mathbb{Z}_\ell, \xi)$ be the twisted connective K-theory groups. We refer to Stolz [17] for the proof of the following result.

Theorem 2.2. $\mathrm{ko}(B\mathbb{Z}_\ell, \xi)_{(2)} \cong \{\Omega(B\mathbb{Z}_\ell, \xi)/T(B\mathbb{Z}_\ell, \xi)\}_{(2)}$.

Let $\Omega^+(B\mathbb{Z}_\ell, \xi)$ denote the subgroup of $\Omega(B\mathbb{Z}_\ell, \xi)$ generated by manifolds which admit metrics of positive scalar curvature. Let $D : \Omega_m(B\mathbb{Z}_\ell, \xi) \to \mathrm{ko}_m(B\mathbb{Z}_\ell, \xi)$ be the natural projection from bordism to connective K-theory. Since the standard metric on $\mathbb{H}P^2$ has positive scalar curvature, the total space E of such a fibration admits a metric of positive scalar curvature; see Besse [2, (9.70)] for details. Thus $T_m(B\mathbb{Z}_\ell, \xi) \subset \Omega_m^+(B\mathbb{Z}_\ell, \xi)$. Since the α invariant vanishes on $\Omega_m^+(B\mathbb{Z}_\ell, \xi)$, it vanishes on $T_m(B\mathbb{Z}_\ell, \xi)$ and extends to K-theory. Let

$$\mathrm{ko}_m^+(B\mathbb{Z}_\ell, \xi) := D(\Omega_m^+(B\mathbb{Z}_\ell, \xi)) \subseteq \ker(\alpha) \cap \mathrm{ko}_m(B\mathbb{Z}_\ell; \xi).$$

Since the groups in question are 2-primary, we can suppress localization at the prime 2 and use the geometric characterization of Theorem 2.2. The following reformulation of the Gromov–Lawson–Rosenberg conjecture in terms of bordism and connective K-theory is now immediate.

Theorem 2.3. *Let $m \geq 5$. The following assertions are equivalent.*

(a) *Conjecture 1.2 holds for the fundamental group $\pi = \mathbb{Z}_\ell$, for the dimension m, and for the twisting bundle ξ.*

(b) $\Omega_m^+(B\mathbb{Z}_\ell, \xi) = \ker(\alpha) \cap \Omega_m(B\mathbb{Z}_\ell, \xi)$.

(c) $\mathrm{ko}_m^+(B\mathbb{Z}_\ell, \xi) = \ker(\alpha) \cap \mathrm{ko}_m(B\mathbb{Z}_\ell, \xi)$.

The connective K-theory groups $\mathrm{ko}(\cdot)$ are much smaller and easier to work with than are the bordism groups $\Omega(\cdot)$ so (c) is a significant improvement over (b) in Theorem 2.3. What is important for our purposes is that the order of $\ker(\alpha) \cap \mathrm{ko}_m(B\mathbb{Z}_\ell, \xi)$ can be computed using the Adams spectral sequence, which is a tool of algebraic topology. Let $0 \neq x$ generate $H^1(B\mathbb{Z}_\ell; \mathbb{Z}_2) = \mathbb{Z}_2$ and let $0 \neq y$ generate $H^2(B\mathbb{Z}_\ell; \mathbb{Z}_2) = \mathbb{Z}_2$. We refer to [5, 6] for the proof of the following result.

Theorem 2.4. *Let $\epsilon(m, \xi) := |\ker(\alpha) \cap \mathrm{ko}_m(B\mathbb{Z}_\ell, \xi)|$. Let $\vec{u} = (w_1(\xi), w_2(\xi))$.*

(a) *If m is even and if $w_1(\xi) = 0$ or if m is odd and if $w_1(\xi) \neq 0$, $\epsilon(m, \xi) = 1$.*

(b) *If $\vec{u} = (0, 0)$, $\epsilon(8k + 1, \xi) = (\ell/2)^{2k+1}$, $\epsilon(8k + 3, \xi) = 2(2\ell)^{2k+1}$,*

$$\epsilon(8k + 5, \xi) = (\ell/2)^{2k+2},\ \epsilon(8k + 7, \xi) = (2\ell)^{2k+2}.$$

(c) If $\vec{u} = (0, y)$, $\epsilon(8k + 1, \xi) = (2\ell)^{2k+1}$, $\epsilon(8k + 3, \xi) = (\ell/2)^{2k+1}$,

$$\epsilon(8k + 5, \xi) = (2\ell)^{2k+2},\ \epsilon(8k + 7, \xi) = (\ell/2)^{2k+2}.$$

(d) If $\vec{u} = (x, 0)$, $\epsilon(8k+2, \xi) = \epsilon(8k+4, \xi) = \epsilon(8k+6, \xi) = \epsilon(8k+8, \xi) = 2^{2k+2}$.

(e) If $\vec{u} = (x, y)$, $\epsilon(8k, \xi) = \epsilon(8k + 2, \xi) = 2^{2k+1}$,

$$\epsilon(8k + 4, \xi) = \epsilon(8k + 6, \xi) = 2^{2k+2}.$$

Since $\ker(\alpha) \cap \mathrm{ko}_m(B\mathbb{Z}_\ell, \xi) = 0$ if $w_1(\xi) = 0$ and if m is even or if $w_1(\xi) \neq 0$ and if m is odd, Theorem 1.3 follows in these two cases trivially. The case $w_1(\xi) = 0$ and m odd comprises the remainder of this paper; the case $w_1(\xi) \neq 0$ and m even is still under investigation.

3. The eta invariant

Let $\mathbb{Z}_\ell = \{\lambda \in \mathbb{C} : \lambda^\ell = 1\}$ be the cyclic group of order $\ell = 2^q$ for $q > 1$. Let ξ_0 be the trivial line bundle over $B\mathbb{Z}_\ell$ and let ξ_1 be underlying real 2-plane bundle of the classifying line bundle over $B\mathbb{Z}_\ell$. Then $\vec{u}(\xi_0) = (0, 0)$ and $\vec{u}(\xi_1) = (0, y)$. Note that $\Omega_m(B\mathbb{Z}_\ell) = \Omega_m(B\mathbb{Z}_\ell, \xi_0)$ and $\mathrm{ko}_m(B\mathbb{Z}_\ell) = \mathrm{ko}_m(B\mathbb{Z}_\ell, \xi_0)$. We restrict to $\xi = \xi_0$ or $\xi = \xi_1$ and assume that m is odd henceforth. We refer to [8] for details concerning the analytic facts used in this section.

If $f : M \to B\mathbb{Z}_\ell$ gives M a $B\mathbb{Z}_\ell$ structure and if ρ is a representation of $\mathbb{Z}_\ell$, let $\rho(M, f)$ be the associated flat vector bundle over M. If $[(M, s, f)] \in \Omega_m(B\mathbb{Z}_\ell, \xi)$, then M admits a natural spin^c structure with flat associated determinant line bundle. Let g be a Riemannian metric on M and let $P_\rho = P(s, f, g, \rho)$ denote the tangential operator of the spin^c complex on M with coefficients in $\rho(M, f)$. Let

$$\eta(P_\rho) := \tfrac{1}{2}\{\mathrm{Tr}_{L^2}(P_\rho(P_\rho^2)^{-(s+1)/2}) + \dim\ker(P_\rho)\}|_{s=0} \in \mathbb{R}$$

be the eta invariant of Atiyah, Patodi, and Singer [1]. Since $\eta(P_\rho)$ is additive in ρ, it extends to the group representation ring $R(\mathbb{Z}_\ell)$. Let $R_0(\mathbb{Z}_\ell)$ denote the augmentation ideal of virtual representations of virtual dimension 0.

Theorem 3.1. *Let $\xi = \xi_0$ or let $\xi = \xi_1$.*

(a) *If $m \equiv 3 \bmod (4)$ and if $\rho \in R_0(\mathbb{Z}_\ell)$, then the map $(M, \cdot) \to \eta(P_\rho)$ extends to homomorphisms $\eta_\rho : \Omega_m(B\mathbb{Z}_\ell, \xi) \to \mathbb{R}/\mathbb{Z}$ and $\eta_\rho : \mathrm{ko}_m(B\mathbb{Z}_\ell, \xi) \to \mathbb{R}/\mathbb{Z}$.*

(b) *If $m \equiv 1 \bmod (4)$ and if $\rho \in R(\mathbb{Z}_\ell)$, then the map $(M, \cdot) \to \eta(P_\rho)$ extends to homomorphisms $\eta_\rho : \Omega_m(B\mathbb{Z}_\ell, \xi) \to \mathbb{R}/\mathbb{Z}$ and $\eta_\rho : \mathrm{ko}_m(B\mathbb{Z}_\ell, \xi) \to \mathbb{R}/\mathbb{Z}$.*

(c) *If $m \equiv 3 \bmod (8)$ and if $\rho = \rho_0 - \rho_{\ell/2}$, then the map $(M, \cdot) \to \eta(P_\rho)$ extends to homomorphisms $\eta_\rho : \Omega_m(B\mathbb{Z}_\ell) \to \mathbb{R}/2\mathbb{Z}$ and $\eta_\rho : \mathrm{ko}_m(B\mathbb{Z}_\ell) \to \mathbb{R}/2\mathbb{Z}$.*

Proof. We first show η extends to bordism. Suppose that $(M, s, f) = \partial(W, s_W, f_W)$. This means that W is a compact Riemannian manifold of dimension $m+1$ with smooth boundary M and that the structures on M extend over W. Let D_ρ be the operator of the spin^c complex on W with coefficients in $\rho(W, f_W)$. Since the determinant line bundle of the spin^c structure on W is flat, the Atiyah–Patodi–Singer index theorem for manifolds with boundary yields

$$\mathrm{index}(D_\rho) = \dim(\rho)\textstyle\int_W \hat{A}(W) - \eta(P_\rho).$$

If $\dim(\rho) = 0$, then $\eta(P_\rho) = -\mathrm{index}(D_\rho) \in \mathbb{Z}$ which proves the first assertion. If $m \equiv 1 \bmod (4)$, the interior integral vanishes for dimensional reasons and $\eta(P_\rho) = -\mathrm{index}(D_\rho) \in \mathbb{Z}$ and assertion (b) follows. If ρ is of real type and if $\xi = \xi_0$, then the spin^c structure is a spin structure. If $m \equiv 3 \bmod (8)$, then the spin bundle with coefficients in ρ admits a quaternion structure; D_ρ is $\mathbb{H}$ linear so index(D_ρ) is even and (c) follows. This proves the assertions concerning bordism. To show that η extends to connective K-theory, it suffices to show that η vanishes on geometric fiber bundles $\mathbb{H}P^2 \to E \to B$. If $\xi = \xi_0$, we are dealing with a spin structure and this follows from the Adiabatic limit theorem of Bismut and Cheeger [3]; if $\xi = \xi_1$, we can give a direct argument; see [6]. □

We define linear representations of $\mathbb{Z}_\ell$ by setting $\rho_\mu(\lambda) = \lambda^\mu$. Let $\vec{a} = (a_1, \ldots, a_k)$ be a collection of odd indices. Let $\rho_{\vec{a}} = \rho_{a_1} \oplus \cdots \oplus \rho_{a_k} : \mathbb{Z}_\ell \to U(k)$. This representation is fixed point free and we let $L^{2k-1}(\ell; \vec{a}) := S^{2k-1}/\rho_{\vec{a}}(\mathbb{Z}_\ell)$ be the resulting quotient manifold. Let $H^{\otimes 2}$ be the tensor square of the Hopf line bundle over the sphere S^2. Let $\mathcal{S}(\mathcal{W}_k)$ be sphere bundle of the complex vector bundle $\mathcal{W}_k := H^{\otimes 2} \oplus (k-1)\epsilon$ over S^2. We extend $\rho_{\vec{a}}$ to act without fixed points on the fiber spheres of $\mathcal{S}(\mathcal{W}_k)$ and define $X^{2k+1}(\ell; \vec{a}) := \mathcal{S}(\mathcal{W}_k)/\rho_{\vec{a}}(\mathbb{Z}_\ell)$.

Definition 3.2. Let $\lambda \in \mathbb{Z}_\ell$ and let $\lambda \neq 1$. Let $|\vec{a}| = a_1 + \cdots + a_k$.

(a) If k is even, let $\mathcal{G}_L(\vec{a})(\lambda) := \lambda^{|\vec{a}|/2} \det(I - \rho_{\vec{a}}(\lambda))^{-1}$.

(b) If k is odd, let $\mathcal{G}_L(\vec{a})(\lambda) := \lambda^{(|\vec{a}|+1)/2} \det(I - \rho_{\vec{a}}(\lambda))^{-1}$.

(c) Let $\mathcal{G}_X(\vec{a})(\lambda) = \mathcal{G}_L^k(\vec{a})(\lambda)(1 + \lambda^{a_1})/(1 - \lambda^{a_1})$.

(d) Let $\tilde{\Sigma}_\lambda = \Sigma_{\lambda^\ell = 1, \ell \neq 1}$ and $\Sigma_\lambda = \Sigma_{\lambda^\ell = 1}$.

Lemma 3.3. *Give $L^m(\ell; \vec{a})$ and $X^m(\ell; \vec{b})$ the natural $\mathbb{Z}_\ell$ structures.*

(a) *If $m \geq 3$, $L^m(\ell; \vec{a})$ and $X^m(\ell; \vec{b})$ admit metrics of positive scalar curvature.*

(b) *If k is even, then $w_2(L^{2k-1}(\ell; \vec{a})) = 0$ and $w_2(X^{2k+1}(\ell; \vec{a})) = 0$.*

(c) *If k is odd, then $w_2(L^{2k-1}(\ell; \vec{a})) \neq 0$ and $w_2(X^{2k+1}(\ell; \vec{a})) \neq 0$.*

(d) $\eta(L^{2k-1}(\ell; \vec{a}))(\rho) = \ell^{-1} \tilde{\Sigma}_\lambda \operatorname{Tr}(\rho(\lambda)) \mathcal{G}_L(\vec{a})(\lambda)$.

(e) $\eta(X^{2k-1}(\ell; \vec{a}))(\rho) = \ell^{-1} \tilde{\Sigma}_\lambda \operatorname{Tr}(\rho(\lambda)) \mathcal{G}_X(\vec{a})(\lambda)$.

(f) *If $\rho \in R_0(\mathbb{Z}_\ell)^{k+1}$, then $\eta(L^{2k-1}(\ell; \vec{a}))(\rho) \in \mathbb{Z}$ and $\eta(X^{2k-1}(\ell; \vec{a}))(\rho) \in \mathbb{Z}$.*

Proof. The first three assertions are immediate from the definition. Donnelly [7] has generalized the Atiyah–Patodi–Singer index theorem [1] to the equivariant setting. Let the sphere S^{2k-1} bound the disk D^{2k} and extend the $\mathbb{Z}_\ell$ action to D^{2k} with an isolated fixed point at the origin; (d) now follows from Donnelly's work. Let the sphere bundle $\mathcal{S}(\mathcal{W}_k)$ bound the disk bundle $\mathcal{D}(\mathcal{W}_k)$ and extend the $\mathbb{Z}_\ell$ action to $\mathcal{D}(\mathcal{W}_k)$ with fixed point set S^2; (e) now follows. Let $\rho \in R_0(\mathbb{Z}_\ell)^{k+1}$. If a is odd, then $R_0(\mathbb{Z}_\ell) = (\rho_0 - \rho_a)R(\mathbb{Z}_\ell)$. Let $M = L^{2k-1}(\ell; \vec{a})$ or $M = X^{2k-1}(\ell; \vec{a})$. Then there exists $\gamma \in R_0(\mathbb{Z}_\ell)$ so that $\mathrm{Tr}(\gamma(\lambda)) = \mathcal{G}_M(\vec{a})(\lambda)\,\mathrm{Tr}(\rho(\lambda))$ for $\lambda \neq 1$. Then

$$\eta(M)(\rho) = \ell^{-1}\tilde{\Sigma}_\lambda \mathrm{Tr}(\gamma(\lambda)) = \ell^{-1}\Sigma_\lambda \mathrm{Tr}(\gamma(\lambda)) \in \mathbb{Z}.$$

□

Definition 3.4. In the free Abelian group generated by lens spaces and lens space bundles, define:

(a) $\mathcal{B} : L^m(\ell; \vec{a}) \to L^{m+4}(\ell; \vec{a}, 1, 1) - 3L^{m+4}(\ell; \vec{a}, 1, 3)$.

(b) $\mathcal{B} : X^m(\ell; \vec{a}) \to X^{m+4}(\ell; \vec{a}, 1, 1) - 3X^{m+4}(\ell; \vec{a}, 1, 3)$.

(c) $K_L^5 := L^5(\ell; 1, 1, 1) - 3L^5(\ell; 1, 1, 3)$ and $K_L^{4j+1} := \mathcal{B}K_L^{4j-3}$ for $j > 1$.

(d) $K_X^5 := X^5(\ell; 1, 1) - 3X^5(\ell; 1, 3)$ and $K_X^{4j+1} = \mathcal{B}K_X^{4j-3}$ for $j > 1$.

Lemma 3.5. *Let* $\epsilon = \rho_{-1}(\rho_0 - \rho_1)^2$ *and let* $j \geq 1$. *Then*

(a) *There exists* $\rho \in R(\mathbb{Z}_\ell)$ *so* $\eta(L^5(\ell; 1, 1, 1))(\epsilon\rho)$ *has order* 2ℓ *in* $\mathbb{R}/\mathbb{Z}$.

(b) *There exists* $\rho \in R(\mathbb{Z}_\ell)$ *so* $\eta(X^5(\ell; 1, 1))(\epsilon\rho)$ *has order* $\ell/2$ *in* $\mathbb{R}/\mathbb{Z}$.

(c) *If* $\rho \in R(\mathbb{Z}_\ell)$, *then* $\eta(K_L^{4j+1})(\epsilon\rho) = 0$ *and* $\eta(K_X^{4j+1})(\epsilon\rho) = 0$.

(d) *There exists* $\gamma_{L,j} \in R(\mathbb{Z}_\ell)$ *so that* $\eta(K_L^{4j+1})(\gamma_{L,j})$ *has order* 2ℓ *in* $\mathbb{R}/\mathbb{Z}$.

(e) *There exists* $\gamma_{X,j} \in R(\mathbb{Z}_\ell)$ *so that* $\eta(K_X^{4j+1})(\gamma_{X,j})$ *has order* $\ell/2$ *in* $\mathbb{R}/\mathbb{Z}$.

(f) $\eta(L^3(1, 1) - 3L^3(1, 3))(\rho_0 - \rho_{\ell/2}) \neq 0$ *in* $\mathbb{R}/2\mathbb{Z}$.

Proof. We use Lemma 3.3 to prove (a) and (b) by computing

$$\begin{aligned}
\eta(L^5(\ell; 1, 1, 1))(\rho_{-1}\epsilon) &= \ell^{-1}\tilde{\Sigma}_\lambda(1-\lambda)^{-1}\\
&= (2\ell)^{-1}\tilde{\Sigma}_\lambda\{(1-\lambda)^{-1} + (1-\bar{\lambda})^{-1}\}\\
&= (2\ell)^{-1}\tilde{\Sigma}_\lambda 1 = (2\ell - 1)/(2\ell)\\
\eta(X^5(\ell; 1, 1))(\epsilon(\rho_0 - \rho_1)) &= \ell^{-1}\tilde{\Sigma}_\lambda(1+\lambda) = (\ell - 2)/\ell.
\end{aligned}$$

It is immediate from the definition that

$$\begin{aligned}
&\mathcal{G}_L(\vec{a}, 1)(\lambda) - 3\mathcal{G}_L(\vec{a}, 3)(\lambda) = \epsilon(\lambda)\mathcal{G}_L(\vec{a}, 3)(\lambda)\\
&\mathcal{G}_X(\vec{a}, 1)(\lambda) - 3\mathcal{G}_X(\vec{a}, 3)(\lambda) = \epsilon(\lambda)\mathcal{G}_X(\vec{a}, 3)(\lambda)\\
&\eta(K_L^{4j+1})(\gamma) = \eta(L^{4j+1}(\ell; (j+1)\cdot 1, j\cdot 3))(\epsilon^j\gamma)\\
&\eta(K_X^{4j+1})(\gamma) = \eta(X^{4j+1}(\ell; j\cdot 1, j\cdot 3))(\epsilon^j\gamma).
\end{aligned}$$

If $\gamma = \epsilon\rho$, then $\epsilon^j\gamma \in R_0(\mathbb{Z}_\ell)^{2j+2}$ so (c) follows from Lemma 3.3 (f). Since $\epsilon^j R(\mathbb{Z}_\ell) = R_0(\mathbb{Z}_\ell)^{2j}$, we can choose $\gamma_{L,j}$ and $\gamma_{X,j}$ so that

$$\begin{aligned}
&\gamma_{L,j}\epsilon^j\mathcal{G}_L(\ell; (j+1)\cdot 1, j\cdot 3)(\lambda) = (1-\lambda)^{-1}, \text{ and}\\
&\gamma_{X,j}\epsilon^j\mathcal{G}_X(\ell; j\cdot 1, j\cdot 3)(\lambda) = 1+\lambda.
\end{aligned}$$

The same calculation made to prove (a) and (b) then establishes (d) and (e); the proof of (f) is similar and is omitted. □

Definition 3.6. Let $k \geq 1$, and let $\xi = \xi_0$ or $\xi = \xi_1$.

(a) Let $\mathcal{M}_{4k+1}(\xi_0) := \operatorname{span}\{[X^{4k+1}(\ell; \vec{a})]\} \subset \Omega_{4k+1}(B\mathbb{Z}_\ell)$.

(b) Let $\mathcal{M}_{4k+3}(\xi_0) := \operatorname{span}\{[L^{4k+3}(\ell; \vec{a})]\} \subset \Omega_{4k+3}(B\mathbb{Z}_\ell)$.

(c) Let $\mathcal{M}_{4k+1}(\xi_1) := \operatorname{span}\{[L^{4k+1}(\ell; \vec{a})]\} \subset \Omega_{4l+1}(B\mathbb{Z}_\ell, \xi_1)$.

(d) Let $\mathcal{M}_{4k+3}(\xi_1) := \operatorname{span}\{[X^{4i+3}(\ell; \vec{a})]\} \subset \Omega_{4l+3}(B\mathbb{Z}_\ell, \xi_1)$.

If A is an Abelian group, the Poincaré dual A^* is the group of homomorphisms from A to $\mathbb{R}/\mathbb{Z}$. Let $\tilde{\eta}(M)$ be the homomorphism which sends ρ to $\eta(M)(\rho)$. By Theorem 3.1, the map which sends M to $\tilde{\eta}(M)$ extends to homomorphisms $\tilde{\eta} : \mathcal{M}_{4k+1}(\xi) \to R(\mathbb{Z}_\ell)^*$ and $\tilde{\eta}_0 : \mathcal{M}_{4k+3}(\xi) \to R_0(\mathbb{Z}_\ell)^*$.

Lemma 3.7. *Let* $k \geq 1$.

(a) $|\tilde{\eta}(\mathcal{M}_{4k+1}(\xi_0))| \geq (\ell/2)^{k+1}$ *and* $|\tilde{\eta}(\mathcal{M}_{4k+1}(\xi_1))| \geq (2\ell)^{k+1}$.

(b) $|\tilde{\eta}_0(\mathcal{M}_{4k+3}(\xi_0))| \geq (2\ell)^{k+1}$ *and* $|\tilde{\eta}_0(\mathcal{M}_{4k+3}(\xi_1)| \geq (\ell/2)^{k+1}$.

(c) $|\operatorname{ko}^+_{4k+1}(B\mathbb{Z}_\ell, \xi_0)| \geq (\ell/2)^{k+1}$ *and* $|\operatorname{ko}^+_{4k+1}(B\mathbb{Z}_\ell, \xi_1))| \geq (2\ell)^{k+1}$.

(d) $|\operatorname{ko}^+_{4k+3}(B\mathbb{Z}_\ell, \xi_0)| \geq (2\ell)^{k+1}$ *and* $|\operatorname{ko}^+_{4k+3}(B\mathbb{Z}_\ell, \xi_1)| \geq (\ell/2)^{k+1}$.

(e) $|\operatorname{ko}^+_{8k+3}(B\mathbb{Z}_\ell, \xi_0) \geq 2(2\ell)^{2k+1}$.

Proof. Let $\epsilon := \rho_{-1}(\rho_0 - \rho_1)^2$. Then the homomorphism which sends ρ to $\epsilon\rho$ defines a dual map ϵ^* from $R(\mathbb{Z}_\ell)^*$ to $R(\mathbb{Z}_\ell)^*$ so that $\epsilon^*(\tilde{\eta}(M))(\rho) = \eta(M)(\epsilon\rho)$. If $k \geq 2$, then $\eta(M^{4k+1}(\ell; \vec{a}, 1, 1))(\epsilon\rho) = \eta(M^{4k-3}(\ell; \vec{a}))(\rho)$ where M is a lens space or lens space bundle. Consequently $\tilde{\eta}\mathcal{M}_{4k-3}(\xi) \subset \epsilon^*\mathcal{M}_{4k+1}(\xi)$. By Lemma 3.5, $\tilde{\eta}(K_L^{4k+1})$ and $\tilde{\eta}(K_X^{4k+1})$ are elements of $\ker(\epsilon^*)$ of order at least 2ℓ and $\ell/2$ respectively. Thus $|\ker(\epsilon^*) \cap \tilde{\eta}\mathcal{M}_{4k+1}(\xi_1)| \geq 2\ell$ and $|\ker(\epsilon^*) \cap \tilde{\eta}\mathcal{M}_{4k+1}(\xi_0)| \geq \ell/2$. By Lemma 3.5, $\epsilon^*\tilde{\eta}(L^5(\ell; 1, 1, 1))$ and $\epsilon^*\tilde{\eta}(X^5(\ell; 1, 1))$ are elements of orders at least 2ℓ and $\ell/2$ in $R(\mathbb{Z}_\ell)^*$. This completes the proof of (a) if $m = 5$; the general case of (a) now follows by induction.

Let $\delta = (\rho_0 - \rho_1)$. Then $\eta(M^{4k+3}(\ell; \vec{a}, 1))(\delta\rho) = \eta(M^{4k+1}(\ell; \vec{a}))(\rho)$ for all ρ in $R(\mathbb{Z}_\ell)$ and for M either a lens space or a lens space bundle. Thus

$$\tilde{\eta}\mathcal{M}_{4k+1}(\xi_0) \subseteq \delta^*\tilde{\eta}_0\mathcal{M}_{4k+3}(\xi_1) \text{ and } \tilde{\eta}\mathcal{M}_{4k+1}(\xi_1) \subseteq \delta^*\tilde{\eta}_0\mathcal{M}_{4k+3}(\xi_0).$$

Thus (b) follows from (a). Assertions (c) and (d) follow from (a) and (b) since the eta invariant extends to connective K-theory and since $\mathcal{M}_m(\xi_0)$ and $\mathcal{M}_m(\xi_1)$ are generated by bordism classes of manifolds which admit metrics of positive scalar curvature. To prove (e), we must squeeze out a single factor of 2 if $m = 8k + 3$ and if ξ is trivial. Let $M = (L^3(1, 1) - 3L^3(1, 3)) \times B^j$ where B is the Bott manifold; B *is* an 8 dimensional spin manifold with $\hat{A}(B) = 1$. Then by Lemma 3.5 (f) and using the multiplicative nature of the eta invariant, we compute

$$\eta(M)(\rho_0 - \rho_{\ell/2}) = \eta(L^3(1,1) - 3L^3(1,3))(\rho_0 - \rho_{\ell/2}) \neq 0 \text{ in } \mathbb{R}/2\mathbb{Z}.$$

Thus in particular, the refined eta invariant of Theorem 3.1 (c) is non-trivial and generates the missing factor of 2. □

We compare these lower bounds of Lemma 3.7 with the upper bounds of Theorem 3.4 to see that $\mathrm{ko}_m^+(B\mathbb{Z}_\ell, \xi) = \ker(\alpha) \cap \mathrm{ko}_m(B\mathbb{Z}_\ell, \xi)$ if m is odd and if $w_1(\xi) = 0$. This completes the proof of the Gromov–Lawson–Rosenberg conjecture in these cases. It also shows that the eta invariant completely detects $\ker(\alpha) \cap \mathrm{ko}_m(B\mathbb{Z}_\ell, \xi)$.

In the non-orientable case, the basic strategy is the same. We must find manifolds of positive scalar curvature, compute their eta invariants, and show that the order of the range of the eta invariant agrees with the estimate of Theorem 2.4. This has been done for $\ker(\alpha) \cap \mathrm{ko}_m(B\mathbb{Z}_\ell, u_1 \neq 0, u_2 = 0)$ if $m \equiv 2 \bmod 4$ and hence the Gromov–Lawson–Rosenberg conjecture is established in these cases. Our research is continuing in the remaining cases.

References

[1] M. F. Atiyah, V. K. Patodi, and I. M. Singer, Spectral asymmetry and Riemannian geometry I, Math. Proc. Camb. Phil. Soc. 77 (1975), 43–69.

[2] A. L. Besse, Einstein manifolds, Ergeb. Math. Grenzgeb. (3) 10, Springer-Verlag, 1987.

[3] J. Bismut and J. Cheeger, η invariants and their adiabatic limits, J. Amer. Math. Soc. 2 (1989), 33–70.

[4] B. Botvinnik and P. Gilkey The eta invariant, equivariant spin bordism, and metrics of positive scalar curvature, in: Partial Differential Operators and Mathematical Physics (Demuth, M., Schulze, B.-W., ed.), Oper. Theory Adv. Appl. 78, Birkhäuser, Basel 1995, 141–152.

[5] B. Botvinnik, P. Gilkey, and S. Stolz, The Gromov–Lawson–Rosenberg conjecture for groups with periodic cohomology, J. Differential Geom., to appear.

[6] B. Botvinnik and P. Gilkey, The Gromov–Lawson–Rosenberg conjecture: the twisted case, submitted.

[7] H. Donnelly, Eta invariants for G spaces, Indiana Univ. Math. J. 27 (1978), 889–918.

[8] P. Gilkey, The Geometry of Spherical Space Form Groups, Ser. Pure Math. 7, World Scientific Press, Singapore 1989.

[9] M. Gromov and H. B. Lawson, Spin and scalar curvature in the presence of a fundamental group I, Ann. of Math. 111 (1980), 209–230; see also: The classification of simply connected manifolds of positive scalar curvature, Ann. of Math. 111 (1980), 423–434.

[10] S. Kwasik and R. Schultz, Positive scalar curvature and periodic fundamental groups, Comment. Math. Helv. 65 (1990), 271–286.

[11] A. Lichnerowicz, Spineurs harmoniques, C. R. Acad. Sci. Paris 257 (1963), 7–9.

[12] T. Miyazaki, On the existence of positive curvature metrics on non simply connected manifolds, J. Fac. Sci. Univ. Tokyo Sect. IA 30 (1984), 549–561.

[13] J. Rosenberg, C^* algebras, positive scalar curvature, and the Novikov conjecture, II, in: Geometric Methods in Operator Algebras, Pitman Res. Notes Math. Ser. 123, Longman Sci. Techn., Harlow 1986., 341–374; III, Topology 25 (1986) 319–336.

[14] J. Rosenberg and S. Stolz, Manifolds of positive scalar curvature, in: Algebraic Topology and its applications (G. Carlsson, R. Cohen, W. C. Hsiang and J. D. S. Jones, ed.), Math. Sci. Res. Inst. Publ. 27 Springer-Verlag, New York 1994, 241–267; see also: A stable version of the Gromov–Lawson conjecture, to appear in the Proceedings of the Centennial Czech conference; The stable classification of manifolds of positive scalar curvature, preprint; A general approach to the Gromov–Lawson–Rosenberg conjecture, preprint.

[15] R. Schoen and S. T. Yau, The structure of manifolds with positive scalar curvature, Manuscripta Math. 28 (1979), 159–183.

[16] S. Stolz, Simply connected manifolds of positive scalar curvature, Ann. of Math. 136 (1992), 511–540.

[17] —, Splitting certain MSpin module spectra, Topology 33 (1994), 159–180.

Geometry of the Radon–Penrose transform and interpretation of singularities

*Hoàng Lê Minh**

Abstract. A recent work of D'Agnolo–Schapira on general Radon–Penrose transform for sheaves and $\mathcal{D}$-Modules shows that, starting from a complex analytic vector bundle, one can get an analytic $\mathcal{D}$-Module which admits only regular singularities. In this paper we present two examples extending this fact. Firstly, we interpret singular solutions of a class of linear equations on the Yang–Mills background in a region of the complexified compactified space-time in terms of cohomology classes of a singular set of corresponding vector bundles on the generalized twistor manifold. Then, when the Yang–Mills background is self-dual, by taking the Radon–Penrose transform of a class of vector bundles on the twistor manifold we obtain a class of singular analytic connections which admits only first-order poles along a noncharacteristic hypersurface.

1991 Mathematics Subject Classification: 14F05, 32L25.

1. Introduction

The original concept of Radon–Penrose transform is based on the integral formulas of Radon for recovering the values of a function on $\mathbf{R}^3$ by its integrals along all two-dimensional planes, and the Penrose correspondence between analytic solutions of some massless field equations in mathematical physics and their contour integral presentations (in geometric language: Čech cohomology classes) by the theory of twistors ([1]).

To be precise, let us consider the complexified, compactified model **CM** of the (flat) Minkowski space-time. There are exactly two families of complex totally null-planes (so-called α and β-planes) passing through each point $x \in \mathbf{CM}$. By assigning $x \in \mathbf{CM}$ to the α-planes through x, one can define the three-dimensional complex projective twistor space $\mathbf{CP}^3$ and relate these objects in a diagram of double fibration

*Department of Mathematics and Informatics, University of Hochiminh City, Vietnam

$$\begin{array}{ccc} & \mathbf{F} & \\ g \swarrow & & \searrow f \\ \mathbf{CP}^3 & & \mathbf{CM} \end{array} \tag{1.1}$$

Geometrically speaking, this diagram can be considered as a standard diagram of complex flag manifolds with the natural "forgetful" projections, namely $\mathbf{CM} = Gr(2; \mathbf{C}^4)$, $F = Fl(1, 2; \mathbf{C}^4)$ and its local parts, i.e. subsets $U \subset \mathbf{CM}$, their inverse images $F(U) = f^{-1}(U)$ and $P(U) = g(F(U)) \subset \mathbf{CP}^3$. Again, the set $P(U)$ can be considered as the set of all α-planes having non-empty intersections with U. By introducing the reality structure to the diagram, R. Penrose and his colleagues were able to interpret analytic (singular) solutions of a class of linear massless field equations of mathematical physics in terms of (local) analytic cohomology classes on $P(U)$ (see [2] for details).

After the success of the Penrose correspondence, several ambitious programs were pushed ahead during the last twenty years with the goal of interpreting and solving many linear and non-linear partial equations on **CM** (or more generally, on a complex manifold X) in terms of complex geometric objects on $\mathbf{CP}^3$ (on an appropriate "twistor" manifold Z).

The mechanism of all known interpretations is based on the so called diagram of the Radon–Penrose Transform

$$\begin{array}{ccc} & Y & \\ g \swarrow & & \searrow f \\ Z & & X \end{array}$$

where X, Y, Z are complex manifolds; f, g surjective holomorphic morphisms such that (f, g) yields a closed embedding $Y \hookrightarrow Z \times X$.

The geometrical objects under consideration can be holomorphic vector bundles, cohomological classes and, more generally, objects of the derived category of $\mathcal{D}$-modules and sheaves. By means of algebraic and geometric methods, many interesting results were achieved with the help of the Radon–Penrose transform (see [3]).

Recently, A. D'Agnolo and P. Schapira have proved some results on the interpretation of singular solutions for linear partial equations. By considering additional topological conditions, these authors have shown that there is a correspondence between a class of complex analytic vector bundles on Z and a certain class of $\mathcal{D}$-modules, which admits only regular singularities on X (in terms of microlocal analysis and the derived category, see [4], Theorem 3.14).

The aim of this work is to give a systematic study of the complex geometry of the underlying Radon–Penrose diagram and to extend related interpretations of singular solutions in terms of cohomology groups and classes.

2. Conformally flat Radon–Penrose transform

Following R. Penrose, we consider the flat compactified complexified space-time **CM** as the Grassmannian manifold of all complex two-dimensional planes in a four-dimensional complex vector space T (and its dual T^*). We denote the corresponding tautological two-dimensional bundles over **CM** by $S_\pm$. There is a canonical isomorphism between the holomorphic cotangent bundle $T^*\mathbf{CM}$ and the four-dimensional bundle $S_+ \otimes S_-$. It defines a complex conformally flat structure on **CM** representing the one-dimensional subbundle of the complex Riemannian metric $\wedge^2 S_+ \otimes \wedge^2 S_-$ in $\mathbf{S}^2(T^*\mathbf{CM})$. The complex projective lines in $\mathbf{P}(T)$, lying on **CM** under Klein–Plucker embedding, are complex null-lines (complex null-geodesics with respect to the complex Riemannian conformal metric).

All null-lines passing through a point $x \in \mathbf{CM}$ form a complex null-cone C_x with vertex at this point. Then the space L of all null-lines is identified canonically with the flag manifold of $(1, 3)$-dimensional flags in T, and therefore it can be embedded in the direct product $P^\alpha \times P^\beta$, where $P^\alpha = \mathbf{P}(T)$ and $P^\beta = \mathbf{P}(T^*)$. Such three-dimensional projective spaces can be interpreted as the spaces of α- and β-null-planes in **CM**. Note that L is defined in the product as a set of zeros of the unit section $s \in H^0(P^\alpha \times P^\beta, \mathcal{O}_{P^\alpha \times P^\beta}(-1,-1)) = T^* \otimes T$ corresponding to the identity endomorphism id_T. We relate L and **CM** by means of the flag manifold F of complete $(1, 2, 3)$-dimensional flags in T together with the natural projections g onto L and f onto **CM**, and also by the projections μ_α and μ_β from F onto P^α and P^β.

By a conformally flat Radon–Penrose transform we mean the following diagram of flag manifolds and its localized portions to open subsets $U \subset \mathbf{CM}$. This diagram enables us to interpret the general complex (non)-self-dual Yang–Mills field equations.

$$\begin{array}{ccccc} & & F(U) & & \\ & g \swarrow & & \searrow f & \\ P^\alpha(U) \times P^\beta(U) & \overset{(\rho_\alpha, \rho_\beta)}{\hookleftarrow} & L(U) & & U \subset \mathbf{CM} \end{array} \tag{2.1}$$

Here $F(U) = f^{-1}(U)$, $P^\alpha(U) = \mu_\alpha F(U)$, $P^\beta(U) = \mu_\beta F(U)$, $L(U) = g(F(U))$ and ρ_α, ρ_β are canonical projections of $(1, 3)$-flags onto the projective spaces. $F(U)$ can be interpreted geometrically as the space of the incidence relation

$$(\text{null-line; point on the line}) \in L(U) \times U$$

and $L(U)$ as the space of all null lines intersecting U.

The base of the null-cone C_x of a point $x \in U$ is a two-dimensional quadric $L_x = gf^{-1}(x)$ contained entirely in $L(U)$, while $L(U)$ is generated by such quadrics. The projective lines $P_x^\alpha = \mu_\alpha g^{-1}(x)$ and $P_x^\beta = \mu_\beta g^{-1}(x)$ are contained in $P^\alpha(U)$ and $P^\beta(U)$ respectively, and there is a canonical identification

$$L_x = P_x^\alpha \times P_x^\beta \subset P^\alpha \times P^\beta.$$

Let $L^{(k)}$ be the kth infinitesimal neighborhood of L in $P^\alpha \times P^\beta$, $k \geq 0$. Then there are standard invertible sheaves

$$\mathcal{O}_L^{(k)}(a,b), \quad a,b \in \mathbf{Z}$$

induced by the embedding of $L^{(k)}$. The analytic spaces

$$(L(U), \mathcal{O}_L^{(k)}(a,b)|_{L(U)})$$

are denoted by $L^{(k)}(U)$. Since $L^{(k)}$ is defined by the analytic equation $s^{k+1} = 0$, there are the following exact sequences of sheaves:

$$0 \to \mathcal{O}_L(-k-1, -k-1) \to \mathcal{O}_{L^{(k+1)}} \to \mathcal{O}_{L^{(k)}} \to 0$$

$$0 \to \mathcal{O}_{P^\alpha \times P^\beta}(-k-1, -k-1) \to \mathcal{O}_{P^\alpha \times P^\beta} \to \mathcal{O}_{P^\alpha \times P^\beta} \to 0.$$

3. Geometry of complex characteristic hypersurfaces

Let C be a complex hypersurface, defined locally in a neighborhood of its nonsingular point $x \in C$ by the analytic equation $u(x^a) = 0$. Then C is called characteristic (or null) at x if and only if the conormal vector to C at x; locally $\dfrac{\partial u}{\partial x^a}$ is a null-vector. It is not hard to see that locally, C is generated in a neighborhood of its nonsingular point by a two-parameter family Z_C of null-lines which are contained in C. It follows from the flat Radon–Penrose diagram (2) that Z_C can be identified with a two-dimensional submanifold in L, such that $fg^{-1}(Z_C) = C$ holds locally. If a point x is nonsingular, then $L_x \cap Z_C$ contains a single point γ_x corresponding to the generator passing through x. If the intersection contains more then one point, then x should be singular. We wish to describe the characteristic complex hypersurface $C \subset \mathbf{CM}$ in terms of the geometry on L. For this purpose we define the following set:

$$Z'_C = \{l \in L \text{ , the null line } gf^{-1}(l) \text{ is tangent to } C\ \}.$$

We have the following criterion:

Proposition 3.1. *A complex hypersurface C is characteristic in a neighborhood of a nonsingular point $x \in C$ if and only if the following conditions are satisfied:*

a) *There is a two-dimensional submanifold Z_C in L such that $fg^{-1}(Z_C) = C$ (at least locally).*

b) *A point $x \in C$ is nonsingular if and only if $Z_C \cap L_x$ contains a single point γ_x.*

c) *$Z'_C \cap L_x$ is a cone with vertex at γ_x.*

Proof. Let $x \in C$ be a nonsingular point and let U be a neighborhood of x such that $C \cap U$ does not contain any singular points in C. Assigning to each point of $C \cap U$ the null vector normal to C at this point, we obtain a nondegenerate vector field on $C \cap U$

since its normal null vector belongs to tangent spaces. This vector field is self-parallel with respect to the induced Riemannian connection. Therefore its integral curves are null-geodesic on C, i.e., they are its generators. Obviously a single generator passes through each point and all generators together form a two-dimensional submanifold Z_C in L such that $fg^{-1}(Z_C) = C$ and $Z_C \subset Z'_C$. Condition b) is clearly satisfied. If for a nonsingular point $x \in C \cap U$ there are $\gamma_x \neq \gamma'_x \in L_x \cap Z'_C$, then the tangent plane to C at x spanned by the null lines corresponding to γ_x and γ'_x is a null-plane. It follows that all the projective lines contained in this plane and passing through x are tangent to C at x. They correspond to a line in L_x passing through γ_x and γ'_x. This proves condition c).

Conversely, let conditions a)–c) be satisfied. The hypersurface $C = fg^{-1}(Z_C)$ is then generated by the null-lines $fg^{-1}(\gamma)$, $\gamma \in Z_c$. We shall denote by ν_x the null-vector in the space tangent to C at a nonsingular point x which indicates the direction of the generator $fg^{-1}(\gamma_x)$. Consider an arbitrary vector v_x tangent to C at x and the tangent plane π_x, spanned by ν_x and v_x. If π_x does not contain any other null-vectors apart from ν_x, then ν_x is orthogonal to the plane π_x and, in particular, to v_x (with respect to the induced inner product). If there is another null-vector in π_x, then condition c implies that all the tangent vectors in π_x should be null-vectors, i.e. π_x is a null-plane and, therefore, ν_x is orthogonal to v_x. We have proved that ν_x is a null-vector normal to c at x, thus C is characteristic at any nonsingular point. This completes the proof of the theorem. □

The null-cone C_x is an example of the simplest characteristic hypersurface in **CM** with a single singular point x. In fact, C_x is generated by null-lines passing through x, i.e. parametrized by a two-dimensional quadric L_x. If $x \neq y \in C_x$, then $L_x \cap L_y$ corresponds to the null-line passing through x and y. An arbitrary line which has more then one intersection point with C_x or is tangent to C_x is contained in C_x. It follows that the set Z'_{C_x} defined above reduces to the set $\tilde{L}_x$ of all the null-lines contained in C_x. It is easy to see that $\tilde{L}_x$ is the union of two three-dimensional extended quadrics L_x^α and L_x^β corresponding, respectively, to the null-lines in the α- and β-planes in C_x. Such a geometry of complex null-cones is reflected in the diagram (2.1) by the following relations

$$L_x = L_x^\alpha \cap L_x^\beta \subset L_x^{\alpha,\beta} = \rho_{\alpha,\beta}^{-1} P_x^{\alpha,\beta} \subset \tilde{L}_x = L_x^\alpha \cup L_x^\beta$$

where $\rho_{\alpha,\beta} : L \to P^{\alpha,\beta}$ is a canonical projection. We can also identify L_x^α with the intersection of L and $H_x^\alpha = P_x^\alpha \times P^\beta$, which are embedded in $P^\alpha \times P^\beta$; a similar identification can be made for L_x^β. Since P_x^α is a complete intersection of two hypersurfaces in P^α defined by $z_1 = z_2 = 0$, where $z_1, z_2 \in H^0(P^\alpha, \mathcal{O}_{P^\alpha}(1))$, the Koszul complex

$$0 \to \mathcal{O}_{P^\alpha}(-2) \stackrel{\mu_2}{\to} \mathcal{O}_{P^\alpha}(-1) \oplus \mathcal{O}_{P^\alpha}(-1) \stackrel{\mu_1}{\to} J_{P_x^\alpha} \to 0$$

where $\mu_2(p) = (pz_2, pz_1)$, $\mu_1(q_1, a_2) = q_1 z_1 + q_2 z_2$, and p and q are, respectively, local sections of the sheaves $\mathcal{O}(-2)$ and $\mathcal{O}(-1)$, is acyclic. We also have the following resolution to the sheaf $\mathcal{O}_{P^\alpha_x}$

$$0 \to \mathcal{O}_{P^\alpha}(-2) \overset{\mu_2}{\to} \mathcal{O}_{P^\alpha}(-1) \oplus \mathcal{O}_{P^\alpha}(-1) \overset{\mu_1'}{\to} \mathcal{O}_{P^\alpha_x} \overset{\mu_0}{\to} \mathcal{O}_{P^\alpha_x} \to 0.$$

The projection $\pi_\alpha : H = P^\alpha \times P^\beta \to P^\alpha$ is a flat morphism and, therefore, the functor π^*_α is exact on $\mathcal{O}_{P^\alpha}$-modules and sends the last complex to the resolution of the sheaf $\mathcal{O}_{H^\alpha_x}$

$$0 \to \mathcal{O}_H(-2, 0) \overset{\mu_2}{\to} \mathcal{O}_H(-1, 0) \oplus \mathcal{O}_H(-1, 0) \overset{\mu_1'}{\to} \mathcal{O}_H \overset{\mu_0}{\to} \mathcal{O}_{H^\alpha_x} \to 0.$$

We set

$$\mathcal{O}^{(k)}_{L^\alpha_x} = \mathcal{O}_{L^{(k)}} \otimes_{\mathcal{O}_H} \mathcal{O}_{H^\alpha_x}, \quad J^{(k)}_{L^\alpha_x} = \mathcal{O}_{L^{(k)}} \otimes_{\mathcal{O}_H} J_{H^\alpha_x}.$$

The above sequence remains true when we are restricted on kth-infinitesimal neighborhood of L, i.e. there is an analogous exact sequence

$$0 \to \mathcal{O}_{L^{(k)}}(-2, 0) \overset{\mu_2}{\to} \mathcal{O}_{L^{(k)}}(-1, 0) \oplus \mathcal{O}_{L^{(k)}}(-1, 0) \overset{\mu_1'}{\to}$$

$$\overset{\mu_1'}{\to} \mathcal{O}_{L^{(k)}} \overset{\mu_0}{\to} \mathcal{O}_{L^{(k)\alpha}_x} \to 0.$$

We note, for $k = 0$, the sheaf $J_{L^\alpha_x} \hookrightarrow \mathcal{O}_L$ is the sheaf of ideals which determines the submanifold L^α_x in L.

4. Non-self-dual Yang–Mills backgrounds and singular solutions

We now assume that arbitrary nonempty intersection of null-lines with U is connected and simply connected. Then the Radon–Penrose transform establishes an equivalence between the category of holomorphic vector bundles with a holomorphic connection (E, ∇) on U and the category of holomorphic vector bundles E_L over $L(U)$ which are trivial for all quadrics L_x, $x \in U$. The bundle E_L has a unique extension $E^{(1)}_L$ to the first infinitesimal neighborhood $L^{(1)}(U)$ which admits a further second extension to $L^{(2)}(U)$. We set $E^{(1)}_L(a, b) = E^{(1)}_L \otimes_{\mathcal{O}_L} \mathcal{O}_{L^{(1)}}(a, b)$. Then the cohomological classes in $H^1(L(U), E^{(1)}_L(-2, 0))$ classify the holomorphic solutions of the Laplace equation, defined in U, on the Yang–Mills background (E, ∇). In fact, the cohomological interpretation of the sheaf of local holomorphic solutions of the equation can be described as follows. Let us consider a complex version of the conformally invariant Laplace operators on the Yang–Mills background

$$D_\pm : E \otimes \Lambda^2 S_\pm \to E \otimes (\Lambda^2 S_\pm)^{\otimes 2} \otimes \Lambda^2 S_\mp$$

(the operators $D_\pm$ are identified in the domain over which the bundles $\Lambda^2 S_\pm$ have nonvanishing sections, i.e., where it is possible to define a global nondegenerate Riemannian metric representing a conformal structure). We denote by Ker D the sheaf of local holomorphic sections of the bundle $E \otimes \Lambda^2 S_\pm$ which are annulled by D_+. We then obtain the following isomorphism of sheaves

$$f_* g^{-1} E_L^{(1)}(-2,0) = 0, \quad \mathbf{R}^1 f_* g^{-1} E_L^{(1)}(-2,0) = \operatorname{Ker} D$$

where the left-hand sides contain f-images of the inverse g-image of the sheaf of local holomorphic sections of the bundle $E_L^{(1)}(-2,0)$ over $L(U)$.

One should note that there are more general forms of the correspondence using the general theory of $\mathcal{D}$-modules and sheaves on manifolds. But then some geometric properties of the constructions are not clear. In particular, one has to consider only the trivial extensions of the linear equations, and not the cases of (anti)-self-dual or non-self-dual Yang–Mills backgrounds. Also the generalized construction requires strictly topological conditions on U (more precisely, on fibres of the canonical surjective mapping f and g). In our case, many nice topological properties follow from the geometry of characteristic hypersurfaces.

We are interested in an interpretation of singular solutions of linear field equations (and the holomorphic connections themselves) which can admit singularities in U along a complex hypersurface. In the case of the Laplace equation, the only possible pole singularities can occur along a characteristic hypersurface C, outside C they are complex analytic and are defined up to the modulus of global holomorphic solutions (over U). In other words, these solutions should belong to the factor group

$$\operatorname{Coker}[H^0(U, \operatorname{Ker} D) \overset{\text{res}}{\to} H^0(U - C, \operatorname{Ker} D)]$$

or be represented by a class of the local cohomology group $H_C^1(U, \operatorname{Ker} D)$.

The following Proposition gives a cohomological interpretation of these solutions.

Proposition 4.1. *One has the following commutative diagram with isomorphisms in each column:*

$$\begin{array}{ccccc}
H^1(L(U), E_L^{(1)}(-2,0)) & \to & H^1(L(U) - \tilde{Z}_C, E_L^{(1)}(-2,0)) & \to \\
\tau \downarrow & & \tau \downarrow & \\
H^0(U, \operatorname{Ker} D) & \to & H^0(U - C, \operatorname{Ker} D) & \to
\end{array}$$

$$\begin{array}{cc}
\to & H^2_{\tilde{Z}_C}(L(U), E_L^{(1)}(-2,0)) \\
& \tau \downarrow \\
\to & H_C^1(U, \operatorname{Ker} D).
\end{array}$$

Proof. The proof of these isomorphisms follows from standard arguments with the help of the two Leray spectral sequences related to the morphisms f and g. Details can be seen in ([5]). □

The next step is the construction of a local cohomology class on $L(U)$ which yields the singular solution of the Laplace equation on U with minimal singularities along C in the form of a simple pole. In view of the preceding isomorphisms, the corresponding representing class should lie in the local cohomology group. We first consider the scalar case (without any Yang–Mills background). There is the following exact sequence related to the spectral sequence for the sheaves Ext and Ext-groups

$$0 \to H^1(L(U), \mathcal{O}_{L^{(1)}}(-2,0)) \to \mathrm{Ext}^1(J^{(1)}_{L^\alpha_x}, \mathcal{O}_{L^{(1)}}(-2,0))$$
$$\stackrel{\gamma}{\to} H^2(L(U), \mathcal{O}_{L^{(1)}} \otimes \mathcal{O}^{(1)}_{L^\alpha_x}) \to H^2.$$

We denote by $\mathrm{id}_{\mathcal{O}^{(1)}_{L^\alpha_x}}$ the unit section of the sheaf $\mathcal{O}^{(1)}_{L^\alpha_x}$.

Then if $H^2(L(U), \mathcal{O}_{L^{(1)}} \otimes \mathcal{O}^{(1)}_{L^\alpha_x}) = 0$, it is always possible to find a class

$$\epsilon_x \in \mathrm{Ext}^1(J^{(1)}_{L^\alpha_x}, \mathcal{O}_{L^{(1)}}(-2,0))$$

such that

$$\gamma(\epsilon_x) = \mathrm{id}_{\mathcal{O}^{(1)}_{L^\alpha_x}} \in H^0(L(U), \mathcal{O}^{(1)}_{L^\alpha_x})$$

and if

$$H^1(L(U), \mathcal{O}_{L(1)}(-2,0)) = 0$$

then this class is defined uniquely. These cohomology groups are trivial if, for example, U contains the real 4-dimensional sphere S^4. However, a class ϵ_x corresponding to the unit section can be constructed also for other regions. In fact, let us first choose the characteristic class ϵ of the following extension

$$0 \to \mathcal{O}_{L^{(1)}}(-2,0) \stackrel{\mu_2}{\to} \mathcal{O}_{L^{(1)}}(-1,0) \oplus \mathcal{O}_{L^{(1)}}(-1,0) \stackrel{\mu_1}{\to} J^{(1)}_{L^\alpha_x} \to 0.$$

Since this extension is locally free, it follows from the Serre lemma that $\epsilon|_l$ for a point $l \in L(U) \cap L^\alpha_x$ generates a group

$$\mathrm{Ext}^1_{L(U),l}(J^{(1)}_{L^\alpha_x,l}, \mathcal{O}^{(1)}_{L^\alpha_x,l}(-2,0))$$

as an $\mathcal{O}^{(1)}_{L,l}$-module and, therefore, γ maps $\epsilon|_l$ to a nonzero value on the fibre of $\mathcal{O}^{(1)}_{L^\alpha_x,l}$. The section $\gamma(\epsilon)$ is nonzero everywhere on L^α_x and, after suitable renormalization, we obtain a class ϵ_x satisfying the condition $\gamma(\epsilon) = \mathrm{id}_{\mathcal{O}^{(1)}_{L^\alpha_x}}$.

We shall denote by res the restriction functor

$$\mathrm{res} : \mathrm{Ext}^1_{L(U)}(J^{(1)}_{L^\alpha_x}, \mathcal{F}) \to \mathrm{Ext}^1_{L(U)-\tilde{L}_x}(J^{(1)}_{L^\alpha_x}, \mathcal{F}) = H^1(L(U) - \tilde{L}_x, \mathcal{F})$$

where $\mathcal{F}$ is a sheaf of $\mathcal{O}_{L^{(1)}}$-module over $L(U)$. The above isomorphism follows from the explicit calculations of cohomology groups and from the inclusion

$$\operatorname{Supp}\mathcal{O}^{(1)}_{L^\alpha_x} \subset L^\alpha_x \subset \tilde{L}_x.$$

Then the class ϵ_x represents the singular solution of the scalar Laplace equation on U, which admits only a first order pole at x (and therefore along the characteristic null-cone C_x).

More generally, assume that the domain of definition U of a Yang–Mills field (E, ∇) is such that the parallel translation of fibres of the bundle E, induced by the connection ∇ and by the Riemannian connection along paths passing through two points $x, y \in U$ and lying on a complex geodesic containing these points is independent of the choice of a particular path. The parallel translation then induces mappings of the fibres into one another. Fixing the point x and varying the point y, we obtain the canonical homomorphism between $\tilde{E}_x$ and E. They are horizontal with respect to the induced canonical connection on $\operatorname{Hom}(\tilde{E}_x, E)$ and, therefore they generate a homomorphism of the corresponding fibrations over $L(U)$

$$\tilde{E}_{x,L} \longleftrightarrow E_L,$$

and, as in the scalar case, there is the following exact sequence of cohomology groups on $L(U)$

$$0 \to H^1(L(U), \operatorname{End} E^{(1)}(-2,0)) \to \operatorname{Ext}^1(J^{(1)}_{L^\alpha_x}, \operatorname{End} E^{(1)}(-2,0))$$

$$\xrightarrow{\gamma} H^0(L(U), \operatorname{End} E^{(1)} \otimes \mathcal{O}^{(1)}_{L^\alpha_x}).$$

The homomorphisms between fibres of E induce the mapping

$$\Phi_x : \operatorname{Ext}^1(J^{(1)}_{L^\alpha_x}, \operatorname{End} E^{(1)}_L(-2,0)) \to$$

$$\operatorname{Ext}^1(J^{(1)}_{L^\alpha_x}, \operatorname{Hom}(\tilde{E}^{(1)}_{x,L}, E(1)_{L^\alpha_x}(-2,0))).$$

This mapping can be interpreted as a fixing of the variable x for functions of two variables with values in the groups of homomorphisms of fibres of E.

Let us consider the class

$$\epsilon^E_x \in \operatorname{Ext}^1_{L(U)}(J^{(1)}_{L^\alpha_x}, \operatorname{End} E^{(1)}_L(-2,0))$$

such that ϵ^E_x is sent by γ to the unit section of the fibration $\operatorname{End} E^{(1)}_L$ on L^α_x with respect to the natural identification

$$\gamma(\epsilon^E_x) = \operatorname{id}_{E^{(1)}_L|_{L^\alpha_x}} \in \operatorname{Hom}_{L(U)}(E^{(1)}_L|_{L^\alpha_x}, E^{(1)}_L|_{L^\alpha_x}) = H^0(L(U), \operatorname{End} E^{(1)}_L \otimes \mathcal{O}^{(1)}_{L^\alpha_x}).$$

We prove the following theorem which gives an interpretation of the singular solutions for the Laplace equation on the Yang–Mills background with minimal singularities along the null-cones.

Theorem 4.1. *The class*

$$\epsilon_x^E \in \mathrm{Ext}^1(J^{(1)}_{L_x^\alpha}, \mathrm{End}\, E_L^{(1)}(-2,0))$$

corresponding to the identity morphism $\mathrm{id}_{E_L^{(1)}}$ *restricted to* L_x^α *,i.e. in*

$$\mathrm{Hom}(E_L^{(1)}|_{L_x^\alpha}, E_L^{(1)}|_{L_x^\alpha}) = H^0(L(U), \mathrm{End}\, E_L^{(1)} \otimes_{\mathcal{O}_L^{(1)}} \mathcal{O}^{(1)}_{L_x^\alpha})$$

under the map γ *is the local cohomology class representing the Radon–Penrose transform of a singular solution for the Laplace equation on* U *with simple poles along the null cone* C_x.

Proof. We first consider the scalar case. We show that $\mathrm{res}(\epsilon_x)$ determines a holomorphic section of the bundle $\wedge^2 S_+$over $U - C_x$ and satisfies the Laplace equation there. Suppose that $y \notin C_x$. This is equivalent to supposing that y does not lie on any null line contained in C_x, i.e. $L_y \cap \tilde{L}_x = \emptyset$. Hence $L_y \subset L(U) - \tilde{L}_x$ and it follows from the exact sequence

$$0 \to \mathcal{O}_L(-3,-1) \to \mathcal{O}_L^{(1)}(-2,0) \to \mathcal{O}_L(-2,0) \to 0$$

restricted to L_y that there is a canonical identification

$$H^1(L_y, \mathcal{O}_L^{(1)}(-2,0)) = H^1(L_y, \mathcal{O}_L(-2,0)) = \wedge^2 S_+(y).$$

The class $\mathrm{res}(\epsilon_x)$, being restricted to each quadric L_y, $y \in U - C_x$, determines a holomorphic section of the bundle $\wedge^2 S_+$ over $U - C_x$, and clearly satisfies the Laplace equation.

We now claim that ϵ_x defines over U a meromorphic section with a first-order pole on C_x. It follows from Proposition 3.1 that $fg^{-1}(L_x^\alpha) = C_x$. Hence we have the following inclusion on $F(U)$

$$F_x^\alpha = g^{-1}(L_x^\alpha) \subset \tilde{C}_x \tag{4.2}$$

where $\tilde{C}_x := f^{-1}(C_x)$ is a divisor (manifold of codimension one) in $F(U)$. We can assume that C_x is defined locally as the set of zeros of a holomorphic function s in U. Lifting s to $F(U)$ by means of f, we obtain a holomorphic function denoted by $f^{-1}s$. It has constant values on the fibers of the morphism f and vanishes on $\tilde{C}_x$. Let us multiply $f^{-1}s$ by the local sections of the sheaf $g^{-1}\mathcal{O}_{L^{(1)}}$, i.e. by holomorphic functions lifted from $L^{(1)}$, and denote by $J^{(1)}_{\tilde{C}_x}$ the resultant principal ideal in $g^{-1}\mathcal{O}_{L^{(1)}}$. It follows from the inclusion (4.2) that $J^{(1)}_{\tilde{C}_x}$ is contained in $g^{-1}(J^{(1)}_{L_x^\alpha})$. Lifting the class ϵ_x together with the Ext^1-group containing it to $F(U)$ by means of g^{-1}, we obtain the following natural map:

$$g^{-1}(\epsilon_x) \in \mathrm{Ext}^1_{F(U)}(g^{-1}J^{(1)}_{L_x^\alpha}, g^{-1}\mathcal{O}_{L^{(1)}}(-2,0)) \to \mathrm{Ext}^1_{F(U)}(J^{(1)}_{\tilde{C}_x}, g^{-1}\mathcal{O}_{L^{(1)}}(-2,0)) \tag{4.3}$$

where $g^{-1}\epsilon_x$ is mapped to a class φ_x from the right-hand side of (4.3). The restrictions of these classes to $F(U) - \tilde{C}_x$ are identical and define a holomorphic section of the bundle $\wedge^2 S_+$ over $U - C_x$ when dropped to U by means of f. Since $J^{(1)}_{\tilde{C}_x}$ is locally generated by a single section, it is a locally free sheaf of $g^{-1}\mathcal{O}_{L^{(1)}}$-modules. Hence, we have the canonical identification

$$\mathrm{Ext}^1_{F(U)}(J^{(1)}_{\tilde{C}_x}, g^{-1}\mathcal{O}_{L^{(1)}}(-2,0)) = H^1(F(U), \mathrm{Hom}(J^{(1)}_{\tilde{C}_x}, g^{-1}\mathcal{O}_{L^{(1)}}(-2,0))). \tag{4.4}$$

It follows from (4.4) that multiplication of φ_x by a holomorphic function $f^{-1}s$ yields a cohomology class in the group

$$H^1(F(U), g^{-1}\mathcal{O}_{L^{(1)}}(-2,0)))$$

which corresponds to a global holomorphic section of the bundle $\wedge^2 S_+$ over U. Consequently, the section of $\wedge^2 S_+$ over U determined by the class ϵ_x (or φ_x) has a simple pole along C_x. This means that we have proved the theorem in the scalar case.

The case of Yang–Mills background can be proved by the same arguments. Therefore we do not repeat it here. □

5. Self-dual Yang–Mills backgrounds and singular connections

An (anti-)self-dual Yang–Mills field (E, ∇) on $U \subset \mathbf{CM}$ corresponds under the Radon–Penrose transform to the holomorphic vector bundle E_L on $L(U)$ which is the descent of a holomorphic vector bundle E_{P^α} defined on the $P^\alpha(P^\beta)$, which is trivial on P^α_x (P^β_x) with respect to the canonical embedding ρ_α (ρ_β). In this case, the restriction of E_L to $\tilde{L}^\alpha_x$ ($\tilde{L}^\beta_x$) is trivial for all x in U. We shall prove in this section the following result:

Theorem 5.1. *Let C be an analytic hypersurface in U, locally defined by an analytic section of $H^0(U, \mathcal{O})$, and let (E, ∇) be a self-dual Yang–Mills field, defined outside the hypersurface C. Suppose that C is noncharacteristic with respect to the riemannian structure on* **CM** *(that is the tangent vector space to C at every point does not contain any α- or β-plane).*

Then (E, ∇) can be extended to a locally free sheaf $\mathcal{E}$ on U such that in local bases near each point on C, the connection ∇ is given by a matrix of meromorphic sections having only first-order poles along C.

Proof. Since it is concerned with the self-dual case, one can consider the associated self-dual diagram of the Radon–Penrose transform of the form

$$\begin{array}{ccc} & Y \hookrightarrow P^{\alpha}(U) \times U & \\ g \swarrow & & \searrow f \\ Z = P^{\alpha}(U) & & X = U \end{array}$$

where $Y \hookrightarrow P^{\alpha}(U) \times U$ is the graph of the incident relation

$$\{\alpha\text{-plane intersecting with } U; \text{ a point in } U \text{ on this plane }\}.$$

Suppose (E_Z) is the Radon–Penrose transform of the self-dual Yang–Mills field (E, ∇), defined on $Z(U \setminus C)$. Since C is noncharacteristic, from Proposition 3.1 it follows that $Z(U \setminus C) = Z(U)$. Therefore E_Z is defined globally on $Z(U)$, and its inverse image with respect to g defines a locally free sheaf $\mathcal{E}_Y$ on Y with relative f-connection

$$\nabla_Y : \mathcal{E}_Y \to \mathcal{E}_Y \otimes T^*_{Y/X}$$

where $T^*_{Y/X}$ is the sheaf of g-relative 1-forms.

The above diagram is associated with the following exact sequence of locally free sheaves on Y

$$0 \to \mathcal{N}^* \to f^*T^*_X \to T^*_{Y/X} \to 0, \tag{5.5}$$

where $\mathcal{N}^*$ is the conormal sheaf of the induced embedding $Y \hookrightarrow P^{\alpha}(U) \times U$.

Now by tensoring $\mathcal{E}_Y$ with (5.5), followed by taking direct image functor $\mathbf{R}^{\cdot} f_*$, we obtain an exact sequence of coherent sheaves on U

$$0 \to f_*(\mathcal{N}^* \otimes \mathcal{E}_Y) \to \mathbf{T}^*_X \otimes f_*\mathcal{E}_Y \xrightarrow{\sigma} f_*(\mathbf{T}^*_{Y/Z} \otimes \mathcal{E}_Y) \to \mathbf{R}^1 f_*(\mathcal{N}^* \otimes \mathcal{E}_Y) \tag{5.6}$$

It is clear that $f_*\mathcal{E}_Y \equiv E$ outside C and C consists of the points $x \in X$ such that the restriction of $\mathcal{E}_Y$ on $f^{-1}(x)$ is not trivial. Since $f^{-1}(x) \equiv \mathbf{CP}^1$ and the field (E, ∇) is self-dual, it follows from cohomological criteria and Serre's duality theorem that C is also defined by the non-triviality of the cohomology groups $H^0(f^{-1}\mathcal{E}_Y(-1))$ and $H^1(f^{-1}\mathcal{E}_Y(-1))$. We have the isomorphism $\mathcal{N}^* = \mathcal{O}_Y(-1) \otimes f^*S_+$, and thus we can consider C as the support of the sheaves $f_*(\mathcal{N}^* \otimes \mathcal{E}_Y)$ and $\mathbf{R}^1 f_*(\mathcal{N}^* \otimes \mathcal{E}_Y)$. Note that the inverse Radon–Penrose transform of E_Z is the first-order differential operator $D = f_*(\nabla_{Y/Z})$, acting on the coherent sheaf $\mathcal{E} = f_*\mathcal{E}_Y$ with values in the coherent sheaf $\mathcal{E}' = f_*(\mathcal{E}_Y \otimes \mathbf{T}^*_{Y/Z})$, whose symbol is isomorphic to the mapping σ in (5.6). These sheaves are the coherent extensions of the Yang–Mills field E (considered as a locally free sheaf outside C) and $E \otimes \mathbf{T}^*_X$. In general $\mathcal{E}$ and $\mathcal{E}'$ may not be locally free. But the field E can be extended to a locally free sheaf $\bar{\mathcal{E}}$ globally on X, and in local bases near each point of C, and the operator D is given by a matrix with holomorphic coefficients. For we take a point $p \in C$ and a sufficiently small neighborhood $\mathcal{V}$ of a point on $f^{-1}(p)$ in Y such that $\mathcal{E}_Y$ is trivial in it. By extending $\mathcal{E}_Y$ together with $\nabla_{Y/Z}$ in a trivial way to $\mathcal{V}$ on the whole $f^{-1}f(\mathcal{V})$, and dropping all this onto X using f^*, we obtain locally trivial extension of E (and $E \otimes \mathbf{T}^*_X$) together with the operator $\bar{D}$

near to p. Set $\bar{\sigma} = \sigma|_{U-C}$. Obviously, the composite

$$\bar{\nabla} = \bar{\sigma}^{-1} \cdot \bar{D} : \bar{\mathcal{E}} \to \bar{\mathcal{E}} \otimes \mathbf{T}_X^*$$

coincides with the holomorphic connection ∇ outside C, in other words, it is an extension of the analytic connection for the Yang–Mills Field. We now prove that $\bar{D}$ is represented in local bases of the sheaves $\bar{\mathcal{E}}$ and $\bar{\mathcal{E}} \otimes \mathbf{T}_X^*$ by a matrix of local sections, having first-order poles along C. Indeed, the sheaves $\mathcal{E}$ and $\mathcal{E}'$ are reflexive on X, being direct images of locally free sheaves under f, and the ranks of $\mathcal{E} \otimes \mathbf{T}_X^*$ and $\mathcal{E}'$ are equal. Hence the sheaves of determinants $\det(\mathcal{E} \otimes \mathbf{T}_X^*)$ and $\det \mathcal{E}$ are of rank one, and the induced morphism $\det \sigma$ between them yields the divisor

$$\mathrm{Supp}(\mathrm{Ker} \det \sigma) = \mathrm{Supp}(\mathrm{Coker} \det \sigma).$$

In other words, this is the hypersurface determined by the zeros of a section of the rank one sheaf $\det(\mathcal{E} \otimes \mathbf{T}_X^*)^* \otimes \det \mathcal{E}'$ over X. The surjectivity of σ clearly implies that of $\det \sigma$. Conversely, let $\det \sigma$ be surjective at every point. Then the corresponding vector spaces induced by the fibres of the sheaves $\mathcal{E} \otimes \mathbf{T}_X^*$ and $\mathcal{E}'$ over the same point have the same dimension. Hence, by Nakayama's lemma, the morphism σ between these fibres must be surjective. Thus we have

$$\mathrm{Supp}(\mathrm{Ker} \det \sigma) = \mathrm{Supp}(\mathrm{Coker} \det \sigma) = C.$$

This shows that $\det \sigma$ is the local section defining C locally in X. In local bases near C, the matrices M_∇, representing the connection ∇ and the terms of zero order of the operator $\bar{D}$ are connected by the relation

$$M_\nabla = \Sigma^{-1} \cdot M_{\bar{D}}$$

where Σ is the matrix of the morphism σ. It is now easy to see that the coefficients of the matrix Σ^{-1}, and also those of M_∇, have only first-order poles along $C = \{\det \Sigma = 0\}$, since the coefficients of $M_{\bar{D}}$ can be holomorphically extended globally on X. This completes the proof of the Theorem. □

Acknowledgements. I would like to thank the Department of Mathematics, National University of Singapore, for the invitation to give a talk during the Pacific Rim Geometric Conference, 12–17 Dec, 1994, and their hospitality during the visit.

References

[1] R. Penrose, The twistor programme, Rep. Math. Phys. 12 (1977), 65–76.

[2] M. G. Eastwood, R. Penrose, and R. O. Wells, Jr., Cohomology and massless fields, Comm. Math. Phys. 78 (1981), 305–351.

[3] M. F. Atiyah, Geometry of Yang–Mills Fields, Lezioni Fermiane, Pisa 1979.

[4] A. D'Agnolo and P. Schapira, La transforme de Radon–Penrose des $\mathcal{D}$-modules, C. R. Acad. Sci. Paris 319 (1994), 461–466.

[5] Hoang Le Minh, Green Functions and Cohomology, Math.-USSR Izv. 26 (1986), 405–421.

[6] Hoang Le Minh, Singularities of complex analytic solutions of the self-dual Yang–Mills equations, Russian Math. Surveys, 43 (1988), 207–208.

Realization in the large in $\mathbb{R}^3$ of surfaces

Hong Jiaxing

Given a smooth two-dimensional Riemannian manifold (M^2, g), and the standard metric h on $\mathbb{R}^3$, can we find a map

$$\phi : M^2 \longrightarrow \mathbb{R}^3$$

such that $\phi^* h = g$? This is a long standing problem in Differential Geometry. In this talk we restrict ourselves to global results. As is well known, most global results are very closely related to the sign of the curvature of the given surface.

For the positive curvature case the first result is due to H. Weyl and H. Lewy. Their result states:

Theorem (Weyl–Lewy). *Any analytic positive curvature metric defined on S^2 always admits an analytic isometric embedding in $\mathbb{R}^3$.*

It was extended to the smooth case independently by A. V. Pogorelov [PO1] and L. Nirenberg [NR].

Theorem (Pogorelov–Nirenberg). *Any smooth positive curvature metric defined on S^2 always admits a smooth isometric embedding in $\mathbb{R}^3$.*

For the noncompact case we have:

Theorem (Olovjanisnikov–Pogorelov). *Any smooth, complete positive curvature metric defined on $\mathbb{R}^2$ always admits a smooth isometric embedding in $\mathbb{R}^3$.*

Based on Aleksandrov's theory on convex surfaces Olovjanisnikov obtained the weak solution for the complete positively curved $\mathbb{R}^2$ and Pogorelov proved that this weak solution is smooth if the metric is smooth. For the nonnegative curvature case we have, in [HZ]:

Theorem A. *Any C^4-nonnegative curvature metric defined on S^2 always admits a $C^{1,1}$ isometric embedding.*

It should be mentioned that Theorem A was also independently obtained by Guan and Li by a different approach in [GL]. For the noncompact nonnegatively curved plane the result remains true. We have, in [HO3]:

Theorem B. *Any complete C^4-nonnegative curvature metric defined on $\mathbb{R}^2$ always admits a $C^{1,1}$ isometric embedding in $\mathbb{R}^3$. Moreover it is smooth where the metric is smooth and the curvature positive.*

Remark 1. According to Olovjanisnikov's result in [PO1] any convex graph is completely determined by its metric, its limit cone and a prescribed limit generate of the limit cone. From this one infers that the solution in Theorem B is not unique. Indeed we have obtained more. Given a convex region $\Sigma \subset S^2$ whose area equals the total curvature of the above given surface one can find a $C^{1,1}$ isometric embedding in $\mathbb{R}^3$ such that its spherical image is contained in $\overline{\Sigma}$.

Remark 2. Alexsandrov gave a C^2 convex surface which is not C^3-continuous but realizes an analytic metric on S^2 with positive curvature except at one point. On the other hand, Pogorelov gave in [PO2] a C^{2+1} geodesic disk with nonnegative curvature not even admitting a C^2 local isometric embedding in $\mathbb{R}^3$ at the center of the disk. Therefore a natural open question is:

> Does there exist a $C^{2+\alpha}$ $(0 < \alpha < 1)$ isometric embedding in $\mathbb{R}^3$ for any sufficiently smooth (even analytic) nonnegatively curved sphere or plane?

For the negative curvature case most global results give a negative answer. The famous ones are the Hilbert theorem and Efimov's generalization [EF].

Theorem (Hilbert–Efimov). *Any complete surface with negative constant curvature (with curvature bounded above by a negative constant) has no C^2 isometric immersion in $\mathbb{R}^3$.*

Indeed, Efimov's result provides a necessary condition for a complete negatively curved surface to embed isometrically in $\mathbb{R}^3$, that is

$$\inf_M |k| = 0.$$

To the author's knowledge, for a complete negatively curved surface, no affirmative answer has been known. The following problem posed by Yau in [Y] should be mentioned:

> Find a nontrivial sufficient condition for a complete negatively curved surface to embed isometrically in $\mathbb{R}^3$.

He also pointed out that such a nontrivial condition might be the rate of decay of the curvature at infinity. Recently some development in this direction has been made in [HO4]. Let M be a simply connected noncompact complete surface of negative curvature k. By Hadamard's theorem $\exp : T_p(M) \longrightarrow M$ is a global diffeomorphism for each point $p \in M$ and it induces global geodesic polar coordinates (ρ, θ) on M.

Theorem C. *Suppose that*
(1) *for some $\delta > 0$, $\rho^{2+\delta}|k|$ is decreasing in ρ outside a compact set, and that*
(2) $\partial_\theta^i \ln |k|$, $i = 1, 2$ *and* $\rho\partial_\rho\partial_\theta \ln |k|$ *are bounded on M.*
Then M always admits a smooth isometric immersion in $\mathbb{R}^3$.

Remark 1. If $M \in C^{s,1}$ $(s \geq 4)$ and (1) and (2) are satisfied, then it admits a $C^{s-1,1}$ isometric immersion in $\mathbb{R}^3$.

Remark 2. Assumption (1) implies that

$$|k| \leq \frac{A}{\rho^{2+\delta}},$$

outside a compact set for some constant A. It is not clear now what will happen if $\delta = 0$. Moreover, there is a gap between Efimov's necessary condition and the sufficient condition here for a complete negatively curved surface to embed isometrically in $\mathbb{R}^3$. How can one fill this gap? Maybe this is a more interesting open problem.

For the mixed curvature case the situation is more complicated even for the local isometric embedding. The first local result is due to C. S. Lin. In [LI] he obtained a local isometric embedding in $\mathbb{R}^3$ for any metric with curvature clearly changing its sign at the point considered. Recently some semiglobal isometric embeddings for the metric with the curvature clearly changing its sign has been obtained in [HO2]. Let M be a smooth surface of curvature k. Set

$$M^+ = \{p \in M \mid k(p) > 0\}.$$

Theorem D. *Let M^+ be simply connected and bounded. Suppose that $dk|_{\partial M^+} \neq 0$ and ∂M^+ has positive geodesic curvature to M^+. Then there is a neighborhood of M^+ admitting a smooth isometric embedding in $\mathbb{R}^3$.*

Remark 1. The assumption that ∂M^+ has positive geodesic curvature cannot be removed. Gromov and Rochlin in [GR] gave an analytic geodesic disk of positive curvature not admitting a C^2 isometric embedding whose boundary is just of negative geodesic curvature.

Remark 2. In the domain of the isometric embedding obtained in Theorem D the region where the curvature is positive is global and the region where the curvature is negative is local. This is the reason why we call it a semiglobal result.

Sketch of the proof for Theorem B. If the curvature k is identically zero, it is easy to see that the exponential map centered at any point gives the required isometric embedding. Therefore we always assume that $k > 0$ in $B_R(p)$ for some point $p \in \mathbb{R}^2$. Under the present circumstances the proof is split into three steps.

Step 1. Applying Proposition 8.5 in [CE] to the present complete nonnegatively curved plane one can find a sequence of simply connected totally convex bounded regions $\widetilde{D}_j$ monotone-increasing and exhausting $\mathbb{R}^2$. Then for each j, approximate $\partial\widetilde{D}_j$ by a C^4-continuous convex curve which is composed of geodesic segments and several convex arcs. On each $\widetilde{D}_j$, without difficulty, one can find a C^4-continuous positive curvature metric $\tilde{g}_j$ conformal to g such that $\partial\widetilde{D}_j$ has positive geodesic curvature. Finally the mollifying technique provides at once a sequence of smooth bounded domains D_j and a sequence of smooth metrics g_j defined on $\overline{D}_j$ such that

(1) $p \in D_1 \subset D_2 \subset \cdots \subset D_j \subset \cdots$ and $\bigcup D_j = \mathbb{R}^2$,

(2) g_j has positive curvature and ∂D_j is of positive geodesic curvature in g_j,

(3) for each compact set ω, $|g_j - g|_{C^4(\omega)} \longrightarrow 0$ as $j \longrightarrow \infty$.

Step 2. To find a smooth graph $\vec{r}_j$ for each j which is a smooth isometric embedding of (D_j, g_j). The existence of these isometric embeddings comes from the following theorem of Pogorelov in [PO1]:

Theorem. *Let g be a smooth positive curvature metric defined in a simply connected domain $\overline{D} \subset \mathbb{R}^2$ and let ∂D be of nonnegative geodesic curvature. Then (D, g) admits a smooth isometric embedding in $\mathbb{R}^3$ which is a convex cap.*

By a convex cap we mean a convex surface whose boundary lies on a plane and the projection on this plane of this surface is a one-to-one mapping.

Step 3. For each compact set ω, find a constant C_ω such that

$$|D\vec{r}_j|,\ |D^2\vec{r}_j| \le C_\omega \quad \text{on } \omega \text{ for all } j.$$

The first inequality follows immediately from the fact that $\vec{r}_j$ realizes the metric g_j and g_j tends to g in $C^4(\omega)$ for any compact set ω. The proof of the second inequality is the crucial part of the proof of the present theorem. Since all the surfaces $\vec{r}_j$ are convex it suffices to evaluate the above bound of their mean curvature. Set

$$\rho = -\frac{1}{2}(\vec{r}, \vec{r}).$$

A direct computation gives

$$H = \frac{\Delta_g \rho + 2}{2\sqrt{-2\rho - |\nabla\rho|^2}}.$$

Thus the C^2 estimates are reduced to the proofs for the following lemmas if we choose the origin suitably.

Lemma 1. $-2\rho - |\nabla\rho|^2 \geq \frac{1}{C}$, *for some positive constant* C.

Lemma 2. *For each compact set* ω, $\Delta_g\rho + 2 \leq C_\omega$ *on* ω *for some constant* C_ω *independent of* j.

Let us first look at Lemma 1. $-2\rho - |\nabla\rho|^2$ is nothing else but the square of the Minkowsky function $p = \langle\vec{r}, \vec{n}\rangle$ of the surface in question, where $\vec{n}$ is the unit normal at $\vec{r}$. To prove Lemma 1, it suffices to show that there is a ball centered at the origin inscribed in the above region of the convex graph with the radius of positive lower bound depending only on the intrinsic quantities of $\vec{r}$. Such an intrinsic estimation was obtained in [CY] for a closed convex surface. Following the idea of [CY] the lower bound for the radii of balls inscribed in a closed convex surface is completely determined by the lower bound of the areas of the projections on any hyperplane of the considered convex surface and the intrinsic diameter of the surface. Now for convex graphs a similar estimation can be obtained.

Lemma 3. *Let* B_R *be a geodesic disk of positive curvature* k *where* R *is less than the convex radius at the center. Suppose that* $\vec{r} = (x, y, z) : B_R \longrightarrow \mathbb{R}^3$ *is the isometric embedding of* B_R *which is a* C^2*-convex graph with respect to the z-axis. Then the projection onto the* x-y *plane of the convex graph contains a disk with radius*

$$r_0 \geq \frac{1}{C}$$

where $C = C(\max k, L, 1/(\inf \int_{HB_R} k\,d\sigma))$, L *is the intrinsic diameter of* B_R *and* HB_R *denotes some half geodesic disk of* B_R*, and where the infimum is taken over all possible half geodesic disks.*

Proof. Suppose the center of the given geodesic disk is mapped to the origin in $\mathbb{R}^3$. Denote by Σ the level set $\{\vec{r}(p) = (x, y, z) \mid z(p) = 0\}$.

Case 1. $\Sigma \cap \vec{r}(\partial B_R) = \emptyset$. In the present case we can show the projection Π onto the x-y plane of $\vec{r}(B_R)$ is convex too. In view of the fact that R is less than the convex radius at the center, we can use the geodesic polar coordinates on B_R and the metric is of the form

$$g = d\rho^2 + G^2\,d\theta^2, \quad 0 \leq \rho \leq R.$$

It turns out that

$$dx^2 + dy^2 = g - dz^2 = (1 - z_\rho^2)\,d\rho^2 - 2z_\rho z_\theta\,d\rho\,d\theta + (G^2 - z_\theta^2)\,d\theta^2.$$

The curvature of $\partial\Pi$, as a plane curve, equals

$$\tilde{k} = \frac{G^2G_\rho(1 - |\nabla z|^2) + Gz_\rho(z_{\theta\theta} - (G_\theta/G)z_\theta + GG_\rho z_\rho)}{\sqrt{1 - |\nabla z|^2}(G^2 - z_\theta^2)^{3/2}}.$$

Since G_ρ and $(1 - |\nabla z|^2)$ are positive and the Hessian of z is positive definite, it follows that the second term of the numerator is also positive and hence Π is convex. Next we turn to compute its area $|\Pi|$.

$$|\Pi| = \int_{\vec{r}(B_R)} \cos(\vec{n}, \vec{k})\, d\sigma.$$

The worst case occurs when all the spherical images $G(\vec{r}(B_R))$ are distributed and concentrated on a strip $\theta \in [\frac{\pi}{2} - \delta, \frac{\pi}{2})$ near the equator, where $\delta = \sin^{-1}((\int k)/2\pi)$. It turns out that

$$|\Pi| = \int_{G(\vec{r}(B_R))} \cos(\vec{n}, \vec{k})\, \frac{d\omega}{k} \geq \frac{(\int k)^2}{4\pi \max k}.$$

For each direction on the x-y plane, one can find a pair of supporting straight lines parallel to this direction. Minkowsky defined the minimum of the distances of all pairs of the parallel supporting straight lines as the *width* d of this convex domain Π. Next, we proceed to evaluate the lower bound of d. Indeed the area of Π is less than the diameter of Π times d. Consequently, in view of the fact that the diameter of Π is less than the intrinsic diameter of $\vec{r}(B_R)$, it turns out that

$$d\big(\text{the intrinsic diameter of } \vec{r}(B_R)\big) \geq |\Pi| \geq \frac{(\int k)^2}{4\pi \max k}.$$

From this it follows immediately that $d \geq C^{-1}$ for some constant C depending only on the intrinsic quantities of the metric g. In the sequel, unless otherwise stated, all such constants are always called the constants under control. Let L denote the intrinsic diameter of $\vec{r}(\partial B_R)$. Now, without difficulty, one can find a triangle $p_1 p_2 p_3$ contained in Π such that all the sides are less than or equal to L, the side $p_2 p_3 \geq d$ and the perpendicular segment from p_1 to the side $p_2 p_3$ is bigger than or equal to $d/2$. It turns out that its interior tangent circle is of radius $r_0 \geq d^2/32L$.

Case 2. $\vec{r}(\partial B_R) \cap \Sigma \neq \emptyset$. Under the present circumstances, generally speaking, the projection mentioned above ceases to be convex. Instead of the whole graph we consider the part $\vec{r}(B_R) \cap \{z > 0\}$ denoted by σ. Note that σ contains $\vec{r}(HB_R)$ for some half geodesic disk HB_R since from $z_{\rho\rho} > 0$ we infer that the geodesic tangent to Σ at $\vec{r}(0)$ always lies in the set $\{z \geq 0\}$. Hence the area of the projection Π of σ satisfies

$$|\Pi| \geq \int_{\vec{r}(HB_R)} \langle \vec{n}, \vec{k} \rangle\, d\sigma \geq \frac{\left(\inf \int_{HR} k\right)^2}{4\pi \max k}.$$

Although the projection of σ may not be convex, one can find a convex subregion of Π with area bigger than or equal to $|\Pi|$ times a universal constant. It is evident that the diameter of this convex subregion is also less than the intrinsic diameter of $\vec{r}(\partial B_R)$. A similar argument to that in Case 1 will soon complete the proof of the present lemma. □

Lemma 3 enables us to reduce the proof of Lemma 1 to a local investigation. Indeed, set h equal to the maximum of the projections onto the z axis of the surface $\vec{r}$. Lifting the inscribed disk constructed in Lemma 3 along the unit vector of the z axis produces a circular cylinder ω. If we take as the origin the center of the disk which is the intersection of the plane $z = h + r_0$ and ω, Lemma 1 is true on $\vec{r}(B_R)$ for the constant $C = 1/r_0^2$ and hence, continues to be true on any convex graph with respect to the same z axis containing $\vec{r}(B_R)$. This proves Lemma 1.

By means of Lemma 1 and Heinz's interior estimates [HE2] on convex surfaces it is easy to get the following C^2-interior estimates for convex graphs with positive curvature.

Lemma 4. *Let B_R be a C^3-geodesic disk of positive curvature and let its isometric embedding $\vec{r} : B_R \longrightarrow \mathbb{R}^3$ be a C^2-convex graph. Then*

$$|D^2\vec{r}|_{C(B_{R-\delta})} \leq C\Big(L, |g|_{C^3(B_R)}, \frac{1}{\min k}, \frac{1}{\min \det(g_{ij})}, \frac{1}{\delta}\Big).$$

Proof. According to Heinz's interior estimates, we have

$$|D^2\vec{r}|_{C(B_{R-\delta})} \leq C\Big(L, |g|_{C^3(B_R)}, \frac{1}{\min k}, \frac{1}{\min \det(g_{ij})}, \frac{1}{\delta}, \int H\Big).$$

The last quantity of the right hand side of the last inequality involves extrinsic quantities of the convex surface in question and hence it is inconvenient for applications. With the aid of Lemma 1 for convex graphs we can directly evaluate the above bound of the total mean curvature and hence, remove this extrinsic quantity. Let us now compute its total mean curvature:

$$\begin{aligned}\int_{\vec{r}(B_R)} H\, d\sigma &= \int_{\vec{r}(B_R)} \frac{\Delta_g \rho + 2}{2\sqrt{-2\rho - |\nabla\rho|^2}}\, d\sigma \\ &\leq C\Big[\int_{B_R} d\sigma + \int_{\partial B_R} |\nabla\rho|\, dl\Big].\end{aligned}$$

In getting the last inequality we have used Lemma 1 and integration by parts. Thus we have proved that the total mean curvature is bounded above by a constant depending only on the intrinsic quantities of $\vec{r}(B_R)$ if we note that Lemma 1 and $|\nabla\rho|^2 \leq 2|\rho| \leq \sup|\vec{r}|^2$, where the supremum is taken over the whole geodesic disk B_R. □

Thus Lemma 4 implies Lemma 2 if the curvature of the given surface is positive everywhere. Under the present circumstances the curvature may be zero somewhere. Therefore we have to turn to prove Lemma 2 from the equation satisfied by ρ which is as follows:

$$\det(\nabla\rho_{ij} + g_{ij}) = k \det(g_{ij})(-2\rho - |\nabla\rho|^2).$$

Following Pogorelov and Cafarelli–Nirenberg–Spruck's technique we take, as an auxiliary function,

$$W = (\Delta_g \rho + 2)\exp\Big(\frac{\lambda}{2}|\nabla\rho|^2 + \phi\Big),$$

where λ is a constant to be fixed and ϕ is a function smooth inside the domain considered, tending to $-\infty$ as points approach the boundary of the domain. Suppose that W attains its maximum at some point p^* inside. We shall take an orthonormal moving frame near p^* and without loss of generality, after a rotation, we may assume that the Hessian of ρ at p is diagonal. Hence, at p^*

$$W_i = 0: \quad (\Delta\rho)_i + (\Delta\rho + 2)(\lambda\rho_i\rho_{ii} + \phi_i) = 0,$$

and with $F^{ii} = \partial\det(\nabla\rho_{ij} + g_{ij})/\partial\rho_{ij}$,

$$\begin{aligned} 0 &\geq \exp(-(\lambda/2)|\nabla\rho|^2 - \phi)F^{ii}W_{ii} \\ &= (\Delta\rho + 2)F^{ii}(\phi_{ii} + \phi_i^2) + 2F^{ii}\phi_i((\Delta\rho)_i + (\Delta\rho + 2)\lambda\rho_i\rho_{ii}) \\ &\quad + F^{ii}[(\Delta\rho)_{ii} + 2(\Delta\rho)_i\lambda\rho_i\rho_{ii} + \lambda(\rho_{ii}^2 + \rho_l\rho_{lii}) + \lambda^2\rho_i^2\rho_{ii}^2] \end{aligned}$$

or

$$\begin{aligned} 0 &\geq \exp(-(\lambda/2)|\nabla\rho|^2 - \phi)F^{ii}W_{ii} \\ &= (\Delta\rho + 2)F^{ii}(\phi_{ii} - \phi_i^2) + F^{ii}[(\Delta\rho)_{ii} - 2\lambda\phi_i\rho_i\rho_{ii}(\Delta\rho + 2) + \lambda\rho_l\rho_{lii}] \\ &\quad + \lambda(\Delta\rho + 2)F^{ii}(1 - \lambda\rho_i^2)\rho_{ii}^2 \\ &= I_1 + I_2 + I_3. \end{aligned}$$

Obviously,

$$I_3 \geq \lambda(\Delta\rho + 2)^2(1 - \lambda|\nabla\rho|^2) - C(\Delta\rho + 3).$$

Now we are in a position to fix the function ϕ. Translate the origin to $\vec{r}(p)$ and rotate the axis such that the interior normal at $\vec{r}(p)$ is the axis $\tilde{z}$. In the new coordinates, the position vector, denoted by $\vec{\tilde{r}} = (\tilde{x}, \tilde{y}, \tilde{z})$, is the rotation of $\vec{r} - \vec{r}(p)$. For each $h \in (R, \infty)$, set

$$D_h = \{q \in \mathbb{R}^2 \mid \tilde{z}(q) < h\}.$$

We can show that

$$B_h \subset D_h \subset D_{j(h)}$$

for some integer $j(h)$. This implies that for each h, $\overline{D}_h$ is compact. Define

$$\phi = -\frac{4}{1 - (\tilde{z}/h)}.$$

A computation gives

$$I_1 \geq -C_h\phi^2(\Delta\rho + 2) - (3/2)F^{ii}\phi_i^2(\Delta\rho + 2).$$

Let us focus our attention on I_2. Differentiation of the original equation using the Ricci identity gives

$$I_2 \geq 2(\rho_{12l}\rho_{12l} - \rho_{11l}\rho_{22l}) - 2\lambda F^{ii}\phi_i\rho_i\rho_{ii}(\Delta\rho + 2) - C_h(1+\phi^2)(\Delta\rho + 3).$$

Set $f = k\det(g_{ij})(-2\rho - |\nabla\rho|^2)$. Solving the third derivatives from the system $F^{ii}\rho_{iil} = f_l$ and $\rho_{iil} = (\Delta\rho)_l$, $l = 1, 2$ and combining the expression $W_l = 0$, $l = 1, 2$, we have,

$$\begin{aligned} 2(\rho_{12l}\rho_{12l} - \rho_{11l}\rho_{22l}) &\geq (3/2)F^{ii}\phi_i^2(\Delta\rho + 2) + (1/2)F^{ii}(\phi_i + \lambda\rho_i\rho_{ii})^2(\Delta\rho + 2) \\ &\quad + 2\lambda F^{ii}\phi_i\rho_i\rho_{ii}(\Delta\rho + 2) - 8\lambda L^2 k(\Delta\rho + 2)^2 - C_h(1+\phi^2)(\Delta\rho + 3), \end{aligned}$$

where L is the intrinsic diameter of D_h. It is bounded since D_h is compact for each $h \in (R, \infty)$. Summing up we obtain

$$\begin{aligned} 0 &\geq \exp(-(\lambda/2)|\nabla\rho|^2 + \phi)F^{ii}W_{ii} \\ &\geq \lambda e^{2\phi}(\Delta\rho + 2)^2(1 - \lambda L^2 - 8L^2 k) - C_h(1+\phi^2)e^{\phi}(\Delta\rho + 3)e^{\phi}. \end{aligned}$$

If we choose $\lambda L^2 < 1/4$ and assume $L^2 k(p^*) < 1/32$, it is easy to get the estimation for the above bound of W. Suppose that $L^2 k(p^*) \geq 1/32$. Then Lemma 4 provides the estimation of $W(p^*)$, hence also the above bound of W and the above bound of $(\Delta_g\rho + 2)$ on $D_{h/2}$. This completes the proof of Lemma 2. □

References

[AL] Aleksandrov, A. D., Intrinsic geometry of convex surfaces, Moscow, 1948, German transl., Academie Verlag, Berlin 1955.

[CE] Cheeger, J. and Ebin, D., Comparison theorems in Riemannian geometry, North-Holland Publishing Company, 1975.

[CY] Cheng, S. Y. and Yau, S. T., On the regularity of the n-dimensional Minkowski problem, Comm. Pure Appl. Math. 29 (1976), 495–516.

[EF] Efimov, N. E., Generation of singularities on surface of negative curvature, Mat. Sb. 64 (1964), 286–320.

[GL] Guan, P. and Li, Y. The Weyl problem with nonnegative Gauss curvature. J. Differential Geom. 39 (1994), 331–342.

[GR] Gromov, M. L. and Rochlin, V. A., Embeddings and Immersions in Riemannian Geometry, Russian Math. Surveys (1971), 1–57; Uspekhi Mat. Nauk 25:5 (1970), 3–62.

[HE1] Heinz, E., On elliptic Monge–Ampere equations and Weyl's embedding problem, J. Analyse Math. 7 (1959–1960), 1–52.

[HE2] —, Neue a priori Abschätzungen für den Ortsvektor einer Fläche positiver Gaussscher Krümmung durch ihr Linienelement, Math. Z. 74 (1960), 129–157.

[HO1] Hong, J. X., Dirichlet problems for general Monge-Ampere equations, Math. Z. 209 (1992), 289–306.

[HO2] —–, Darboux equations and isometric embedding in $\mathbb{R}^3$ of Riemannian manifolds with nonnegative curvature I, to appear in Math. Ann.

[HO3] —–, Isometric embedding in $\mathbb{R}^3$ of complete noncompact nonnegatively curved surface, to appear in Manuscripta Math.

[HO4] —–, Realization in $\mathbb{R}^3$ of the complete Riemannian manifolds with negative curvature, Comm. Anal. Geom. 1 (1993), 487–514.

[HZ] Hong, J. X. and Zuily, C., Isometric embedding in $\mathbb{R}^3$ of the 2-sphere with non-negative curvature, Math. Z. 219 (1995), 323–334.

[IA] IAIA, J. A., Isometric embeddings of surfaces with nonnegative curvature in $\mathbb{R}^3$, Duke Math. J. 67 (1992), 423–459.

[LI] Lin, C. S., The local isometric embedding in $\mathbb{R}^3$ of two dimensional Riemannian manifolds with Gaussian curvature changing sign clearly, Comm. Pure Appl. Math. 39 (1986), 307–326.

[NI1] Nirenberg, L., The Weyl and Minkowski problems in Differential Geometry in the large, Comm. Pure Appl. Math. 6 (1953).

[NI2] —–, On nonlinear elliptic partial differential equations and Hölder continuity, Comm. Pure Appl. Math. 6 (1953), 103–156.

[PO1] Pogorelov, A. V., Extrinsic geometry of convex surfaces, Transl. Math. Monographs 35, Amer. Math. Soc., Providence, RI, 1973.

[PO2] —–, An example of a two dimensional Riemannian metric not admitting a local realization in $\mathbb{R}^3$, Dokl. Akad. Nauk. USSR. 198 (1971), 42–43.

[Y] Yau, S. T., Seminar on differential geometry, Princeton University Press, 1982.

Borsuk–Ulam type theorems and systems of bilinear equations

*Kee Yuen Lam**

Abstract. It is shown that the Borsuk–Ulam theorem, particularly the "equal height" case of this theorem, can be used to provide concise and uniform derivation of several major results in the classical problem of sectioning vector bundles over real projective spaces, or in the almost equivalent problem of nontrivial solvability of systems of bilinear equations over the field of real numbers.

1991 Mathematics Subject Classification: 57N10

1. Introduction

Let us begin by considering a system of equations

$$(*) \qquad \begin{cases} f_1(x_1, \dots, x_m) = 0 \\ \quad\vdots \\ f_n(x_1, \dots, x_m) = 0. \end{cases}$$

When each f_j is a homogeneous linear form in $x_1, \dots, x_m$ (say with real coefficients), Gaussian elimination confirms the validity of the following assertion:

(A) If the number m of unknowns exceeds the number n of equations, then $(*)$ admits nontrivial solutions in $(x_1, \dots, x_m)$.

It turns out, however, that Gaussian elimination isn't quite relevant here. Much more relevant is perhaps the Borsuk–Ulam theorem (B-U), which states that

Theorem 1.1. *For $m > n$, there is no continuous map $G : S^{m-1} \longrightarrow S^{n-1}$ satisfying $G(-x) = -G(x)$ for all $x \in S^{m-1}$.*

Here S^{m-1} denotes the unit sphere in the Euclidean space $\mathbb{R}^m$, and the condition $G(-x) = -G(x)$ is referred to as the "antipodal" condition. Using (B-U) one can

*Research supported by a grant from the National Science and Engineering Research Council of Canada.

easily prove, by contradiction, that assertion (A) remains true if each f_j in $(*)$ is a continuous odd function in $(x_1, \ldots, x_m)$. Here again a non-trivial solution means one for which $x_1, \ldots, x_m$ are not all zero.

Another similar situation is dealt with in classical algebraic geometry. There, assertion (A) is confirmed as a "rule of thumb" in the case when each f_j is a (homogeneous) polynomial function in the *complex* variables $x_1, \ldots, x_m$. Our present article, however, will concentrate on cases over $\mathbb{R}$, so unless otherwise stated the f_j's from now on are assumed to be real-valued continuous functions, and nontrivial *real* solutions in the unknowns are to be sought. A moment's reflection shows that in this generality assertion (A) cannot be expected to serve as a rule of thumb; its inadequacy is clearly seen in the system consisting of one single equation

$$x_1^2 + x_2^2 + \cdots + x_m^2 = 0$$

with $m > 1$. Indeed this example shows that for systems of quadratic equations there is no reason to expect any general principle resembling (A) to hold. Rather, the most immediate generalization beyond linear or odd systems is to the case of *bilinear systems*, in which each f_j $(1 \leq j \leq n)$ is a real bilinear form

$$f_j = f_j(x_1, \ldots, x_r; y_1, \ldots, y_s) = \sum_{1 \leq p \leq r,\ 1 \leq q \leq s} a_j^{pq} x_p y_q, \quad (a_j^{pq} \in \mathbb{R})$$

in two sets of unknowns $x_1, \ldots x_r$ and $y_1, \ldots, y_s$. Of course, a nontrivial solution to $(*)$ now means one for which $x_1, \ldots, x_r$ are not all zero *and* $y_1, \ldots, y_s$ are not all zero. For such systems it is certainly an interesting problem to discover various numerical conditions on r, s and n, like those in (A), that can guarantee the existence of nontrivial solutions. From the remarks above it isn't surprising that the Borsuk–Ulam theorem ought to play a major rôle in this problem, even though its relevance has hardly been observed in the literature. Part of the purpose of this article is to bring out this major rôle. We shall show that the core results in the papers of James [7] and Davis [4], when interpreted in the context of bilinear systems, are actually direct consequences of Borsuk–Ulam type theorems and Adams' classical result on vector fields on spheres [1]. We emphasize in this article the novelty of the proof rather than of the results, leaving generalizations and refinements of results to a forthcoming paper.

2. Bilinear systems of (real) equations

For systems of real bilinear equations

$$(\#) \qquad \begin{cases} f_1(x_1, \ldots, x_r; y_1, \ldots, y_s) = 0 \\ \vdots \\ f_n(x_1, \ldots, x_r; y_1, \ldots, y_s) = 0 \end{cases}$$

two famous and classical theorems are known, both of which being highly nontrivial. The first is the Stiefel–Hopf theorem (1941), whose statement requires a numerical function $r \circ s$. Recall that the real projective space $\mathbb{R}P^{m-1}$ has cohomology ring with mod 2 coefficients equal to

$$H^*(\mathbb{R}P^{m-1}; \mathbb{Z}_2) = \mathbb{Z}_2[w]/(w^m)$$

where w denotes the unique nonzero cohomology class in dimension one, and (w^m) is the ideal generated by w^m in the polynomial ring $\mathbb{Z}_2[w]$. By definition, $r \circ s$ is the *height* of the cohomology class $\theta = w \otimes 1 + 1 \otimes w$ in the cohomology ring

$$H^*(\mathbb{R}P^{r-1} \times \mathbb{R}P^{s-1}; \mathbb{Z}_2) = H^*(\mathbb{R}P^{r-1}; \mathbb{Z}_2) \otimes H^*(\mathbb{R}P^{s-1}; \mathbb{Z}_2),$$

namely, it is equal to the positive integer h such that $\theta^h = 0$ but $\theta^{h-1} \neq 0$. As an alternative definition, $r \circ s$ is the smallest integer h for which the binomial coefficients $\binom{h}{t}$ are even for all t satisfying $h - r < t < s$. Note that $h = r + s - 1$ is one such integer, for then the binomial conditions become void. It is also clear that the smallest such h gives the height of θ. Further, in addition to the obvious inequality

$$\mathrm{Max}\{r, s\} \leq r \circ s \leq r + s - 1,$$

a little checking using standard facts on the parity of binomial coefficients will confirm that $r \circ s (= s \circ r)$ has the following recursive properties: $1 \circ s = s$ and

$$r \circ s = \begin{cases} 2^t & \text{if } 2^{t-1} < r, s \leq 2^t \\ 2^{t-1} + r \circ (s - 2^{t-1}) & \text{if } r \leq 2^{t-1} < s. \end{cases}$$

These allows for quick evaluation of any desired value. For example:

$$14 \circ 37 = 32 + 14 \circ (37 - 32) = 32 + 5 \circ 14 = 32 + 8 + 5 \circ 6 = 32 + 8 + 8 = 48.$$

The statement of the classical Stiefel–Hopf theorem (S-H) is simply

Theorem 2.1 ([13], [6]). *The system (#) always admits nontrivial solutions if* $n < r \circ s$.

To compare with assertion (A), let us write the numerical condition $n < r \circ s$ as

$$r + s > n + (r + s - r \circ s).$$

What (S-H) asserts on real bilinear systems is that for nontrivial solutions to exist it suffices to replace condition (A) of Section 1 by the condition that the number $r + s$ of unknowns exceed the number n of equations by a gap wider than $r + s - r \circ s$!

Unfortunately, rather than being the final word on bilinear systems, the Stiefel–Hopf theorem is only a beginning. For many values of r, s and n, the "gap" it requires is definitely too wide! A good case in point is Adams' vector field theorem, for whose statement we shall need another numerical function $\rho(n)$, first considered by A. Hurwitz and J. Radon. Write a positive integer n as $n =$ an odd number $\times 2^{a+4b}$, where $0 \leq a \leq 3$. Now define $\rho(n) = 2^a + 8b$. The first few values of ρ are given in the following table

n	1	2	3	4	5	6	7	8	9	10	11	12	13	14	15	16	17	⋯
$\rho(n)$	1	2	1	4	1	2	1	8	1	2	1	4	1	2	1	9	1	⋯

and the general value is given periodically by $\rho(16n) = \rho(n)+8$. What Adams proved via homotopy theory in [1] is that $\rho(n) - 1$ is the maximal number of everywhere independent vector fields tangential to the sphere S^{n-1}. When applied to bilinear systems, this theorem can be paraphrased as

Theorem 2.2. *The system (#) always admits nontrivial solutions if the following numerical conditions are satisfied: $n = r$ and $\rho(n) < s$.*

Thus, *in the special case $r = n$*, nontrivial solvability is guaranteed by requiring $r+s > n+\rho(n)$, namely that the number of unknowns exceed the number of equations by a gap wider than $\rho(n)$, which is much narrower than the gap required by (S-H).

It is now clear that one should try to determine, for various kinds of r, s and n, necessary and sufficient conditions to guarantee the nontrivial solvability of an arbitrary bilinear system (#). The joint paper [10], for example, offered a small extension of Adams' theorem, addressing the case when r, while not equal to n, is no more than 5 units behind it. In fact, there are many papers on homotopy theory, almost too numerous to cite, whose results can be interpreted as work along this direction. But up to this day, the ultimate necessary and sufficient conditions still elude us.

3. Extending (S-H) as well as Adams' theorem

The numerical condition $n = r$ in Adams' theorem is clearly undesirable: it limits itself to the special case when the number of equations equals the number of x-variables. It is interesting, therefore, to observe that Adams' geometric and homotopical considerations, when combined suitably with a Borsuk–Ulam type argument, can yield the much more general result below.

Theorem 3.1. *A bilinear system (#) always admits nontrivial solutions if the following numerical conditions are satisfied*

$$\binom{n-1}{r-1} \equiv 1 \pmod 2 \textit{ and } r + s > n + \rho(n).$$

Notice that the condition $n = r$ is now drastically relaxed to the condition that $\binom{n-1}{r-1}$ be odd.

To see a proof as well as the connection with (S-H), assume on the contrary that the system (#) in the variables $x_1, \ldots, x_r$ and $y_1, \ldots, y_s$ has no nontrivial solutions. With this assumption, one can define a map

$$F : S^{r-1} \times S^{s-1} \longrightarrow S^{n-1}$$

given by $F(x, y) = \frac{1}{N}(f_1(x, y), f_2(x, y), \dots, f_n(x, y))$, where $x = (x_1, \dots, x_r)$ and $y = (y_1, \dots, y_s)$ are unit vectors and N is the norm of the nonzero vector $(f_1(x, y), \dots, f_n(x, y))$ in $\mathbb{R}^n$. Clearly $F(-x, y) = F(x, -y) = -F(x, y)$, so that F factors through the quotient space $S^{r-1} \times_T S^{s-1}$ obtained by identifying each point (x, y) in $S^{r-1} \times S^{s-1}$ with its "antipodal point" $T(x, y) = (-x, -y)$. One may think of $S^{r-1} \times_T S^{s-1}$ as a kind of "projective torus", and note that it has on its own self a fixed-point-free involution τ given by

$$((x, y)) \longleftrightarrow \tau((x, y)) \stackrel{\text{defn}}{=} ((-x, y)) = ((x, -y)).$$

When corresponding points under τ are further identified, $S^{r-1} \times_T S^{s-1}$ clearly becomes $\mathbb{R}P^{r-1} \times \mathbb{R}P^{s-1}$, the Cartesian product of two real projective spaces. The entire situation can be summarized by the diagram (†) below, where vertical arrows denote identification maps, which are also double covering maps, and where $\widetilde{F}$ and PF are maps induced by F.

$$(\dagger)\qquad \begin{array}{ccc} S^{r-1} \times S^{s-1} & \xrightarrow{\quad F \quad} & S^{n-1} \\ \downarrow & & \\ S^{r-1} \times_T S^{s-1} & \nearrow \widetilde{F} & \Big\downarrow \\ \downarrow & & \\ \mathbb{R}P^{r-1} \times \mathbb{R}P^{s-1} & \xrightarrow[\quad PF \quad]{} & \mathbb{R}P^{n-1} \end{array}$$

We shall call PF the "projectivisation" of F, a map which exists under the assumption that (#) is only trivially solvable. One might recall here that Hopf [6] obtained theorem (S-H) by simply considering the homomorphism induced by PF on mod 2 cohomology ring, *i.e.*,

$$(PF)^* : H^*(\mathbb{R}P^{n-1}; \mathbb{Z}_2) \longrightarrow H^*(\mathbb{R}P^{r-1} \times \mathbb{R}P^{s-1}; \mathbb{Z}_2).$$

In fact, by restricting PF to $\mathbb{R}P^0 \times \mathbb{R}P^{s-1}$ and $\mathbb{R}P^{r-1} \times \mathbb{R}P^0$, Hopf first reasoned that $(PF)^*(w) = w \otimes 1 + 1 \otimes w$. Since a ring homomorphism cannot increase height, he further concluded that $n \geq r \circ s$, which contradicts the hypotheses of (S-H). This argument by contradiction is clearly the same as that for deducing theorem (B-U): the projectivisation of an antipodal map $G : S^{m-1} \longrightarrow S^{n-1}$ produces a map $PG : \mathbb{R}P^{m-1} \longrightarrow \mathbb{R}P^{n-1}$ whose induced morphism on cohomology rings cannot increase height!

No contradiction will arise in Hopf's proof if the elements $w \in H^1(\mathbb{R}P^{n-1}; \mathbb{Z}_2)$ and $\theta = w \otimes 1 + 1 \otimes w \in H^1(\mathbb{R}P^{r-1} \times \mathbb{R}P^{s-1}; \mathbb{Z}_2)$ were to have *equal height*. This is the case when $\theta^{n-1} \neq 0, \theta^n = 0$. But

$$\theta^{n-1} = \binom{n-1}{r-1} w^{r-1} \otimes w^{n-r} + \binom{n-1}{r-2} w^{r-2} \otimes w^{n-r+1} + \cdots + \binom{n-1}{n-s} w^{n-s} \otimes w^{s-1},$$

and

$$\theta^n = \theta^{n-1}\theta = \theta^{n-1}(1 \otimes w + w \otimes 1) = \sum_{r-1 \leq j \leq n-s+1} \left[\binom{n-1}{j} + \binom{n-1}{j-1}\right] w^j \otimes w^{n-j}.$$

So "equal height" occurs precisely when

$$\binom{n-1}{r-1} \equiv \binom{n-1}{r-2} \equiv \cdots \equiv \binom{n-1}{n-s} \equiv 1 \pmod 2.$$

Theorem 3.1 can thus be interpreted also as a supplement to (S-H). For it says that, in the case when the latter's hypothesis marginally fails to hold for n, namely when $r \circ s = n$, nontrivial solution to (#) can still be guaranteed by the extra numerical condition $r + s > n + \rho(n)$. Also, to prove Theorem 3.1 there is now no harm to assume $r \leq s$, because its conclusion is already secured by (S-H) when $r \circ s > n$, and as we have just seen, $\binom{n-1}{r-1}$ has the same parity as $\binom{n-1}{n-s}$ or as $\binom{n-1}{s-1}$, for the remaining case $r \circ s = n$.

The proof of Theorem 3.1 now proceeds by analysing the lower half of diagram (†). The elements $w \in H^1(\mathbb{R}P^{n-1}; \mathbb{Z}_2)$ and $\theta \in H^1(\mathbb{R}P^{r-1} \times \mathbb{R}P^{s-1}; \mathbb{Z}_2)$ correspond under the induced homomorphism $(PF)^*$, and they both have height $r \circ s = n$. This, then, is the case of "equal height".

Recall that the classical Borsuk–Ulam theorem has a sharper version which asserts that if a map $PG : \mathbb{R}P^{m-1} \longrightarrow \mathbb{R}P^{m-1}$ is such that under $(PG)^*$ the one dimensional cohomology generators—which are of equal height—correspond to each other, then PG must be covered by a map $\widetilde{G} : S^{m-1} \longrightarrow S^{m-1}$ of *odd* degree. In the similar but more general case of PF, there is a clever spectral sequence argument which owes its origin to D. Randall [12], proving that

$$\widetilde{F^*} : H^{n-1}(S^{n-1}; \mathbb{Z}_2) \longrightarrow H^{n-1}(S^{r-1} \times_T S^{s-1}; \mathbb{Z}_2)$$

must be nontrivial, or that $S^{r-1} \times_T S^{s-1}$ must have a nonzero cospherical cohomology class in dimension $n-1$. We postpone this argument to the next section, and continue on to discuss the geometry of $\widetilde{Y} = S^{r-1} \times_T S^{s-1}$. If ξ denotes the canonical Hopf line bundle over $\mathbb{R}P^{r-1}$ and $s\xi = \xi \oplus \ldots \oplus \xi$ (s times), then the vector bundle $s\xi$ has total space $S^{r-1} \times_T \mathbb{R}^s$, and $\widetilde{Y}$ is its corresponding unit sphere bundle. Since we are assuming $r \leq s$, $s\xi$ admits a unit section $\sigma : \mathbb{R}P^{r-1} \longrightarrow \widetilde{Y}$. It thus splits up into a Whitney sum $\eta \oplus \epsilon$ where ϵ denotes a trivial line bundle whose fiber over $[x] \in \mathbb{R}P^{r-1}$ is spanned by $\sigma[x]$, and η is a vector bundle of dimension $s-1$ whose Thom space, or one-point compactification, shall be denoted by $T(\eta)$. By elementary point set topology, it is easy to verify, using "stereographic projection", that $T(\eta)$ is homeomorphic to $\widetilde{Y}/\sigma(\mathbb{R}P^{r-1})$, the space obtained from $\widetilde{Y}$ by pinching the section image $\sigma(\mathbb{R}P^{r-1})$ to a point. By standard properties of Thom complexes, we now have homeomorphisms

$$T(s\xi) \cong T(\eta \oplus \epsilon) \cong \sum T(\eta) \cong \sum(\widetilde{Y}/\sigma(\mathbb{R}P^{r-1})),$$

exhibiting $\widetilde{Y}/\sigma(\mathbb{R}P^{r-1})$ as a *de*suspension of the Thom space $T(s\xi)$, which is well-known by [2] to be homeomorphic to the truncated projective space $\mathbb{R}P_s^{r+s-1} = \mathbb{R}P^{r+s-1}/\mathbb{R}P^{s-1}$.

Returning to diagram (†), we note that the map $\widetilde{F} : \widetilde{Y} \longrightarrow S^{n-1}$ is homotopically trivial when restricted to the section image $\sigma(\mathbb{R}P^{r-1})$. This is simply for reasons of dimension, because $r \leq s < n$. (The case $s = n$ is already covered by Theorem 2.1). Thus $\widetilde{F}$ factors up to homotopy through a map $\widehat{F}$ as in the diagram

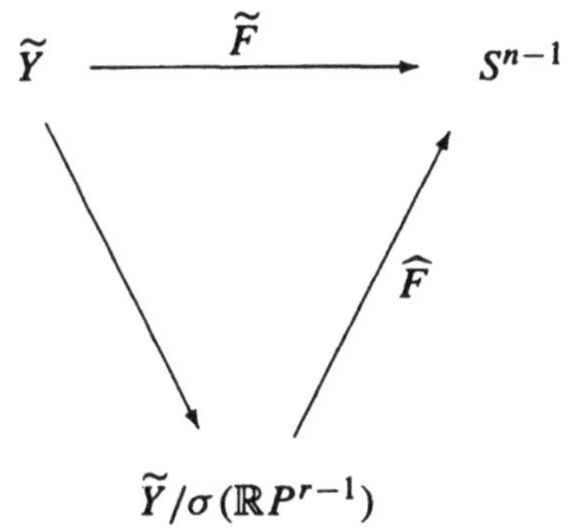

Suspending this diagram once and using the homeomorphisms above, we get a homotopy commutative diagram

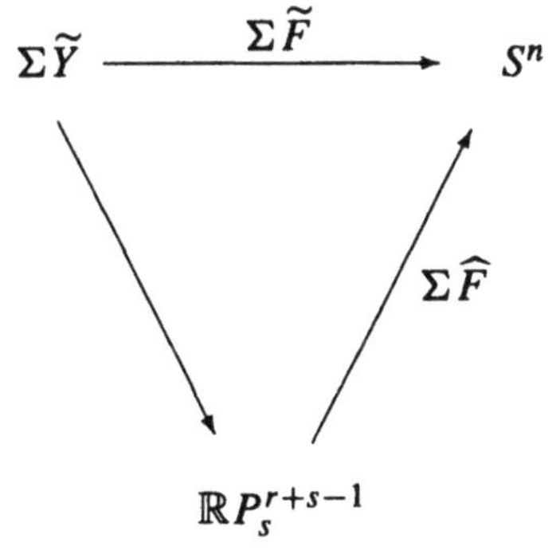

in which $\sum(\widetilde{F})^*$ is nontrivial in n-dimensional cohomology. The same must be true, by commutativity, for $\sum(\widehat{F})^*$, and we can now use cellularity to pinch the $(n-1)$-skeleton of $\mathbb{R}P_s^{r+s-1}$ down to a point, obtaining from $\sum \widehat{F}$ a quotient map $\overline{F} : \mathbb{R}P_n^{r+s-1} \longrightarrow S^n$ which exhibits the cosphericity of the nonzero cohomology class in $H^n(\mathbb{R}P_n^{r+s-1}; \mathbb{Z}_2)$. In other words, $\overline{F}$ shows, in Bott's terminology [3], that the bottom cell of $\mathbb{R}P_n^{r+s-1}$ is co-reducible. By Adams' celebrated result [1, Theorem 1.2 and Section 8], this can happen when and only when the vector bundle $n\xi$ over $\mathbb{R}P^{r+s-n-1}$ is J-trivial, or when $r + s - n \le \rho(n)$.

Thus, the assumption that (#) has only trivial solutions leads to a contradiction of the numerical hypotheses $r + s > n + \rho(n)$, and Theorem 3.1 is thereby proved.

Remark. Theorems 1.1, 2.1 and 3.1 are all valid for biskew systems, namely for the case when $f_j(x, y)$ are continuous functions satisfying $f_j(-x, y) = f_j(x, -y) = -f_j(x, y)$. This is clear from the proofs and discussions given above.

4. A general Borsuk–Ulam type theorem

In this section we establish a Borsuk–Ulam type Theorem 4.1 general enough for use in Section 3. Let $\widetilde{W}$ be a connected cell-complex with a fixed-point-free involution simply denoted by $\widetilde{w} \longleftrightarrow -\widetilde{w}$, and let W be the quotient space obtained by identifying antipodal points. Denote the "first Stiefel–Whitney class" associated with the double covering $\pi : \widetilde{W} \longrightarrow W$ by $\theta \in H^1(W; \mathbb{Z}_2)$. As is well-known, π as a fibration is classified by a map $\beta : W \longrightarrow \mathbb{R}P^\infty$ with $\beta^*(w_\infty) = \theta$, where w_∞ denotes the generator of $H^1(\mathbb{R}P^\infty; \mathbb{Z}_2)$. Furthermore, up to homotopy type

$$\widetilde{W} \xrightarrow{\pi} W \xrightarrow{\beta} \mathbb{R}P^\infty$$

can be viewed as a fibration with base $\mathbb{R}P^\infty$, fiber $\widetilde{W}$, and total space W. When $\widetilde{W}$ is S^{n-1} with the usual antipodal map, we have $W = \mathbb{R}P^{n-1}, \theta = w \in H^1(\mathbb{R}P^{n-1}; \mathbb{Z}_2)$, and the corresponding maps are denoted by π_0 and β_0 respectively.

We can now state

Theorem 4.1. *Let $\widetilde{W}$ be such that its first Stiefel–Whitney class $\theta \in H^1(W; \mathbb{Z}_2)$ is of height n. Suppose $\widetilde{F} : \widetilde{W} \longrightarrow S^{n-1}$ is an antipodal map so that under its projectivisation $P\widetilde{F} : W \longrightarrow \mathbb{R}P^{n-1}$ the equal height classes $w \in H^1(\mathbb{R}P^{n-1}; \mathbb{Z}_2)$ and θ correspond. Suppose further that a certain technical assumption (T) below is satisfied. Then*

$$\widetilde{F}^* : H^{n-1}(S^{n-1}; \mathbb{Z}_2) \longrightarrow H^{n-1}(\widetilde{W}; \mathbb{Z}_2)$$

is a nonzero homomorphism.

For the proof of such a theorem note that an antipodal $\widetilde{F}$ induces a map of fibrations

$$\begin{array}{ccc}
\widetilde{W} & \xrightarrow{\widetilde{F}} & S^{n-1} \\
\downarrow \pi & & \downarrow \pi_0 \\
W & \xrightarrow{F} & \mathbb{R}P^{n-1} \\
\downarrow \beta & & \downarrow \beta_0 \\
\mathbb{R}P^{\infty} & = & \mathbb{R}P^{\infty}
\end{array}$$

In the Serre spectral sequence of the left fibration, the E_2-term is

$$H^*(\mathbb{R}P^{\infty}, \{H^*(\widetilde{W}; \mathbb{Z}_2)\}),$$

where $\{H^*(\widetilde{W}; \mathbb{Z}_2)\}$ denotes a possibly twisted system of coefficients over $\mathbb{R}P^{\infty}$. The technical condition (T) we want to impose is simply that

(T) $\{H^*(\widetilde{W}; \mathbb{Z}_2)\}$ is a *simple* system of coefficients over $\mathbb{R}P^{\infty}$.

This holds, for example, when $H^q(\widetilde{W}; \mathbb{Z}_2)$ has no more than 2 elements for each q, and in particular for $\widetilde{W}$ = the space $\widetilde{Y}$ in Section 3.

Under assumption (T), we can now compare the Serre spectral sequences of the right and left fibrations through the respective induced homomorphisms in cohomology. Since $w = \beta_0^*(w_\infty)$ and $\theta = \beta^*(w_\infty)$ have the same height, n, in both spectral sequences the element w_∞^{n-1} survives to E_∞, while w_∞^n must be killed by the nth differential d_n. Clearly, for the spectral sequence of the fibration on the right, $d_n[S^{n-1}] = w_\infty^n$, where $[S^{n-1}]$, the generator of $H^{n-1}(S^{n-1}; \mathbb{Z}_2)$, is situated at $E_2^{0,n-1}$. By naturality then, we have $d_n(\widetilde{F}^*[S^{n-1}]) = w_\infty^n$. In particular, $\widetilde{F}^*[S^{n-1}] \neq 0$ in $H^{n-1}(\widetilde{W}; \mathbb{Z}_2)$. □

There is a dual theorem which we record as

Theorem 4.2. *Let $\widetilde{W}$ be as in Theorem* 4.1 *satisfying the technical condition* (T). *If θ has height n, then any antipodal map $G : S^{n-1} \longrightarrow \widetilde{W}$ induces a nonzero homomorphism $G^* : H^{n-1}(\widetilde{W}; \mathbb{Z}_2) \longrightarrow H^{n-1}(S^{n-1}; \mathbb{Z}_2)$.*

The proof, based on equal height and spectral sequence arguments, is the same as that of Theorem 4.1.

5. Another extension of (S-H) and Adams' theorem

Similar to Theorem 3.1 we would like to present in this section

Theorem 5.1. *A bilinear system (#) always admits nontrivial solutions if the following numerical conditions are satisfied:*

$$\binom{n}{r} \equiv 1 \pmod 2 \quad \textit{and} \quad r + s > n + \rho(r).$$

This theorem reduces to Adams' when $r = n$, and is kind of dual to Theorem 3.1. Again we can confine ourselves to the case $r \circ s \leq n$, otherwise the conclusion is already immediate by (S-H). The hypothesis $\binom{n}{r} \equiv 1 \pmod 2$ signifies that the element $w \otimes 1 + 1 \otimes w$ in $H^1(\mathbb{R}P^r \times \mathbb{R}P^{s-1}; \mathbb{Z}_2)$ has height $> n$, or that $(r+1) \circ s > n$. In this sense Theorem 5.1 is just another supplement to (S-H), addressing the case when the latter's hypothesis marginally fails to hold for r.

Let $V_{n,s}$ be the Stiefel manifold of orthonormal s-frames in $\mathbb{R}^n$, with antipodal map defined by reversing the sign of all the vectors constituting a given frame. When antipodal frames are identified, one obtains the projective Stiefel manifold $PV_{n,s}$, whose detailed study is provided by Gitler and Handel [5]. In particular, if $y \in H^1(PV_{n,s}; \mathbb{Z}_2)$ denotes the first Stiefel–Whitney class of the double cover $V_{n,s} \longrightarrow PV_{n,s}$ as in [5], its height N in $H^*(PV_{n,s}; \mathbb{Z}_2)$ is given in Theorem 1.6 of [5], which can be succinctly translated into the following: N is the largest integer q satisfying $q \circ s \leq n$. To prove Theorem 5.1 now we again assume that (#) had only trivial solutions, and from this define a map $G : S^{r-1} \longrightarrow V_{n,s}$ by sending $x = (x_1, \ldots, x_r)$ in S^{r-1} to the s-frame in $\mathbb{R}^n$ consisting of the vectors

$$(f_1(x_1, \ldots, x_r; 1, 0, \ldots, 0), \ldots, f_n(x_1, \ldots, x_r; 1, 0, \ldots, 0),$$

$$(f_1(x_1, \ldots, x_r; 0, 1, \ldots, 0), \ldots, f_n(x_1, \ldots, x_r; 0, 1, \ldots, 0),$$

$$\vdots$$

$$(f_1(x_1, \ldots, x_r; 0, 0, \ldots, 1), \ldots, f_n(x_1, \ldots, x_r; 0, 0, \ldots, 1).$$

These vectors are linearly independent under our assumption, and are understood to have been orthonormalized via Gram–Schmidt into an orthonormal frame $G(x) \in V_{n,s}$. Clearly $G(-x) = -G(x)$ so that G projectivises into a map $PG : \mathbb{R}P^{r-1} \longrightarrow PV_{n,s}$ satisfying $(PG)^*(y) = w$, while by hypothesis w and y have equal height r. By the Borsuk–Ulam type Theorem 4.2,

$$G^* : H^{r-1}(V_{n,s}; \mathbb{Z}_2) \longrightarrow H^{r-1}(S^{r-1}; \mathbb{Z}_2)$$

must be onto, showing that the unique $(r-1)$-dimensional generator of $H^*(V_{n,s}; \mathbb{Z}_2)$ is a spherical cohomology class. The same statement is also true, then, for $V_{r,r+s-n}$ which is standardly embedded in $V_{n,s}$. This is because one can deform G homotopically into $V_{r,r+s-n}$ by cellular approximation. We are now back to the vector field situation, for such a deformed G furnishes a "mod 2" section for the standard fibration of $V_{r,r+s-n}$ over the sphere S^{r-1}. Invoking Adams [1] again, one gets $r+s-n \leq \rho(r)$, contradicting the numerical hypothesis on r, s and n. This proves Theorem 5.1 completely. □

6. Miscellaneous remarks

Remark 1. The dual Theorem 5.1, like its analogs in Sections 2 and 3, also holds for biskew systems of equations. In fact, there is a standard technique to reduce the discussion of biskew systems to bilinear systems, using corollary 6.6 of [8] and the stablisation procedures developed in [9] . But we shall omit the details.

Remark 2. One advantage of formulating and highlighting Theorem 3.1 as in Section 3 is the simplicity of its statement. Comparing with other works, one will note that this Theorem is a tiny bit stronger than Theorem 6.2 of James [7]. On the other hand, it is slightly weaker than the results of D. Davis, who in his thesis [4] obtained theorems to the effect that when $\binom{n-1}{r}$ is odd, (#) must have nontrivial solution if $r + s > n + \rho(n) - \lambda(n, r)$, where $\lambda = \lambda(n, r)$ is some integer between 0 and 3, its exact value being dependent on the mod 4 values of n and r. The situation is the same for Theorem 5.1. Under the same binomial hypotheses Davis obtained $r + s > n + \rho(r) - \mu(n, r)$ as sufficient condition for nontrivial solvability, where $\mu(n, r) \geq 0$ is again small and quite similar to $\lambda(n, r)$. Davis's techniques, however, are much more involved, invoking a new homology theory **bo** as well as heavy computations with Adams spectral sequences. Still, it would be interesting to understand the basic geometry which allows for the small improvements $\lambda(n, r)$ and $\mu(n, r)$ to occur. Even more tantalisingly, it would be wonderful to find out (*cf.* [10]) whether these improvements are already optimal under the stated hypotheses on binomial coefficients. We hope to return to questions of this sort in a future article.

The author would like to express his thanks to professors S. Gitler and D. Randall for numerous fruitful discussions, and to his colleagues E. Luft, A. Melzak, L. Roberts, D. Rolfsen, D. Sjerve and P. Yiu for generous help. Thanks are also due to J. Sung, and to the author's children Duhane and Selena, for their understanding and assistance.

References

[1] Adams, J. F., Vector fields on spheres, Ann. of Math. 75 (1962), 603–632.

[2] Atiyah, M., Thom complexes, Proc. London Math. Soc. 11 (1961), 291–310.

[3] Bott, R., Lectures on K-theory, Benjamin, New York 1969.

[4] Davis, D., Generalized homology and the generalized vector field problem, Quart. J. Math. Oxford Ser. (2) 25 (1974), 169–193.

[5] Gitler, S. and Handel, D., The projective Stiefel manifolds I, Topology 7 (1967), 39–46.

[6] Hopf, H., Ein topologischer Beitrag zur reellen Algebra, Comment. Math. Helv. 13 (1941), 219–239.

[7] James, I. M., On the immersion problem for real projective spaces, Bull. Amer. Math. Soc. 69 (1963), 231–238.

[8] James, I. M., The space of bundle maps, Topology 2 (1963), 45–59.

[9] Lam, K. Y., On bilinear and skew-linear maps that are nonsingular, Quart. J. of Math. Oxford Ser. (2) 19 (1968), 281–288.

[10] Lam, K. Y. and Randall, D., Low dimensional spinor representations, Adams maps and geometric dimension, London Math. Soc. Lecture Note Ser. 175 (1992), 89–102.

[11] Lam, K. Y. and Yiu, P., Sums of squares formulae near the Hurwitz–Radon range, Contemp. Math. Part II, 58 (1987), 51–55.

[12] Randall, D., On equivariant maps of Stiefel manifolds, Contemp. Math. 36 (1985), 145–149.

[13] Stiefel, E., Über Richtungsfelder in den projektiven Räumen und einen Satz aus reellen Algebra, Comment. Math. Helv. 13 (1941), 201–218.

Generalized affine rotation surfaces with geodesic orbits

Isaac Chaujun Lee

Abstract. Generalized affine rotation surfaces in affine 3-space have been investigated in [L2] from a group theoretic point of view. Using this, we discover, from the view point of shadow boundaries, two well-known propositions about geodesic orbits of rotation surfaces in euclidean 3-space and those in Lorentzian 3-space. Furthermore, we classify generalized affine rotation surfaces with geodesic orbits as follows:

$$x^2 + \varepsilon y^2 = (\cos z)^2, \quad x^2 + \varepsilon y^2 = (\cosh z)^2, \quad x^2 + \varepsilon y^2 = (\sinh z)^2,$$
$$y^2 = z(x + \varepsilon z^3), \quad y^2 = \ (x + \varepsilon z^2),$$

where ε represents 1 or -1 in this paper.

1991 Mathematics Subject Classification: 53A15.

1. Preliminaries

1.1. Basic structures

For an affine space $\mathbb{R}^{n+1}$ with the usual flat connection D and a volume element ω defined by

$$\omega(\mathcal{Y}_1, \mathcal{Y}_2, \ldots, \mathcal{Y}_{n+1}) = \det[\mathcal{Y}_1\, \mathcal{Y}_2\, \ldots\, \mathcal{Y}_{n+1}],$$

we see that such an affine space $(\mathbb{R}^{n+1}, D, \omega)$ has an *equiaffine structure*, that is, $D\omega = 0$. For a given immersion

$$f : \mathcal{M}^n \longrightarrow \mathbb{R}^{n+1}$$

with a transversal vector field η, we can define the *induced connection* ∇^η, the *affine fundamental form* h^η, the *shape operator* S^η and the *transversal connection* τ^η as follows:

$$\begin{cases} D_X f_*(Y) = f_*(\nabla^\eta{}_X Y) + h^\eta(X, Y)\eta \\ \quad D_X \eta = -f_*(S^\eta(X)) + \tau^\eta(X)\eta. \end{cases}$$

* This paper is part of the author's Ph.D. dissertation at Brown University, 1993

An *induced volume element* θ^η is defined by

$$\theta^\eta(Y_1, Y_2, \ldots, Y_n) = \omega(f_*(Y_1), f_*(Y_2), .., f_*(Y_n), \eta).$$

We have

(1) ∇^η is a torsion free affine connection;

(2) h^η is a symmetric tensor;

(3) $\nabla^\eta_X \theta^\eta = \tau^\eta(X)\theta$.

We say the affine immersion f is *nondegenerate* if h^η is nondegenerate, and we know that this condition is independent of the choice of the transversal vector η. From now on, we always assume the surface is *nondegenerate.*

We say a transversal vector field η is *equiaffine* (or η is *relative normal*) if

$$\nabla^\eta \theta^\eta = 0,$$

i.e., τ^η is identically zero. We then have a unique choice (up to a sign) of a transversal vector field ξ, called *the affine normal* which is equiaffine and satisfies

$$\theta^\xi = \omega_h,$$

where ω_h is the volume element of the affine fundamental form h^ξ. In this case, h^ξ is called the *affine metric* and is denoted by g if necessary. An affine immersion with affine normal is called a *Blaschke immersion.*

We have the following fundamental equations for a relative normal:

Gauss Eq.	$R(X, Y)Z$	$=$	$h(Y, Z)SX - h(X, Z)SY$
Codazzi Eq. for h	$(\nabla_X h)(Y, Z)$	$=$	$(\nabla_Y h)(X, Z)$
Codazzi Eq. for S	$(\nabla_X S)Y$	$=$	$(\nabla_Y S)X$
Ricci Eq.	$h(SX, Y)$	$=$	$h(X, SY)$.

For its shape operator S, we define the *affine mean curvature* $\mathbf{H}$ and the *Gauss-Kronecker curvature* $\mathbf{K}$ to be trace$(\mathbf{S})/n$ and det$(\mathbf{S})$, respectively. A hypersurface is called an *affine minimal* if the affine mean curvature $\mathbf{H}$ is zero.

On the other hand, a hypersurface is called an *affine sphere* if $\mathbf{S} = \lambda \mathbf{I}$ at each point of the hypersurface. If $\lambda \neq 0$, the surface is called a *proper affine sphere* and if $\lambda = 0$, an *improper affine sphere.*

Let $\widehat{\nabla}$ be the Levi-Civita connection relative to the affine metric g. We define the difference tensor of ∇ and $\widehat{\nabla}$ to be $K := \nabla - \widehat{\nabla}$. One has

$$g(K_X Y, Z) = (-\frac{1}{2})(\nabla_X g)(Y, Z).$$

This implies that K is symmetric, that is, $K_X Y = K_Y X$ by the Codazzi equation for g. Therefore, the *cubic form* $C(X, Y, Z) := (\nabla_X g)(Y, Z)$ is related to K. One important condition for Blaschke immersions is the *apolarity condition*, $\nabla \omega_g = 0$. It is equivalent to the condition

$$\text{trace}(K_X) = 0,$$

Furthermore, one can define relative normal maps and the corresponding relative conormal maps [NO] as follows.

Let $f : \mathcal{M}^n \to \mathbb{R}^{n+1}$ be a equiaffine immersion with relative normal ξ. We define the relative normal map φ to be the end point of the relative normal vector ξ when it is displaced parallelly so as to have the starting point at the origin of $\mathbb{R}^{n+1}$. Since $\varphi(X) = \nabla_X \varphi = \nabla_X \xi = -f_*(SX)$, it follows that φ is an immersion if and only if S is nonsingular. In this case, the induced connection ∇' of the centro-affine immersion φ with $-\varphi$ as transversal vector field satisfies

$$\begin{cases} \nabla'_X Y &= S^{-1}(\nabla_X S) + \nabla XY \\ h'(X, Y) &= h(SX, Y). \end{cases}$$

On the other hand, let $\mathbb{R}_{n+1}$ be the dual affine space of $\mathbb{R}^{n+1}$. We have a choice of a relative conormal ζ in $\mathbb{R}_{n+1}$ such that

$$\begin{cases} \zeta(f_*(Z)) = 0, & \text{for all } Z \text{ tangent to } \mathcal{M}^n \\ \zeta(\xi) = 1. \end{cases} \tag{1}$$

The relative conormal map ν is defined to be the end point of the relative conormal vector ζ when it is displaced parallelly so as to have the starting point at the origin of $\mathbb{R}_{n+1}$. In this case, the induced connection ∇^* of the centro-affine immersion ν with $-\nu$ as transversal vector field satisfies

$$\begin{cases} \text{(a) } Xh(Y, Z) = h(\nabla_X Y, Z) + h(Y, \nabla^*_X Z), \text{ and} \\ \text{(b) } \widehat{\nabla} - \nabla^* = \nabla - \widehat{\nabla}, \end{cases} \tag{2}$$

where (b) implies

$$K = \tfrac{1}{2}(\nabla - \nabla^*).$$

A curve $x(t)$ on $\mathcal{M}^n$ is called a *shadow boundary* with parallel light $\vec{a}$ if $\vec{a} \parallel T_{x(t)}\mathcal{M}$ for all t. It is known that

Lemma 1.1.1([NS]). *A curve is a shadow boundary if and only if it is a ∇^*-geodesic.*

In [NO], they proved the following proposition:

For a nondegenerate affine surface $\mathcal{M}$ with relative normal η and nonsingular shape operator, the following conditions are equivalent:

(1) the image $\varphi(\mathcal{M})$ of the normal map is a hyperquadric;

(1′) $\nabla^* h^* = 0$;

(2) the image $\nu(\mathcal{M})$ of the conormal map is a hyperquadric;

(2′) $\nabla' h' = 0$;

(3) $\nabla^* = \nabla'$;

(4) $2K_X Y = -S^{-1}(\nabla_X S)Y$.

1.2. Generalized affine rotation surfaces

In [L2], we defined generalized affine rotation surfaces and classified them up to affine congruence into four standard types, namely,

$$\begin{aligned} lx^2 + y^2 &= g_1(z)^2, \\ lx^2 - y^2 &= g_2(z)^2, \\ y^2 &= z(x - g_3(z)), \\ y^2 &= (x - g_4(z)), \end{aligned}$$

with $g_i''(z) \neq 0$.

From a group theoretic point of view, each of them can be generated by ${}^t(g(v), 0, v)$, a profile curve on the xz-plane, under the action of a generalized affine rotation groups, namely,

$$\left\{ \begin{array}{ll} Q_1(t) = \begin{pmatrix} \cos t & -\sin t & 0 \\ \sin t & \cos t & 0 \\ 0 & 0 & 1 \end{pmatrix}, & Q_2(t) = \begin{pmatrix} \cosh t & \sinh t & 0 \\ \sinh t & \cosh t & 0 \\ 0 & 0 & 1 \end{pmatrix}, \\ Q_3(t) = \begin{pmatrix} 1 & -t & t^2/2 \\ 0 & 0 & -t \\ 0 & 0 & 1 \end{pmatrix}, & Q_4(t) = \begin{pmatrix} 1 & -t & 0 & t^2/2 \\ 0 & 1 & 0 & -t \\ 0 & 0 & 1 & 0 \\ 0 & 0 & 0 & 1 \end{pmatrix}. \end{array} \right. \tag{3}$$

Note that $\{Q_4(t)\}$ denotes a subgroup of $\mathbf{AS}(3, \mathbb{R})$ with ${}^t(t^2/2, -t, 1)$ as the translation factor.

There are wonderful dualities between ARS with elliptic orbits and that with hyperbolic orbits [L2]. Some common expressions of affine quantities among generalized affine rotation surfaces are listed as follows:

$$\left\{ \begin{aligned} \text{shape operator } S &= \begin{pmatrix} -T_2\delta'/\delta & 0 \\ 0 & -(T_2\mu)'/\mu \end{pmatrix} \\ \Gamma_{11}^1 &= 0 \\ \Gamma_{11}^2 &= (-\tfrac{1}{4})(\delta/g'')' = -{}^E\Gamma_{11}^2 \\ \Gamma_{12}^1 &= \delta'/\delta \\ \Gamma_{12}^2 &= 0 \\ \Gamma_{22}^1 &= 0 \\ \Gamma_{22}^2 &= (-\tfrac{1}{4})(\delta/g'')'/(\delta/g'') \\ h_{11} &= -\delta^2/\phi = -{}^H h_{11} \\ h_{12} &= 0 \\ h_{22} &= \delta g''/\phi = -{}^P h_{22} = -{}^I h_{22}, \end{aligned} \right. \tag{4}$$

where

$$\begin{cases} \phi & := |\delta^3 g''|^{\frac{1}{4}} \\ T_2 & := (-\phi\tau_2 - \phi')/(\delta g'') = -\,{}^P T_2 = -\,{}^I T_2 \\ \tau_2 & := -(\delta\mu)'/(\delta\mu) \end{cases}$$

and

$$(\delta, \mu, \psi) := \begin{cases} (g, g', 1) \text{ for elliptic/hyperbolic ARS} \\ (v, 1, g') \text{ for parabolic ARS} \\ (1, 1, g') \text{ for improper ARS.} \end{cases}$$

Note that $\delta' = \mu$ for ARS. More details can be found in [L2].

2. Shadow boundaries and geodesics

2.1. GARS with shadow boundary orbits

We would like to give an affine geometric interpretation for two propositions that are known for both euclidean and Lorentzian rotation surfaces. Recall that the euclidean space $\mathbf{E}^3$ is defined as $(\mathbb{R}^3, ds_{\mathrm{E}})$ with metric $ds_{\mathrm{E}}^2 = dx^2 + dy^2 + dz^2$. The Lorentz space $\mathbf{L}^3$ is defined as $(\mathbb{R}^3, ds_{\mathrm{L}})$ with metric $ds_{\mathrm{L}}^2 = dx^2 + dy^2 - dz^2$. Rotation surfaces in these two spaces are defined as follows. An *euclidean rotation surface* (ERS) is a surface invariant under a 1-parameter subgroup of SO(3,$\mathbb{R}$) with a line of fixed points. On the other hand, a *Lorentzian rotation surface* (LRS) is a surface invariant under a 1-parameter subgroup of the isometry group SO(2,1,$\mathbb{R}$) with a line of fixed points [HN]. It is well-known that any ERS is congruent to

$$\begin{pmatrix} \cos t & -\sin t & 0 \\ \sin t & \cos t & 0 \\ 0 & 0 & 1 \end{pmatrix} \cdot \begin{pmatrix} g(v) \\ 0 \\ f(v) \end{pmatrix}$$

and any LRS is congruent to one of the following surfaces [HN]

$$\begin{pmatrix} \cos t & -\sin t & 0 \\ \sin t & \cos t & 0 \\ 0 & 0 & 1 \end{pmatrix} \cdot \begin{pmatrix} g(v) \\ 0 \\ f(v) \end{pmatrix}, \quad \begin{pmatrix} 1 & 0 & 0 \\ 0 & \cosh t & \sinh t \\ 0 & \sinh t & \cosh t \end{pmatrix} \cdot \begin{pmatrix} f(v) \\ 0 \\ g(v) \end{pmatrix},$$

$$\begin{pmatrix} 1 & -t & t \\ t & 1 - t^2/2 & t^2/2 \\ t & -t^2/2 & 1 + t^2/2 \end{pmatrix} \cdot \begin{pmatrix} 0 \\ f(v) \\ g(v) + f(v) \end{pmatrix}.$$

The following two propositions are easy to prove.

Proposition 2.1.1. *For an ERS (or LRS), the image of the Gauss map is also an ERS (or LRS) of the same type with the same* 1*-parameter group.*

Proposition 2.1.2. *For a standard ERS (or a standard LRS), an orbit passing through a point p_0 of the profile curve is a geodesic if and only if $g'(p_0) = 0$.*

We now give an affine version of Proposition 2.1.1 for GARS.

Proposition 2.1.3. *For a GARS $\mathcal{M}$ with 1-parameter group $\{\mathcal{Q}(t)\}$, we have:*

(1) *If $\mathcal{M}$ is an ARS, then $\varphi(\mathcal{M})$ and $\nu(\mathcal{M})$ are ARS with 1-parameter group $\{\mathcal{Q}(t)\}$ and $\{\mathcal{Q}(t)^*\}$ respectively;*

(2) *If $\mathcal{M}$ is an improper ARS, then $\varphi(\mathcal{M})$ and $\nu(\mathcal{M})$ are degenerate surfaces invariant under the 1-parameter groups which are the linear part of $\{\mathcal{Q}(t)\}$ and $\{\mathcal{Q}(t)^*\}$ respectively.*

Proof. We know that any affine transformation A is obtained by a linear transformation $A_{\mathcal{L}}$ plus a parallel translation $A_{\mathcal{P}}$. Let ξ_x be the affine normal at the point x, we have in general

$$\xi_{A(x)} = \left(\frac{1}{\sqrt{\det(A_{\mathcal{L}})}}\right) \cdot (A_{\mathcal{L}} \cdot \xi_x).$$

Now, the condition $\det(\mathcal{Q}_i(t)) = 1$ gives statement (1).

As for statement (2), we can consider standard improper ARS. The linear part of $\mathcal{Q}_4$ (see Eq. (3)) is

$$\begin{pmatrix} 1 & -t & 0 \\ 0 & 1 & 0 \\ 0 & 0 & 1 \end{pmatrix}. \tag{5}$$

The surface generated by this 1-parameter group is a degenerate surface [L2]. □

In the remainder of the paper, we deal with a nondegenerate surface with a relative normal and give affine geometric interpretation of Proposition 2.1.2. We provide two different view points, one with relative conormal, and the other with relative normal. We first discuss things with relative conormal.

We would like to show that the condition $g'(p_0) = 0$ in Proposition 2.1.2 comes from the conditions for ∇^*-geodesics rather than those of ∇-geodesics. Note that a ∇^*-geodesic can be visualized as a shadow boundary by Lemma 1.1.1. For an orbit of a GARS, we would like to find the condition for being a shadow boundary. Note that if an ARS (or an improper ARS) has a tangent vector parallel to the line of fixed points (respectively, the plane of fixed points) at one point, then the orbit passing through this point is a shadow boundary. Conversely, we have

Proposition 2.1.4. *Let C be an orbit of any ARS (or improper ARS) and assume C is a shadow boundary. Then, along C, the light direction is parallel to the line of fixed points (repectively, the plane of fixed points).*

Proof. Suppose the surface is $f(u, v)$ and C is the orbit $f(u, v_0)$ which is a shadow boundary. Let $\vec{a}$ be the lighting direction. Then, there are two functions $\lambda(u, v_0)$,

$\mu(u, v_0)$ such that

$$\vec{a} = \lambda(u, v_0) f_u(u, v_0) + \mu(u, v_0) f_v(u, v_0) = \lambda(0, v) f_u(0, v_0) + \mu(0, v_0) f_v(0, v_0). \tag{6}$$

Claim. *$\{\mathcal{G}(u) \cdot \vec{a}\}$ is a plane curve parallel to the tangent plane at $f(0, v_0)$.*

A standard GARS invariant under a 1-parameter group $\mathcal{G}(u)$ means

$$f(u, v) = \mathcal{G}(u) \cdot f(0, v). \tag{7}$$

We want to differentiate this equation in order to rewrite Eq. (6). Differentiating Eq. (7) with respect to u, we have $f_u(u, v_0) = \mathcal{G}'(u) \cdot f(0, v_0)$, which gives us two pieces of information. One is

$$f_u(0, v_0) = \mathcal{G}'(0) \cdot f(0, v_0), \tag{8}$$

which is obtained by substituting $u = 0$. The other is

$$f_u(u, v_0) = (\mathcal{G}(u) \cdot \mathcal{G}'(0)) \cdot f(0, v_0) \tag{9}$$

since $\mathcal{G}(u)$ is a 1-parameter group and thus $\mathcal{G}'(u) = \mathcal{G}(u) \cdot \mathcal{G}'(0) = \mathcal{G}'(0) \cdot \mathcal{G}(u)$. Substituting Eq. (8) in Eq. (9), we get

$$f_u(u, v_0) = \mathcal{G}(u) \cdot f_u(0, v_0). \tag{10}$$

On the other hand, by differentiating Eq. (7) with respect to v we have

$$f_v(u, v_0) = \mathcal{G}(u) \cdot (f(0, v_0))' = \mathcal{G}(u) \cdot f_v(0, v_0). \tag{11}$$

We substitute both Eq. (10) and Eq. (11) into Eq. (6) and get

$$\begin{aligned}\vec{a} &= \lambda(u, v_0)\mathcal{G}(u) \cdot f_u(0, v_0) + \mu(u, v_0)\mathcal{G}(u) \cdot f_v(0, v_0) \\ &= \lambda(0, v_0) f_u(0, v) + \mu(0, v_0) f_v(0, v_0),\end{aligned}$$

in other words,

$$\begin{aligned}\vec{a} &= \mathcal{G}(u) \cdot (\lambda(u, v_0) f_u(0, v_0) + \mu(u, v_0) f_v(0, v_0)) \\ &= \lambda(0, v_0) f_u(0, v_0) + \mu(0, v_0) f_v(0, v_0).\end{aligned}$$

This proves what is claimed.

For a standard ARS with elliptic orbits (or hyperbolic orbits), we know that, for any point $p \in \mathbb{R}^3$, $\{\mathcal{G}(u) \cdot p\}$ is a plane curve parallel to the xy-plane unless $p \in z$-axis.

Combining this fact and what we just claimed, we conclude that $\vec{a}$ has to be parallel to the z-axis, namely, the line of fixed points of $\mathcal{G}(u)$.

A similar argument can be used for the other two types of GARS. In particular, we have

Corollary 2.1.5.

(1) *For a parabolic type ARS, no orbit is a shadow boundary.*

(2) *For an improper ARS, every orbit is a shadow boundary.*

Proof. (1) For the 1-parameter group $\mathcal{G}_3$ (see Eq. (3)) of the standard parabolic type, the x-axis is the eigenspace. But the profile curve ${}^t(g(v), 0, v)$ is on the xz-plane. It is impossible for the tangent line of the profile curve $x = g(z)$ to be parallel to the x-axis. Thus statement (1) follows from Lemma 1.1.1 and Proposition 2.1.4.
(2) For a standard improper ARS, the linear part of $\mathcal{Q}_4$ (see Eq. (3)) has the xz-plane as eigenspace. So, any shadow boundary has its light direction on the xz-plane by view of Proposition 2.1.4. Now, the profile curve lies in exactly the same plane, and so does its tangent vector. This proves statement (2). □

Using Lemma 1.1.1 and Proposition 2.1.4, the following corollary gives us one affine geometric interpretation for Proposition 2.1.2. Recall that the unit normals of surfaces in $\mathbf{E}^3$ (or $\mathbf{L}^3$) and affine normals are examples of relative normals.

Corollary 2.1.6. *For any standard elliptic or a hyperbolic ARS, an orbit passing through a point p_0 of the profile curve is a geodesic with respect to a relative conormal if and only if $g'(p_0) = 0$.*

This corollary implies Proposition 2.1.2 once we realize that the unit normal and unit conormal of any euclidean (or Lorentzian) rotation surface coincide.

Secondly, we would like to discuss the affine geometric interpretation of Proposition 2.1.2 from the view point of relative normals.

Lemma 2.1.7. *For a nondegenerate equiaffine immersion $(\mathcal{M}^n, \nabla)$ of $\mathbb{R}^{n+1}$ with relative normal ξ, let $x(t)$ be a curve on $\mathcal{M}^2$. The following conditions are equivalent:*

(1) *$x(t)$ is a ∇-pre-geodesic;*

(2) *the relative normal $\xi(p)$ is co-planar with the osculating plane at every point p of $x(t)$;*

(3) *the relative co-normal vector ζ (in $\mathbb{R}_{n+1}$) satisfies*

$$(D_Y\zeta)(D_t f_*(X_t)) = 0$$

for all tangent vectors Y in $T(M)$ such that $h(Y, X_t) = 0$, where $X_t := dx(t)/dt$.

Proof. Denoting D_t for D_{X_t} and ∇_t for ∇_{X_t}, we have

$$D_t f_*(X_t) = f_*(\nabla_t X_t) + h(X_t, X_t)\xi. \tag{12}$$

We see that $x(t)$ is ∇-pre-geodesic if and only if

$$\xi \in \mathrm{Span}\{D_t f_*(X_t), f_*(X_t)\}. \tag{13}$$

Thus, the equivalence between (1) and (2) follows.

As for (3), we take the derivative D_Y on Eq. (1) and get

$$\begin{cases} (D_Y\zeta)(\xi) = 0 \\ (D_Y\zeta)(Z) = -h(Y, Z). \end{cases} \tag{14}$$

Applying $D_Y\zeta$ onto Eq. (12) we have

$$(D_Y\zeta)(D_t f_*(X_t)) = -h(Y, f_*(\nabla_t X_t))$$

by virtue of Eq. (14). Then the equivalence of (1) and (3) follows. □

One can check that all orbits of any GARS are plane curves and these planes are parallel to each other (see [L1] for more geometric characterization). The *base of orbits* of a GARS is defined to be the plane passing through the origin 0 to which any orbit plane is parallel. Then, we have:

Corollary 2.1.8. *For an orbit C of a GARS with a relative normal η, the following conditions are equivalent:*

(1) *C is a geodesic relative to η;*

(2) *$\varphi(C)$ lies on the the base of orbits.*

And, if we exclude improper ARS,

(3) *$\nu(C)$ is a shadow boundary with parallel lighting in the direction dual to the direction of the axis of rotation.*

Proof. For a standard ARS,an orbit is on the plane parallel to the xy-plane. Applying Eq. (13), we get the equivalence among (1) , (2) and (3). □

Condition (2) of this proposition implies Proposition 2.1.2 and thus gives an affine geometric interpretation from the view point of relative normals.

Since the affine normals are relative normals, we have

Corollary 2.1.9. *For any standard GARS, an orbit through p_0 of the profile curve is a geodesic relative to the affine normal if and only if*

$$(\delta/g'')'(p_0) = 0. \tag{15}$$

Proof. Applying Lemma 2.1.7 to the standard GARS we have the condition $\Gamma^2_{11} \equiv 0$. From Eq. (4), the equivalence follows. □

Remark. It can be proved that, by excluding improper affine spheres, the focal points of the affine normals along an orbit are concurrent at a point on the axis of rotation.

For elliptic (or hyperbolic, parabolic) type ARS, the image by the normal map of an orbit is an *ellipse* (respectively, *hyperbola, parabola*) on a plane parallel to the base of orbits as we know from Proposition 2.1.3, whereas that for an improper ARS is a *point* on a fixed plane. In particular, if $\mathcal{M}$ is a standard GARS with geodesic orbit C, then the affine normal image $\varphi(C)$ satisfies one of the following properties:

i) For elliptic or hyperbolic type ARS, the image is $x^2 \pm y^2 = \sqrt{g''(z_0)/g(z_0)}$, where z_0 satisfies $(g''/g)'(z = z_0) = 0$;

ii) For parabolic type ARS, the image is a point on the x-axis $(g''(z_0)/z_0)^{1/4}, 0, 0)$, where z_0 satisfies $(g''/z)'(z_0) = 0$;

iii) For improper type ARS, the image is a point on the x-axis $(g''(z_0))^{1/4}, 0, 0)$, where z_0 satisfies $(g'')'(z_0) = 0$.

2.2. GARS with geodesic orbits

For GARS, we have the notion of meridians and orbits and they provide a natural parametrization for the surface. With the affine normal, a GARS whose meridians and orbits are geodesics relative to the induced connection is called *a GARS with geodesic parameters*.

From Eqs. (4) and (2), we have $\Gamma_{22}^1 = 0$, and $\Gamma_{22}^{1\,*} = 0$, which immediately implies:

Proposition 2.2.1. *All meridians of any GARS are geodesics relative to the induced connection of the affine normal or conormal.*

We are ready to classify all GARS with geodesic parameters.

Theorem 2.2.2. *Any GARS with geodesic parameters is affine congruent to one of the following surfaces.*

(1) $x^2 + \varepsilon y^2 = (\cos z)^2$,

(2) $x^2 + \varepsilon y^2 = (\cosh z)^2$, $x^2 + \varepsilon y^2 = (\sinh z)^2$,

(3) $y^2 = z(x + \varepsilon z^3)$,

(4) $y^2 = (x + \varepsilon z^2)$.

Proof. From Corollary 2.1.9, we have to solve $(\delta/g'')' \equiv 0$, where

$$\delta = \begin{cases} g, & \text{for elliptic or hyperbolic type ARS.} \\ v, & \text{for parabolic type ARS.} \\ 1, & \text{for improper ARS.} \end{cases}$$

The solutions are

$$g(v) = \begin{cases} \begin{cases} (1)\ m_1 \cos(k_1 v) \pm m_2 \sin(k_1 v), & \text{or} \\ (2)\ n_1 \cosh(k_2 v) \pm n_2 \sinh(k_2 v), & \text{both for elliptic/hyperbolic.} \end{cases} \\ (3)\ p_1 v^3 + p_2 v + p_3, \ \text{for proper parabolic.} \\ (4)\ q_1 v^2 + q_2 v + q_3, \ \text{for improper parabolic.} \end{cases}$$

where m_i, k_i, n_i, p_i, and q_i's are arbitary constants. They generate the following surfaces respectively

(1) $x^2 \pm y^2 = (m_1 \cos(k_1 z) \pm m_2 \sin(k_1 z))^2$

(2) $x^2 \pm y^2 = (n_1 \cosh(k_2 z) \pm n_2 \sinh(k_2 z))^2$

(3) $y^2 = z(x - (p_1 z^3 + p_2 z + p_3))$

(4) $y^2 = x - (q_1 z^2 + q_2 z + q_3)$.

Classification is done up to affine congruences. □

We would like to remark that the last surfaces, $y^2 = (x + \varepsilon z^2)$, have geodesic parameters only when they are considered as improper ARS, although they could be generated as ARS with elliptic/hyperbolic, namely, $x^2 + \varepsilon y^2 = z$.

Corollary 2.2.3. *In the class of GARS, any affine sphere with geodesic parameters has to be an improper affine sphere with parabolic orbits.*

Proof. We can prove that surfaces in classes (1) and (2) in the theorem are not affine spheres. On the other hand, surfaces in classes (3) and (4) are improper affine spheres. Recall that class (3) is characterized as an improper affine sphere of parabolic type ARS. □

Some further results on generalized affine rotation surfaces are sketched in [L4].

References

[L1] I. Lee, Affine Rotation Surfaces, Master's Thesis, Brown Univ., 1991.

[L2] I. Lee, On Generalized Affine Rotation Surfaces, Results Math. 27 (1995), 63–76.

[L3] I. Lee, Projectively flat affine surfaces that are not locally symmetric, Proc. Amer. Math. Soc. 123 (1995), 237–246.

[L4] I. Lee, On generalized affine rotation surfaces, Geometry and topology of submanifolds, VII (F. Dillen, M. Magid, U. Simon, I. Van de Woestinje, L. Verstraelen, ed.), World Scientific, Singapore 1995.

[NO] K. Nomizu and B. Opozda, On normal and conormal maps for affine hypersurfaces, Tohoku Math. J. 44 (1992), 425–431.

[NS1] K. Nomizu and T. Sasaki, Affine Differential Geometry and Geometry of Affine Immersions, Cambridge University Press, New York 1994.

[NS2] —, Centroaffine immersion of co-dimension two and projective hypersurface theory, Nagoya Math. J. 132 (1993), 63–90.

Lower bounds on some geometric quantities and applications

Man Chun Leung

Abstract. Lower bounds on volumes and $L^{\frac{n}{2}}$-norms of scalar curvatures are discussed for some compact manifolds of negative curvature. The results are applied to an isospectral problem.

1991 Mathematics Subject Classification: Primary 53C20; Secondary 53C21.

1. Introduction

The uniformization theorem for compact surfaces without boundary says that every such surface admits a Riemannian metric of constant curvature. It happens that such a Riemannian metric of constant curvature is an *absolute minimum* for some geometric quantities. For example, the Gauss–Bonnet theorem for two-manifolds shows that, if M is a compact surface and h is a metric on M with constant curvature $S(h)$, then

$$\int_M |S(g)|\, dv_g \geq \int_M |S(h)|\, dv_h \tag{1.1}$$

for any Riemannian metric g on M, where $S(g)$ is the scalar curvature of the metric g. For higher dimensions, results show that "nice" metrics like constant sectional curvature metrics, locally symmetric metrics or Einstein metrics could be minima for some geometric quantities. Gromov had conjectured that if (M, h) is a compact hyperbolic n-manifold with $n \geq 3$, then

$$\mathrm{Vol}(M, g) \geq \mathrm{Vol}(M, h)$$

for any Riemannian metric g on M with sectional curvature $K(g) \geq -1$. Here $\mathrm{Vol}(M, g)$ is the volume of (M, g). Besson, Courtois and Gallot proves this in [5, 6]. In addition, they demonstrate that, up to isometry and scaling, there is just one Einstein metric on a compact hyperbolic 4-manifold. Recently, using the Seiberg–Witten invariant, LeBrun [10] shows that if η is a Kähler metric of constant negative scalar curvature on a compact 4-manifold M, then

$$\int_M |S(g)|^2\, dv_g \geq \int_M |S(\eta)|^2\, dv_\eta \tag{1.2}$$

for any Riemannian metric g that is compatible with a fixed decomposition $H^2(M) = H^+ \oplus H^-$. An interesting question in Riemannian geometry is this: Given a compact n-manifold M without boundary, when can we find a positive constant $c(M)$ and a sufficiently large class of Riemannian metrics $C(M)$ on M such that

$$\int_M |S(g)|^{\frac{n}{2}}\, dv_g \geq c(M)$$

for any $g \in C(M)$? Note that the $L^{n/2}$-norm of scalar curvature is invariant under scaling. As the $L^{n/2}$-norm of the scalar curvature of a flat torus is zero, we will consider "hyperbolic"-type manifolds only, for example, when M admits a rank one locally symmetric metric of non-compact type. In addition to the above-mentioned results, Besson, Courtois and Gallot [4] show that, for $n \geq 3$, if h is a Riemannian metric of constant negative scalar curvature, then for any Riemannian metric g conformal to h (*i.e.*, $g = u^{\frac{4}{n-2}} h$ for some positive smooth function u on M), we have

$$\int_M |S(g)|^{\frac{n}{2}}\, dv_g \geq \int_M |S(h)|^{\frac{n}{2}}\, dv_h. \tag{1.3}$$

Furthermore they show that if h is an Einstein metric of negative sectional curvature, then inequality (1.3) holds for any Riemannian metric g in a neighborhood of h. We will demonstrate this result using the Ricci flow introduced by Hamilton.

Theorem A. *For $n \geq 4$, let (M, g) be a compact Riemannian n-manifold with $S(g) < 0$. Let g_t be the Ricci flow starting at g. If $S(g_t) \leq 0$, then*

$$\frac{d}{dt}\left(\int_M |S(g_t)|^{\frac{n}{2}}\, dv_{g_t}\right) \leq 0.$$

Under the above conditions, if the Ricci flow converges to a smooth metric h on M, then

$$\int_M |S(g)|^{\frac{n}{2}}\, dv_g \geq \int_M |S(h)|^{\frac{n}{2}}\, dv_h.$$

Convergence of Ricci flows under various curvature assumptions is discussed by many authors, including Hamilton [8], Min-Oo [12] and Ye [16]. Under certain conditions, the Ricci flow may blow up and hence has no smooth limits at all. The topology of the underlying manifold plays an important role. As a rather direct result of Seiberg–Witten theory, one can obtain the following.

Theorem B. *Let (M, η) be a compact Kähler–Einstein surface of negative scalar curvature. Then for any Riemannian metric g on M, we have*

$$\int_M |S(g)|^2 \, dv_g \geq \int_M |S(\eta)|^2 \, dv_\eta. \tag{1.4}$$

It follows from Theorem B that such a Kähler–Einstein metric η achieves the supremum of Yamabe invariants on the complex surface M. We then apply some of the results to an isospectral problem.

Theorem C. *Let (M, h) be a compact rank one locally symmetric 4-manifold of non-compact type. If g is a Riemannian metric on M having the same spectrum as h, then g is isometric to h.*

2. Ricci curvature flow and some estimates

Let (M, g_o) be a compact Riemannian manifold. Hamilton introduced the Ricci curvature flow and it can be defined by the following equation [16].

$$\frac{\partial g}{\partial t} = -2\mathbf{z}(g) - \frac{2\delta S(g)}{n} g, \quad g(0) = g_o, \tag{2.1}$$

where $\mathbf{z}(g) = \mathrm{Ric}(g) - [S(g)/n]g$ is the trace free Ricci tensor and

$$\delta S(g) = S(g) - \frac{\int_M S(g)\, dv_g}{\mathrm{Vol}(M, g)}.$$

Lemma 2.2. *Let M be a compact n-manifold with $n \geq 4$. Let g be a solution to the Ricci curvature flow equation (2.1) on the time interval $(0, t')$. Here t' is either a positive number or equal to infinity. Assume that $\lim_{t \to t'} g = h$ is a smooth Riemannian metric on M. If $S(g) < 0$ for $t \in (0, t')$, then*

$$\frac{d}{dt} \int_M |S(g)|^{\frac{n}{2}} \, dv_g \leq 0 \quad \text{for all } t \in (0, t').$$

Hence

$$\int_M |S(g_o)|^{\frac{n}{2}} \, dv_{g_o} \geq \int_M |S(h)|^{\frac{n}{2}} \, dv_h.$$

Proof. From (2.1) we have [16]

$$\frac{dS(g)}{dt} = \Delta S(g) + 2|\mathbf{z}|^2 + \frac{2\delta S(g)}{n} S(g). \tag{2.3}$$

As $\mathbf{z}(g)$ is trace-free, we have

$$(dv_g)' = \frac{1}{2}\mathrm{tr}_g(\frac{dg}{dt})\, dv_g = -\delta S(g). \tag{2.4}$$

Therefore

$$\begin{aligned}
&\frac{d}{dt}\int_M |S(g)|^{\frac{n}{2}}\, dv_g \\
&= \int_M (\frac{n}{2}|S(g)|^{\frac{n}{2}-1}\frac{d}{dt}|S(g)|)\, dv_g + \int_M |S(g)|^{\frac{n}{2}}(dv_g)' \\
&= \int_M (\frac{n}{2}|S(g)|^{\frac{n}{2}-1}(-\frac{d}{dt}S(g))\, dv_g - \int_M |S(g)|^{\frac{n}{2}}(\delta S(g))\, dv_g \quad (\text{as } S(g) < 0) \\
&= \int_M \frac{n}{2}|S(g)|^{\frac{n}{2}-1}(-\Delta S(g) - 2|\mathbf{z}|^2 - \frac{2\delta S(g)}{n}S(g))\, dv_g - \int_M |S(g)|^{\frac{n}{2}}(\delta S(g))\, dv_g \\
&= -\int_M (\frac{n}{2})(\frac{n}{2}-1)|S(g)|^{\frac{n}{2}-2}|\nabla|S(g)||^2\, dv_g - 2\int_M \frac{n}{2}|S(g)|^{\frac{n}{2}-1}|\mathbf{z}|^2\, dv_g \le 0,
\end{aligned}$$

as $-\Delta S(g) = \Delta|S(g)|$. □

Theorem 2.5 (Besson-Gallot-Courtois [4]). *For $n \ge 4$, let (M, h) be a compact Einstein n-manifold of strictly negative sectional curvature. Then there exists an open neighborhood of h in the space of smooth metrics on M with C^∞-norm such that for any metric g in that open neighborhood we have*

$$\int_M |S(g)|^{\frac{n}{2}}\, dv_g \ge \int_M |S(h)|^{\frac{n}{2}}\, dv_h.$$

Proof. It is shown in [12] and [16] that there exists an open neighborhood of h for which the Ricci curvature flow (2.1), starting with any Riemannian metric on that open neighborhood, converges to h. Moreover, we may choose the open neighborhood such that the scalar curvature remains negative during the Ricci curvature flow [16]. Then we can apply Lemma 2.2. □

It follows that the Einstein metric h gives a local minimum on $L^{\frac{n}{2}}$-norms of scalar curvatures. The Gauss–Bonnet theorem for two-manifolds shows that any metric of constant curvature is the *absolute* minimum of L^1-norms of scalar curvatures. Another example is Kähler–Einstein metrics of negative scalar curvature on compact complex surfaces ($n = 4$).

Theorem 2.6. *Let η be a Kähler–Einstein metric of negative scalar curvature on a compact complex surface M. Then for any Riemannian metric g on M, we have*

$$\int_M |S(g)|^2\, dv_g \ge \int_M |S(\eta)|^2\, dv_\eta.$$

Proof. This follows rather directly from the Seiberg–Witten theory [15, 9]. The main point is this, for the Kähler–Einstein surface (M, η), the first Chern class is given by $[S(\eta)\omega/8\pi]$, where ω is the Kähler form of (M, η). Hence

$$c_1(L)^2 = 2\chi(M) + 3\tau(M) = \frac{1}{32\pi^2}\int_M |S(\eta)|^2\, dv_\eta,$$

where $\chi(M)$ and $\tau(M)$ are the Euler characteristic and signature of M respectively. It follows from Seiberg–Witten theory that for any Riemannian metric g on M, there exists a solution to the Seiberg–Witten equations [15] and we have [9]:

$$c_1(L)^2 \le \frac{1}{32\pi^2}\int_M |S(g)|^2\, dv_g.$$

That is,

$$\int_M |S(g)|^{\frac{n}{2}}\, dv_g \ge \int_M |S(\eta)|^{\frac{n}{2}}\, dv_\eta.$$

This completes the proof. □

Remark. Using the Seiberg–Witten invariant, LeBrun [9] is able to show that, up to isometries and scalings, there is just one Einstein metric on a compact complex hyperbolic 4-manifold.

Remark. In the two-step variational program proposed to obtain an Einstein metric on a compact manifold M, one considers the quotient

$$Q(g) = \frac{\int_M S(g)\, dv_g}{[\mathrm{Vol}(M, g)]^{\frac{n-2}{n}}}$$

and finds a minimizing metric g_o in the conformal class of a metric g, denoted by $C(g)$, that is,

$$\lambda(M, g_o) = \inf_{g' \in C(g)} Q(g') = Q(g_o).$$

($\lambda(M, g_o)$ is known as the Yamabe invariant of M for the conformal class $C(g)$.) Let

$$\Lambda(M) = \sup_{g} \inf_{g' \in C(g)} Q(g').$$

Then one tries to find a metric h such that

$$Q(h) = \lambda(M, h) = \Lambda(M).$$

It is known that if such a Riemannian metric h exists and $\Lambda(M) \le 0$, then h is an Einstein metric [13]. Let M be a compact manifold which admits a metric of negative sectional curvature. Then by the solution to the Yamabe problem, on each conformal class $C(g)$ of Riemannian metrics on M, there is a unique metric $g_o \in C(g)$ with $S(g_o) = -1$ and

$$\lambda(M, g_o) = -[\mathrm{Vol}(M, g_o)]^{\frac{2}{n}}.$$

Hence in this situation $\Lambda(M)$ is related to a lower bound on $\mathrm{Vol}(M, g)$ for any Riemannian metric with $S(g) = -1$. This is equivalent to a lower bound on the $L^{n/2}$-norm of scalar curvature, as it follows from [4] that

$$\int_g |S(g)|^{\frac{n}{2}}\, dv_g \geq \int_{g_o} |S(g_o)|^{\frac{n}{2}}\, dv_{g_o}$$

for any $g \in C(g_o)$ and $S(g_o) \equiv -1$.

Theorem 2.7. *Let M be a compact complex surface which admits a Kähler–Einstein metric η of negative scalar curvature. Then up to scaling and isometry, η is the unique metric on M such that*

$$Q(\eta) = \lambda(M, \eta) = \Lambda(M).$$

Proof. From Theorem 2.6 and the above remark, it follows that

$$Q(\eta) = \lambda(M, \eta) = \Lambda(M).$$

Suppose that g is another metric on M such that

$$Q(g) = \lambda(M, g) = \Lambda(M).$$

Then g is an Einstein metric with negative scalar curvature as $\Lambda(M)$ is negative. We may assume that, without loss of generality, $S(\eta) = S(g) = -1$. Then

$$Q(\eta) = \Lambda(M) = Q(g)$$

implies that

$$\int_M |S(g)|^2\, dv_g = \int_M |S(\eta)|^2\, dv_\eta$$

and

$$c_1(L)^2 = \frac{1}{32\pi^2} \int_M |S(g)|^2\, dv_g.$$

A result in [9] implies that g is a Kähler–Einstein metric. Hence g is isometric to η up to a scaling factor [3]. □

Remark. For a compact hyperbolic 4-manifold (M, h), the hyperbolic metric h is the only candidate to achieve

$$Q(h) = \lambda(M, h) = \Lambda(M).$$

For if there is another Riemannian metric g such that

$$Q(g) = \lambda(M, g) = \Lambda(M),$$

then g is an Einstein metric [13]. Hence g is isometric to h up to scaling by a result of Besson–Courtois–Besson [5, 6]. However, it is unsettled whether $Q(h) = \Lambda(M)$.

For real dimension two, a manifold has both complex and conformally flat structures on it. We observe the following:

Theorem 2.8. *Let (M, h) be a compact hyperbolic 4-manifold. Then for any locally conformally flat metric g on M, we have*

$$\int_M |S(g)|^2\, dv_g \geq \int_M |S(h)|^2\, dv_h.$$

Proof. In dimension four, the Gauss–Bonnet formula takes the form [2, 3]

$$\chi(M) = \frac{1}{8\pi^2}\int_M (|U(g)|^2 - |Z(g)|^2 + |W(g)|^2)\, dv_g, \tag{2.9}$$

where $W(g)$ is the Weyl curvature tensor and

$$U(g)_{ijkl} = \frac{S(g)}{n(n-1)}(g_{ik}\, g_{jl} - g_{il}\, g_{jk}), \tag{2.10}$$

$$Z(g)_{ijkl} = \frac{1}{n-2}(\mathbf{z}(g)_{ik}\, g_{jl} + \mathbf{z}(g)_{jl}\, g_{ik} - \mathbf{z}(g)_{il}\, g_{jk} - \mathbf{z}(g)_{jk}\, g_{il}). \tag{2.11}$$

We have

$$|U(g)|^2 = \frac{2S(g)^2}{n(n-1)}, \qquad |Z(g)|^2 = \frac{4}{(n-2)}|\mathbf{z}(g)|^2 \tag{2.12}$$

and

$$|\operatorname{Ric}(g)|^2 = |\mathbf{z}(g)|^2 + \frac{S(g)^2}{n}. \tag{2.13}$$

Let h be a hyperbolic metric on M, then $W(h) = Z(h) = 0$ and hence

$$\chi(M) = \frac{1}{48\pi^2}\int_M S(h)^2\, dv_h. \tag{2.14}$$

For dimension bigger than or equal to four, a Riemannian metric g is conformally flat if and only if $W(g) \equiv 0$. Then (2.9), (2.12), (2.13) and (2.14) show that if g is any conformally flat metric on M we have

$$\int_M S(g)^2\, dv_g \geq \int_M S(h)^2\, dv_h.$$

This completes the proof of the theorem. □

Let $W^+(g)$ and $W^-(g)$ be the irreducible components of the Weyl curvature tensor $W(g)$ of an oriented Riemannian 4-manifold (M, g). (M, g) is said to be half-conformally flat if $W^-(g) \equiv 0$. Our next result shows that half-conformally flat Einstein metrics minimize the L^2-norm of scalar curvatures among half-conformally flat metrics on M.

Theorem 2.15. *Let (M, h) be an oriented half-conformally flat Einstein manifold. Then for any half-conformally flat Riemannian metric g on M, we have*

$$\int_M S(g)^2\, dv_g \geq \int_M S(h)^2\, dv_h.$$

Proof. Let $\tau(M)$ be the signature M. Then we have

$$\tau(M) = \frac{1}{12\pi^2}\int_M (|W^+(g)|^2 - |W^-(g)|^2)\, dv_g.$$

In particular, if h and g are half-conformally flat metrics on M, then we have

$$\tau(M) = \frac{1}{12\pi^2}\int_M |W^+(h)|^2\, dv_h = \frac{1}{12\pi^2}\int_M |W^+(g)|^2\, dv_g. \tag{2.16}$$

The Gauss–Bonnet formula for half-conformally flat manifolds is given by

$$\chi(M) = \frac{1}{8\pi^2}\int_M (|U(g)|^2 - |Z(g)|^2 + |W^+(g)|^2)\, dv_g. \tag{2.17}$$

If in addition h is Einstein then

$$\chi(M) = \frac{1}{8\pi^2}\int_M (|U(h)|^2 + |W^+(h)|^2)\, dv_g. \tag{2.18}$$

It follows from (2.16), (2.17) and (2.18) that

$$\int_M S(g)^2\, dv_g \geq \int_M S(h)^2\, dv_h.$$

This completes the proof. □

In [3] chapter 13, section F, a construction of half-conformally flat Einstein manifolds is discussed. Examples include the Fubini–Study metric on $\mathbf{CP}^2$ and irreducible locally symmetric 4-manifolds. We refer to [11] for further applications of the Gauss–Bonnet formula to obtain lower bounds on the $L^{n/2}$-norms of scalar curvatures.

Remark. As the scalar curvature is a summation of sectional curvatures, a positive lower bound on the $L^{n/2}$-norms of scalar curvatures roughly says that if the volume of the manifold is small, then the absolute value of the scalar curvature of the manifold is large. Using results in [4, 5, 6], it can be shown that if (M, h) is a compact Riemannian manifold with strictly negative sectional curvature, then for any Riemannian metric g on M with $\mathrm{Ric}(g)_{ij} \geq -(n-1)a^2 g_{ij}$, one has

$$a^n \cdot \mathrm{Vol}(M, g) \geq C(M, h),$$

where $C(M, h)$ is a positive constant that depends on the manifold M and the Riemannian metric h. This estimate makes use of a global lower bound on the Ricci curvature. The scalar curvature may still be close to zero. Suppose that one replaces the $L^{n/2}$-norms of scalar curvatures by the $L^{n/2}$-norms of full curvature tensors, then it follows rather easily from the Gauss–Bonnet formula that if (M, h) is a compact hyperbolic n-manifold, then there exists a positive constant $c(M, n)$ which depends on M and n, such that

$$\int_M |R_g|^{\frac{n}{2}}\, dv_g \geq c(M, n)$$

for any Riemannian metric g on M, where R_g is the curvature tensor of (M, g). If n is an odd integer then one considers $M \times M$.

3. Isospectral metrics

Let M be a compact 4-manifold. Suppose that g and h are two Riemannian metrics on M having the same spectra, then g and h are called isospectral metrics on M. If g and h are isospectral metrics on M, then it is known that [1, 2, 7]

$$\mathrm{Vol}(M, h) = \mathrm{Vol}(M, g); \tag{3.1}$$

$$\int_M S(h)\, dv_h = \int_M S(g)\, dv_g; \tag{3.2}$$

$$\int_M (6|\operatorname{Ric}(h)|^2 + 3|S(h)|^2)\, dv_h = \int_M (6|\operatorname{Ric}(g)|^2 + 3|S(g)|^2)\, dv_g. \tag{3.3}$$

(3.1) and (3.2) hold for all dimensions while (3.3) is valid in dimension four only.

Lemma 3.4. *Let (M, h) be a Riemannian metric of constant scalar curvature. If g and h are isospectral metrics on M, then*

$$\int_M S(g)^2\, dv_g \geq \int_M S(h)^2\, dv_h.$$

Proof. We may assume that $\mathrm{Vol}(N, h) = \mathrm{Vol}(M, g) = 1$ by scaling on both g and h by a positive constant. From (3.2) we have

$$\begin{aligned} \left|\int_M S(h)\, dv_h\right| &= \left|\int_M S(g)\, dv_g\right| \\ &\leq \int_M |S(g)|\, dv_g \\ &\leq \left(\int_M |S(g)|^2\, dv_g\right)^{\frac{1}{2}} \left(\int_M 1\, dv_g\right)^{\frac{1}{2}} \\ &= \left(\int_M |S(g)|^2\, dv_g\right)^{\frac{1}{2}}, \end{aligned} \tag{3.5}$$

as $\mathrm{Vol}(M,g)=1$. Since $S(h)$ is a constant, we have

$$\left|\int_M S(h)\,dv_h\right| = \int_M |S(h)|\,dv_h. \tag{3.6}$$

From (3.5) and (3.6) we have

$$\begin{aligned}|S(h)|\int_M |S(h)|\,dv_h &\le \left(\int_M |S(g)|^2\,dv_g\right)^{\frac{1}{2}}\cdot|S(h)|\\ &= \left(\int_M |S(g)|^2\,dv_g\right)^{\frac{1}{2}}\left(|S(h)|^2\cdot\mathrm{Vol}(M,h)\right)^{\frac{1}{2}}\\ &= \left(\int_M |S(g)|^2\,dv_g\right)^{\frac{1}{2}}\left(\int_M |S(h)|^2\,dv_h\right)^{\frac{1}{2}}.\end{aligned}$$

That is,

$$\left(\int_M |S(h)|^2\,dv_h\right)^{\frac{1}{2}} \le \left(\int_M |S(g)|^2\,dv_g\right)^{\frac{1}{2}},$$

or

$$\int_M |S(h)|^2\,dv_h \le \int_M |S(g)|^2\,dv_g.$$

This completes the proof of the lemma. □

Theorem 3.7. *Let (M,h) be a rank one compact locally symmetric space of non-compact type. If g is another Riemannian metric on M that is isospectral to h, then g is isometric to h.*

Proof. As (M,h) is a rank one compact locally symmetric space, it is an Einstein manifold. Lemma 3.4 and equation (3.3) show that

$$\int_M |\,\mathrm{Ric}(g)|^2\,dv_g \le \int_M |\,\mathrm{Ric}(h)|^2\,dv_h.$$

As (cf. (2.11))

$$|\,\mathrm{Ric}(g)|^2 = |\mathbf{z}(g)|^2 + \frac{|S(g)|^2}{4} \quad\text{and}\quad |\,\mathrm{Ric}(h)|^2 = \frac{|S(h)|^2}{4},$$

where $\mathbf{z}_{ij} = \mathrm{Ric}_{ij} - [S(g)/4]g_{ij}$, we obtain

$$\int_M |\mathbf{z}|^2\,dv_g + \int_M \frac{|S(g)|^2}{4}\,dv_g \le \int_M \frac{|S_o|^2}{4}\,dv_h.$$

From Lemma 3.4 we obtain

$$\int_M |\mathbf{z}(g)|^2\,dv_g = 0.$$

Hence (M, g) is an Einstein manifold with $\mathrm{Ric}(g) = -3g$. A rank one symmetric spaces of non-compact type in dimension four is either a real hyperbolic space or a complex hyperbolic space. If (M, h) is a compact real hyperbolic manifold, then up to isometries, it has a unique Einstein metric with scalar curvature equal to $-4 \cdot 3$ [5, 6]. The corresponding statement for complex hyperbolic manifolds is proved in [9]. □

Remark. Berger proves that if two compact Riemannian manifolds of dimension less than or equal to five are isospectral, and if one of the Riemannian manifold has constant sectional curvature, then the other Riemannian manifold also has constant sectional curvature [1]. Tanno extends the result to dimension six [14]. Berger's result and Mostow's rigidity theorem provide a more direct proof to the above theorem if (M, h) is a compact hyperbolic 4-manifold.

Acknowledgement. The author would like to thank Professor Peter Gilkey for helpful conversations, and the referee for useful suggestions.

References

[1] M. Berger, Le spectre des variétés riemanniennes, Rev. Roum. Math. Pure Appl. 13 (1969), 915–931.

[2] M. Berger, P. Gauduchon, E. Mazet, Le Spectre d'Une Variété Riemannienne, Lecture Notes in Math. 194, Springer-Verlag, Berlin–Heidelberg–New York 1971.

[3] A. Besse, Einstein Manifolds, Springer-Verlag, Berlin–Heidelberg–New York 1987.

[4] G. Besson, G. Courtois, S. Gallot, Volume et entropie minimale des espaces localement symétriques, Invent. Math. 103 (1991), 417–445.

[5] G. Besson, G. Courtois, S. Gallot, Les variétés hyperboliques sont des minima locaux de l'entropie topologique, Invent. Math. 117 (1994), 403–445.

[6] G. Besson, G. Courtois, S. Gallot, Entropies et Rigidités des Espaces Localement Symétriques de Courbure Strictement Négative, Prépublication de l'Institut Fourier, Grenoble, No. 281, (1994).

[7] R. Brooks, P. Perry, P. Yang, Isospectral sets of conformally equivalent metrics, Duke Math. J. 58 (1989), 131–150.

[8] R. Hamilton, Three-manifolds with positive Ricci curvature, J. Diff. Geom. 17 (1982), 225–306.

[9] C. LeBrun, Einstein metrics and Mostow rigidity, Math. Research Letters 2 (1995), 1–8.

[10] C. LeBrun, Polarized 4-manifolds, extremal Kähler metrics, and Seiberg–Witten theory, Math. Research Letters 2 (1995) 653–662.

[11] M.-C. Leung, On the $L^{\frac{n}{2}}$-norm of scalar curvature, Illinois J. of Math. 40 (1996), 606–631.

[12] M. Min-Oo, Almost Einstein manifolds of negative Ricci curvature, J. Diff. Geom. 32 (1990), 457–472.

[13] R. Schoen, Variational theory for the total scalar curvature functional for Riemannian metrics and related topics, in: Topics in Calculus of Variations (M. Giaquinta, ed.), Lecture Notes in Math. 1365, Springer-Verlag, Berlin–Heidelberg–New York, 1989, 120–154.

[14] S. Tanno, Eigenvalues of the Laplacian of Riemannian manifolds, Tohoku Math. J. 25 (1973), 391–403.

[15] E. Witten, Monopoles and four-manifolds, Math. Res. Lett. 1 (1994), 769–796.

[16] R. Ye, Ricci Flow, Einstein metrics and space forms, Trans. Amer. Math. Soc. 338 (1993), 871–896.

Wavelet geometric analysis of the Dirichlet problem

En-Bing Lin and Xiaolin Zhou

Abstract. We prove a generalized convergence theorem of Wavelet-Galerkin approximation for the Dirichlet problem.

1991 Mathematics Subject Classification: 42C20, 35J35, 65N30

1. Introduction

Wavelets have been used for solutions of differential equations directly as bases with Galerkin's method [12], for algebraic manipulations to simplify the results using existing discretization methods [14] and after adaptation to specific differential operators when Galerkin's method is used [4]. Numerical results on the accuracy and complexity of algorithms are also discussed in several papers [2, 18, 19]. The wavelet point of view for approaching boundary value problems in partial differential equations was introduced in [14], in which the boundary data, and the unknown solution (of a boundary value problem) on the interior of domain are all uniformly represented in terms of compactly supported wavelet functions in an extrinsic ambient Euclidean space. Subsequently, the localization property of wavelets was used to extend Daubechies' orthonormal representation to boundary data and geometric boundaries of the Dirichlet problem [19]. On the other hand, several variations are given on the construction of Daubechies' wavelet [8]. In this paper, we consider one of theses variations, namely, coiflets. We generalize the result on the rate of convergence of wavelet solutions of Dirichlet problem by considering generalized coiflets. We organize our paper as follows. In Section 2, we outline some basic ingredients about wavelets which we need in this paper. In Section 3, we give a precise description of a wavelet series interpolation, and show that this interpolation approximates functions in $L^2(R^2)$ and in the Sobolev space $H^1(R^2)$ with convergence rates $(\frac{1}{2})^{jN}$ and $(\frac{1}{2})^{j(N-1)}$, respectively. In section 4, we review some basic geometric measure theory and provide some formulas to calculate boundary integrals. In Section 5, we prove a convergence theorem for wavelet-Galerkin solution in the sense of fictitious domain schemes. In Section 6, we generalize the convergence property of wavelet-Galerkin approximation in [19]. Here we formulate the results in two dimensions; higher dimensional cases can also be derived similarly.

2. Daubechies wavelets and coiflets

In this section, we will briefly review the basic ideas and descriptions of wavelets, multiresolution analysis, Daubechies wavelets, Mallat algorithm and coiflets.

Construction of wavelet functions can start from the building of scaling function, $\phi(x)$, and a set of related coefficients, $\{a_k\}_{k\in Z}$, which satisfy the *two-scale relation* or *refinement equation*,

$$\phi(x) = \sum_k a_k \phi(2x - k) \tag{2.1}$$

and some additional conditions. The scaling function $\phi(x)$ has compact support if and only if finitely many coefficients a_k are non-zero.

Translations of the scaling function, $\{\phi(x - k)\}$, form a Riesz or unconditional basis of a subspace $V_0 \subset L^2(R)$. Furthermore, through translation and dilation of ϕ, a Riesz basis $\{\phi_{n,k}(x)\}_{k\in Z}$ is obtained for the subspace $V_n \subset L^2(R)$, where

$$\phi_{n,k}(x) = 2^{\frac{n}{2}}\phi(2^n x - k) \tag{2.2}$$

If translation of the scaling function, $\{\phi(x)\}$, is orthonormal at the same level n, *i.e.*,

$$\int_{-\infty}^{\infty} \phi_{n,k}(x)\,\phi_{n,l}(x)\,dx = \delta_{k,l}, \quad n, k, l \in Z, \tag{2.3}$$

then the best approximation of an $L^2(R)$ function, $f(x)$, by a function $P_n f(x)$ in the subspace V_n of $L^2(R)$ is given by the orthogonal projection of f on V_n, as follows:

$$P_n f(x) = \sum_k \alpha_{n,k}\,\phi_{n,k}(x) \tag{2.4}$$

where $\alpha_{n,k}$ is given by the inner product of $f(x)$ and $\phi_{n,k}(x)$,

$$\alpha_{n,k} = \langle f, \phi_{n,k}\rangle = \int_{-\infty}^{\infty} f(x)\phi_{n,k}(x)\,dx \tag{2.5}$$

The operator $P_n : L^2(R) \to V_n$ is a linear orthogonal projection resulting in the best approximation of $f(x)$ in V_n.

Approximation of a function, $f(x)$, can be conducted at different resolution levels, and the approximations in the subspaces, $\ldots, V_{n-1}, V_n, V_{n+1}, \ldots$, have the following properties:

(i) $V_n \subset V_{n+1}$,

(ii) $f(x) \in V_n \Leftrightarrow f(2x) \in V_{n+1}$,

(iii) $f(x) \in V_n \Leftrightarrow f(x + 2^{-n}k) \in V_n$, $\forall k \in Z$,

(iv) $\lim_{n\to\infty} V_n = \cup_n V_n$ is dense in $L^2(R)$,

(v) $\cap_{n\in Z} V_n = \{0\}$,

(vi) $\{\phi(x - n)\}_{n\in Z}$ is an orthonormal basis of V_0.

Here, we call $\phi(x) \in L^2(R)$ a scaling function that generates a *multiresolution analysis* (MRA) with the above properties.

For every $j \in Z$, define W_j to be the orthogonal complement of V_j in V_{j+1}. We have

$$V_{j+1} = V_j \oplus W_j \tag{2.6}$$

and

$$W_j \perp W_{j'} \quad \text{if} \quad j \neq j' \tag{2.7}$$

(If $j > j'$,*e.g.*, then $W_{j'} \subset V_j \perp W_j$.) It follows that, for $j > J$,

$$V_j = V_J \oplus (\oplus_{k=0}^{J-j+1} W_{J-k}) \tag{2.8}$$

where all these subspaces are orthogonal. By properties (iv) and (v) above, we have

$$L^2(R) = \oplus_{j\in Z} W_j, \tag{2.9}$$

a decomposition of $L^2(R)$ into mutually orthogonal subspaces. It turns out [6,12] that a basis for W_n can be obtained by dilating and translating a single function, $\psi(x)$, called a *basic (mother, analyzing)* wavelet, which is defined by

$$\psi(x) = \sum_k b_k \phi(2x - k) \tag{2.10}$$

where b_k are a set of coefficients for the two-scale relationship of wavelet basis. In fact, $\{\psi_{j,k}(x) = 2^{\frac{j}{2}}\psi(2^j x - k)\ ;\ j, k \in Z\}$ is an orthonormal basis for $L^2(R)$.

We can define an operator $Q_n : L^2(R) \to W_n$ similar to P_n. Then

$$Q_n f(x) = \sum_k \beta_{n,k}\psi_{n,k}(x) \tag{2.11}$$

where $\{\beta_{n,k}\}$ are coefficients that can be determined through an equation analogous to (2.5), *i.e.*

$$\beta_{n,k} = \langle f, \psi_{n,k}\rangle = \int_{-\infty}^{\infty} f(x)\psi_{n,k}(x)\,dx \tag{2.12}$$

Thus, for a function, $f(x)$, we have

$$\begin{aligned} P_n f(x) &= \textstyle\sum_k \alpha_{n,k}\phi_{n,k}(x) \\ &= \textstyle\sum_k \alpha_{n-1,k}\phi_{n-1,k}(x) + \sum_k \beta_{n-1,k}\psi_{n-1,k}(x) \end{aligned} \tag{2.13}$$

where the first sum is $P_{n-1}f(x) \in V_{n-1}$ and the second sum is $Q_{n-1}f(x) \in W_{n-1}$.

The relationships between the approximation coefficients at different levels, *e.g.* $\{\alpha_{n,k}\}$ at level n, and $\{\alpha_{n-1,k}\}$, $\{\beta_{n-1,k}\}$ at level $n-1$, are determined through some sequences of constants which depend only on the wavelet basis being used. This is the *Mallat algorithm* [13], and the formulas will be given below.

Recall that $\{\phi_{n-1,k}(x)\}_k$ and $\{\psi_{n-1,k}(x)\}_k$ are bases for V_{n-1} and W_{n-1} respectively. Thus a scaling function $\phi_{n,l}(x) \in V_n = V_{n-1} \oplus W_{n-1}$ can be represented in terms of these bases, using coefficients $\frac{1}{2}g(\ell - 2k)$, and $\frac{1}{2}h(\ell - 2k)$, as

$$\phi_{n,\ell}(x) = \sum_k [\tfrac{1}{2}g(\ell - 2k)\phi_{n-1,k}(x) + \tfrac{1}{2}h(\ell - 2k)\psi_{n-1,k}(x)], \quad \ell \in Z \tag{2.14}$$

where the sequences $\{\frac{1}{2}g(k)\}$ and $\{\frac{1}{2}h(k)\}$ depend only on the mother wavelet. From this equation and (2.13), we get

$$\alpha_{n-1,\ell} = \tfrac{1}{2}\textstyle\sum_k g(\ell - 2k)\alpha_{n,k} \tag{2.15}$$

$$\beta_{n-1,\ell} = \tfrac{1}{2}\textstyle\sum_k h(\ell - 2k)\alpha_{n,k} \tag{2.16}$$

Thus, if we know $\{\alpha_{n,k}\}$, we can go from a higher level n to a lower level $n - \ell$.

By applying the two-scale relations of scaling and wavelet functions to the right hand side of equation (2.13), we obtain the following reconstruction algorithm,

$$\alpha_{n+1,\ell} = \sum_k a_{\ell-2k}\alpha_{n,k} + b_{\ell-2k}\beta_{n,k} \tag{2.17}$$

where $\{a_k\}$ and $\{b_k\}$ are called *reconstruction sequences*. So, if we know $\{\alpha_{n-1,k}\}$ and $\{\beta_{j,k}\}_{j=n-\ell}^{n-1}$, we can construct the approximation at level n from the lower level $n - l$.

The construction of *Daubechies wavelets* [6] starts from finding a sequence with finite non-zero terms a_k, $k = 0, \ldots, 2N - 1$, for the two-scale relation of the scaling function (2.1), and the coefficients $\{a_0, a_1, \ldots, a_{2N-1}\}$ satisfy

$$\sum_{k=0}^{2N-1} a_k = 2 \tag{2.18}$$

$$\sum_{k=0}^{2N-1} a_k a_{k+2\ell} = 2\delta_{0,\ell}, \quad \ell \in Z \tag{2.19}$$

The vector $(a_0, \ldots, a_{2N-1})$ is called the *scaling vector* and we define the associated wavelet vector $(b_0, \ldots, b_{2N-1})$ by defining

$$b_k = (-1)^{k+1} a_{2N-1-k}. \tag{2.20}$$

Then wavelet function $\psi(x)$ is

$$\psi(x) = \sum_{k=0}^{2N-1} b_k \phi(2x - k). \tag{2.21}$$

With this convention, both the scaling function $\phi(x)$ and the corresponding wavelet function $\psi(x)$ have the same compact support $[0, 2N-1]$. Moreover, we have the following properties of moments of these functions.

$$\int \phi(x)\,dx = 1, \tag{2.22}$$

$$\int x\phi(x)\,dx = \tfrac{1}{2}\textstyle\sum_{k=0}^{2N-1} ka_k, \tag{2.23}$$

$$\int x^l\psi(x)\,dx = 0, \quad l = 0, \dots, N-1. \tag{2.24}$$

Let us denote by

$$c := \tfrac{1}{2}\textstyle\sum_{k=0}^{2N-1} ka_k, \tag{2.25}$$

the *first moment* of the function $\phi(x)$.

For $j, k \in Z$, define

$$\phi_k^j(x) = 2^{\frac{j}{2}}\,\phi(2^jx - k), \tag{2.26}$$

$$\psi_k^j(x) = 2^{\frac{j}{2}}\,\psi(2^jx - k), \tag{2.27}$$

where $x \in R$, and as usual we define $\phi_k(x) := \phi_k^0(x)$. The following properties of these functions are valid and easy to verify:

$$\mathrm{supp}(\phi_k^j) = \mathrm{supp}(\psi_k^j) = [\frac{k}{2^j}, \frac{k+2N-1}{2^j}], \tag{2.28}$$

$$\|\phi_k^j\|_{L^2(R)} = 1. \tag{2.29}$$

Smooth scaling functions arise as a consequence of the degree of approximation of the translates. The conditions

$$\sum_k (-1)^k k^m a_k = 0 \quad m = 0, 1, \dots, N-1 \tag{2.30}$$

imply that $\{1, x, \dots, x^{N-1}\}$ can be expressed as linear combinations of the translates of $\phi(x-k)$ [17], which has the following equivalent expressions.

Proposition 2.1. *The following are equivalent results:*

(i) $\{1, x, \dots, x^{N-1}\}$ *are linear combinations of* $\{\phi(x-k)\}_{k\in Z}$.

(ii) $\|f - \sum c_k\phi(2^jx-k)\| \le C2^{-jN}\|f^{(N)}\|$ *where* $c_k = 2^j\int f(x)\phi(2^jx-k)\,dx$.

(iii) $\int x^m\psi(x)\,dx = 0$ *for* $m = 0, 1, \dots, N-1$.

(iv) *The decay of the wavelet coefficients of a smooth function is:*

$$\int f(x)\psi(2^jx)\,dx \le C2^{-jN}.$$

The following lemma will be needed in next section.

Lemma 2.2. *Let $\phi, \psi \in C^M$ be the orthonormal multiresolution analysis satisfying*

$$\int x^l \psi(x)\,dx = 0, \quad l = 0, 1, \ldots, M. \tag{2.31}$$

Then for $k = 0, \ldots, M$, $x \in R$,

$$\sum_l l^k \phi(l + x) = \int (-x + y)^k \phi(y)\,dy. \tag{2.32}$$

Proof. The condition of the compact support along with vanishing moment assumption imply (see the proof of Corollary 5.5.4 of [7]) that ϕ and $\hat{\phi}$ are rapidly decreasing, and

$$\hat{\phi}(2\pi l) = \hat{\phi}'(2\pi l) = \cdots = \hat{\phi}^{(M)}(2\pi l) = 0 \tag{2.33}$$

for $l \neq 0$. We use the Poisson formula

$$\sum_l f(l) = \sum_l \hat{f}(2\pi l) \tag{2.34}$$

and $f(t) = t^k \phi(t + x)$ which give rise to the following:

$$\begin{aligned} \textstyle\sum_l l^k \phi(l + k) &= \textstyle\sum_l i^k \partial^k (e^{iwx} \hat{\phi}(w))|_{w=2\pi l} \\ &= i^k \partial^k (e^{iwx} \hat{\phi}(w))|_{w=0} \\ &= \textstyle\sum_{r \le k} \binom{k}{r} i^{k-r} (-x)^r \partial^{k-r} \hat{\phi}(0) \\ &= \textstyle\sum_{r \le k} \binom{k}{r} i^{k-r} (-x)^r \int (-iy)^{k-r} \phi(y)\,dy \\ &= \textstyle\int (-x + y)^k \phi(y)\,dy. \end{aligned}$$

□

It is easy to verify the following corollary by changing variables.

Corollary 2.3. *For $c, x \in R$*

$$\sum_p \left(\frac{p + c}{2^j} - x\right)^k \phi_p^j(x) = \int \left(\frac{t + c}{2^j} - x\right)^k 2^{\frac{j}{2}} \phi(2^j x - t)\,dt.$$

We will use this formula in the next section.

Depending on the application one had in mind, Daubechies constructed coiflets of order L [8]. Namely, to construct ϕ, ψ, *coiflet* of order L which satisfy

$$\begin{aligned} \textstyle\int x^m \psi(x)\,dx &= 0, \quad m = 0, 1, \ldots, L - 1, \\ \textstyle\int x^m \phi(x)\,dx &= 0, \quad m = 0, 1, \ldots, L - 1. \end{aligned}$$

The need for orthonormal bases with this property first came up in the application of wavelet bases to numerical analysis in the work of [2]. Imposing such vanishing moments on ϕ also increases its symmetry, but there is a price to pay, namely, a coiflet with $2k$ vanishing moments typically has support width $6k-1$, as compared to $4k-1$ for Daubechies wavelet. Because these orthonormal wavelet bases with vanishing moments for both ϕ and ψ were requested by Coifman, these wavelets were name coiflets by Daubechies, [7,8].

3. Wavelet interpolation

We prove the following interpolation theorem in R^2 which is also true in R^n. Let ϕ, ψ be the orthonormal multiresolution analysis with compact support and let ψ satisfy (2.24). Let $\{M_l\}$ denote the moments of the scaling function ϕ, *i.e.*,

$$M_l := \int x^l \phi(x)\, dx \quad l = 1, 2, \ldots. \tag{3.1}$$

In particular,

$$M_1 = c = \frac{1}{2}\sum_k k a_k. \tag{3.2}$$

Theorem 3.1. *Assume the function $f \in C^N(\overline{\Omega})$, where Ω is a bounded open set in R^2, $N \geq 2$. For $j \in Z$, let*

$$f^j(x,y) := \frac{1}{2^j} \sum_{p,q\in\wedge} f(\tfrac{p+c}{2^j}, \tfrac{q+c}{2^j})\phi_p^j(x)\phi_q^j(y), \quad (x,y) \in \Omega, \tag{3.3}$$

where the index set $\wedge = \{(p,q) \mid (\mathrm{supp}(\phi_p^j) \otimes \mathrm{supp}(\phi_q^j)) \cap \Omega \neq \phi\}$. In addition, suppose that the moments M_l satisfy

$$M_l = (c)^l \quad l = 1, 2, \ldots, N-1. \tag{3.4}$$

Then

$$\|f - f^j\|_{L^2(\Omega)} \leq C\|f^{(N)}\|_\infty (\frac{1}{2^j})^N, \tag{3.5}$$

$$\|f - f^j\|_{H^1}(\Omega) \leq C\|f^{(N)}\|_\infty (\frac{1}{2^j})^{N-1}, \tag{3.6}$$

where C is a constant depending only on N, the diameter of Ω and

$$\|f^{(N)}\|_\infty := \max_{(x,y)\in\Omega,\ m=0,1,\ldots,N} \left|\frac{\partial^N f}{\partial x^m \partial y^{N-m}}(x,y)\right|. \tag{3.7}$$

Proof. Without loss of generality, we can ssume that Ω is a square with edge width L. Then the number of p's (and q's) in the summation in (3.3) is no more than $2^j L + W$, where W denotes the support width of ϕ.

For each $(p, q) \in \wedge$, using the Taylor expansion for f at a point $(x, y) \in \Omega$, we have

$$f(\tfrac{p+c}{2^j}, \tfrac{q+c}{2^j}) = f(x, y) + \textstyle\sum_{n=1}^{N-1} \frac{1}{n!} \sum_{m=0}^{n} \binom{n}{m} \frac{\partial^n f(x,y)}{\partial x^m \partial y^{n-m}} (\frac{p+c}{2^j} - x)^m (\frac{q+c}{2^j} - y)^{n-m}$$

$$+ \tfrac{1}{N!} R_{p,q}(x, y) \tag{3.8}$$

where

$$R_{p,q}(x, y) = \sum_{m=0}^{N} \binom{N}{m} \frac{\partial^N f(\xi_p, \eta_q)}{\partial x^m \partial y^{N-m}} (\tfrac{p+c}{2^j} - x)^m (\tfrac{q+c}{2^j} - y)^{N-m} \tag{3.9}$$

for some (ξ_p, η_q) on the segment connecting (x, y) and $(\frac{p+c}{2^j}, \frac{q+c}{2^j})$. Therefore

$$2^j (f^j - f)(x, y) = \textstyle\sum_{p,q\in\wedge} \Big\{ \sum_{n=1}^{N-1} \frac{1}{n!} \sum_{m=0}^{n} \binom{n}{m} \frac{\partial^n f(x,y)}{\partial x^m \partial y^{n-m}} (\frac{p+c}{2^j} - x)^m$$

$$\times (\tfrac{q+c}{2^j} - y)^{n-m} + \tfrac{1}{N!} \textstyle\sum_{p,q\in\wedge} R_{p,q}(x, y) \Big\} \phi_p^j(x) \phi_q^j(y). \tag{3.10}$$

Now we have, by Lemma 2.1 and (3.2),

$$\begin{aligned} \textstyle\sum_p (\frac{p+c}{2^j} - x) \phi_p^j(x) &= \textstyle\int (\frac{t+c}{2^j} - x) 2^{\frac{j}{2}} \phi(2^j x - t)\, dt \\ &= 2^{-\frac{j}{2}} \textstyle\int (t - c) \phi(t)\, dt = 0. \end{aligned} \tag{3.11}$$

For $m > 1$, if $\phi, \psi \in C^{N-1}$, then

$$\begin{aligned} \textstyle\sum_p (\frac{p+c}{2^j} - x)^m \phi_p^j(x) &= \textstyle\int (\frac{t+c}{2^j} - x)^m 2^{\frac{j}{2}} \phi(2^j x - t)\, dt \\ &= 2^{(\frac{1}{2} - m)j} \textstyle\int (t - c)^m \phi(t)\, dt. \end{aligned} \tag{3.12}$$

By (3.4) and induction, we have

$$\int (t - c)^m \phi(t)\, dt = 0, \quad m = 2, \ldots, N - 1. \tag{3.13}$$

Hence

$$\sum_p (\tfrac{p+c}{2^j} - x)^m \phi_p^j(x) = 0, \quad m = 1, 2, \ldots, N - 1. \tag{3.14}$$

This implies that the first summation in (3.10) vanishes. Without the assumption of $\phi, \psi \in C^{N-1}$, (3.14) is still valid, see Remark (3.1) and we obtain

$$2^j (f^j - f)(x, y) = \frac{1}{N!} \sum_{p,q\in\wedge} R_{p,q}(x, y) \phi_p^j(x) \phi_q^j(y). \tag{3.15}$$

But for $x \in \operatorname{supp} \phi_p^j$, $y \in \operatorname{supp} \phi_q^j$

$$\left| \frac{p + c_1}{2^j} - x \right| \le \frac{C}{2^j}, \qquad \left| \frac{q + c_1}{2^j} - y \right| \le \frac{C}{2^j}, \tag{3.16}$$

where C depends only on W. From (3.9):

$$|\tfrac{1}{N!}R_{pq}(x,y)| \leq C\|f^{(N)}\|_\infty(\tfrac{1}{2^j})^N. \tag{3.17}$$

Therefore,

$$2^{2j}|f^j - f|^2 \leq C\|f^{(N)}\|_\infty^2 \left(\frac{1}{2^j}\right)^{2N} \left(\sum_{p,q\in\wedge} \phi_p^j(x)\phi_q^j(y)\right)^2, \tag{3.18}$$

$$\begin{aligned} 2^{2j}\|f^j - f\|_{L^2(\Omega)}^2 &\leq \|f^{(N)}\|_\infty^2(\tfrac{1}{2^j})^{2N} \sum_{p,q\in\wedge}\sum_{p',q'\in\wedge} \iint |\phi_p^j\phi_q^j\phi_{p'}^j\phi_{q'}^j|\,dx\,dy \\ &= C\|f^{(N)}\|_\infty^2(\tfrac{1}{2^j})^{2N} \sum_{p,q\in\wedge}\sum_{|p'-p|\leq W} \int |\phi_p^j\phi_{p'}^j|\,dx \sum_{|q'-q|\leq W} \int |\phi_q^j\phi_{q'}^j|\,dy \\ &\leq C\|f^{(N)}\|_\infty^2(\tfrac{1}{2^j})^{2N}2^j2^j. \end{aligned} \tag{3.19}$$

Hence (3.5) is established.

The proof of (3.6) can be obtained by showing the following inequality

$$\|\nabla f - \nabla f^j\|_{L^2(\Omega)} \leq C\|f^{(N)}\|_\infty(\tfrac{1}{2^j})^{N-1}. \tag{3.20}$$

We again use the Taylor expansion (3.8).

From (3.3) and (3.14), we have

$$\begin{aligned} 2^j \cdot \tfrac{\partial f^j}{\partial x}(x,y) &= \sum_{p,q\in\wedge} f\left(\tfrac{p+c}{2^j}, \tfrac{q+c}{2^j}\right) \tfrac{\partial\phi_p^j(x)}{\partial x}\phi_q^j(y) \\ &= \sum_{p,q\in\wedge} \left\{ \sum_{n=1}^{N-1} \tfrac{1}{n!}\tfrac{\partial^n f(x,y)}{\partial x^n}\left(\tfrac{p+c}{2^j} - x\right)^n \right\} \tfrac{\partial\phi_p^j(x)}{\partial x}\phi_q^j(y) \\ &\quad + \tfrac{1}{N!}\sum_{p,q\in\wedge} R_{p,q}(x,y)\tfrac{\partial\phi_p^j(x)}{\partial x}\phi_q^j(y). \end{aligned} \tag{3.21}$$

Now

$$\begin{aligned} \textstyle\sum_p \left(\tfrac{p+c}{2^j} - x\right)^n \tfrac{\partial\phi_p^j(x)}{\partial x} &= \textstyle\sum_p n\left(\tfrac{p+c}{2^j} - x\right)^{n-1}\phi_p^j(x) \\ &= \begin{cases} 2^{\frac{j}{2}}, & n = 1 \\ 0, & n > 1. \end{cases} \end{aligned} \tag{3.22}$$

Substituting (3.22) into (3.21) one has

$$2^j(\frac{\partial f^j}{\partial x} - \frac{\partial f}{\partial x})(x,y) = \frac{1}{N!}\sum_{(p,q)\in\wedge} R_{p,q}(x,y)\frac{\partial\phi_p^j}{\partial x}\phi_q^j. \tag{3.23}$$

Therefore by (3.17) and (3.23),

$$2^{2j}\left|\frac{\partial f^j}{\partial x} - \frac{\partial f}{\partial x}\right|^2 \leq C\|f^{(N)}\|_\infty^2(\tfrac{1}{2^j})^{2N}(\sum_{p,q\in\wedge}\frac{\partial\phi_p^j}{\partial x}\phi_q^j)^2, \tag{3.24}$$

$$2^{2j}\left\|\frac{\partial f^j}{\partial x}-\frac{\partial f}{\partial x}\right\|^2_{L^2(\Omega)} \le C\|f^{(N)}\|^2_\infty(\tfrac{1}{2^j})^{2N}\sum_{p,q\in\wedge}\sum_{|p'-p|\le W}\int|\tfrac{\partial\phi^j_p(x)}{\partial x}\cdot\tfrac{\partial\phi^j_{p'}(x)}{\partial x}|\,dx\sum_{|q'-q|\le W}\int|\phi^j_q\phi^j_{q'}|\,dy. \tag{3.25}$$

Note that

$$\int\left|\frac{\partial\phi^j_p(x)}{\partial x}\right|^2 dx \le 2^{2j}\int|\phi'|^2\,dx = C2^{2j}. \tag{3.26}$$

Then combining (3.25) and (3.26) and using Hölder's inequality, we obtain

$$2^{2j}\left\|\frac{\partial f}{\partial x}-\frac{\partial f^j}{\partial x}\right\|^2_{L^2(\Omega)} \le C\|f^{(N)}\|^2_\infty(\tfrac{1}{2^j})^{2N}2^j2^{2j}2^j, \tag{3.27}$$

$$\left\|\frac{\partial f}{\partial x}-\frac{\partial f^j}{\partial x}\right\|_{L^2(\Omega)} \le C\|f^{(N)}\|_\infty(\tfrac{1}{2^j})^{N-1}. \tag{3.28}$$

Similarly, we can derive

$$\left\|\frac{\partial f}{\partial y}-\frac{\partial f^j}{\partial y}\right\|_{L^2(\Omega)} \le C\|f^{(N)}\|_\infty(\tfrac{1}{2^j})^{N-1}. \tag{3.29}$$

Then (3.20) follows and the Theorem is proved. □

Remark 3.1. In order to derive (3.14) without exploiting the formula (2.32), one can use Proposition 2.1.

In (ii) of Proposition 2.1, we set $f(x) = x$, then $x = \sum_k c_k\phi(2^jx - k)$ and $c_k = 2^j\int x\phi(2^jx-k)\,dx = 2^{-j}(c+k)$. This implies (3.11). Letting $f(x) = x^2,\ldots,x^{N-1}$, by (3.4) and induction, one can obtain (3.14). This is an alternative way to prove the above theorem.

Remark 3.2. The coiflets of order N satisfies (3.4) since $M_1 = M_2 = \ldots = M_{N-1} = 0$, and hence the theorem holds for coiflets.

Remark 3.3. Under the assumption $\phi, \psi \in C^N$, one can derive the following estimates:

$$\|f - f^j\|_{H^m(\Omega)} \le C\|f^{(N)}\|_\infty(\frac{1}{2^j})^{N-m}, \qquad m = 0,1,\ldots,N-1. \tag{3.30}$$

4. Wavelet geometric measure theory

In this section, we will briefly review some basic geometric measure theory which we will need in the following sections. Readers are referred to [9, 10, 19] for more details.

We will say that a Lebesgue integrable function u is of *bounded variation* over a domain $U \subset R^2$ if

$$\sup_{g \in C^1(U,R^2),|g|\leq 1} \int_U u \operatorname{div} g \, dx \, dy \leq +\infty. \tag{4.1}$$

Suppose that u is of bounded variation in $U \subset R^2$. Then, by the Riesz representation theorem, there exists a Radon measure μ and a μ-measurable function ν with $|\nu(x, y)| = 1$, for μ-almost every $(x, y) \in U$, so that, for any $g \in C_0^1(U, R)$.

$$\int_U u \operatorname{div} g \, dx \, dy = \int_U g \cdot \nu \, d\mu. \tag{4.2}$$

We will denote by $|\nabla u|$ the Radon measure μ and by $\nabla\mu$ the vector-valued measure ν. Note that if μ is continuously differentiable, then μ is the 2-dimensional Lebesgue measure weighted by the norm of the gradient of u and $\nu = \frac{\nabla u}{|\nabla u|}$.

Assume Ω is a bounded domain in R^2 whose boundary $\partial\Omega$ admits a representation of the form

$$\partial\Omega = \{(x, y) \in R^2 \ : \ F(x, y) = c\}, \tag{4.3}$$

for some Lipschitz function F and constant c. Then the unit normal vector along the boundary $\partial\Omega$, $\vec{n}$, can be written as

$$\vec{n} = \frac{\nabla F}{|\nabla F|}. \tag{4.4}$$

Extending $\vec{n}$ to R^2 smoothly, we have the arclength formula of $\partial\Omega$:

$$L(\partial\Omega) = \int_{\partial\Omega} ds = -\int_{R^2} \nabla x_\Omega \cdot \vec{n} \, dx \, dy,$$

where χ_Ω is the characteristic function of Ω. In general, one has for any integrable function f defined on $\partial\Omega$, after extending f to R^2,

$$\int_{\partial\Omega} f \, ds = -\int_{r^2} f \|\partial\Omega\|, \tag{4.5}$$

where we use $\|\partial\Omega\|$ to denote this numerical boundary measure $-\nabla\chi_\Omega \cdot \vec{n}$.

Interpolate the characteristic function χ_Ω at a certain level j by using the interpolation given in (2.26), (3.3), we have:

$$\chi_\Omega^j (x, y) = \sum_{p,q} \chi_{p,q}^j \, \phi_p^j (x) \phi_q^j (y), \tag{4.6}$$

where

$$\chi_{p,q}^j = \frac{1}{2^j} \chi_\Omega \left(\frac{p+c}{2^j}, \frac{q+c}{2^j}\right). \tag{4.7}$$

Using the connection coefficients

$$\Gamma_{i,j}^k =: \int_R \phi_k \phi_i \phi_j \, dx, \tag{4.8}$$

and (4.5), one obtains that

$$\|\partial\Omega\|^j = \sum_{p,q\in\wedge} (\|\partial\Omega\|)^j_{p,q}\phi^j_p(x)\phi^j_q(y)\,dx\,dy, \tag{4.9}$$

where for each $(p,q) \in \wedge^j_\Omega$,

$$(\|\partial\Omega\|)^j_{p,q} = \sum_{k,l,s,t} \frac{((\frac{\partial\chi}{\partial x})_{k,l}(\frac{\partial F}{\partial x})_{s,t} + (\frac{\partial\chi}{\partial y})_{k,l}(\frac{\partial F}{\partial y})_{s,t})\Gamma^p_{k,s}\Gamma^q_{l,t}}{2^j\sqrt{(\frac{\partial F}{\partial x})^2_{s,t} + (\frac{\partial F}{\partial y})^2_{s,t}}}, \tag{4.10}$$

where F is the defining function in (4.3).

5. Convergence of the penalized Dirichlet problem

Let Ω be a bounded domain in R^2 with Lipschitz boundary $\partial\Omega$. Given functions $f \in L^2(\Omega)$, $g \in H^{\frac{1}{2}}(\partial\Omega)$, a well-known Dirichlet problem is to find a function $u \in H^1(\Omega)$ so that

$$\Delta u + u = f, \quad \text{in}\,\Omega, \tag{5.1}$$

$$u = g, \quad \text{on}\,\partial\Omega. \tag{5.2}$$

Here $H^p(\Omega)$ denotes the Sobolev space of L^2 functions in Ω whose p-th distribution derivative is also in L^2 (see[11]).

There are a number of different ways of treating such a problem numerically. In [19], a penalty method combined with wavelet-Galerkin methods (see[5],[16]) was formulated and a convergence theorem was proved. In this section we use Theorem 3.1 to derive a better convergent error estimate.

Suppose that $f \in H^{N+1}(\Omega)$, (N was given by (3.4)), and g can be extended to be a function in $H^{N+1}(\Omega)$, *i.e.*, there exists a $\tilde{g} \in H^{N+1}(\Omega)$, so that $g = \gamma_0\tilde{g}$, where $\gamma_0 : H^1(\Omega) \to H^{\frac{1}{2}}(\partial\Omega)$ is the continuous trace operator. Let D be a square containing Ω. Extend f, g to H^{N+1} D by letting $f = 0, g = 0$ on $D - \bar{\Omega}$. Then for any positive number ϵ, consider the following penalized variational problem

$$\int_D \nabla u_\epsilon \cdot \nabla v\,dx\,dy + \int_D u_\epsilon v\,dx\,dy + \frac{1}{\epsilon}\oint_{\partial\Omega} u_\epsilon v\,ds = \int_D fv\,dx\,dy + \frac{1}{\epsilon}\oint_{\partial\Omega} gv\,ds, \tag{5.3}$$

where u_ϵ is the desired solution and (5.3) is satisfied for all $v \in H^1(D)$.

For each $\epsilon > 0$, there is a unique solution $u_\epsilon \in H^1(D)$ satisfying (5.3). This follows from the "direct method" of calculus of variations for its equivalent problem [9].

$$\inf_{v\in H^1(D)}\Big\{\tfrac{1}{2}\int_D |\nabla v|^2\,dx\,dy + \tfrac{1}{2}\int_D v^2\,dx\,dy - \int_D fv\,dx\,dy + \tfrac{1}{2\epsilon}\oint_{\partial\Omega} |v-g|^2\,ds\Big\}. \tag{5.4}$$

We have the following theorem on the relation between the solutions of above problems.

Theorem 5.1. *Let u_ϵ be the solution to* (5.3), *and $u \in H^2(\Omega)$ be the solution to* (5.1), (5.2). *Extend u to $H^2(D)$ by letting $u = 0$ in $D - \bar{\Omega}$. Then*

$$\|u_\epsilon - u\|_{H^1(D)} = O(\epsilon). \tag{5.5}$$

Proof. We define u^1 as the solution of the Dirichlet problem:

$$\Delta u^1 = 0 \quad \text{in} \quad \Omega, \qquad u^1 = -\frac{\partial u}{\partial n} \quad \text{on} \quad \partial\Omega. \tag{5.6}$$

Let $u^1 = 0$ on $D - \overline{\Omega}$, we then define $w_\epsilon \in H^1(\Omega)$ by

$$w_\epsilon = u + \epsilon u^1. \tag{5.7}$$

We clearly have

$$\|w_\epsilon - u\|_{H^1(D)} = \epsilon\|u^1\|_{H^1(D)}. \tag{5.8}$$

From (5.8) it follows that to prove (5.5) it suffices to prove that

$$\|w_\epsilon - u_\epsilon\|_{H^1(D)} = O(\epsilon). \tag{5.9}$$

Multiplying (5.1) by $v = w_\epsilon - u_\epsilon$ and using Green's theorem we obtain

$$\begin{aligned}\textstyle\int_\Omega \nabla u \cdot \nabla(w_\epsilon - u_\epsilon)\,dx\,dy + \int_\Omega u\cdot(w_\epsilon - u_\epsilon)\,dx\,dy \\ = \textstyle\int_\Omega f(w_\epsilon - u_\epsilon)\,dx\,dy + \oint_{\partial\Omega} \frac{\partial u}{\partial n}(w_\epsilon - u_\epsilon)\,ds,\end{aligned} \tag{5.10}$$

that is,

$$\begin{aligned}\textstyle\int_D \nabla u \cdot \nabla(w_\epsilon - u_\epsilon)\,dx\,dy + \int_D u\cdot(w_\epsilon - u_\epsilon)\,dx\,dy \\ = \textstyle\int_D f(w_\epsilon - u_\epsilon)\,dx\,dy + \oint_{\partial\Omega} \frac{\partial u}{\partial n} v\,ds.\end{aligned} \tag{5.11}$$

Taking $v = w_\epsilon - u_\epsilon$ in (5.3) we have

$$\begin{aligned}\textstyle\int_D \nabla u_\epsilon \cdot \nabla(w_\epsilon - u_\epsilon)\,dx\,dy + \int_D u_\epsilon\cdot(w_\epsilon - u_\epsilon)\,dx\,dy \\ + \textstyle\frac{1}{\epsilon}\oint_{\partial\Omega}(u_\epsilon - g)(w_\epsilon - u_\epsilon)\,ds = \int_D f(w_\epsilon - u_\epsilon)\,dx\,dy,\end{aligned} \tag{5.12}$$

which we rewrite as

$$\begin{aligned}\textstyle\int_D \nabla(w_\epsilon - u_\epsilon)\cdot\nabla(w_\epsilon - u_\epsilon)\,dx\,dy + \int_D (w_\epsilon - u_\epsilon)^2\,dx\,dy + \frac{1}{\epsilon}\oint_{\partial\Omega}(w_\epsilon - u_\epsilon)^2\,ds \\ = \textstyle\int_D \nabla w_\epsilon\cdot\nabla(w_\epsilon - u_\epsilon)\,dx\,dy + \int_D w_\epsilon(w_\epsilon - u_\epsilon)\,dx\,dy \\ + \textstyle\frac{1}{\epsilon}\oint_{\partial\Omega}(w_\epsilon - g)(w_\epsilon - u_\epsilon)\,ds - \int_D f(w_\epsilon - u_\epsilon)\,dx\,dy.\end{aligned} \tag{5.13}$$

Combining (5.6), (5.7) and (5.8) we obtain

$$\begin{aligned}
&\int_D \nabla(w_\epsilon - u_\epsilon)\cdot\nabla(w_\epsilon - u_\epsilon)\,dx\,dy + \int_D (w_\epsilon - u_\epsilon)^2\,dx\,dy + \tfrac{1}{\epsilon}\oint_{\partial\Omega}(w_\epsilon - u_\epsilon)^2\,ds \\
&= \epsilon\int_D[\nabla u^1\cdot\nabla(w_\epsilon - u_\epsilon) + u^1\cdot(w_\epsilon - u_\epsilon)]\,dx\,dy + \int_D \nabla u\cdot\nabla(w_\epsilon - u_\epsilon)\,dx\,dy \\
&\quad + \int_D u\cdot(w_\epsilon - u_\epsilon)\,dx\,dy - \int_D f(w_\epsilon - u_\epsilon)\,dx\,dy - \oint_{\partial\Omega}\tfrac{\partial u}{\partial n}(w_\epsilon - u_\epsilon)\,ds.
\end{aligned} \tag{5.14}$$

From (5.11), the second term in the right-hand side of (5.14) vanishes; we then have

$$\begin{aligned}
\|w_\epsilon - u_\epsilon\|^2_{H^1(D)} &+ \tfrac{1}{\epsilon}\oint_{\partial\Omega}(w_\epsilon - u_\epsilon)^2\,ds \\
&= \epsilon\int_D[\nabla u^1\cdot\nabla(w_\epsilon - u_\epsilon) + u^1\cdot(w_\epsilon - u_\epsilon)]\,dx\,dy,
\end{aligned} \tag{5.15}$$

which implies (by Schwarz's inequality):

$$\|w_\epsilon - u_\epsilon\|_{H^1(D)} \le \epsilon\|u^1\|_{H^1(D)}, \tag{5.16}$$

and

$$\|w_\epsilon - u_\epsilon\|_{L^2(\partial\Omega)} \le \epsilon\|u^1\|_{H^1(D)}. \tag{5.17}$$

Hence (5.5) follows from (5.8) and (5.16)

$$\|u - u_\epsilon\|_{H^1(D)} \le 2\epsilon\|u^1\|_{H^1(D)}. \tag{5.18}$$

□

Remark 5.1. Since $\partial\Omega$ is Lipschitz, we have

$$\|w_\epsilon - u\|_{L^2(\partial\Omega)} \le C\|w_\epsilon - u\|_{H^1(\Omega)} \le C\,\epsilon\,\|u^1\|_{H^1(D)}. \tag{5.19}$$

Hence,

$$\|w_\epsilon - u\|_{L^2(\partial\Omega)} \le C\,\epsilon\,\|u^1\|_{H^1(D)}. \tag{5.20}$$

6. Convergence of the wavelet-Galerkin approximation

Next we want to derive an error estimate for the wavelet-Galerkin approximation to the solution of the penalized problem (5.3). To this we will consider a wavelet basis consisting of the special Daubechies scaling function ϕ with the condition (3.4) and consider the space $V_j(R^2)$ define by the multiresolution given in Section 2. Let f^j, g^j be the wavelet interpolations for f and g as in the Theorem 3.1.

$$f^j(x,y) := \sum_{p,q\in\wedge_D^j} f^j_{p,q}\phi^j_p(x)\phi^j_q(y), \tag{6.1}$$

$$g^j(x,y) := \sum_{p,q\in\wedge_D^j} g^j_{p,q}\phi^j_p(x)\phi^j_q(y), \tag{6.2}$$

where $\wedge_D^j$ is the index set defined as

$$\wedge_D^j = \{(p,q) \in Z \otimes Z \ : \ \text{supp}(\phi_p^J(X)\phi_q^J(Y)) \cap D \neq \phi\}.$$

Replace f, g in (5.3) by f^j, g^j and let

$$u_\epsilon^j(x,y) := \sum_{p,q \in \wedge_D^j} u_{p,q}^j \phi_p^j(x)\phi_q^j(y) \tag{6.3}$$

be the wavelet-Galerkin solution to

$$\begin{aligned} \int_D \nabla u_\epsilon^j \cdot \nabla v \, dx\, dy + \int_D u_\epsilon^j v \, dx\, dy + \tfrac{1}{\epsilon} \oint_{\partial\Omega} u_\epsilon^j v \, ds \\ = \int_D f^j v \, dx\, dy + \tfrac{1}{\epsilon} \oint_{\partial\Omega} g^j v \, ds, \quad \forall v \in H^1(D), \end{aligned} \tag{6.4}$$

where $u_{p,q}^j$ is to be determined by a system of linear equations and the boundary integrals are calculated by (4.5), (4.9).

Theorem 6.1. *If $u_\epsilon \in H^{N+1}(D)$ is the solution to* (5.3), *and u_ϵ^j is the wavelet-Galerkin solution to* (6.4), *then*

$$\|u_\epsilon - u_\epsilon^j\|_{H^1(D)} \le C(\tfrac{1}{2^j})^N, \tag{6.5}$$

where C depends on the diameter of D, on N and on $\|f\|_{H^{N+1}(D)}, \|g\|_{H^{N+1}(D)}$.

Proof. From, (5.3) and (6.4) we obtain

$$\begin{aligned} \int_D \nabla(u - u_\epsilon^j) \cdot \nabla v \, dx\, dy + \int_D (u - u_\epsilon^j) v \, dx\, dy + \tfrac{1}{\epsilon} \oint_{\partial\Omega} (u - u_\epsilon^j) \, ds \\ = \int_D (f - f^j) v \, dx\, dy + \tfrac{1}{\epsilon} \oint_{\partial\Omega} (g - g^j) v \, ds, \quad \forall v \in H^1(D). \end{aligned} \tag{6.6}$$

Let $\tilde{u}_\epsilon^j$ be the wavelet interpolations for u_ϵ as in the Theorem:

$$\tilde{u}_\epsilon^j(x,y) := \sum_{p,q \in \wedge_D^j} \tilde{u}_{p,q}^j \phi_p^j(x)\phi_q^j(y), \tag{6.7}$$

where

$$\tilde{u}_{p,q}^j = u_\epsilon(\tfrac{p+c}{2^j}, \tfrac{q+c}{2^j}).$$

From Theorem and (6.1), (6.2) and (6.7) we have

$$\|f - f^j\|_{H^1(D)} \le C(\tfrac{1}{2^j})^N, \tag{6.8}$$

$$\|g - g^j\|_{H^1(D)} \le C(\tfrac{1}{2^j})^N, \tag{6.9}$$

$$\|u_\epsilon - \tilde{u}_\epsilon^j\|_{H^1(D)} \le C(\tfrac{1}{2^j})^N. \tag{6.10}$$

Taking $v = u_\epsilon - \tilde{u}_\epsilon^j$ in (6.6) we obtain by Schwarz's inequality

$$\begin{aligned}
\|u_\epsilon &- u_\epsilon^j\|_{H^1(D)}^2 \\
&= \int_D \nabla(u_\epsilon - u_\epsilon^j)\cdot\nabla(u_\epsilon - u_\epsilon^j - v)\,dx\,dy \\
&\qquad + \int_D (u_\epsilon - u_\epsilon^j)\cdot(u_\epsilon - u_\epsilon^j - v)\,dx\,dy \\
&= \int_D \nabla(u_\epsilon - u_\epsilon^j)\cdot\nabla(u_\epsilon - \tilde{u}_\epsilon^j)\,dx\,dy + \int_D (u_\epsilon - u_\epsilon^j)\cdot(u_\epsilon - \tilde{u}_\epsilon^j)\,dx\,dy \\
&\qquad - \{\int_D \nabla(u_\epsilon - u_\epsilon^j)\cdot\nabla v\,dx\,dy + \int_D (u_\epsilon - u_\epsilon^j)\cdot v\,dx\,dy\} \\
&= \int_D \nabla(u_\epsilon - u_\epsilon^j)\cdot\nabla(u_\epsilon - \tilde{u}_\epsilon^j)\,dx\,dy + \int_D (u_\epsilon - u_\epsilon^j)\cdot(u_\epsilon - \tilde{u}_\epsilon^j)\,dx\,dy \\
&\qquad - \int_D (f - f^j)v\,dx\,dy + \tfrac{1}{\epsilon}\oint_{\partial\Omega}((u_\epsilon - u_\epsilon^j) - (g - g^j))v\,ds \\
&\le \|u_\epsilon - u_\epsilon^j\|_{H^1(D)}\|u_\epsilon - \tilde{u}_\epsilon^j\|_{H^1(D)} + \|f - f^j\|_{L^2(\partial\Omega)}\|v\|_{L^2(D)} \\
&\qquad + \tfrac{1}{\epsilon}\|u_\epsilon - u_\epsilon^j\| - \|g - g^j\|_{L^2(\partial\Omega)}\|v\|_{L^2(D)}.
\end{aligned} \tag{6.11}$$

Let w be the solution to the Dirichlet problem

$$-\Delta w + w = f - f^j \quad \text{in } \Omega, \tag{6.12}$$

$$w = g - g^j \quad \text{on } \partial\Omega, \tag{6.13}$$

and let w^1 be the solution of

$$\Delta w^1 = 0 \quad \text{in } \Omega, \quad w^1 = -\frac{\partial w}{\partial n} \quad \text{on } \partial\Omega, \tag{6.14}$$

then we have from Remark 5.1

$$\|(u_\epsilon - u_\epsilon^j) - (g - g^j)\|_{L^2(\partial\Omega)} \le C\epsilon\|w^1\|_{H^1(\Omega)} \le C\epsilon(\tfrac{1}{2^j})^N, \tag{6.15}$$

(see Remark 6.1). Also we note that $v = u_\epsilon^j - \tilde{u}_\epsilon^j = u_\epsilon^j - u_\epsilon + u_\epsilon - \tilde{u}_\epsilon^j$, so that

$$\|v\|_{L^2(\partial\Omega)} \le C\|v\|_{H^1(\Omega)} \le \|u_\epsilon - u_\epsilon^j\|_{H^1(D)} + \|u_\epsilon - \tilde{u}_\epsilon^j\|_{H^1(D)}. \tag{6.16}$$

Combining (6.11), (6.3–6.5) and (6.15, 6.16) we finally have

$$\|u_\epsilon - u_\epsilon^j\|_{H^1(D)}^2 \le C(\tfrac{1}{2^j})^N\|u_\epsilon - u_\epsilon^j\|_{H^1(D)} + C(\tfrac{1}{2^j})^{2N}. \tag{6.17}$$

Then (6.5) follows from solving the above inequality. □

Remark 6.1. Without loss of generality, we assume that

$$\|w^1\|_{H^1(\Omega)} \le C(\|f - f^j\|_{H^1(\Omega)} + \|g - g^j\|_{H^1(\Omega)}), \tag{6.18}$$

i.e., the solution of the Dirichlet problem (6.12–6.14) depends continuously on the data $f - f^j$ and $g - g^j$ as $\|f - f^j\|_{H^1(\Omega)} \to 0$, $\|g - g^j\|_{H^1(\Omega)} \to 0$.

Remark 6.2. We have proved that the wavelet-Galerkin approximation u_ϵ^j differs from the solution u of the original differential equation by an estimate of the following type

$$\|u - u_\epsilon^j\|_{H^1(D)} \leq O(\epsilon) + O(2^{-Nj}). \tag{6.19}$$

This is what the numerical evidence from a number of sources have indicated [19].

References

[1] R. A. Adams, Sobolev spaces. Academic Press, New York 1975.

[2] E. Bacry, S. Mallat and G. Papanicolaou, A wavelet based space-time adaptive numerical method for partial differential equations, Math. Modelling Numerical Analysis 26 (1992), 793–834.

[3] G. Beylkin, R. Coifman, and V. Rokhlin, Fast wavelet transforms and numerical algorithms. I, Comm. Pure Appl. Math. 44 (1991), 141–183.

[4] G. Beylkin, On wavelet-based algorithms for solving differential equations, in: Wavelets: Mathematics and Applications (J. J. Benedetto and M. Frazier, ed.), CRC Press, Boca Raton, FL (1993).

[5] Phillipe G. Ciarlet, The finite element methods for elliptic problems, North-Holland, 1987.

[6] I. Daubechies, Orthonormal basis of compactly supported wavelets, Comm. Pure Appl. Math. 41 (1988), 909–996.

[7] —, Ten lectures on wavelets, CBMS-NSF Regional Conf. Ser. in Appl. Math., SIAM, Philadelphia, PA, 1992.

[8] —, Orthonormal bases of compactly supported wavelets II. Variations on a theme, SIAM J. Math. Anal. 24 (2) (1993), 499–519.

[9] Herbert Federer, Geometric measure theory, Springer-Verlag, New York 1969.

[10] Enrico Giusti, Minimal surfaces and functions of bounded variation, Birkhäuser, Boston 1984.

[11] Roland Glowinski, Numerical methods for nonlinear variational problems, Springer Ser. Comput. Phys. Springer-Verlag, New York 1984.

[12] R. Glowinski, W. Lawton, M. Ravachol, and E. Tenenbaum, in: Proceedings, 96th International Conference on Numerical Methods in Applied Sciences and Engineering, SIAM, Philadelphia, PA, 1990, 55–119.

[13] S. Mallat, Multiresolution approximation and wavelet orthonormal bases of $L^2(R)$, Trans. Amer. Math. Soc. 315 (1989), 69–87.

[14] S. Qian and J. Weiss, Wavelets and the numerical solution of partial differential equations, J. Comput. Phys. 106 (1993), 155–175.

[15] H. L. Resnikoff and R. O. Wells, Jr., Wavelet analysis and the geometry of euclidean domains, J. Geom. Phys. 8 (1–4) (1992), 273–282.

[16] Gilbert Strand and George J. Fix, An analysis of the finite element method, Prentice-Hall, Englewood Cliffs, NJ, 1973.

[17] G. Strang, Wavelets and dilation equations: a brief introduction, SIAM Review 31 (4) (1989), 614–627.

[18] R. O. Wells, Jr. and X. Zhou, Wavelet interpolation and approximate solutions of elliptic partial differential equations, NATO Adv. Sci. Inst. Ser. C Math. Phys. Sci. 429, Kluwer Acad. Publ. 1994, 349–366.

[19] R. O. Wells, Jr. and X. Zhou, Wavelet solutions for the Dirichlet problem, preprint, Rice University.

[20] Xu, J. C. and W. C. Shann, Galerkin-wavelet methods for two-point boundary value problems, Numer. Math. 63 (1) (1992), 123–142.

Moser's inequality and n-Laplacian

Kai-Ching Lin

Abstract. Let Ω be a bounded smooth domain in $\mathbb{R}^n$, and $u(x)$ a C^1 function with compact support in Ω. Moser's inequality states that there is a constant c_0, depending only on the dimension n, such that

$$\frac{1}{|\Omega|}\int_\Omega e^{n\omega_{n-1}^{1/(n-1)}u^{n/(n-1)}}\,dx \leq c_0,$$

where $|\Omega|$ is the Lebesgue measure of Ω, and ω_{n-1} the surface area of the unit ball in $\mathbb{R}^n$. It has been proved that there are extremal functions for this inequality. In other words, the

$$\sup\left\{\frac{1}{|\Omega|}\int_\Omega e^{n\omega_{n-1}^{1/(n-1)}u^{n/(n-1)}}\,dx \mid u \in W_0^{1,n},\ \|\nabla u\|_n \leq 1\right\}$$

is attained. This is a survey on the existence of extremal functions, including results from Carleson–Chang (1986), Flucher (1992), and Lin (1995).

1991 Mathematics Subject Classification: 14P10, 14P15

1. Introduction

Let Ω be a bounded smooth domain in $\mathbb{R}^n$, and $u(x)$ a C^1 function supported in Ω with $\|\nabla u\|_q < \infty$. Sobolev's Imbedding Theorem says that if $1 \leq q < \infty$, then

$$\|u\|_p \leq C(n,q), \tag{1.1}$$

where $C(n,q)$ is a constant independent of the function u, as well as the domain Ω. The imbedding is no longer valid when $q = n$. Indeed, there are unbounded functions whose gradients are in L^n. However, in 1967 Trudinger [11] proved that if $\|\nabla u\|_n \leq 1$, then u is in an exponential class. More precisely, the integral

$$\int_\Omega e^{\beta_0 u^{n/(n-1)}}\,dx,$$

is uniformly bounded for some positive β_0 depending only on the dimension. Then in 1971, Moser [10] found the best exponent β_0. He showed that if $\|\nabla u\|_n \leq 1$, then

$$\frac{1}{|\Omega|}\int_\Omega e^{n\omega_{n-1}^{1/(n-1)}u^{n/(n-1)}}\,dx \leq c_0, \tag{1.2}$$

where c_0 is a constant depending only on n. (ω_{n-1} is the surface area of the unit ball in $\mathbb{R}^n$.)

This paper surveys the results on the existence of extremal functions of Moser's inequality (1.2). In other words, what we are concerned with is whether or not the

$$\sup\left\{\frac{1}{|\Omega|}\int_\Omega e^{n\omega_{n-1}^{1/(n-1)}u^{n/(n-1)}}\,dx \mid u \in W_0^{1,n},\ \|\nabla u\|_n \leq 1\right\}$$

is attained.

The first result in this direction is due to Carleson–Chang [1], who proved in 1986 that there are extremals when Ω is a ball in any dimension. Their result came as a surprise, since it was known at that time that no extremals exist for Sobolev's inequality (1.1) when Ω is a ball. (See below for an account of this.) Two years ago, M. Flucher [3] proved the same existence for any bounded smooth domain in 2-dimensions. Then Lin [5] did it for any smooth bounded domain in any dimension.

2. Concentration-compactness

By $W_0^{1,n}(\Omega)$ we mean the Sobolev space of functions vanishing on the boundary $\partial\Omega$ with $\|\nabla u\|_n < \infty$, and we denote by $F_\Omega(u)$ the Moser functional

$$\int_\Omega \left(e^{n\omega_{n-1}^{1/(n-1)}u^{n/(n-1)}} - 1\right) dx = \int_\Omega f(u)\,dx,$$

(where the '-1' term in the integrand is introduced for convenience). We now describe the difficulties that will occur and the ideas used in [1], [3], and [5] to resolve them. Let $\{u_j\}$ be a maximizing sequence for (1.2), that is, $\{u_j\} \subset W_0^{1,n}(\Omega)$, $\|\nabla u_j\|_n \leq 1$, and $F_\Omega(u_j)$ tends to the supremum. From functional analysis we obtain immediately that we can extract a subsequence, still denoted by $\{u_j\}$, which satisfies

$$\|\nabla u_j\|_n \leq 1, \quad u_j \rightharpoonup u \text{ weakly, and } |\nabla u(x)|^n\,dx \rightharpoonup d\mu \text{ weakly,} \tag{2.3}$$

where u is a fuction in $W_0^{1,n}(\Omega)$, and $d\mu$ is a finite measure on Ω. Our goal is to prove that $F_\Omega(u_j) \to F_\Omega(u)$. ($u$ will then be an extremal.)

The main difficulty in this type of problems is that the Moser functional $F_\Omega(u)$ is not compact. In other words, there exists a sequence of functions $\{u_j\}$ which satisfies all the conditions in (2.3), but $F_\Omega(u_j)$ fails to converge to $F_\Omega(u)$. Here is an example.

Take Ω to be the unit ball in $\mathbb{R}^n$, and define u_a to be $c_a \log \frac{1}{|x|}$ for $a \le |x| \le 1$, and choose a constant d_a for $0 \le |x| \le a$, so that the functions are continuous, and where c_a is chosen so that $\|\nabla u_a\|_n = 1$. It is easy to see that as $a \to 0, u_a \rightharpoonup u = 0$ weakly, $|\nabla u_a(x)|^n dx \rightharpoonup \delta_0$ weakly where δ_0 is the Dirac measure at 0, and that $\limsup F_\Omega(u_a) > F_\Omega(0)$.

All is not lost, however. If we are willing to put more restrictions on $\{u_j\}$, then we have compactness:

Lemma 1. *Suppose* $\{u_j\}$ *satisfies*

$$\|\nabla u_j\|_n \le \delta < 1,\ u_j \rightharpoonup u \text{ weakly, and } |\nabla u(x)|^n\, dx \rightharpoonup d\mu \text{ weakly,} \tag{2.4}$$

then $F_\Omega(u_j) \to F_\Omega(u)$.

Proof. By Moser's inequality (1.2) the condition $\|\nabla u_j\|_n \le \delta \le 1$ implies that

$$\frac{1}{|\Omega|}\int_\Omega e^{n\omega_{n-1}^{1/(n-1)}\delta^{-n/(n-1)}u_j^{n/(n-1)}}\, dx \le c_0.$$

In other words, u_j is actually in a better exponential class than condition (2.3) would have promised. This says that the distribution function of u_j would decrease in a faster exponential manner. More precisely,

$$|\{u_j > t\}| \le c_0|\Omega| e^{n\omega_{n-1}^{1/(n-1)}\delta^{-n/(n-1)}t^{n/(n-1)}}.$$

Therefore, we have

$$\begin{aligned} F_\Omega(u_j) + |\Omega| &= c_0 n\omega_{n-1}^{\frac{1}{n-1}} \int_0^\infty n(n-1)t^{\frac{1}{n-1}} e^{n\omega_{n-1}^{1/(n-1)}t^{n/(n-1)}} |\{u_j > t\}|\, dt \\ &\le c_0 n^2(n-1)\omega_{n-1}^{\frac{1}{n-1}} \int_0^\infty t^{\frac{1}{n-1}} e^{n\omega_{n-1}^{1/(n-1)}t^{n/(n-1)}[1-\delta^{-n/(n-1)}]}\, dt, \end{aligned}$$

which converges, since $\delta < 1$. By the Dominated Convergence Theorem we have that $F_\Omega(u_j) \to F_\Omega(u)$. □

Due to the homogeneity of the L^p norm, condition (2.4) will *not* put u_j in a better L^p space. So the above argument will not work. This partially explains why the Sobolev inequality (1.1) does not have extremals. Indeed, one can argue that, by the invariance of (1.1) under dilations $u \mapsto a^{-n/q}u(\cdot/a)$, the best constant for (1.1) is the same, regardless of domains. The reason that no extremals exist for (1.1) on any bounded domain for $q > 1$ is that any extremal on $\mathbb{R}^n$ must be of the form $(u(x) = (\lambda + |x - x_0|^{\frac{q}{q-1}})^{1-(n/q)}$, which is not compactly supported.

With the help of this lemma, one can give, as Flucher [3] did, an easier proof of the following result of P. L. Lions, which says that the only case that can go wrong for sequences $\{u_j\}$ satisfying the more relaxed condition (2.3) is the case of the example above—it concentrates at some point.

Theorem 1 (P. L. Lions). *Suppose* $\{u_j\}$ *is a sequence satisfying* (2.3). *Then, either*

(a) compactness *holds, i.e.,* $F_\Omega(u_j) \to F_\Omega(u)$; *or*

(b) $\{u_j\}$ concentrates *at some point* x_0, *i.e.,* $u_j \rightharpoonup u = 0$ *weakly, and* $|\nabla u(x)|^n dx \rightharpoonup \delta_{x_0}$.

This is the so-called the *concentration-compactness principle* for the Moser functional. See [9] for other functionals as well. We will present here Flucher's proof of the special case when u is bounded.

Proof of Theorem 1. ($n = 2$, u is bounded.) Suppose μ is not a Dirac measure of mass 1. Then there is an $R > 0$ such that $\mu(B(x, R)) < 1$ for every x in Ω. For this R, choose r with $0 < r < R$ so that the capacity of $B(0, r)$ with respect to $B(0, R)$ is less then ϵ, *i.e.*, there is a function ϕ which is harmonic on $B(0, R) \setminus B(0, r)$ with $\phi = 0$ on $\mathbb{R}^2 \setminus B(0, R)$, $\phi = 1$ on $B(0, r)$, and $\int_{\mathbb{R}^2} |\nabla\phi|^2\, dx < \epsilon$. Set $\phi_{x_0}(x) = \phi(x - x_0)$. We have

$$\begin{aligned}\limsup_{j\to\infty} \int_\Omega |\nabla(\phi_{x_0} u_j)|^2\, dx &\le (1+\epsilon)\limsup_{j\to\infty}\int_\Omega |\phi_{x_0}|^2 |\nabla u_j|^2\, dx \\ &\quad + C(\epsilon)\limsup_{j\to\infty}\int_\Omega |\nabla\phi_{x_0}|^2 |u_j|^2\, dx \\ &\le (1+\epsilon)\mu(B(x_0, R)) + C(\epsilon)\int_\Omega |\nabla\phi_{x_0}|^2 |u|^2\, dx.\end{aligned}$$

By choosing r sufficiently small, we can make the above expression less than 1. We then apply Lemma 1 to $\{\phi_{x_0} u_j\}$, so that $f(u_j) \to f(u)$ in L^1 on $B(x_0, r)$. Covering $\overline{\Omega}$ by finitely many balls gives $F_\Omega(u_j) \to F_\Omega(u)$. □

So far, we have not used the condition that u_j is *maximizing*. In the following, we will show that maximizing sequences *never* concentrate. To do this, we quantify the concentration phenomenon:

Definition 1. Let x_0 be a point in $\overline{\Omega}$. The *concentration function* at x_0 is defined to be

$$C_\Omega(x_0) = \sup\{\limsup F_\Omega(u_j) : \|\nabla u\|_n \le 1,\ \{u_j\} \text{ concentrates at } x_0\}.$$

Obviously, we have $\sup_u F_\Omega(u) \ge \sup_x C_\Omega(x)$. In view of Lions's concentration-compactness principle (Theorem 1), it now suffices to prove

$$\sup_u F_\Omega(u) > \sup_x C_\Omega(x). \tag{2.5}$$

In the subsequent sections, we will explain the ideas of proving this inequality on various domains.

3. Ω is a ball in $\mathbb{R}^n$

The existence in this case is due to Carleson–Chang[1]. They showed

Theorem 2 (Carleson–Chang). *Let B be the unit ball in $\mathbb{R}^n$. Then*

(a) $\sup_x C_B(x) = C_B(0) = e^{1+\frac{1}{2}+\frac{1}{3}+\cdots+\frac{1}{n-1}}|B|$,

(b) $\sup_u F_B(u) > e^{1+\frac{1}{2}+\frac{1}{3}+\cdots+\frac{1}{n-1}}|B|$.

The first equality in (a) says that the concentration achieves its maximum at the origin. This is a consequence of the following result on *classical reaarangement*. Recall that if u is a function on Ω, then the classical reaarangement u^* is a radial, decreasing function defined on the symmetrized domain Ω^* (where Ω^* is a ball centered at the origin, $|\Omega^*| = |\Omega|$) such that $|\{u^* > t\}| = |\{u > t\}|$. Recall also (cf. [5]):

Theorem 3. *For $u \in W_0^{1,n}(\Omega)$, we have*

(a) $\|\nabla u^*\|_{L^n(\Omega^*)} \leq \|\nabla u\|_{L^n(\Omega)}$,

(b) $F_{\Omega^*}(u^*) = F_\Omega(u)$.

This theorem also enables us to treat the extremal problem on a ball in a one-variable setting since we may assume all the functions we deal with are radial and decreasing. The difficult part of this theorem lies in finding the *exact* value of $C_B(0)$, the second part of (a). The key ingredient is the following result on the 'linearized Moser inequality':

Lemma 2. *Let K_δ denote the class of functions $w(t)$ defined on $(0,\infty)$ such that $w(0) = 0$, $\int_0^\infty \dot{w}^n\,dt \leq \delta < 1$. Then for every $c > 0$, we have*

$$\sup_{w\in K_\delta}\int_0^\infty e^{cw(t)-t}\,dt < e^{((n-1)/n)^{n-1}c^n/n}e^{1+\frac{1}{2}+\frac{1}{3}+\cdots+\frac{1}{n-1}}.$$

Moreover, when $c^n\delta \to \infty$, the inequality tends to an equality.

As in Lemma 1, the condition on K_δ ensures the existence of extremals for the linearized functional $\int_0^\infty e^{cw(t)-t}\,dt$. They show that the Euler equation satisfied by the extremal can be solved by an ODE method. (Such a method is not available for the original Moser inequality.) After proving (a), they then proceed to exhibit a function whose functional value is bigger than $C_B(0)$.

When Ω is a domain other than a ball, we shall compare the concentration function and the functional on these two domains.

4. $n = 2$, Ω is simply-connected

If Ω is a simply-connected domain in the complex plane, and x_0 a point in Ω, then there is a Riemann mapping $h_{x_0}(z)$ from the unit disc D onto Ω so that $h_{x_0}(0) = x_0$. It is easy to see that extremals are *not* invariant under conformal mappings: take Ω to be the unit disc D, and $h(z)$ a Möbius transform. If $v(z)$ is an extremal, then $v \circ h_{x_0}^{-1}$ cannot be an extremal, since an extremal on D has to be radial (Theorem 3). However, concentrating sequences are invariant. To prove this, let $\{v_j\}$ be a concentrating (radial, decreasing) sequence at 0 on D which realizes $C_B(0)$. The functions $v_{j,x_0} = v_j \circ h_{x_0}^{-1}$ satisfy $\|\nabla v_{x_0}\|_{L^2(\Omega)} = \|\nabla v_0\|_{L^2(D)}$. In fact, more is true:

$$\int_{\{v_{j,x_0}<t\}} |\nabla v_{j,x_0}|^2\,dx = \int_{\{v_j<t\}} |\nabla v_j|^2\,dz,$$

which tends to 0, as $j \to \infty$ for every t. So, $\{v_{j,x_0}\}$ concentrates at x_0. Furthermore, we have

$$\begin{aligned} F_\Omega(v_{j,x_0}) &= \int_\Omega f(v_{j,x_0})\,dx = \int_\Omega f(v_j((h_{x_0}^{-1}(x)))\,dx \\ &= \int_D f(v_j(z))|h'_{x_0}(z)|^2\,dz = \int_0^1 f(v_j(r)) \int_0^{2\pi} |h'_{x_0}(re^{i\theta})|^2\,d\theta\,r\,dr \\ &\geq 2\pi|h'_{x_0}(0)|^2 \int_0^1 f(v_j(r))\,r\,dr = |h'_{x_0}(0)|^2 F_D(v_j). \end{aligned}$$

(The inequality used is the mean-value inequality for analytic functions.) Taking $j \to \infty$, we obtain $C_\Omega(x_0) \geq |h'_{x_0}(0)|^2 C_D(0)$. A similar argument would give us the reverse inequality. So, we have

$$C_\Omega(x_0) = |h'_{x_0}(0)|^2 C_D(0). \tag{4.6}$$

Consequently,

$$\sup_x C_\Omega(x) = (\sup_x |h'_x(0)|^2) C_D(0). \tag{4.7}$$

Notice that the above inequality also yields, by taking the supremum over v_j in $W_0^{1,2}$,

$$\sup_u F_\Omega(u) \geq |h'_x(0)|^2 \sup_u F_D(u), \tag{4.8}$$

for every x. Therefore,

$$\sup_u F_\Omega(u) \geq (\sup_x |h'_x(0)|^2) \sup_u F_D(u). \tag{4.9}$$

Combining (4.7) and (4.9), we obtain

$$\frac{\sup_u F_\Omega(u)}{\sup_x C_\Omega(x)} \geq \frac{\sup_u F_D(u)}{\sup_x C_D(x)}. \tag{4.10}$$

By Theorem 2 of Carleson–Chang, we have proved the case when Ω is a simply-connected domain in the plane.

5. $n = 2$, Ω is arbitrary

When Ω is an arbitrary domain in $\mathbb{R}^2$, Riemann mappings are no longer available. To overcome this difficulty, we will use *Green's functions* instead. We now re-do the previous section (where Ω is simply-connected) in order to explain the ideas. In particular, we will reformulate the conformal factor $|h'_{x_0}(0)|^2$, and re-define the functions v_{Ω,x_0} using Green's functions.

Let x_0 be a point in the domain Ω, and let $G = G_{x_0} = G_{\Omega,x_0}$ be Green's function on Ω with the pole at x_0. It is well-known that this Green function G has the following asymptotic expansion:

$$G = -\frac{1}{2\pi}\log|x - x_0| - H_{x_0}(x), \tag{5.11}$$

where $H_{x_0}(x)$ is a harmonic function on Ω. By the invariance of Green's functions,

$$G_{\Omega,x_0}(x) = G_{D,0}(h_{x_0}^{-1}(x)), \tag{5.12}$$

we have, setting $z = h_{x_0}^{-1}(x)$,

$$\begin{aligned} -\frac{1}{2\pi}\log|h_{x_0}(z) - h_{x_0}(0)| - H_{x_0}(h_{x_0}(z)) &= -\frac{1}{2\pi}\log|z|, \\ \frac{1}{2\pi}\log\left|\frac{h_{x_0}(z) - h_{x_0}(0)}{z}\right| &= -H_{x_0}(h_{x_0}(z)). \end{aligned}$$

Letting $z \to 0$, we obtain $|h'_{x_0}(0)| = e^{-2\pi H_{x_0}(x_0)}$. This leads to the following definition:

Definition 2. The *conformal radius* at x_0 is defined to be

$$r_\Omega(x_0) = e^{-2\pi H_{x_0}(x_0)}.$$

Thus equation (4.6) can now be re-written as $C_\Omega(x_0) = r_\Omega^2(x_0)C_D(0)$ when Ω is simply-connected. We can also define the function v_{x_0} in the previous section by using Green's function only. Again, the invariance of Green's functions implies that the conformal mapping $h_{x_0}(z)$ takes the level curves of G_0 to those of G_{x_0}. Therefore, when v_0 is a radial, decreasing function on D, the function v_{x_0} defined earlier by $v_0 \circ h_{x_0}^{-1}$ has the same level sets as G_{x_0}, and v_{x_0} and v_0 agree on the corresponding level sets of the two Green's functions. In fact, we can make the following definition:

Definition 3. Let x_0 be a point in Ω, and let v_0 be a decreasing, radial function on D. The *conformal reaarrangement* of v_0 on Ω at x_0 is defined to be $v_{x_0} = v_{\Omega,x_0} = v_0 \circ G_{D,0}^{-1} \circ G_{\Omega,x_0}$.

The following theorem is what we need to know about the conformal reaarangement on a general domain:

Theorem 4. *For a radial, decreasing function v_0 in $W_0^{1,n}(D)$, we have*

(a) $\|\nabla v_{x_0}\|_{L^2(\Omega)} = \|\nabla v_0\|_{L^2(D)}$;

(b) $F_\Omega(v_{x_0}) \geq r_\Omega^2(x_0)F_D(v_0)$;

(c) *if $\{v_j\}$ concentrates at 0, then $\{v_{j,x_0}\}$ concentrates at x_0; and* $\lim_{j\to\infty} F_\Omega(v_{j,x_0}) = \lim_{j\to\infty} F_D(v_j)$.

The main tool to prove Theorem 4 is the co-area formula [2]. Also, somewhere in the proof we will need a replacement of the mean-value inequality for analytic functions which we used in Section 4. See Flucher [3] for details. From (a) and (b) of Theorem 4 we obviously have that

$$\sup_u F_\Omega(u) \geq (\sup_x r_\Omega^2(x)) \sup_u F_D(u), \tag{5.13}$$

which is a replacement of (4.9). It is also not hard to show from (c) that

$$C_\Omega(x) = r_\Omega^2(x)C_D(0), \tag{5.14}$$

which replaces (4.6). We thus arrive at the situation of equation (4.10) as before. This proves the case when Ω is a planar domain.

6. Ω is an arbitrary domain in $\mathbb{R}^n$

The key properties of Green's functions that we used in Section 5 are:

(a) *Green's functions are invariant under conformal mappings*, and

(b) *Green's functions have logarithmic poles.*

None of these properties is satisfied by the usual Green's functions in higher dimensions. We shall, instead, work with Green's functions with respect to the *n-Laplacian*.

Definition 4. Let x_0 be a point in Ω. The *n-Green's function* $G = G_{x_0} = G_{\Omega,x_0}$ on Ω with a pole at x_0 is the singular solution to the *n-Laplacian*:

$$\Delta_n G = \mathrm{Div}(|\nabla G|^{n-2}\nabla G) = \delta_{x_0} \quad \text{in } \Omega, \tag{6.15}$$

with $G = 0$ on $\partial\Omega$.

In terms of distributions, equation (6.15) means

$$\int_\Omega |\nabla G|^{n-2}\nabla G \cdot \nabla\phi \, dx = \phi(x_0), \tag{6.16}$$

for every compactly supported smooth function $\phi(x)$ on Ω. Of course, when $n = 2$, the 2-Laplacian is the usual Laplacian, and the 2-Green's function is the usual Green's function. In higher dimensions, the existence and uniqueness of this G is also well-known, see [6] and [7], for example. The n-Green's function on the ball B with a pole at the origin is $G_0 = -\omega_{n-1}^{-1/(n-1)} \log |x|$, and for general domain we have the following asymptotic expansion:

$$G_{x_0}(x) = -\omega_{n-1}^{-1/(n-1)} \log |x - x_0| - H_{x_0}(x), \tag{6.17}$$

where $H_{x_0}(x)$ is a continuous function on Ω and is $C^{1,\alpha}$ in $\Omega \setminus \{x_0\}$. Furthermore, n-Green's functions are invariant under Möbius transforms. Therefore, the machinery described in Section 5 can be carried out without much difficulty. We refer the reader to [8] for details.

References

[1] Carleson, L., and Chang, S.-Y. A., On the existence of an extremal function for an inequality of J. Moser, Bull. Sci. Math. Astro. (2) 110 (1986), 113–127.

[2] Federer, H., Geometric measure theory, Springer-Verlag 1969.

[3] Flucher, M., Extremal functions for the Trudinger–Moser inequality in 2 dimensions, Comment. Math. Hel. 67 (1992), 471–497.

[4] Heinonen J., Kilpelainen, T, and Martio, O., Non-linear potential theory of degenerate elliptic equations, Oxford Sci. Publ. 1993.

[5] Kawohl, B., Rearangements and convexity of level sets in PDE, Lecture Notes in Math. 1150, Springer-Verlag 1985.

[6] Kichenassamy, S., and Veron L., Singular solutions of the p-Laplace equation, Math. Ann. 275 (1986), 599–615.

[7] Kilpelainen, T., and Maly, J., Degenerate elliptic equations with measure data and non-linear potentials, Ann. Scuola Norm. Sup. Pisa. Cl. Sci. (4) 19 (1992), 591–613.

[8] Lin, K., Extremal functions for Moser's inequality, Trans. Amer. Math. Soc. 348 (1996), 2663–2671.

[9] Lions, P. L., The concentration-compactness principle in the calculus of variation, the limit case, Part I, Rev. Mat. Iberoamericana 1 (1985), 145–201.

[10] Moser, J., A sharp form of an inequality by N. Trudinger, Indiana Univ. Math. J. 20 (1971), 1077–1092.

[11] Trudinger, N. S., On embeddings into Orlicz spaces and some applications, J. Math. Mech. 17 (1967), 473–484.

Submanifold of a Riemannian manifold with semisymmetric recurrent connections

Ma Chuanyu

0. Introduction

K. Yano [4] studied semisymmetric metric connections in a Riemannian manifold in 1971. In [8], T. Imai discussed the hypersurfaces of a Riemannian manifold with semisymmetric metric connection. In 1993, Chuanyu Ma and Jianming Liu [1] introduced semisymmetric recurrent connections. In 1996, Chuanyu Ma [2] studied the hypersurfaces of a Riemannian manifold with semisymmetric recurrent connections and obtained Gauss's equation and Weingarten's equation. In this paper, we discuss the submanifolds of a Riemannian manifold with semisymmetric recurrent connections. We give the Gauss formula, the Gauss equation, the Theorem Egregium of Gauss and applications to the conformal change of metric in a Riemannian manifold.

Let M' be an $(n+p)$-dimensional Riemannian manifold (with $n > 2$). A linear connection ∇' on M' is said to be a semisymmetric recurrent connection if its torsion tensor T' satisfies the following relations:

$$T'(X', Y') = \pi(Y')X' - \pi(X')Y', \tag{1}$$

$$\nabla_{X'} g(Y', Z') = -2\omega(X')g(Y', Z'), \tag{2}$$

where π and ω are 1-forms and X', Y', Z' are arbitrary C^∞ vector fields.

Ma-Liu proved that

$$\nabla'_{X'} g(Y', Z') = \overset{o}{\nabla}'_{X'} Y' + \pi(Y')X' - g(X', Y')P' + \omega(X')Y' + \omega(Y')X' - g(X', Y')V' \tag{3}$$

where $\overset{o}{\nabla}'$ is the Levi-Civita connection on M', and P' and V' are defined by

$$g(X', P') = \pi(X'), \tag{4}$$

$$g(X', V') = \omega(X'). \tag{5}$$

Let M be an n-dimensional Riemannian manifold which is immersed in a Riemannian manifold M' of dimension $n + p$. We denote by g the induced metric on M.

Suppose that X', Y', Z', W' are C^∞ vector fields on M' and X, Y, Z, W are C^∞ vector fields on M. For the normal bundle $T(M)^\perp$ of M in M', we choose p orthonormal fields $\xi_i, i = 1, \ldots, p$, of normal vectors to M. We decompose $P' + V'$ into its unique tangential and normal components

$$P' + V' = W + \sum_{i=1}^{p} \lambda_i \xi_i, \tag{6}$$

where W is a vector field on M and $\lambda_i, i = 1, \ldots, p$ are real-valued functions on M.

1. Gauss's formula and Weingarten's formula

Gauss's formula with respect to the Levi-Civita connection $\overset{o}{\nabla}{}'$ is given by

$$\overset{o}{\nabla}{}'_X Y = \overset{o}{\nabla}_X Y + \sum_{i=1}^{p} \overset{o}{h}_i(X, Y)\xi_i.$$

To the semisymmetric recurrent connection ∇', we have a similar Gauss formula:

$$\nabla'_X Y = \nabla_X Y + \sum_{i=1}^{p} h_i(X, Y)\xi_i, \tag{7}$$

where

$$h_i = \overset{o}{h}_i - \lambda_i g. \tag{8}$$

A straightforward calculation shows that ∇ is also a semisymmetric recurrent connection on M, that is:

Theorem 1. *The connection ∇ which is induced by* (7) *on a submanifold of a Riemannian manifold with a semisymmetric recurrent connection ∇' is also a semisymmetric recurrent connection.*

The Weingarten equation with respect to $\overset{o}{\nabla}{}'$ is given by

$$\overset{o}{\nabla}{}'_X \xi_i = -\overset{o}{A}_i(X) + D_X \xi_i, \tag{9}$$

where $\overset{o}{A}_i$ are tensors of type (1,1) on M and D is a (metric) connection in the normal bundle $T(M)^\perp$ with respect to the fibre metric induced from g.

Analogously for ∇' we have

Proposition 1. *The Weingarten formula with respect to ∇' is*

$$\nabla'_X \xi_i = -A_i(X) + \bar{D}_X \xi_i, \tag{10}$$

where

$$A_i(X) = (\overset{o}{A}_i - \lambda_i I)(X), \tag{11}$$
$$\bar{D}_X \xi_i = D_X \xi_i + \omega(X)\xi_i. \tag{12}$$

Using $\overset{o}{H} = \frac{1}{n}\sum_{i=1}^{p}(\text{trace } \overset{o}{A}_i)\xi_i$, we can define similarly the mean curvature normal H of M with respect to ∇ by

$$H = \frac{1}{n}\sum_{i=1}^{p}(\text{trace } A_i)\xi_i.$$

Let X_j, $j = 1, 2, \ldots, n$, be n orthonormal local vector fields on M. Note that

i) $\overset{o}{h}_i(X, Y) = g(\overset{o}{A}_i(X), Y) = g(X, \overset{o}{A}_i(Y))$ and

ii) $h_i(X, Y) = g(A_i(X), Y) = g(X, A_i(Y))$

so that $\overset{o}{H} = \frac{1}{n}\sum_{i=1}^{p}\{\overset{o}{h}_i(X_i, X_i)\}\xi_i$, $H = \frac{1}{n}\sum_{i=1}^{p}\{h_i(X_i, X_i)\}\xi_i$. From (8), we get

Proposition 2. *A submanifold M of a Riemannian manifold M' is totally umbilical if and only if it is totally umbilical with respect to the semisymmetric recurrent connection ∇.*

2. Equations of Gauss and Codazzi

Let $\overset{o}{R}'$ and $\overset{o}{R}$ be the Riemannian-Christoffel curvature tensors of type (0,4) on M and M' respectively. The equation of Gauss is given by

$$\overset{o}{R}'(W, Z, X, Y) = \overset{o}{R}(W, Z, X, Y) + \sum_{i=1}^{p}\{\overset{o}{h}_i(X, Z)\overset{o}{h}_i(Y, W) - \overset{o}{h}_i(Y, Z)\overset{o}{h}_i(X, W)\}. \tag{13}$$

The equation of Codazzi–Mainardi is given by

$$\begin{aligned}\overset{o}{R}'(\xi_i, Z, X, Y) = {} & (\overset{o}{\nabla}_X \overset{o}{h}_i)(Y, Z) - (\overset{o}{\nabla}_Y \overset{o}{h}_i)(X, Z) \\ & + \sum_{j=1}^{p} g\left(\{\overset{o}{h}_j(Y, Z)D_X\xi_j - \overset{o}{h}_j(X, Z)D_Y\xi_j\}, \xi_j\right).\end{aligned} \tag{14}$$

We define similarly the curvature tensors of semisymmetric recurrent connections ∇' and ∇ by

$$R'(X', Y')Z' = \nabla'_{X'}\nabla'_{Y'}Z' - \nabla'_{Y'}\nabla'_{X'}Z' - \nabla'_{[X', Y']}Z'$$

and

$$R(X,Y)Z = \nabla_X \nabla_Y Z - \nabla_Y \nabla_X Z - \nabla_{[X,Y]} Z.$$

Using straightforward calculation, we get the following theorem:

Theorem 2.

$$R'(W,Z,X,Y) = R(W,Z,X,Y) + \sum_{i=1}^{p} \{h_i(X,Z)h_i(Y,W) - h_i(Y,Z)h_i(X,W)\}. \tag{15}$$

$$R'(\xi_i, Z, X, Y) = h_i(\pi(Y)X - \pi(X)Y, Z)(\nabla_X h_i)(Y,Z) - (\nabla_Y h_i)(X,Z) + \sum_{j=1}^{p} g\left(\{h_j(Y,Z)\bar{D}_X \xi_j - h_j(X,Z)\bar{D}_Y \xi_j\}, \xi_j\right). \tag{16}$$

(15) is called the equation of Gauss, and (16) is called the equation of Weingarten with respect to ∇' and ∇. From (15), we obtain

$$R'(X,Y,X,Y) = R(X,Y,X,Y) + \sum_{i=1}^{p} \{g(A_i(X),Y)^2 - g(A_i(X),X)g(A_i(Y),Y)\}.$$

Therefore we have

Theorem 3. *Let P be a 2-dimensional subspace of $T(M)$, and let $R'(P)$ and $R(P)$ be, respectively, the sectional curvatures of P in M' and M with respect to the semisymmetric recurrent connections ∇' and ∇. If X and Y form an orthonormal basis of P, then*

$$R'(P) = R(P) + \sum_{i=1}^{p} \{g(A_i(X),Y)^2 - g(A_i(X),X)g(A_i(Y),Y)\}.$$

If $\dim M' = 3$ and M is a surface of M', then the determinant of A_i are exactly $\{g(A_i(X),Y)^2 - g(A_i(X),X)g(A_i(Y),Y)\}$, so we have

Corollary. *If $\dim M' = 3$ and M is a surface in M', then the determinant of A_i (where there is now only one such map) is independent of A_i but depends only on the Riemannian metric g and the semisymmetric recurrent connections ∇' and ∇.*

3. Conformal change of Riemannian metric

We consider a conformal change of Riemannian metric g:

$$g^* = e^{2\rho} g. \tag{17}$$

Using notation as in [1] and (17), we have

Proposition 3. *Let* $\{\nabla; \pi, \omega\}$ *be a semisymmetric recurrent connection on a Riemannian manifold* (M, g) *and let* $g^* = e^{2\rho} g$ *be a conformal change of Riemannian metric* g. *Then* ∇ *is also a semisymmetric recurrent connection which is determined by* π *and* $\omega^* = \omega - d\rho$ *with respect to the metric* g^*.

For Riemannian manifolds (M, g) and (M, g^*), we obtain a relation of Riemannian curvature tensors between the two manifolds:

Theorem 4. *By a conformal change* (17), *the Riemannian curvature tensor* $\overset{o}{R}$ *of* (M, g) *and* $\overset{o}{R}{}^*$ *of* (M, g^*) *satisfies*

$$\overset{o}{R}{}^*(X, Y)Z = \overset{o}{R}(X, Y)Z - \alpha(Y, Z)X + \alpha(X, Z)Y - g(Y, Z)AX + g(Y, Z)AY. \quad (18)$$

References

[1] Ma Chuanyu and Liu Jianming, Semisymmetric recurrent connection on Riemannian manifold, J. Nanjing University 29 (4) 1993,10 (in Chinese).

[2] Ma Chuanyu, Hypersurfaces of a Riemannian manifold with semisymmetric recurrent connections, J. Nanjing University 32 (1) 1996, 1 (in English).

[3] Z. Nakao, Submanifold of Riemannian manifold with semisymmetric metric connection, Proc. Amer. Math. Soc. 54 (1976), 261–266.

[4] K. Yano, On semisymmetric metric connection, Rev. Roumaine Math. Pure Appl. 15 (1970), 1579–1586.

[5] N. J. Hicks, Submanifold of semi-Riemannian manifolds, Rend. Circ. Mat. Palermo (2) 12 (1963), 137–149.

[6] N. J. Hicks, Notes on differential geometry, Van Nostrand Math. Studies 3, Van Nostrand, Princeton, N.J., 1965.

[7] S. Kobayashi and K. Nomizu, Foundations of differential geometry, Vol. II, Interscience Tracts in Pure and Appl. Math. Vol. II, Interscience, New York 1969.

[8] T. Imai, Hypersurfaces of Riemannian manifold with semisymmetric connection, Tensor 32, 1978.

[9] On the geometry of differentiable manifolds, Astérisque 163/164 (1988).

[10] S. Gallot, D. Hulin and J. Lafontaine, Riemannian Geometry, Springer-Verlag, 1987.

Frénet theorem for spaces of constant curvature

J. Muñoz Masqué and G. Rodríguez Sánchez

Abstract. A classical result by Frénet states that two curves placed in general position in the Euclidean m-space are congruent if and only if their $m-1$ curvatures are equal. We prove that Frénet's theorem is characteristic of the spaces of constant curvature; *i.e.*, Frénet's theorem holds true for a connected Riemannian manifold (M, g), if and only if (M, g) is of constant curvature.

1991 Mathematics Subject Classification: 53B20, 53A35

1. Frénet curves and Frénet frames

Definition 1. Let (M, g) be an m-dimensional Riemannian manifold and let ∇ be its Levi-Civita connection. A curve $\sigma : (a, b) \to M$ is said to be a *Frénet curve* at a point $t \in (a, b)$ if the tangent vectors

$$T_t, (\nabla_T T)_t, \ldots, (\nabla_T^{m-2} T)_t$$

are linearly independent, where $T = \sigma_*(d/dt)$.

Definition 2. If σ is a Frénet curve at every point $t \in (a, b)$, we can define an orthonormal system of vector fields $X_1, \ldots, X_{m-1}$ along σ by imposing the following two conditions:

1) For every $i = 0, 1, \ldots, m-2$, $a < t < b$,

$$\langle T_t, \ldots, (\nabla_T^i T)_t \rangle = \langle X_1(t), \ldots, X_i(t) \rangle;$$

i.e., the systems of vectors $\left(T_t, \ldots, (\nabla_T^i T)_t\right)$ and $(X_1(t), \ldots, X_i(t))$ span the same subspace.

2) $\det_{(X_1(t), \ldots, X_i(t))} \left(T_t, \ldots, (\nabla_T^i T)_t\right) > 0$.

Once one of the two Riemannian volume m-forms induced by g on $\sigma^* T(M)$ has been fixed, say ω_m, we can complete the above system to an orthonormal basis $(X_1, \ldots, X_{m-1}, X_m)$, called the *Frénet frame* of σ, by setting

$$\omega_m(X_1, \ldots, X_{m-1}, X_m) = 1.$$

In that case, there exist unique differentiable functions, called the *curvature functions* of σ, $\kappa_i : (a, b) \to \mathbb{R}^+$, $1 \leq i \leq m-2$, $\kappa_{m-1} : (a, b) \to \mathbb{R}$, such that the following *Frénet formulas* hold true:

$$\nabla_{X_1} X_1 = \kappa_1 X_2, \tag{1.1}$$

$$\nabla_{X_1} X_i = -\kappa_{i-1} X_{i-1} + \kappa_i X_{i+1}, \quad 2 \leq i \leq m-1, \tag{1.2}$$

$$\nabla_{X_1} X_m = -\kappa_{m-1} X_{m-1}. \tag{1.3}$$

For the sake of simplicity we also denote by κ_0 the *velocity* of a Frénet curve σ; *i.e.*, $\kappa_0(t) = |T_t|$, $t \in (a, b)$.

Remark 1. There is a functional argument to justify the above definition of Frénet curves. It is not difficult to see that Frénet curves are an open dense subset in $C^\infty((a, b), M)$ with respect to the strong W_∞ topology, but curves having T, $\nabla_T T, \dots$, $\nabla_T^{m-2} T$, $\nabla_T^{m-1} T$ linearly independent are not dense since inflection points are unavoidable.

Definition 3. Two curves $\sigma, \sigma' : (a, b) \to M$ are said to be *congruent* if there exists an open isometric embedding $\phi : U \to M$ such that $\sigma(t) \in U$ and $\sigma'(t) = \phi(\sigma(t))$ for every $t \in (a, b)$.

Remark 2. Two curves $\sigma, \sigma' : (a, b) \to M$ with values in a connected and simply connected complete Riemannian manifold of constant curvature are congruent if and only if there exists an isometry $\phi : M \to M$ such that $\sigma' = \phi \circ \sigma$, since M is endowed with a canonical analytic structure ([1], VI. Theorem 7.7) and consequently every isometric embedding on a connected open subset can be uniquely extended to an isometry of M ([1], VI. Corollary 6.4).

2. Congruent curves in spaces of constant curvature

Theorem 1. *Let $\sigma, \sigma' : (a, b) \to M$ be two curves with values in an m-dimensional connected Riemannian manifold (M, g) of constant curvature, let T, T' be their tangent vectors, respectively, and let ∇ be the Levi-Civita connection of (M, g).*

(i) *Assume σ and σ' are Frénet curves at the points $x_0 = \sigma(t_0)$, $x'_0 = \sigma'(t_0)$, $a < t_0 < b$. Then, σ and σ' are congruent on a neighbourhood of t_0 if and only if $\kappa_i(t) = \kappa'_i(t)$, for $i = 0, 1, \dots, m-1$, and for sufficiently small $|t - t_0|$.*

(ii) *Assume σ, σ' and (M, g) are analytic. Then, σ and σ' are congruent on a neighbourhood of $t_0 \in (a, b)$ if and only if*

$$g\left((\nabla_T^i T)_{t_0}, (\nabla_T^j T)_{t_0}\right) = g\left((\nabla_{T'}^i T')_{t_0}, (\nabla_{T'}^j T')_{t_0}\right), \quad \forall i, j \in \mathbb{N}. \tag{2.4}$$

(iii) *Conversely, assume either* (i) *or* (ii) *hold true for each pair of curves in an m-dimensional, connected Riemannian manifold* (M, g). *Then,* (M, g) *is of constant curvature.*

Proof. We first introduce some preliminary formulas. Let $\sigma : (a, b) \to M$ be a Frénet curve and let $(X_1, \ldots, X_m)$ be its Frénet frame. From Frénet's formulas we obtain:

$$\nabla_T^{j-1} T = \sum_{i=1}^{m} f_{ij} X_i, \quad 1 \leq j \leq m, \tag{2.5}$$

for some differentiable functions $f_{ij} : (a, b) \to \mathbb{R}$ which only depend on $\kappa_0, \kappa_1, \ldots, \kappa_{m-1}$ and their derivatives, such that

$$f_{ij} = 0 \text{ for } i > j, \quad f_{ii} = \kappa_0^i \kappa_1 \ldots \kappa_{i-1}.$$

Let us denote by Δ_k the determinant of the matrix $\left(g(\nabla_T^{i-1} T, \nabla_T^{j-1} T)\right)_{1 \leq i, j \leq k}$. Then, we have:

$$\kappa_0 = \sqrt{\Delta_1}, \quad \kappa_1 = \frac{\sqrt{\Delta_2}}{\Delta_1^{3/2}}, \tag{2.6}$$

$$\kappa_i = \frac{\sqrt{\Delta_{i-1} \Delta_{i+1}}}{\sqrt{\Delta_1} \Delta_i}, \quad 2 \leq i \leq m-2, \tag{2.7}$$

$$\kappa_{m-1} = \frac{\omega_m \left(T, \nabla_T T, \ldots, \nabla_T^{m-1} T\right) \sqrt{\Delta_{m-2}}}{\sqrt{\Delta_1} \Delta_{m-1}} = \frac{\pm\sqrt{\Delta_{m-2} \Delta_m}}{\sqrt{\Delta_1} \Delta_{m-1}}. \tag{2.8}$$

Hence,

(a) Let $\sigma, \sigma' : (a, b) \to M$ be two Frénet curves. Then, $\kappa_i(t) = \kappa_i'(t)$ for all $i = 0, 1, \ldots, m-1$, $a < t < b$, if and only if

$$g\left((\nabla_T^i T)_t, (\nabla_T^j T)_t\right) = g\left((\nabla_{T'}^i T')_t, (\nabla_{T'}^j T')_t\right), \quad 0 \leq i \leq j \leq m-1, \ a < t < b.$$

Furthermore if σ, σ' and (M, g) are analytic then the above conditions are equivalent to (2.4).

Moreover, assume σ, σ' are two congruent curves, say $\sigma' = \phi \circ \sigma$. In that case, T and T' are ϕ-related and taking into account that every open isometric embedding is an affine transformation we conclude ([1], VI.Prop.1.2-(1)) that $\phi_*(\nabla_T^i T) = \nabla_{T'}^i T'$, for all $i \in \mathbb{N}$. As ϕ is an isometry, we obtain:

(b) If $\sigma, \sigma' : (a, b) \to M$ are congruent, for all $i, j \in \mathbb{N}$, $t \in (a, b)$,

$$g\left((\nabla_T^i T)_t, (\nabla_T^j T)_t\right) = g\left((\nabla_{T'}^i T')_t, (\nabla_{T'}^j T')_t\right).$$

Let $(U; x_1, \dots, x_m)$ be an open coordinate neighbourhood of the point x_0. From formulas (2.5) and the definition of the Frénet frame of a curve it follows that there exist universal differentiable functions F_1(i), ..., F_m defined on an open subset of $\mathbb{R}^{m^2+m(m+1)/2}$, such that for every Frénet curve $\sigma : (a, b) \to M$, $\sigma(t_0) = x_0$, $\sigma_h = x_h \circ \sigma$ and for sufficiently small $|t - t_0|$ we have:

$$\frac{d^m \sigma_h}{dt^m} = F_h\left(\frac{d^\alpha \sigma_i}{dt^\alpha}, \frac{d^\beta \kappa_j}{dt^\beta}\right), \ 0 \le \alpha \le m-1, \ 0 \le \beta + j \le m-1, \ 1 \le h, i \le m. \tag{2.9}$$

Let σ, σ' be as in (i). By composing with a suitable local isometry we can assume $x_0 = x_0'$. By virtue of the hypothesis each of the systems

$$\left(T_{t_0}, (\nabla_T T)_{t_0}, \dots, \left(\nabla_T^{m-2} T\right)_{t_0}\right), \quad \left(T_{t_0}', (\nabla_{T'} T')_{t_0}, \dots, \left(\nabla_{T'}^{m-2} T'\right)_{t_0}\right),$$

is linearly independent. Let X_m, X_m' be unit tangent vectors orthogonal to the hyperplanes spanned by the above systems, respectively. We can thus define a linear isomorphism $A : T_{x_0}(M) \to T_{x_0}(M)$, by setting

$$A\left(\nabla_T^i T\right)_{t_0} = \left(\nabla_{T'}^i T'\right)_{t_0}, \quad 0 \le i \le m-2, \quad A(X_m) = X_m'.$$

Taking into account (a) and our hypothesis, it follows that A preserves the scalar product on $T_{x_0}(M)$. As is well-known, manifolds of constant curvature are characterized as the isotropic spaces in the sense of [2] ([2], (4.3.4) and (4.3.11)). Therefore, there exists an isometric embedding ϕ defined on a neighbourhood of x_0 and keeping x_0 fixed such that $\phi_* = A$. We set $\sigma'' = \phi \circ \sigma$. From the very definition of A we deduce that $\left(\nabla_{T''}^i T''\right)_{t_0} = \left(\nabla_{T'}^i T'\right)_{t_0}$, for $i = 0, 1, \dots, m-2$; or equivalently, $\left(d^\alpha \sigma_i'/dt^\alpha\right)(t_0) = \left(d^\alpha \sigma_i''/dt^\alpha\right)(t_0)$, for $\alpha = 0, \dots, m-1$. Since σ and σ' are congruent, from (a) and (b) we obtain $\kappa_i''(t) = \kappa_i'(t)$, for sufficiently small $|t - t_0|$. Accordingly, we can view (2.9) as a system of ordinary differential equations of order m admitting σ_h , σ_h' as solutions. Since they satisfy the same initial conditions, $\sigma' = \phi \circ \sigma$, and part (i) is proved.

As for (ii), we first remark that according to (b) if σ and σ' are congruent, then (2.4) holds for every $t_0 \in (a, b)$. To prove the converse, let us denote by K the curvature of M and let us fix a point $x_0 \in M$. We know there is an open neighbourhood U of x_0 isometric to an open subset of S^m, $\mathbb{R}^m$ or H^m according as $K > 0$, $K = 0$ or $K < 0$. Let k be the biggest of the integers not greater than $m - 1$ for which a value $t_0 \in (a, b)$ exists such that the tangent vectors $T_{t_0}, (\nabla_T T)_{t_0}, \dots, (\nabla_T^{k-1} T)_{t_0}$ are linearly independent. If $k = m - 1$, σ is a Frénet curve at t_0 and the same holds true for σ', since $\Delta_{m-1}' = \Delta_{m-1} > 0$.

As M is of constant curvature, an open isometric embedding ϕ exists, defined on a neighbourhood of $\sigma(t_0)$ such that $\phi(\sigma(t_0)) = \sigma'(t_0)$ and we conclude by simply applying (a) and (i). Therefore, we can assume $k \leq m-2$, $\sigma(t_0) = \sigma'(t_0) = x_0$. In that case, as $k+1 \leq m-1$, there exist analytic functions $f_0, \ldots, f_{k-1}$ on a neighbourhood of t_0 such that

$$\nabla_T^k T = \sum_{i=0}^{k-1} f_i(\nabla_T^i T). \tag{2.10}$$

Firstly, let us assume $K > 0$, so that we shall work in the sphere $S^m \subset \mathbb{R}^{m+1}$, $x_0^2 + \cdots + x_m^2 = K^{-1}$. Gauss' formula reads as

$$\overline{\nabla}_X Y = \nabla_X Y - \sqrt{K}\bar{g}(X, Y)N,$$

where $X, Y \in \mathfrak{X}(S^m)$, $\bar{g} = dx_0^2 + \cdots + dx_m^2$, $\overline{\nabla}$ is the Levi-Civita connection of $\bar{g}$ and N stands for the unit normal vector field; *i.e.*,

$$N_x = |x|^{-1} \sum_{i=0}^{m} x_i(\partial/\partial x_i)_x, \quad x \neq 0.$$

From the equations $\bar{g}(T, N) = 0$, $\bar{g}(N, N) = 1$, we obtain $\overline{\nabla}_T N = \sqrt{K}T$, and by applying successively Gauss' formula we have

$$\nabla_T^k T = \overline{\nabla}_T^k T + \sum_{i=0}^{k-2} \phi_i \overline{\nabla}_T^i T + \phi_{k-1} N, \quad k \geq 2, \tag{2.11}$$

for some analytic functions $\phi_0, \ldots, \phi_{k-2}, \phi_{k-1}$.

Let us denote by $\sigma_0, \sigma_1, \ldots, \sigma_m$ the components of σ in $\mathbb{R}^{m+1}$, so that

$$\overline{\nabla}_T^k T = \sum_{i=0}^{m} (d^{k+1}\sigma_i/dt^{k+1})(\partial/\partial x_i).$$

From (2.10) and (2.11) we obtain

$$\frac{d^{k+1}\sigma}{dt^{k+1}} = \sum_{i=1}^{k} g_i \frac{d^i\sigma}{dt^i} + g_0 N, \tag{2.12}$$

for some analytic functions $g_0, g_1, \ldots, g_k$.

Hence,

$$\frac{d^{k+2}\sigma}{dt^{k+2}} - g_0\sqrt{K}\frac{d\sigma}{dt} - \sum_{i=1}^{k}\left(g_i' \frac{d^i\sigma}{dt^i} + g_i \frac{d^{i+1}\sigma}{dt^{i+1}}\right) = g_0' N. \tag{2.13}$$

We claim that for every $\epsilon > 0$ there exist $t_1 \in (t_0 - \epsilon, t_0 + \epsilon)$ and analytic functions

$h_1, \ldots, h_{k+1}$ on a neighbourhood of t_1 such that

$$\frac{d^{k+2}\sigma}{dt^{k+2}} = \sum_{i=1}^{k+1} h_i \frac{d^i\sigma}{dt^i}. \tag{2.14}$$

This is obvious if g_0 vanishes on a neighbourhood of t_0; otherwise, since g_0 is analytic, we can always pick a t_1 value arbitrarily close to t_0 such that $g_0(t_1) \neq 0$. Eliminating N between (2.12) and (2.13) on a neighbourhood of t_1 we obtain (2.14).

Let $s_1(t), \ldots, s_{k+2}(t)$ be a fundamental system of solutions to the scalar linear differential equation

$$\frac{d^{k+2}u}{dt^{k+2}} = \sum_{i=1}^{k+1} h_i \frac{d^i u}{dt^i}.$$

Hence, there exist real numbers $(\lambda_{ij})_{0\le i\le m,\, 1\le j\le k+2}$, such that $\sigma_i = \sum_{j=1}^{k+2} \lambda_{ij} s_j$. We set $v_j = (\lambda_{0j}, \ldots, \lambda_{mj}) \in \mathbb{R}^{m+1}$. Then, $\sigma(t) = \sum_{j=1}^{k+2} s_j(t) v_j$, thus showing that on a neighbourhood of t_1, σ takes values in

$$\Sigma = S^m \bigcap \langle v_1, \ldots, v_{k+2} \rangle,$$

which is a sphere of the same radius and dimension $\le k+1$. As Σ is a totally geodesic submanifold of S^m, we conclude that σ is a Frénet curve at t_1 with values in Σ. Because of (2.4) the same argument can be applied to σ', thus concluding that σ' is a Frénet curve at a point t_1' arbitrarily close to t_0 taking values in an sphere Σ' of the same radius as Σ, and dimension $\le k+1$. By composing with a suitable isometry of S^m, we can assume $\Sigma = \Sigma'$, $\sigma(t_1) = \sigma'(t_1')$, and the result follows from (i).

In the case $K < 0$, we have $H^m = \{x \in \mathbb{R}^m \; ; \; x_0 > 0, \; -x_0^2 + x_1^2 + \cdots + x_m^2 = K^{-1}\}$, $\bar{g} = -dx_0^2 + dx_1^2 + \cdots + dx_m^2$, and Gauss' formula reads as

$$\overline{\nabla}_X Y = \nabla_X Y + \sqrt{-K}\,\bar{g}(X, Y) N,$$

where N is the vector field on the open neighbourhood of H^m, $x_0^2 > x_1^2 + \cdots + x_m^2$, defined by

$$N_x = (x_0^2 - x_1^2 - \cdots - x_m^2)^{-1/2} \left(\sum_{i=0}^{m} x_i (\partial/\partial x_i)_x \right).$$

Furthermore, $\bar{g}(N, N) = 1$, $\overline{\nabla}_T N = \sqrt{-K}\,T$, and the rest of the proof goes as in the previous case.

For $K = 0$, we can work directly in $\mathbb{R}^m$ and the proof is much easier, since ∇ is flat and equation (2.10) leads us to equation (2.14).

Finally, we shall prove part (iii) of the statement. Let $(X_1, \ldots, X_m)$ be an orthonormal basis for $T_{x_0}(M)$. From the uniqueness and existence theorem for ordinary differential equations we deduce the existence of a curve $\sigma : (t_0 - \epsilon, t_0 + \epsilon) \to M$ such that:

(i) $\sigma(t_0) = x_0$,

(ii) $(\nabla_T^i T)_{t_0} = X_{i+1}, 0 \le i \le m-1$,

(iii) $(\nabla_T^m T)_t = 0$, for every $|t - t_0| < \epsilon$.

Let $A : T_{x_0}(M) \to T_{x_0}(M)$ be an arbitrary $\mathbb{R}$-linear isomorphism preserving the scalar product induced by the Riemannian metric g on $T_{x_0}(M)$, and let $\sigma' : (t_0 - \epsilon, t_0 + \epsilon) \to M$ be the curve determined by:

(i′) $\sigma'(t_0) = x_0$,

(ii′) $(\nabla_{T'}^i T')_{t_0} = X_{i+1}, 0 \le i \le m-1$,

(iii′) $(\nabla_{T'}^m T')_t = 0$, for every $|t - t_0| < \epsilon$.

Taking sufficiently small ϵ, we can also assume that σ and σ' are Frénet curves. Set $h_{ij} = g\left((\nabla_T^i T)_t, (\nabla_T^j T)_t\right) - g\left((\nabla_{T'}^i T')_t, (\nabla_{T'}^j T')_t\right)$. We have:

$$\frac{d^r h_{ij}}{dt^r} = \sum_{k=0}^{r} \binom{r}{k} h_{i+r-k,j+k}, \tag{2.15}$$

and from the definitions of σ and σ',

$$h_{ij}(t) = 0 \quad \text{for } i \ge m, \text{ or for } j \ge m, \tag{2.16}$$

$$h_{ij}(t_0) = 0 \quad \text{for } 0 \le i, j \le m-1. \tag{2.17}$$

Moreover, equations (2.15), (2.16) yield $d^{2m}h_{ij}/dt^{2m} = 0$; hence, h_{ij} is a polynomial of degree $\le 2m-1$. From (2.15) and (2.17) we obtain for every $r \in \mathbb{N}$, $\left(d^r h_{ij}/dt^r\right)(t_0) = 0$. Thus, h_{ij} vanishes identically. By virtue of the hypothesis, there exists an isometric embedding ϕ on an open neighbourhood of x_0, keeping x_0 fixed, such that $\sigma' = \phi \circ \sigma$ and hence $\phi_* = A$. Consequently, M is isotropic in the sense of [2], and therefore of constant curvature (see [2], (4.3.11)). □

References

[1] S. Kobayashi and K. Nomizu, Foundations of Differential Geometry, Volume I, John Wiley & Sons, Inc., New York 1963.

[2] W. Thirring, A Course in Mathematical Physics 2, Classical Field Theory, Springer-Verlag, New York 1979.

The generalized theorem of Castelnuovo–de Franchis for unitary representations

Ngaiming Mok

In the classification theory of compact complex manifolds, varieties of the general type play a central rôle. We are interested in finding analytic and ultimately topological conditions which guarantee that a compact Kähler manifold admits a surjective holomorphic (or even meromorphic) map onto a compact complex space of the general type. In this direction the classical theorem of Castelnuovo–de Franchis states that, given a compact Kähler manifold X with at least 2 linearly independent holomorphic 1-forms ω_1 and ω_2 such that $\omega_1 \wedge \omega_2 \equiv 0$, there exists a surjective holomorphic map onto a compact Riemann surface of genus ≥ 2. Higher dimensional generalizations were obtained by Ran [R] and Catanese [C]. In this article we are interested in twisted holomorphic 1-forms taking values in locally flat bundles arising from unitary representations.

Let X be a compact Kähler manifold, $\Phi : \pi_1(X) \to U(r)$ be a unitary representation and $E_\Phi \mapsto X$ be the associated locally flat vector bundle. Given an E_Φ-valued holomorphic 1-form ω, which is automatically closed, we study the holomorphic foliation defined by $\mathrm{Ker}(\omega)$ on $X - V$ for some subvariety $V \subset X$ of codimension ≥ 2. Under a non-degeneracy assumption on ω, we prove that the leaves of the holomorphic foliation are closed on $X - V$ such that their topological closures are complex-analytic in X. The usual proof of the (generalized) theorem of Castelnuovo–de Franchis using foliations (as in Catanese [C]) consists of showing that the leaves agree with irreducible components of level sets of meromorphic functions. In the presence of twisting given by unitary transformations, such an argument is in general not possible. Our input consists of introducing Differential Geometry into the problem by considering semi-norms on X defined by $\|\eta\| = |\omega(\eta)|$. This allows us to consider metrized holomorphic foliations, which are Riemannian foliations outside some complex-analytic subvariety, and consider pseudogroups G of holomorphic isometries on local complex submanifolds transverse to the foliation. The use of such semi-norms will only be implicit in case $\mathrm{Ker}(\omega)$ is of corank 1 generically, but will be essential in the general case. To prove Castelnuovo–de Franchis for unitary representations we argue by contradiction. The idea is to produce, in the presence of one single transcendental leaf of $X - V$, a positive-dimensional pseudogroup G. From the way that G arises from the unitary representation Φ we prove that the image of the universal cover $\tilde{X}$ under the integral $F : \tilde{X} \to \mathbb{C}^r$ of $\tilde{\omega}$ (lifting of ω) is invariant under a family of Euclidean translations, contradicting with the non-degeneracy assumption on ω.

The nature of the pseudogroup G will play an important rôle. It will be shown that G can be taken to be compact and semisimple, or to be abelian. In the last section we will give a counter-example to the analogue of Castelnuovo–de Franchis for infinite-dimensional unitary representations. In this example the same procedure as in the finite-dimensional case will produce a noncompact simple pseudogroup of affine isometries.

I would like to thank the organizers of the Pacific Rim Geometry Conference, especially Professor A. J. Berrick and Dr. Wing-Keung To, for having invited me and making my stay a very pleasant one. The present article is an outgrowth from the proof of a fibration theorem (arising from unitary representations) over varieties of the general type, which was outlined, among other things, in my talk in the conference. The proof of the fibration theorem will be given in a forthcoming article. Finally, I would like to thank the Hong Kong Research Grants Council for the support it has provided through a research grant.

1. Fibrations over compact Riemann surfaces of genus ≥ 2 — formulation of the problem

(1.1) Let X be an n-dimensional compact Kähler manifold and $\Phi : \pi_1(X) \to U(r)$ be a representation of its fundamental group into the group of unitary transformations of $\mathbb{C}^r$. We have an associated locally flat holomorphic vector bundle $E_\Phi \mapsto X$. Let $\omega \in \Gamma(X, \Omega(E_\Phi))$ be a non-zero E_Φ-valued holomorphic 1-form, $\Omega(E_\Phi)$ denoting the holomorphic bundle of E_Φ-valued tangent covectors. At each $x \in X$ we interpret $\omega(x)$ as being in $\mathrm{Hom}(T_x(X), E_{\Phi,x})$, $T(X)$ denoting the holomorphic tangent bundle. $\mathrm{Ker}\, \omega(x)$ is of constant rank on some non-trivial Zariski-open subset of X. It defines a holomorphic foliation $\mathcal{F}$ on $X - V$ for some complex-analytic subvariety V of codimension ≥ 2.

$\mathcal{F}$ can be described in terms of integrals of the lifting $\tilde{\omega}$ of ω to the universal covering space $\tilde{\omega}$, as follows. On the locally flat vector bundle E_Φ we have a de Rham operator d. Since X is Kähler and Φ is unitary by the Kähler identity any holomorphic E_Φ-valued 1-form ω is d-closed. On the universal covering space $\tilde{X}$ the lifting $\tilde{\omega}$ has a primitive, say $\tilde{\omega} = dF$, where $F : \tilde{X} \to \mathbb{C}^r$ is a vector-valued holomorphic function. The level sets of F define a holomorphic foliation $\tilde{\mathcal{F}}$ on $\tilde{X} - \tilde{V}$ for some complex-analytic subvariety $\tilde{V} \subset \tilde{X}$ of complex codimension ≥ 2. The effect of a deck transformation $\gamma \in \pi_1(X)$ on F is given by

$$F(\gamma\tilde{x}) = \Phi(\gamma)F(\tilde{x}) + c_\gamma$$

for all $\tilde{x} \in \tilde{X}$ and for some $c_\gamma \in \mathbb{C}^r$. The function $c : \pi_1(X) \to \mathbb{C}^r$ given by $c(\gamma) = c_\gamma$ is a 1-cocycle with values in $\mathbb{C}^r$, regarded as a $\pi_1(X)$-representation space via Φ. It follows readily that γ transforms level sets of F to each other, so that $\tilde{V}$ is $\pi_1(X)$-invariant and $\tilde{\mathcal{F}}$ is $\pi_1(X)$-invariant on $\tilde{X} - \tilde{V}$. $\tilde{V}$ then descends to a complex-analytic subvariety $V \subset X$, of codimension ≥ 2, while $\tilde{\mathcal{F}}$ descends to a holomorphic foliation on $X - V$.

(1.2) The classical theorem of Castelnuovo–de Franchis deals with the case where $\Phi : \pi_1(X) \to U(r)$ is the trivial representation and $\mathrm{Ker}(\omega)$ is of corank 1 at generic points. In order to present the basic ideas with a minimum of technicalities we will consider in this section only the case where $\mathrm{Ker}(\omega)$ is of corank 1 at generic points, while $\Phi : \pi_1(X) \to U(r)$ is any unitary representation. Thus, the lifting $\tilde{\omega} = (\tilde{\omega}_1, \ldots, \tilde{\omega}_r)$ satisfies $\tilde{\omega}_i \wedge \tilde{\omega}_j \equiv 0$ for all i, j; $1 \leq i, j \leq r$. We prove

Theorem 1. *Let X be a compact Kähler manifold, $\Phi : \pi_1(X) \to U(r)$ be a unitary representation, and $E_\Phi \mapsto X$ be the associated locally flat vector bundle. Let $\omega \in \Gamma(X, \Omega(E_\Phi))$ be such that* $\mathrm{Ker}(\omega)$ *is of corank* 1 *at generic points of X. Suppose $r \geq 2$ and the lifted holomorphic 1-forms $\{\tilde{\omega}_1, \ldots, \tilde{\omega}_r\}$ on $\tilde{X}$ are linearly independent. Suppose the image $\Phi(\pi_1(X))$ is torsion-free. Then, there exists a compact Riemann surface Z, a unitary representation $\Psi : \pi_1(Z) \to U(r)$ and a holomorphic map $\sigma : X \to Z$, such that*

(a) $\Phi(\gamma) = \Psi(\sigma_*(\gamma))$ *for any $\gamma \in \pi_1(X)$, i.e. $E_\Phi = \sigma^* E_\Psi$;*

(b) *there exists some $\nu \in \Gamma(Z, \Omega(E_\Psi))$ such that $\sigma^*\nu \equiv \omega$.*

Remarks. Since any finitely-generated subgroup of a matrix group contains a torsion-free normal subgroup of finite index, the condition that $\Phi(\pi_1(X))$ is torsion-free is always satisfied if we replace $\pi_1(X)$ by some appropriate normal subgroup $\Gamma \subset \pi_1(X)$ of finite index, X by the corresponding regular finite covering space X' of X with covering map $\alpha : X' \to X$, $\pi_1(X') \cong \Gamma$, Φ by $\Phi|_\Gamma$ and ω by $\alpha^*\omega$.

We proceed with the proof of Theorem 1. As in (1.1) we write $\tilde{\omega} = dF$ on the universal cover $\tilde{X}$, $F : \tilde{X} \to \mathbb{C}^r$ being holomorphic. In what follows we are going to construct complex spaces using holomorphic equivalence relations. For the basic concepts and definitions cf. Kaup [K]. Let $\mathcal{R}_0 \subset \tilde{X} \times \tilde{X}$ be the complex-analytic subvariety defined by $\mathcal{R}_0 = \{(\tilde{x}_1, \tilde{x}_2) \in \tilde{X} \times \tilde{X} : F(\tilde{x}_1) = F(\tilde{x}_2)\}$. $\mathcal{R}_0$ is a holomorphic equivalence relation on $\tilde{X}$. Let $\mathcal{R} \subset \mathcal{R}_0$ be the irreducible component of $\mathcal{R}_0$ containing the diagonal of $\tilde{X}$. Then $\mathcal{R} \subset \tilde{X} \times \tilde{X}$ is the graph of an open holomorphic equivalence relation. Here openness of $\mathcal{R}$ means that the projection map $\mathcal{R} \mapsto \tilde{X}$ onto the first factor is open, which is the case as F is of rank 1 at generic points. By Kaup [K] $\tilde{X}/\mathcal{R}$ can be given the structure of a normal complex space $\tilde{Z}$ such that the canonical map $\pi : \tilde{X} \mapsto \tilde{Z}$ is holomorphic. Since $\tilde{Z}$ is of complex dimension 1 it is a Riemann surface. Here the topology of $\tilde{X}/\mathcal{R}$ is induced by the canonical projection $\tilde{X} \mapsto \tilde{X}/\mathcal{R}$.

As $\mathcal{R} \subset \mathcal{R}_0$ it follows that $(\tilde{x}_1, \tilde{x}_2) \in \mathcal{R}$ implies $F(\tilde{x}_1) = F(\tilde{x}_2)$. Thus the holomorphic map $F : \tilde{X} \to \mathbb{C}^r$ gives rise to a continuous map $f : \tilde{Z} \mapsto \mathbb{C}^r$ such that $F = f \circ \pi$. As $\tilde{Z}$ is normal it follows readily that $f : \tilde{Z} \to \mathbb{C}^r$ is holomorphic. We now define an action of $\pi_1(X)$ on $\tilde{Z}$. From the compatibility condition $F(\gamma\tilde{x}) = \Phi(\gamma)F(\tilde{x}) + c_\gamma$ as in (1.1) we conclude that $F(\tilde{x}_1) = F(\tilde{x}_2)$ implies $F(\gamma\tilde{x}_1) = F(\gamma\tilde{x}_2)$ for any $\gamma \in \pi_1(X)$. It follows that $\mathcal{R}_0 \subset \tilde{X} \times \tilde{X}$ and hence $\mathcal{R} \subset \tilde{X} \times \tilde{X}$ must be invariant under $\pi_1(X)$. As a consequence, given $\gamma \in \pi_1(X)$, the deck transformation $\gamma : \tilde{X} \to \tilde{X}$ descends to a continuous map $\gamma' : \tilde{Z} \to \tilde{Z}$. Since the lifting γ is holomorphic and $\tilde{Z}$ is normal we conclude that $\gamma' : \tilde{Z} \to \tilde{Z}$ is holomorphic. From the uniqueness of γ' it follows readily that the assignment $\gamma \mapsto \gamma'$ is a homomorphism, written $\gamma' = \Theta(\gamma)$, into $\mathrm{Aut}(\tilde{Z})$. We have thus defined an action of $\pi_1(X)$ on $\tilde{Z}$. Under the hypothesis of Theorem 1 we are going to prove that this action $\Theta : \pi_1(X) \to \mathrm{Aut}(\tilde{Z})$ is properly discontinuous.

(1.3) For $p \in \mathbb{C}^r$ and $\gamma \in \pi_1(X)$ consider the affine linear map $\hat{\Phi}(\gamma) : \mathbb{C}^r \to \mathbb{C}^r$ defined by $\hat{\Phi}(\gamma)(p) = \Phi(\gamma)(p) + c_\gamma$. As $\Phi : \pi_1(X) \to U(r)$ is a homomorphism and c_γ is uniquely determined by γ the map $\hat{\Phi} : \pi_1(X) \to \mathrm{Aff}(\mathbb{C}^r)$, the group of affine transformations on $\mathbb{C}^r$, is a homomorphism. In the standard way $\mathrm{Aff}(\mathbb{C}^r)$ is isomorphically a semi-direct product $GL(r, \mathbb{C}) \ltimes \mathbb{C}^r$. With this identification we have $\hat{\Phi}(\pi_1(X)) \subset U(r) \ltimes \mathbb{C}^r$.

The bulk of §1 and §2 will be dedicated to the proof that the action of $\pi_1(X)$ on $\tilde{Z}$ is properly discontinuous. Observe that by the hypothesis of Theorem 1 the linear span of $\{\tilde{\omega}(\tilde{x}) : \tilde{x} \in \tilde{X}\}$ must be $\mathbb{C}^r$, otherwise there would be some non-zero vector $(\alpha_1, \ldots, \alpha_r)$ orthogonal to all $\tilde{\omega}(\tilde{x})$, $\tilde{x} \in \tilde{X}$, so that $\sum_{i=1}^{r} \overline{\alpha}_i\tilde{\omega}_i \equiv 0$ on $\tilde{X}$, contradicting with the linear independence of $\{\tilde{\omega}_1, \ldots, \tilde{\omega}_r\}$. A consequence of this observation is the fact that any affine linear transformation on $\mathbb{C}^r$ fixing every point of $F(\tilde{X})$ must be the identity map. Recall that $\tilde{Z} = \tilde{X}/\mathcal{R}$ and that $\Theta : \pi_1(X) \mapsto \mathrm{Aut}(\tilde{Z})$ is a homomorphism. For any $\gamma_0 \in \mathrm{Ker}\,\Theta$ we know that $F(\gamma_0\tilde{x}) = F(\tilde{x})$ for any $\tilde{x} \in \tilde{X}$. Since $F(\gamma_0\tilde{x}) = \hat{\Phi}(\gamma_0)(F\tilde{x})$ and $\hat{\Phi}(\gamma_0) \in \mathrm{Aff}(\mathbb{C}^r)$ we have $\hat{\Phi}(\gamma_0) = \mathrm{id}$. We may therefore obtain a homomorphism $\hat{\Psi}_0 : \Theta(\pi_1(X)) \to \mathrm{Aff}(\mathbb{C}^r)$ by defining $\hat{\Psi}_0(\gamma') = \hat{\Phi}(\gamma)$ for any $\gamma \in \pi_1(X)$ such that $\Theta(\gamma) = \gamma'$.

To prove that $\Theta(\pi_1(X)) \subset \mathrm{Aut}(\tilde{Z})$ is discrete we argue by contradiction. Since $f : \tilde{Z} \mapsto \mathbb{C}^r$ is a non-trivial holomorphic map, $\tilde{Z}$ cannot be a Riemann sphere, so that there exists an invariant metric h on $\tilde{Z}$. Suppose $\Theta(\pi_1(X)) \subset \mathrm{Aut}(\tilde{Z})$ is not discrete. Then there exists a sequence $\{\gamma_i'\}$ of distinct automorphisms of $\tilde{Z}$ such that $\gamma_i' \in \Theta(\pi_1(X))$ and γ_i' converges to the identity map as isometries of $(\tilde{Z}, h)$. For $\gamma' \in \Theta(\pi_1(X))$ and $\tilde{z} \in \tilde{Z}$

$$f(\gamma'\tilde{z}) = \hat{\Psi}_0(\gamma')(\tilde{z}) := \Psi_0(\gamma')(\tilde{z}) + d_{\gamma'},$$

where $\Psi_0(\gamma') \in U(r)$; $d_{\gamma'} \in \mathbb{C}^r$ are such that $\Psi_0(\gamma') = \Phi(\gamma)$ and $c_{\gamma'} = d_{\gamma'}$ for any $\gamma \in \pi_1(X)$ with $\Theta(\gamma) = \gamma'$. Write θ_i for $\hat{\Psi}_0(\gamma_i')$. We assert first of all that $\operatorname{Ker}\hat{\Psi}_0 \subset \Theta(\pi_1(X)) \subset \operatorname{Aut}(\tilde{Z}) = \operatorname{Isom}(\tilde{Z}, h)$ is a discrete subgroup. To see this take a non-empty open subset $W \subset \tilde{Z}$ such that $f|_W : W \to \mathbb{C}^r$ is a biholomorphism onto its image Σ. If $\mu' \in \Theta(\pi_1(X)) \subset \operatorname{Aut}(\tilde{Z})$ is such that $\mu'(W) \cap W \neq \phi$, then for $\theta = \hat{\Psi}_0(\mu')$, $\theta(\Sigma) \cap \Sigma$ is a non-empty open subset of Σ. Furthermore, $\theta|_\Sigma \equiv \mathrm{id}$ implies that $\theta \equiv \mathrm{id}$ and hence that $\mu' \equiv \mathrm{id}$. We note that $\pi_1(X)$ may switch connected components of level sets of F. In this case there exists $\mu' \in \Theta(\pi_1(X))$, $\mu' \neq \mathrm{id}$, such that $\hat{\Psi}_0(\mu') = \mathrm{id}$. Our assertion is simply that μ' cannot be too close to id (i.e. it cannot happen that $\mu'(W) \cap W \neq \phi$). The non-discreteness of $\Theta(\pi_1(X))$ implies therefore the existence of a countable set $\{\theta_i\}$ of distinct isometries of $\mathbb{C}^r$ such that θ_i converges to the identity map. We proceed to prove

Proposition 1. *Suppose the action of $\Theta(\pi_1(X))$ on the open Riemann surface $\tilde{Z} = \tilde{X}/\mathcal{R}$ is not properly discontinuous. Then, there exists a positive-dimensional (not necessarily closed) connected Lie subgroup $G \subset U(r) \ltimes \mathbb{C}^r$ such that*

(a) *the corresponding Lie subalgebra $\mathbf{g} \subset \mathbf{u}(r) \ltimes \mathbb{C}^r$ satisfies*

$$\mathbf{g} \cap (\{0\} \ltimes \mathbb{C}^r) = \{0\} ,$$

where $\mathbf{u}(r)$ denotes the Lie algebra of $U(r)$;

(b) *for any $\hat{\nu} \in \hat{\Phi}(\pi_1(X))$, we have*

$$\hat{\nu}^{-1} G \hat{\nu} = G .$$

Proof. Introduce a Riemannian metric on $\operatorname{Aff}(\mathbb{C}^r) \cong GL(r, \mathbb{C}) \ltimes \mathbb{C}^r$ and denote by d the distance function thus defined. Denote by $B_\varepsilon^0 \subset GL(r, \mathbb{C}) \ltimes \mathbb{C}^r$ the open ball of radius ε centred at id, measured with respect to d. Without loss of generality we may take $\Sigma \subset W$ to be a closed complex-analytic subvariety of some open subset $W \subset \mathbb{C}^r$ such that Σ is the zero locus of a family of holomorphic functions $\{h_\alpha\}_{\alpha \in A}$. We say that Σ is pseudo-invariant under $\theta \in B_\varepsilon^0$ to mean that $\theta(\Sigma) \cap \Sigma$ is a non-empty open subset of Σ. Choose $\varepsilon > 0$ small enough such that for any $\theta \in B_\varepsilon^0$, $\theta(\Sigma) \cap W \neq \phi$. For such a choice of $\varepsilon > 0$, Σ is pseudo-invariant under θ if and only if $h_\alpha \circ \theta \equiv 0$ for all $\alpha \in A$ on some fixed open polydisk W' on W, centred at some $p \in \Sigma$, chosen so that $\theta(W') \subset W$ for any $\theta \in B_\varepsilon^0$. Write $M(r, \mathbb{C})$ for the Euclidean space of complex r-by-r matrices. Identifying $\operatorname{Aff}(\mathbb{C}^r)$ with $GL(r, \mathbb{C}) \times \mathbb{C}^r$ as complex manifolds in the standard way we may express θ in terms of Euclidean coordinates $(\zeta_1, \ldots, \zeta_s)$, $s = r^2 + r$, arising from the inclusion $GL(r, \mathbb{C}) \times \mathbb{C}^r \subset M(r, \mathbb{C}) \times \mathbb{C}^r \cong \mathbb{C}^{r^2+r}$. Now $\theta \in B_\varepsilon^0$ leaves Σ pseudo-invariant if and only if the power series expansion of $h_\alpha \circ \theta$ on W' has vanishing coefficients. It follows that this is the case if and only if $(\zeta_1, \ldots, \zeta_s)$ lies on the common zero locus of a family of holomorphic functions on B_ε^0. In particular, for $\theta \in B_\varepsilon^0 \cap (U(r) \ltimes \mathbb{C}^r) := B_\varepsilon$, θ leaves Σ pseudo-invariant if and only if θ lies on some real-analytic subvariety $S_0 \subset B_\varepsilon$. Let $S \subset S_0$ be the identity component of S_0. Since there exists a distinct sequence $\{\theta_i\}$ of elements in $U(r) \ltimes \mathbb{C}^r$ converging to id such that $\theta_i \in S_0$, the real-analytic subvariety S must be of positive dimension. Furthermore S is a pseudogroup in the sense that for δ sufficiently small, and for $\alpha, \beta \in B_\delta, \alpha \cdot \beta$ and α^{-1}, β^{-1} are in S. It follows that S must be smooth and that there is an associated non-trivial Lie subalgebra $\mathbf{g} \subset \mathbf{u}(r) \ltimes \mathbb{C}^r$. We write $G \subset U(r) \ltimes \mathbb{C}^r$ for the corresponding (not necessarily closed) Lie subgroup.

Suppose $\mathbf{g} \cap (\{0\} \ltimes \mathbb{C}^r) \neq \{0\}$. Then S contains a positive-dimensional pseudogroup of translations. It follows that Σ is pseudo-invariant under a complex pseudogroup of translations. Since Σ is of complex dimension 1 it follows that Σ and hence $F(\tilde{X})$ lies on some affine line of $\mathbb{C}^r$, contradicting with the assumption that $\{\tilde{\omega}_1, \ldots, \tilde{\omega}_r\}$ are linearly independent.

It remains to prove (ii), i.e., that $\hat{\Phi}(\pi_1(X))$ normalizes G. For this it suffices to show that, given $\hat{\nu} \in \hat{\Phi}(\pi_1(X))$, there exists $\delta > 0$ such that $\hat{\nu}^{-1}(S \cap B_\delta)\hat{\nu} \subset S \subset W$. We are going to prove this by the idea of analytic continuation. Fix a point $p \in \Sigma$ and pick some $\tilde{x} \in \tilde{X}$ such that $F(\tilde{x}) = p$. Write $\hat{\nu} = \hat{\Phi}(\gamma)$ for a choice of $\gamma \in \pi_1(X)$. Let $U \subset\subset \tilde{X}$ be a connected relatively compact open subset of $\tilde{X}$ containing both $\tilde{x}$ and $\gamma(\tilde{x})$. $F(U) \subset\subset \mathbb{C}^r$ is a connected locally closed relatively compact 1-dimensional complex subspace. We may further assume that Σ and U have been chosen so that there exists an open neighborhood D of $\tilde{x}$ in U with the property that $F|_D$ is a submersion of D onto Σ. This implies also that $F|_{\gamma(D)}$ is a submersion of $\gamma(D)$ onto $\hat{\nu}(\Sigma)$. $F(U)$ is the image of some irreducible 1-dimensional complex space T under a holomorphic immersion $\tau : T \to \mathbb{C}^r$. Write $\tau(p_0) = p$, $\tau(q_0) = \hat{\Phi}(\gamma)(p) = \hat{\nu}(p)$, where p_0 and q_0 are chosen such that τ maps the germ of T at p_0 to the germ of Σ at p; and it maps the germ of T at q_0 to the germ of $\hat{\nu}(\Sigma)$ at $\hat{\nu}(p)$. Consider $F(U)$ as a closed complex-analytic subvariety of some open subset V of $\mathbb{C}^r$. Choose now $T_0 \subset\subset T$ to be a relatively compact open subset containing both p_0 and q_0. With $\hat{\nu}$ and T_0 chosen there exists some $\varepsilon > 0$ such that for any $\theta \in B_\varepsilon, \theta(\tau(T_0)) \subset V$. For $\theta \in S \cap B_\varepsilon, \theta(\tau(T_0)) \cap \tau(T_0) \supset \theta(\Sigma) \cap \Sigma$, which is a non-empty open subset of Σ. From the Identity Theorem for holomorphic functions it follows that $\theta(\tau(T_0)) \subset \tau(T) = F(U) \subset V$. For $\varepsilon > 0$ sufficiently small this forces $\theta(\hat{\nu}(\Sigma)) \cap \hat{\nu}(\Sigma)$ to be a non-empty open subset whenever $\theta \in S \cap B_\varepsilon$. It follows that

$$(\hat{\nu}^{-1}\theta\hat{\nu})\Sigma \cap \Sigma = \hat{\nu}^{-1}[(\theta(\hat{\nu}(\Sigma)) \cap \hat{\nu}(\Sigma)]$$

is a non-empty open subset of Σ. In other words, Σ is pseudo-invariant under $\hat{\nu}^{-1}\theta\hat{\nu}$ for any $\theta \in S \cap B_\varepsilon$. From the definition of S it follows readily that $\hat{\nu}$ lies in the normalizer of the Lie algebra $\mathbf{g}$ of the pseudogroup S and hence that $\hat{\nu}$ normalizes the Lie subgroup $G \subset U(r) \ltimes \mathbb{C}^r$, as asserted in (ii). The proof of Proposition 1 is completed. □

2. Contradiction arising from the existence of a non-trivial pseudogroup of isometries

(2.1) We proceed to derive a contradiction from the conclusion of (1.3), Proposition 1. Either $G \subset U(r) \ltimes \mathbb{C}^r$ is semisimple or it has a non-trivial radical. In the first case G is semisimple and of the compact type. In the second case we note that by (ii) of Proposition 1, $\hat{\nu}^{-1} R \hat{\nu} \subset G$ is a solvable normal subgroup. From the maximality of the radical among solvable normal subgroups it follows readily (arguing at the level of ideals of $\mathbf{g}$) that $\hat{\nu}^{-1} R \hat{\nu} = R$, so that replacing G by R we may in the second case assume that G is solvable. For G solvable verifying the conclusions of Proposition 1, its commutator $[G, G]$ also verifies the same conclusions, provided that it is of positive dimension. It follows by considering the derived series of G that there exists a connected abelian subgroup $A \subset G$ of positive dimension for which Proposition 1 holds with G replaced by A. Thus, we have seen that in Proposition 1 we may assume that either (Case I) $G \subset U(r) \ltimes \mathbb{C}^r$ is compact and semisimple or (Case II) G is the homomorphic image of some $\mathbb{R}^k$ such that the kernel is discrete.

Our objective is to analyze in both cases the structure of G. This will allow us to draw conclusion about the subgroup $\hat{\Phi}(\pi_1(X)) \subset U(r) \ltimes \mathbb{C}^r$, finally producing a contradiction to the linear independence of $\{\tilde{\omega}_1, \ldots, \tilde{\omega}_r\}$ on $\tilde{X}$ via the Maximum Principle for plurisubharmonic functions.

(2.2) In Case I, the semisimple Lie group $G \subset U(r) \ltimes \mathbb{C}^r$ is a compact Lie subgroup. Since $G \cap (\{\mathrm{id}\} \ltimes \mathbb{C}^r)$ is a compact subgroup of $\mathbb{C}^r$, it must be trivial, so that the projection map $U(r) \ltimes \mathbb{C}^r \mapsto U(r)$ induces an isomorphism of G onto its image, which we denote by G_0. Identifying $U(r) \ltimes \mathbb{C}^r$ with $U(r) \times \mathbb{C}^r$ as manifolds in the usual way, we have the composition rule for $\hat{U}_i(p) = U_i(p) + \tau_i$; $i = 1, 2$; given by

$$\hat{U}_1(\hat{U}_2(p)) = U_1(U_2(p) + \tau_2) + \tau_1 = (U_1 U_2)(p) + U_1 \tau_2 + \tau_1 ,$$

so that under the identification of $U(r) \ltimes \mathbb{C}^r$ with $U(r) \times \mathbb{C}^r$ we may write

$$(U_1, \tau_1) \cdot (U_2, \tau_2) = (U_1 U_2, U_1 \tau_2 + \tau_1) .$$

Denote by E the complex vector space $\mathbb{C}^r$ considered as a G_0-representation space via the inclusion $G_0 \subset U(r)$ and the standard representation of $U(r)$ on $\mathbb{C}^r$. Write $G = \{\hat{g} = (g, c(g)) : g \in G_0\}$. As is well-known the mapping $c : G_0 \mapsto E$ defines a smooth 1-cocycle in $\tilde{Z}^1(G_0, E)$. In fact, by the composition law

$$c(g_1 g_2) = g_1 c(g_2) + c(g_1)$$

for any $g_1, g_2 \in G_0$. As G_0 is compact, it satisfies Kazhdan's Property (T) (cf. Zimmer [Z, proof of Corollary 7.1.9, p. 132]) so that c is in fact a 1-coboundary, i.e., there exists some $c \in \mathbb{C}^r$ such that for any $g \in G_0$

$$c(g) = c - g(c) .$$

In other words, for $\hat{g} \in G$, $\hat{g}(p) = g(p) + c(g)$ we may write

$$\begin{aligned}\hat{g}(p+c) &= g(p) + c(g) + g(c)\\ &= g(p) + (c - g(c)) + g(c)\\ &= g(p) + c\end{aligned}$$

so that, shifting the origin of $\mathbb{C}^r$ to c the group $G \subset U(r) \ltimes \mathbb{C}^r$ is a group of unitary transformations on $\mathbb{C}^r$. In other words, G is conjugate to a subgroup of $U(r)$ via a translation of $\mathbb{C}^r$.

The proof of the vanishing of $H^1(G_0, E)$ as given in Zimmer [Z] involves knowledge of Kazhdan's Property (T). For the convenience of the reader we include here a self-contained proof as follows.

Lemma 1. *The* 1*-cocycle* $c : G_0 \to E$ *in* $\tilde{Z}^1(G_0, E)$ *is a* 1*-coboundary, i.e., there exists some* $c \in \mathbb{C}^r$ *such that*

$$c(g) = c - g(c)\ .$$

Proof. Decompose $E = E' \oplus E''$ into an orthogonal direct sum such that E'' is fixed pointwise by G_0 and such that there is no non-zero G_0-invariant vector in E'. $c : G_0 \to E$ decomposes accordingly as $c = c' + c''$, where $c'' : G_0 \to E''$ is a homomorphism. Since G_0 is compact, $c'' \equiv 0$. Thus, for the proof of Lemma 1 we may assume without loss of generality that $E = E'$, i.e., that the only vector fixed by G_0 is 0. Let μ be the Haar measure on G_0 such that $\mu(G_0) = 1$. If there exists $c \in E$ (i.e. $\mathbb{C}^r$) such that $c(g) = c - g(c)$, then

$$\int_{G_0} c(g)d\mu = c\mu(G_0) - \int_{G_0} g(c)d\mu = c\ ,$$

where $\int_{G_0} g(c)d\mu = 0$ follows from the invariance of the integral under the action of any $h \in G_0$. It remains to verify that the candidate c as given indeed satisfies $c(h) = c - h(c)$ for any $h \in G_0$. In fact, since μ is invariant under left multiplication and c is a 1-cocycle,

$$\begin{aligned}c - h(c) &= \int_{G_0} [c(g) - h(c(g))]d\mu(g)\\ &= \int_{G_0} [c(hg) - h(c(g))]d\mu(g)\\ &= \int_{G_0} c(h)d\mu(g) = c(h)\ ,\end{aligned}$$

as desired. The proof of Lemma 1 is completed. □

(2.3) In Case II, $G \subset U(r) \ltimes \mathbb{C}^r$ is the image of some $\mathbb{R}^k$, $k \geq 1$, under a homomorpism $\alpha : \mathbb{R}^k \to U(r) \ltimes \mathbb{C}^r$ with discrete fiber. The interesting case is when $\alpha(\mathbb{R}^k)$ is non-compact, otherwise the discussion in Case I applies. We write $\alpha(t)$ as $\hat{U}_t$ defined by

$$\hat{U}_t(p) = U_t(p) + c_t \ .$$

We write E for the Euclidean space $\mathbb{C}^r$, considered as an $\mathbb{R}^k$-representation space via the composition of $\alpha : \mathbb{R}^k \to U(r) \ltimes \mathbb{C}^r$ and the projection $U(r) \ltimes \mathbb{C}^r \to U(r)$. As in Case I we are interested in finding some c such that

$$c_t = c - U_t c$$

for any $t \in \mathbb{R}^k$. Suppose there exists some $s \in \mathbb{R}^k$ such that $1 \notin \mathrm{Spec}(U_s)$, the spectrum of U_s. Then $I - U_s$ is invertible (I denoting the identity transformation, so that for this particular s, $c_s = c - U_s c$ is solvable with the unique solution $c = (I - U_s)^{-1} c_s$. Since $\mathbb{R}^k$ is commutative $U_s U_t = U_t U_s$ for any $t \in \mathbb{R}^k$, so that

$$(I - U_s)(I - U_t)c = (I - U_t)(I - U_s)c = (I - U_t)c_s = c_s - U_t c_s \ .$$

Since $c : \mathbb{R}^k \to E$ defined by $c(t) = c_t$ is a smooth 1-cocycle in $Z^1(\mathbb{R}^k, E)$ we have

$$c_{t+s} - U_t c_s - c_t = 0 = c_{s+t} - U_s c_t - c_s \ ,$$

so that

$$c_s - U_t c_s = c_t - U_s c_t \ .$$

Hence

$$(I - U_s)[(I - U_t)c] = c_t - U_s c_t = (I - U_s)c_t \ , \quad \text{i.e.}$$
$$(I - U_t)c = c_t \ , \quad \text{i.e.} \quad c - U_t c = c_t$$

for all $t \in \mathbb{R}^k$. In other words, $\{c_t : t \in \mathbb{R}^k\}$ constitutes a 1-coboundary, and the Lie subgroup $G \subset U(r) \ltimes \mathbb{C}^r$ is conjugate to a (not necessarily closed) subgroup of $U(r)$ via a translation.

It remains to examine the case where $1 \in \mathrm{Spec}(U_t)$ for any $t \in \mathbb{R}^k$. First of all, since $\{U_t : t \in \mathbb{R}^k\}$ is a commutative family of unitary transformations, they can be simultaneously diagonalized. Thus $E = \oplus E_{\rho_i}$ where $\rho_i : \mathbb{R}^k \to S^1 = \{\zeta \in \mathbb{C} : |\zeta| = 1\}$ is a homomorphism of abelian groups, and E_{ρ_i} is a simultaneous eigenspace for all U_t, $t \in \mathbb{R}^k$, with eigenvalue $\rho_i(t)$. The possibility for $1 \in \mathrm{Spec}(U_t)$ for all $t \in \mathbb{R}^k$ arises only if for some i, $\rho_i(t) = 1$ for all $t \in \mathbb{R}^k$. In this case we decompose $E = E' \oplus E''$ as an orthogonal direct sum, such that E', E'' are both invariant under $\{U_t : t \in \mathbb{R}^k\}$, $U_t|_{E''}$ being the identity map for any $t \in \mathbb{R}^k$, while for some $t \in \mathbb{R}^k$, $1 \notin \mathrm{Spec}(U_t|_{E'})$. It follows by preceding arguments that $G \subset U(r) \ltimes \mathbb{C}^r$ is conjugate via a translation to a group $G' = \alpha'(\mathbb{R}^k)$ whose elements $\hat{U}_t \in G'$, $t \in \mathbb{R}^k$ act in the form

$$\hat{U}_t(p' + p'') = (U'_t p', p'' + c_t) \ ,$$

for $p' \in E'$, $p'' \in E''$ and for some $c_t \in E''$.

(2.4) With the group G in Proposition 1 analyzed in (2.2) and (2.3), we are going to contradict the hypothesis in Theorem 1 of the linear independence of $\{\tilde{\omega}_1, \ldots, \tilde{\omega}_r\}$. By Proposition 1, $\hat{\Phi}(\pi_1(X))$ normalizes G. In Case I, we can assume $G \subset U(r)$. Then, given any $g \in G$, and $\hat{\nu} \in \hat{\Phi}(\pi_1(X))$ we have

$$\hat{\nu}^{-1} g \hat{\nu} = h \quad \text{for some } h \in G ,$$

so that

$$g(\hat{\nu}(0)) = \hat{\nu}(h(0)) = \hat{\nu}(0) .$$

Write $E'' \subset E \cong \mathbb{C}^r$ for the subspace fixed pointwise by G. Then,

$$\hat{\nu}(0) \in E'' \quad \text{for any} \quad \hat{\nu} \in \hat{\Phi}(\pi_1(X)) .$$

Consider the orthogonal decomposition $E = E' \oplus E''$, where $E' \neq \{0\}$. $\hat{\nu}$ acts in the form

$$\hat{\nu}(p', p'') = (\nu'(p'), \nu''(p'') + c(\hat{\nu}))$$

for $p' \in E'$, $p'' \in E''$, where ν' resp. ν'' are unitary transformations on E' resp. E'', and $c(\hat{\nu}) \in E''$. For $\gamma \in \pi_1(X)$ with $\hat{\Phi}(\gamma) = \hat{\nu}$ we have

$$F(\gamma\tilde{x}) = \hat{\Phi}(\gamma)F(\tilde{x}) = \nu'(F'(\tilde{x})) + \nu''(F''(\tilde{x})) + c(\hat{\nu}) ,$$

where $F = F' + F''$ with $F' : \tilde{X} \to E'$ and $F'' : \tilde{X} \to E''$. It follows therefore that the function $\tilde{\varphi} : \tilde{X} \to \mathbb{R}$ defined by

$$\tilde{\varphi}(\tilde{x}) = \|F'(\tilde{x})\|^2$$

is invariant under the action of $\pi_1(X)$ and defines hence a continuous function on the compact Kähler manifold X. But as $\tilde{\varphi}$ and hence φ is plurisubharmonic $\varphi : X \to \mathbb{R}$ must be constant, implying that $F' : \tilde{X} \to E'$ is a constant. If we write $\tilde{\omega} = \tilde{\omega}' + \tilde{\omega}''$ according to the decomposition $E = E' \oplus E''$ then obviously $\tilde{\omega}' \equiv 0$, contradicting with the linear independence of $\{\tilde{\omega}_1, \ldots, \tilde{\omega}_r\}$.

In Case II the only difference arises when we have a similar decomposition $E = E' \oplus E''$; $E', E'' \neq 0$; and G can taken to be a commutative group $\{\hat{U}_t : t \in \mathbb{R}^k\}$ of elements of $U(r) \ltimes \mathbb{C}^r$ of the form

$$\hat{U}_t(p', p'') = (U'_t p', p'' + c_t) .$$

Given $\hat{\nu} \in \hat{\Phi}(\pi_1(X))$, we have $\hat{\nu}^{-1} G \hat{\nu} = G$, i.e., $G\hat{\nu} = \hat{\nu}G$. Recall that G_0 is the image of G under the projection $U(r) \ltimes \mathbb{C}^r \mapsto U(r)$. For $\hat{\nu} \in \hat{\Phi}(\pi_1(X))$ we write $\hat{\nu}(p) = \nu(p) + \tau(\hat{\nu})$ for $\nu \in U(r)$ and $\tau(\hat{\nu}) \in \mathbb{C}^r$. Clearly $G\hat{\nu} = \hat{\nu}G$ implies $G_0\nu = \nu G_0$. Given any vector subspace $A \subset E$ invariant under G_0 we have $G_0\nu(A) = \nu(G_0(A)) = \nu(A)$, so that $\nu(A)$ is invariant under G_0. From the definition of the decomposition $E = E' \oplus E''$, E'' represent an isolated point in the space of G_0-invariant vector subspaces of dimension = $\dim_{\mathbb{C}} E''$. The set of such isolated points being a finite subset of some Grassmannian manifold, we conclude that for some normal subgroup $\Pi \subset \hat{\Phi}(\pi_1(X))$ of finite index, $\hat{\nu} \in \Pi$ implies that $\nu(E'') = E''$. For the purpose of proving that $\hat{\Phi}(\pi_1(X))$ acts properly discontinuously on $\tilde{Z} = \tilde{X}/\mathcal{R}$, we may, replacing X by some finite unramified covering space, assume that E'' is ν-invariant for any $\nu = \Phi(\gamma)$, $\gamma \in \pi_1(X)$. Thus any $\hat{\nu} \in \hat{\Phi}(\pi_1(X))$ acts in the form

$$\hat{\nu}(p', p'') = (\nu' p' + \tau', \nu'' p'' + \tau'')$$

for some $\tau' \in E'$; $\tau'' \in E''$. Since $\hat{\nu}$ normalizes G, given $t \in \mathbb{R}^k$ we have $\hat{U}_t \hat{\nu} = \hat{\nu} \hat{U}_s$ for some $s \in \mathbb{R}^k$. We have

$$\begin{aligned}\hat{U}_t \hat{\nu}(p', p'') &= \hat{U}_t(\nu' p' + \tau', \nu'' p'' + \tau'')\\ &= (U'_t \nu' p' + U'_t \tau', \nu'' p'' + \tau'' + c_t)\,;\\ \hat{\nu}\hat{U}_s(p', p'') &= \hat{\nu}(U'_s p', p'' + c_s)\\ &= (\nu' U'_s p' + \tau', \nu'' p'' + \nu'' c_s + \tau'')\,.\end{aligned}$$

Comparing the E'-components we conclude that $U'_t \tau' = \tau'$ for any $t \in \mathbb{R}^k$. But then $\tau' \in E'' \cap E' = \{0\}$, so that $\tau' = 0$ for any $\hat{\nu} \in \hat{\Phi}(\pi_1(X))$. As a consequence, any $\hat{\nu} \in \hat{\Phi}(\pi_1(X))$ acts in the form

$$\hat{\nu}(p', p'') = (\nu' p', \nu'' p'' + \tau'')\,.$$

The function $\tilde{\varphi} : \tilde{X} \mapsto \mathbb{R}$ defined by $\tilde{\varphi}(\tilde{x}) = \|F(\tilde{x})\|^2$ is then still $\pi_1(X)$-invariant, implying as in Case I that $\tilde{\varphi}$ is a constant, contradicting with the linear independence of $\{\tilde{\omega}_1, \dots, \tilde{\omega}_r\}$.

(2.5) We have proven that the conclusion of [(1.3), Proposition 1] contradicts with the hypothesis of Theorem 1. In other words, under the hypothesis of Theorem 1, we have proven that $\pi_1(X)$ acts properly discontinuously on $\tilde{Z}$ via $\Theta : \pi_1(X) \to \mathrm{Aut}(\tilde{Z})$. We proceed to complete the proof of Theorem 1. To define the compact Riemann surface Z we assume for the time being that $\Theta(\pi_1(X)) \subset \mathrm{Aut}(\tilde{Z})$ is torsion-free, so that $\pi_1(X)$ acts properly discontinuously and without fixed points on $\tilde{Z}$. Define Z to be $\tilde{Z}/\Theta(\pi_1(X))$. The holomorphic map $\tilde{X} \to \tilde{Z} = \tilde{X}/\mathcal{R}$ then descends to a holomorphic map $\sigma : X \to Z$. By the Proper Mapping Theorem Z is a compact Riemann surface. Furthermore, by (1.3) we have a holomorphic map $f : \tilde{Z} \to \mathbb{C}^r$ and a homomorphism $\hat{\Psi}_0 : \Theta(\pi_1(X)) \to U(r) \ltimes \mathbb{C}^r$ such that for any $\tilde{z} \in \tilde{Z}$

$$f(\gamma' \tilde{z}) = \hat{\Psi}_0(\gamma')(f(\tilde{z})) = \Psi_0(\gamma')(\tilde{z}) + d_{\gamma'}\,.$$

$\hat{\Psi}_0$ resp. Ψ_0 may be regarded as being defined on $\pi_1(Z)/\pi_1(\tilde{Z})$, where $\pi_1(\tilde{Z}) \subset \pi_1(Z)$ sits as a normal subgroup. Denoting by $\hat{\Psi}$ resp. Ψ the lifted representations on $\pi_1(Z)$, we have obtained $\Psi : \pi_1(Z) \to U(r)$ such that $\Phi(\gamma) = \Psi(\sigma_* \gamma)$ for any $\gamma \in \pi_1(X)$ and such that ∂f defines $\nu \in \Gamma(Z, \Omega(E_\Psi))$ verifying $\sigma^* \nu \equiv \omega$.

We have now to consider the case where $\Theta(\pi_1(X))$ may have torsion elements. However, since $\Phi(\pi_1(X)) \subset U(r) \ltimes \mathbb{C}^r$ is by assumption torsion-free any element $\gamma' \in \Theta(\pi_1(X))$ of finite order must be mapped to 1 under Ψ_0. We consider a point $\tilde{z}_0 \in \tilde{Z}$ which is a fixed point of some non-trivial element $\gamma' \in \Theta(\pi_1(X))$. By the discreteness of $\Theta(\pi_1(X))$ the subgroup fixing $\tilde{z}_0$ is a finite group H. We have $H \subset \mathrm{Ker}\,\Psi_0$.

We can choose a local holomorphic coordinate ζ at $\tilde{z}_0$ such that $\zeta(\tilde{z}_0) = 0$ and such that H acts as a finite cyclic group of rotations, generated by some $h \in H$, with $h(\zeta) = e^{\frac{2\pi i}{m}}\zeta$ in the ζ-coordinate. Endow $Z = \tilde{Z}/\Theta(\pi_1(X))$ with the usual complex structure, so that the canonical map $\eta : \tilde{Z} \mapsto Z$ is holomorphic and ramified exactly at points $\tilde{z}_0 \in \tilde{Z}$ at which the subgroup $H(\tilde{z}_0) \subset \Theta(\pi_1(X))$ fixing $\tilde{z}$ is non-trivial, the ramification order being exactly $\mathrm{card}(H(\tilde{z}_0)) - 1$. We argue that $f : \tilde{Z} \mapsto \mathbb{C}^r$ descends to some twisted holomorphic 1-form on Z. For $H(\tilde{z}_0) = H$ of cardinality m, as $f(h\zeta) = \hat{\Psi}_0(h)f(\zeta) = f(\zeta)$ for all $h \in H$, expanding in power series $f(\zeta)$ must be a vector-valued holomorphic function in ζ^m, so that it descends locally to a vector-valued holomorphic function on some open neighborhood of $\eta(\tilde{z}_0)$ (with local coordinate $\xi = \zeta^m$). By analytic continuation we can define on the universal cover Z' of Z a holomorphic map $g : Z' \to \mathbb{C}^r$ such that for the canonical map $\eta' : \tilde{Z} \to Z'$, $g \circ \eta' \equiv f$. The vector-valued holomorphic 1-form ∂g on Z' then defines an E_Ψ-valued holomorphic 1-form ν on Z for some unitary representation $\Psi : \pi_1(Z) \to U(r)$. For the holomorphic map $F : \tilde{X} \to \mathbb{C}^r$ we have $F \equiv g \circ \pi'$ where $\pi' : \tilde{X} \to Z'$ is the composition of $\pi : \tilde{X} \to \tilde{Z}$ with $\eta' : \tilde{Z} \mapsto Z'$. It follows that for the holomorphic $\sigma : X \to Z$ induced by $\pi' : \tilde{X} \to Z'$ we have $\sigma^*\Psi = \Phi$ and $\sigma^*\nu = \omega$ as asserted. The proof of Theorem 1 is completed.

(2.6) We have introduced holomorphic foliations $\tilde{\mathcal{F}}$ and $\mathcal{F}$ in (1.1). Given a compact Kähler manifold X together with a twisted holomorphic 1-form $\omega \in \Gamma(X, \Omega(E_\Phi))$, E_Φ arising from a unitary representation $\Phi : \pi_1(X) \to U(r)$, we defined $\mathcal{F}$ on $X - V$ for some complex-analytic subvariety V of codimension ≥ 2. At generic points of $X - V$, $\mathcal{F}$ is defined by the distribution $x \mapsto \mathrm{Ker}\,\omega(x)$. In the case where $\mathrm{Ker}\,\omega(x)$ is of corank 1 generically, which is the case under study, our proof of Theorem 1 makes use of holomorphic equivalence relations. Holomorphic foliations only appear implicitly. In fact, on the universal cover $\tilde{X}$, each fiber of $\tilde{X} - \tilde{V} \mapsto \tilde{X}/\mathcal{R}$ is a disjoint union of an at worst countable union of leaves L of $\tilde{\mathcal{F}}$, where each $L \subset \tilde{X} - \tilde{V}$ is closed and each topological closure $\overline{L} \subset \tilde{X}$ is complex-analytic.

In the case where $\mathrm{Ker}\,\omega$ is generically of corank ≥ 2, we encounter difficulties in trying to adapt arguments using holomorphic equivalence relation. The main difficulty is that the corresponding equivalence relation $\mathcal{R} \subset \tilde{X} \times \tilde{X}$ is no longer open. We can delete a $\pi_1(X)$-equivariant subvariety $\tilde{B} \subset \tilde{X}$ such that the restriction of $\mathcal{R}$ to $\tilde{X} - \tilde{B}$ defines an open equivalence relation. We can then form quotient spaces $(\tilde{X} - \tilde{B})/\mathcal{R} = \tilde{Z}_0$. The problem is that, even if $\Theta(\pi_1(X))$, similarly defined as before, acts properly discontinuously on $\tilde{Z}_0$, it is not clear how one can canonically extend $\tilde{Z}_0/\Theta(\pi_1(X)) = Z_0$ to a compact complex space Z, especially if $\tilde{B} \subset \tilde{X}$ is a non-empty hypersurface. Even if $\mathcal{R} \subset \tilde{X} \times \tilde{X}$ can be shown to be open we have the difficulty of obtaining a pseudogroup of automorphisms of $\tilde{Z}_0 = \tilde{X}/\mathcal{R}$ when $\Theta(\pi_1(X))$ does not act properly discontinuously. The difficulty arises when $\tilde{Z}$ is of complex dimension ≥ 2, in which case we do not know whether $\tilde{Z}$ admits a Hermitian metric invariant under automorphisms. To overcome these and other minor difficulties, in the general case we will make use of the holomorphic foliations $\tilde{\mathcal{F}}$ and $\mathcal{F}$ in place of holomorphic equivalence relations $\mathcal{R}$. In Theorem 1 pseudogroups of automorphisms of $\tilde{Z}$ are identified with pseudogroups of holomorphic isometries of $\mathbb{C}^r$, via the holomorphic map $F : \tilde{X} \to \mathbb{C}^r$. This amounts to studying semi-norms on $\tilde{X}$ which are pull-backs of Euclidean norms on $\mathbb{C}^r$. They are $\pi_1(X)$-invariant and can be defined on X by $\|\eta\| = |\omega(\eta)|$ for complexified tangent vectors η of type $(1, 0)$. Such a semi-norm corresponds to a closed semipositive $(1, 1)$-form Ω on X such that $(X, \Omega, \mathcal{F}, V)$ constitutes a semi-Kähler structure as defined in Mok [M2,4]. In place of considering pseudogroups of automorphisms we consider pseudogroups of holomorphic isometries associated to such semi-Kähler structures.

3. Fibrations over varieties of the general type

(3.1) In this section we are going to prove generalizations of [(1.2), Theorem 1]. Consider $0 \neq \omega \in \Gamma(X, \Omega(E_\Phi))$ such that $\operatorname{Ker}\omega$ is of corank k at generic points with $k \geq 1$ and denote by $\tilde{\omega} = (\tilde{\omega}_1, \ldots, \tilde{\omega}_r)$ its lifting to $\tilde{X}$. For any vector $\eta \in \mathbb{C}^r$, $\eta = (\eta^1, \ldots, \eta^r)$ we denote by $\tilde{\omega}_\eta$ the closed holomorphic 1-form $\sum_{i=1}^{r} \eta^i \tilde{\omega}_i$. Our first generalization is given by

Theorem 2. *Let X be a compact Kähler manifold, $\Phi : \pi_1(X) \to U(r)$ be a unitary representation, and $E_\Phi \mapsto X$ be the associated locally flat vector bundle. Let $\omega \in \Gamma(X, \Omega(E_\Phi))$ be such that* $\operatorname{Ker}(\omega)$ *is generically of* corank k, $1 \leq k < r$, *and such that for any k-dimensional complex vector subspace $V_k \subset \mathbb{C}^r$, $V_k = \operatorname{Span}\{\eta_1, \ldots, \eta_k\}$, we have $\tilde{\omega}_{\eta_1} \wedge \cdots \wedge \tilde{\omega}_{\eta_k} \not\equiv 0$. Then, there exists a nonsingular modification $\rho : \hat{X} \to X$, a projective-algebraic manifold Z of the general type, a unitary representation $\Psi : \pi_1(Z) \mapsto U(r)$ and a holomorphic map $\sigma : \hat{X} \to Z$ with connected fibers such that*

(a) $\Phi(\gamma) = \Psi(\sigma_*(\gamma))$ *for any $\gamma \in \pi_1(\hat{X}) \cong \pi_1(X)$, i.e. $\rho^* E_\Phi = \sigma^* E_\Psi$;*

(b) *there exists some $\nu \in \Gamma(Z, \Omega(E_\Psi))$ such that $\sigma^*\nu \equiv \rho^*\omega$.*

We note that a nonsingular modification $\rho : \hat{X} \to X$ induces an isomorphism $\rho_* : \pi_1(\hat{X}) \cong \pi_1(X)$ on fundamental groups. Theorem 2 is non-trivial even in the case when $k = \dim_{\mathbb{C}} X$. In this case $\hat{X} \cong Z$, $\Psi \equiv \Phi$, and $\sigma : \hat{X} \to Z$ is a biholomorphism. The assertion is that Z and hence X is of the general type. Finally, we note that in Theorem 2 we assert that fibers of $\sigma : \hat{X} \to Z$ are connected, which is not necessarily the case in the construction leading to Theorem 1.

Remarks. In [C, Theorem (1.14), p. 269)] Catanese proved a generalized theorem of Castelnuovo–de Franchis for the trivial representation. A twisted version [C, Theorem (1.18), p. 271] was stated for $\omega_i \in \Gamma(X, \Omega(\mathcal{L}_i))$, where $\mathcal{L}_i$ are holomorphic line bundles belonging to $\operatorname{Pic}^0(X)$. Equivalently, this means that each $\mathcal{L}_i$ arises from a unitary representation $\Phi_i : \pi_1(X) \to U(1) \cong S^1$, the unit circle. The proof of these theorems were based on showing that the leaves L of the holomorphic foliations $\mathcal{F}$ on $X - V$ are closed such that topological closures $\overline{L}$ are complex-analytic. This was done by identifying L with irreducible components of level sets of meromorphic functions. However, in the twisted version there is the difficulty that such functions may not be well-defined on X. In the case where Z is 1-dimensional one considers $\tilde{\omega}_i/\tilde{\omega}_j$ on $\tilde{X}$, which are $\pi_1(X)$-invariant only in the special case when Φ_i and Φ_j are isomorphic. For $\mathcal{L}_i$ and $\mathcal{L}_j$ non-isomorphic $d \log \left(\frac{\tilde{\omega}_i}{\tilde{\omega}_j}\right)$ is indeed $\pi_1(X)$-invariant and corresponds to a meromorphic 1-form μ on X. Unfortunately μ is not the differential of a meromorphic function on X. Arguments using pseudogroups of holomorphic isometries are therefore needed except in the case of scalar representations $\Phi : \pi_1(X) \to U(1) \cdot I_r \subset U(r)$.

For non-trivial unitary representations we have the following consequence of the proof of Theorem 2.

Theorem 3. *Let X be a compact Kähler manifold, $\Phi : \pi_1(X) \to U(r)$ be a non-trivial irreducible unitary representation such that $\Phi(\pi_1(X))$ is torsion-free. Let $E_\Phi \mapsto X$ be the associated locally flat vector bundle. Suppose there exists a non-trivial $\omega \in \Gamma(X, \Omega(E_\Phi))$, then the conclusion of Theorem 2 holds for ω. In particular, X can be meromorphically fibered over a projective-algebraic manifold Z of the general type.*

(3.2) As explained in (2.6), the first difficulty of extending the proof of Theorem 1 to Theorems 2 and 3 is the fact that the corresponding holomorphic equivalence relation $\mathcal{R}$ on $\tilde{X}$ is no longer open. Our principal tool is the notion of a semi-Kähler structure. Details can be found in Mok [M1,3,4].

Definition 1. (Semi-Kähler structures) Let X be a complex manifold. A semi-Kähler structure $(X, \Omega, \mathcal{F}, V)$ consists of

(a) a non-trivial closed positive (1,1)-current Ω on X;

(b) a complex-analytic subvariety $V \subset X$ of codimension ≥ 2, possibly empty, such that Ω is smooth on $X - V$;

(c) a holomorphic foliation $\mathcal{F}$ on $X - V$ such that

(d) the closed semipositive (1,1)-form Ω and $\mathcal{F}$ are compatible on a dense open subset U of $X - V$ in the sense that for any $x \in U$, $\mathrm{Ker}(\Omega(x)) = T_x^{1,0}(\mathcal{F})$.

The subvariety $V \subset X$ will sometimes be left out in the notation for a semi-Kähler structure. $\mathcal{F}$ will also be referred to as a meromorphic foliation. In the situation studied in Theorems 1 and 2, we define $\Omega = \sqrt{-1} \sum_{i=1}^{r} \tilde{\omega}_i \wedge \overline{\tilde{\omega}_i}$ for the lifted closed holomorphic 1-form. Then $d\Omega \equiv 0$ and Ω is invariant under the action of $\pi_1(X)$, since $\Phi : \pi_1(X) \mapsto U(r)$ is a unitary representation. As defined in (1.1) $\tilde{\mathcal{F}}$ will be the meromorphic foliation on $\tilde{X}$ defined by level sets of the holomorphic map $F : \tilde{X} \mapsto \mathbb{C}^r$ with $dF = \tilde{\omega}$. $\tilde{\mathcal{F}}$ descends to a meromorphic foliation $\mathcal{F}$ on X.

For the local structure of semi-Kähler structures we introduce

Definition–Proposition 1. *Let $(X, \Omega, \mathcal{F}, V)$ be a semi-Kähler structure on an m-dimensional complex manifold. We call an open set $U \subset\subset X - V$ a distinguished polydisk if $U \cong D \times D'$ where $D \subset\subset \mathbb{C}^k$, $D' \subset\subset \mathbb{C}^\ell$, $k + \ell = m$, are polydisks centered at 0 such that the leaves of the foliation $\mathcal{F}|_U$ correspond to level sets of the canonical projection $D \times D' \to D$. Then, in terms of the corresponding Euclidean coordinates $(z_1, \ldots, z_k; w_1, \ldots, w_\ell)$ the semi-Kähler form Ω is given by*

$$\Omega = i \sum_{1 \le \alpha, \beta \le k} g_{\alpha\overline{\beta}}(z) dz^\alpha \wedge d\overline{z}^\beta \, .$$

By Definition–Proposition 1 we have an induced quotient closed semipositive $(1, 1)$ form $\overline{\Omega}$ defined on D. Furthermore, $\overline{\Omega}$ is positive definite on a dense open subset of D. The purpose of defining the notion of semi-Kähler structures is to develop sufficient conditions guaranteeing that metrized meromorphic foliations lead at least to meromorphic fibrations. For that purpose we introduce the notion of factorizable semi-Kähler structures, as follows.

Definition–Proposition 2. *For a compact Kähler manifold X we say that a semi-Kähler structure $(X, \Omega, \mathcal{F}, V)$ is factorizable if and only if all leaves L of $\mathcal{F}$ on $X - V$ are closed and the topological closures $\overline{L}$ on X are compact complex-analytic subvarieties of X. In this case there exists a nonsingular modification $\rho : \hat{X} \to X$, a compact Kähler manifold Z and a surjective holomorphic map $\sigma : \hat{X} \to Z$ with generically irreducible fibers such that*

(a) *for all leaves L of $\mathcal{F}$ on $X - V$, $\overline{(\rho^{-1} L)}$ is an irreducible component of a fiber $\sigma^{-1}(z)$, $z \in Z$;*

(b) *$\rho^*\Omega = \sigma^*(\mu)$ for some closed positive $(1, 1)$ form μ on Z which is positive definite on a dense open set.*

The similarity between the approaches using semi-Kähler structures and holomorphic equivalence relations is given by the following result producing pseudogroups of holomorphic isometries.

Proposition 2 (Mok [M1, p. 574]). *Let $(X, \Omega, \mathcal{F}, V)$ be a semi-Kähler structure on a compact complex manifold X. Then, either*

(a) *the semi-Kähler structure is factorizable; or*

(b) *for any distinguished polydisk $U \cong D \times D'$ on $X - V$ such that $\overline{\Omega}$ is positive definite on D, $(D, \overline{\Omega})$ admits a continuous pseudogroup of holomorphic isometries.*

In [M1] in Case (b) we only asserted the existence of some $U \cong D \times D'$ for which (b) is valid. The proof actually yields the stronger statement given here. By abuse of language we will also say, in Case (b) of Proposition 2, that the semi-Kähler structure $(X, \Omega, \mathcal{F})$ admits a continuous pseudogroup of holomorphic isometries.

In our situation where $\overline{\Omega}$ is nothing other than the pull-back of the Euclidean Kähler form on $\mathbb{C}^r$ by $F : \tilde{X} \to \mathbb{C}^r$, in Case (b) of Proposition 2, the pseudogroup can be identified with a pseudogroup of affine isometries in $U(r) \ltimes \mathbb{C}^r$, as in §§1,2. Thus, Proposition 2 allows us to argue as in the case of fibrations over compact Riemann surfaces.

We remark that the crux of the proof of Proposition 2 is the Bishop Extension Theorem on complex-analytic subvarieties. The idea is to measure volumes of leaves of $\mathcal{F}$ on $X - V$. Either leaves are of "bounded recurrence", in which case we prove that most leaves are of bounded volume, thus extendible to complex-analytic subvarieties on X by the Bishop Extension Theorem, or else, they are of "unbounded recurrence", in which case we produce a continuous pseudogroup of holomorphic isometries of $(D, \overline{\Omega})$. See Mok [M1,2] for the precise definitions and the proof of Proposition 2. We note that *a priori* it is possible that every leaf of $\mathcal{F}$ on $X - V$ is closed while $\mathcal{F}$ is of "unbounded recurrence", since for the latter notion we allow the possibility that leaves are recurrent of arbitrarily large finite order on different parts of $X - V$.

(3.3) We consider the semi-Kähler structure $(X, \Omega, \mathcal{F})$ arising from $0 \neq \omega \in \Gamma(X, \Omega(E_\Phi))$, as in the statement of [(3.1), Theorem 1]. The proof of Theorem 2 breaks up into two steps; (I) fibration via the semi-Kähler form Ω and (II) fibration via the Ricci form. From Step (I) we obtain the intermediate conclusion that Theorem 2 holds with the exception that the base space Z is only known to be a compact Kähler manifold. From Step (II) we will deduce that Z is actually of the general type and hence Moišezon. Since Z is also Kähler it is projective-algebraic. For both steps we will argue by contradiction using [(3.2), Proposition 2]. We will use the following simple lemma to contradict the hypothesis of Theorem 2. In what follows we will say that $(X, \Omega, \mathcal{F})$ is pseudo-invariant under a pseudogroup $S \subset U(r) \ltimes \mathbb{C}^r$ if we are in Case (b) of Proposition 2 and the pseudogroup there is naturally identified with S. We have

Lemma 3. *The semi-Kähler structure $(X, \Omega, \mathcal{F})$ arising from the hypothesis of Theorem 2 cannot be pseudo-invariant under a pseudogroup $S \subset \{\mathrm{id}\} \ltimes \mathbb{C}^r$ of translations.*

Proof. . Suppose $(X, \Omega, \mathcal{F})$ is pseudo-invariant under some pseudogroup S of translations. Modulo a unitary transformation of $\mathbb{C}^r$, $(X, \Omega, \mathcal{F})$ is pseudo-invariant under a pseudogroup of translations of the form $(z_1, z_2, \ldots, z_n) \mapsto (z_1 + t; z_1, \ldots, z_n)$,

$|t| < \varepsilon$. Then, $F = (F_1, F')$, where $F' = (F_2, \ldots, F_r) : \tilde{X} \to \mathbb{C}^{r-1}$ is of maximal rank $= k - 1$. In particular $\tilde{\omega}_{i_1} \wedge \cdots \wedge \tilde{\omega}_{i_k} \equiv 0$ whenever $2 \le i_1 < \cdots < i_k \le r$ and the hypothesis of Theorem 2 is violated. □

We proceed with Step (I) of the proof of Theorem 2. By the proof of [(1.2), Theorem 1], whenever $(X, \Omega, \mathcal{F})$ is pseudo-invariant under a pseudogroup S of affine isometries, there must be (by [(1.3), Proposition 1]) a Lie subgroup $G \subset U(r) \ltimes \mathbb{C}^r$ of affine isometries such that either (I) G is semisimple and compact or (II) G is abelian and such that G is normalized by $\hat{\Phi}(\pi_1(X))$, $\hat{\Phi} : \pi_1(X) \to U(r) \ltimes \mathbb{C}^r$ defined by $F : \tilde{X} \to \mathbb{C}^r, dF = \tilde{\omega}$. We note that the arguments as given in (2.2), (2.3) are valid without the assumption that $\operatorname{Ker}\omega$ is generically of corank 1. From the arguments there we can conclude in both cases that $(X, \Omega, \mathcal{F})$ is pseudo-invariant under a group of Euclidean translations, contradicting with Lemma 3. By [(3.2), Proposition 2], the semi-Kähler structure $(X, \Omega, \mathcal{F})$ must be factorizable.

Let $\rho : \hat{X} \to X$ be the nonsingular modification and $\sigma : \hat{X} \to Z$ be the surjective holomorphic map as given in [(3.2), Definition-Proposition 2]. It remains to define a representation $\Psi : \pi_1(Z) \to U(r)$ such that $\Psi \circ \sigma_* = \Phi$, where we have made the identification $\rho_* : \pi_1(\hat{X}) \cong \pi_1(X)$; and to define $\nu \in \Gamma(Z, \Omega(E_\Psi))$ such that $\sigma^*\nu = \rho^*\omega$. Equivalently, we need to prove the existence of a holomorphic map $f : \tilde{Z} \to \mathbb{C}^r$ such that for the lifted map $\tilde{\sigma} : \tilde{\hat{X}} \to \tilde{Z}$ on universal covers $\rho^*F : \tilde{\hat{X}} \to \mathbb{C}^r$ agrees with $f \circ \tilde{\sigma}$.

There is some complex-analytic subvariety $B \subset Z$, such that for $p \in Z - B$, the fiber C_p of $\sigma : \hat{X} \to Z$ over p is irreducible, smooth and reduced. From Definition-Proposition 2, ρ^*F is constant on every connected component of a fiber of $\tilde{\sigma} : \tilde{\hat{X}} \to \tilde{Z}$. Denote by $\pi : \tilde{Z} \to Z$ the universal covering map and choose a simply-connected coordinate chart U on $\tilde{Z} - \pi^{-1}(B)$. Pick a choice of a holomorphic $f : U \to \mathbb{C}^r$ so that the value of $f(\tilde{Z})$ agrees with the value of ρ^*F on some branch of $\tilde{\sigma}^{-1}(\tilde{Z})$ (continuously varying with $\tilde{Z}$). We can extend the domain of definition of f on $\tilde{Z} - \pi^{-1}(B)$ by analytic continuation. The only trouble is that it can become multi-valued as $\tilde{Z} - \pi^{-1}(B)$ is not necessarily simply-connected. To define f it suffices to analytically continue f across each codimension-1 irreducible component of $\pi^{-1}(B)$. Then $f : \tilde{Z} \to \mathbb{C}^r$ can be consistently defined as $\tilde{Z} - A$ is simply-connected for any complex-analytic subvariety A of codimension ≥ 2.

For every $p \in Z - B$ the representation $\Phi : \pi_1(X) \to U(r)$ is trivial on the image of $\pi_1(C_p)$ in $\pi_1(\hat{X}) \cong \pi_1(X)$. Thus $\hat{\Phi}$ induces a representation $\hat{\Psi}_0 : \pi_1(Z - B) \to U(r)$. For each codimension-1 irreducible component B_j of B, $1 \le j \le m$, pick a regular point $p_j \in B_j$ and a holomorphic disk $h_j : \Delta \to Z$, $h_j(0) = p_j$, transverse to B_j. We may choose h_j admitting a multivalent lifting to $\hat{X}$. In other words, there exists an integer $k_j \ge 1$, a holomorphic map $g_j : \Delta \to \hat{X}$, such that

$$\sigma(g_j(0)) = p_j; \quad \sigma(g_j(\zeta)) \notin B \quad \text{for } \zeta \neq 0; \quad \text{and} \quad \sigma(g_j(\zeta)) = h_j(\zeta^{k_j}).$$

Denote by $\mu_j \in \pi_1(Z - B)$ the loop represented by $h_j(\partial\Delta(\frac{1}{2}))$. The argument by analytic continuation leads to a globally defined $f : \tilde{Z} \to U(r)$ if $\hat{\Psi}_0(\mu_j) = \mathrm{id}$ for $1 \leq j \leq m$. Since $\mu_j^{k_j}$ can be lifted via g_j to a loop in $\pi_1(\hat{X})$ it follows that $\hat{\Psi}_0(\mu_j) \in \hat{\Phi}(\pi_1(X))$ must be of finite order. By assumption $\Phi(\pi_1(X))$ and hence $\hat{\Phi}(\pi_1(X))$ is torsion-free. It follows that $\hat{\Psi}_0(\mu_j) = \mathrm{id}$. This justifies the existence of $f : \tilde{Z} \to \mathbb{C}^r$ by analytic continuation and completes the proof of [(3.1), Theorem 2], except for the statement that Z is of the general type.

(3.4) To prove that the base space Z of [(3.1), Theorem 2] is of the general type, we use another semi-Kähler structure, that defined by the Ricci curvature, as in Mok [M1, §6, p. 591ff]. There we dealt with harmonic maps into Riemannian locally symmetric spaces of seminegative sectional curvature. As the arguments here are essentially the same, except that it is simpler, we will only indicate the general scheme of proof and refer the reader to [M1] for details. We note the difference that in [M1] we considered discrete representations whereas here $\hat{\Psi} : \pi_1(X) \to U(r) \ltimes \mathbb{C}^r$ may not be discrete.

The twisted holomorphic 1-form $\nu \in \Gamma(Z, \Omega(E_\Psi))$ defines a closed semipositive $(1,1)$-form θ on Z. By the way that Z is defined via the foliation $\mathcal{F}$, θ arises from a holomorphic map $f : \tilde{Z} \mapsto \mathbb{C}^r$ of maximal rank $k = \dim_{\mathbb{C}} Z$, so that θ is positive definite on a non-trivial Zariski open subset of Z. Where $\theta > 0$, Z can be locally identified with a complex submanifold of $\mathbb{C}^r$, and as such holomorphic bisectional curvatures are seminegative, by the Gauss formula. In particular, where $\theta > 0$ Ricci curvatures are seminegative. If the Ricci curvature form ρ is negative definite at some point, then by the solution to the Kähler case of the Grauert-Riemenschneider Conjecture Z is of the general type (cf. [M1, (8.2), p. 609ff]). If $\mathrm{Ker}\,\rho$ is non-trivial where defined, then there is a compatible meromorphic foliation $\mathcal{E}$ such that $(Z, -\rho, \mathcal{E})$ defines a semi-Kähler structure. While the proof in [M1] applies there is a simple argument in the present special case. Write $f : \tilde{Z} \to \mathbb{C}^r$ as $f = (f_1, \ldots, f_r)$. We denote by $\tilde{\theta}$ the lifting of θ to $\tilde{Z}$, and by $(\tilde{Z}, -\tilde{\rho}, \tilde{\mathcal{E}})$ the lifted semi-Kähler structure. The volume form of $(\tilde{Z}, \tilde{\theta})$ is given by

$$\tilde{\theta}^k = \Big(i \sum_{j=1}^{r} \partial f_j \wedge \overline{\partial f_j}\Big)^k = k! \cdot i^{k^2} {\sum}' (\partial f_{j_1} \wedge \cdots \wedge \partial f_{j_k}) \wedge (\overline{\partial f_{j_1} \wedge \cdots \wedge \partial f_{j_k}}) \,,$$

where Σ' is the summation over $(j_1, \ldots, j_k)$ with $1 \leq j_1 < \cdots < j_k \leq r$. Define $g : \tilde{Z} \to \mathbb{P}^{N-1}$, $N = \binom{r}{k}$, $g = [g_1, \ldots, g_N]$, where each g_i is given by $\partial f_{j_1} \wedge \cdots \wedge \partial f_{j_k} = g_i d\zeta^1 \wedge \cdots \wedge d\zeta^k$, $1 \leq j_1 < \cdots < j_k \leq r$, $(\zeta_1, \ldots, \zeta_k)$ denoting a system of local holomorphic coordinates on Z, then

$$\tilde{\rho} = i\partial\bar{\partial} \log \Big(\sum_{i=1}^{N} |g_i|^2\Big) \,,$$

which is nothing other than the pull-back by g of the Kähler form α of the Fubini–Study metric. The lifted meromorphic foliation $\tilde{\mathcal{E}}$ on $\tilde{Z}$ is thus simply defined by level sets of the meromorphic map $g : \tilde{Z} \to \mathbb{P}^{N-1}$. The leaves of $\tilde{\mathcal{E}}$ on $\tilde{Z}$ defined by the vanishing of the Ricci form $\tilde{\rho}$, have to be in particular of zero holomorphic sectional curvature. It follows that at generic points of $\tilde{Z}$ leaves of $\tilde{\mathcal{E}}$ are defined by affine subspaces $\Sigma \subset \mathbb{C}^r$, say of dimension p.

We argue that the semi-Kähler structure $(Z, -\rho, \mathcal{E})$ is factorizable. Arguing by contradiction using [(3.3), Lemma 3] it suffices to show that when we apply [(3.2), Proposition 2] to $(Z, -\rho, \mathcal{E})$, whenever there is a pseudogroup G as in Case (b) we can always take G to be a pseudogroup of Euclidean translations. To this end we complete p-dimensional affine subspaces Σ of $\mathbb{C}^r$ to p-dimensional projective subspaces of $\mathbb{P}^r$, as the standard compactification of $\mathbb{C}^r$. Then, $\pi_1(Z)$ acts on the Grassmannian $\mathcal{G} = Gr(p+1, r-p)$ of $(p+1)$-planes in $\mathbb{C}^{r+1}$ via the representation $\hat{\Psi} : \pi_1(Z) \to U(r) \ltimes \mathbb{C}^r$. Thus, $U(r) \ltimes \mathbb{C}^r$ embeds into $\mathbb{P}GL(r+1)$ and acts as automorphisms of the Grassmannian $\mathcal{G}$. Assume now that $(Z, -\rho, \mathcal{E})$ is of "unbounded recurrence". Choose a distinguished polydisk $U \cong D \times D'$ adapted to $(\tilde{Z}, -\tilde{\rho}, \tilde{\mathcal{E}})$. As $\tilde{z} \in U$ moves we get a $(k-p)$-dimensional family of affine p-planes in $\mathbb{C}^r$. We may choose U so that this family of affine p-planes yields a holomorphic embedding of D into $\mathcal{G}$, with image some $(k-p)$-dimensional locally closed complex submanifold $S \subset \mathcal{G}$. Analogous to [M1, (7.3), Proposition 3] there is a sequence of distinct elements $\{\hat{\theta}_i\} \subset \hat{\Psi}(\pi_1(Z)) \subset U(r) \ltimes \mathbb{C}^r \subset \mathbb{P}GL(r+1)$ such that $\hat{\theta}_i$ leaves S pseudo-invariant, where $\hat{\theta}_i$ acts on the Grassmannian $\mathcal{G}$ according to the standard action of $\mathbb{P}GL(r+1)$. Furthermore $\hat{\theta}_i$ acts as holomorphic isometries when S is given the Kähler metric α induced by $(D, -\overline{\rho})$.

We claim that $\{\hat{\theta}_i\}$ is a discrete sequence in $U(r) \ltimes \mathbb{C}^r$. Choose a non-empty open subset $U \subset \tilde{Z}$ such that $f|_U : U \to \mathbb{C}^r$ is a holomorphic embedding onto some locally closed complex submanifold Q. If $\{\hat{\theta}_i\}$ contains a bounded subsequence we would be able to produce a continuous pseudogroup of holomorphic isometries of $(Q, \beta|_Q)$, where β denotes the Kähler form of the Euclidean metric. From the arguments of (2.2) and (2.3) we would be able to conclude that Q is pseudo-invariant under a pseudogroup of Euclidean translations, leading to a contradiction of the non-degeneracy assumption on ω in [(3.1), Theorem 2]. We have thus proved the claim that $\{\hat{\theta}_i\}$ is discrete. We note that the proof of discreteness of $\{\hat{\theta}_i\}$ is the only essential difference from that of Mok [M1, §6, p. 591ff.], where the corresponding sequence is discrete by assumption, arising from a discrete representation of the fundamental group.

On the one hand $\{\hat{\theta}_i\} \subset U(r) \ltimes \mathbb{C}^r$ is discrete. On the other hand $\hat{\theta}_i$ act as holomorphic isometries of (S, α), $S \subset \mathcal{G}$ some locally closed complex submanifold. (We recall here that α is not the restriction of some canonical metric on the Grassmannian $\mathcal{G}$, but is induced by $(D, -\overline{\rho})$.). Intuitively this can only occur if there are automorphisms of $\mathcal{G}$ fixing S pointwise and dilating in directions normal to S. We have in fact the following analogue of Mok [M1, (7.6), Proposition 6].

Lemma 4. *There exists a positive-dimensional noncompact algebraic subgroup $R \subset U(r) \ltimes \mathbb{C}^r$ and a projective-algberaic subvariety $T \subset \mathcal{G}$ containing S such that*

(a) *R keeps T invariant as a set,*

(b) *there exists an infinite discrete subset $\{\hat{\theta}_i\}$ of $U(r) \ltimes \mathbb{C}^r$ such that $\{\hat{\theta}_i|_T\}$ constitutes a bounded family of biholomorphisms, and*

(c) *as a consequence there is a noncompact algebraic subgroup $H \subset R$ such that $h(p) = p$ for any $p \in T$.*

Recall that we want to show that $(Z, -\rho, \mathcal{E})$ is pseudo-invariant under a pseudogroup of Euclidean translations, assuming that it is of "unbounded recurrence". An element $h \in H$ fixes each affine p-plane Σ_p represented by some point p of S. As H is noncompact this readily forces H to contain a pseudogroup of Euclidean translations, by the arguments of (2.2) and (2.3). With this we have proven by contradiction that $(Z, -\rho, \mathcal{E})$ is factorizable. To prove [(3.1), Theorem 2] it remains to show that even this possibility leads to a contradiction (unless of course if $\mathcal{E}$ is trivial, so that the Ricci form ρ is generically negative on (Z, θ)). Suppose $\mathcal{E}$ is non-trivial and $(Z, -\rho, \mathcal{E})$ is factorizable. Then the affine subspaces Σ are of dimension $p > 0$ and there exists a subgroup $\Pi \subset \hat{\Psi}(\pi_1(X))$ keeping each of the generic affine subspaces Σ invariant, and such that $\Pi|_\Sigma$ acts as a discrete group of translations so that Σ/Π is a compact complex torus. From this it follows readily that the affine subspaces Σ must be parallel to each other, so that $(Z, -\rho, \mathcal{E})$ is pseudo-invariant under a pseudogroup of Euclidean translations, contradicting with the non-degeneracy assumption on ω in Theorem 2. With this we have completed the proof of [(3.1), Theorem 2]. □

Finally, [(3.1), Theorem 3] is an immediate consequence of the proof of Theorem 2. In fact, as in (2.2) and (2.3), whenever there is a pseudogroup of Euclidean translations there is a non-trivial complex vector subspace $E'' \subset \mathbb{C}^r$ such that $\Psi(\pi_1(X))|_{E''}$ is trivial. In particular $\Psi : \pi_1(X) \to U(r)$ is reducible, contradicting with the hypothesis of Theorem 3.

4. Counter-example for infinite-dimensional unitary representations

(4.1) In this section we give a counter-example to the analogue of Theorem 1 for infinite-dimensional unitary representations. The key point is that the group of unitary transformations of an infinite-dimensional Hilbert space is noncompact and Lie groups G as defined analogously as in [(1.3), Proposition 1] can be noncompact, semisimple and violating Property (T) of Kazhdan. In our example G will be $SU(1,1)/\{\pm I\} \cong SL(2,\mathbb{R})/\{\pm I\}$.

Let Δ be the unit disk, $\Gamma \subset (\mathrm{Aut}(\Delta))^2 = (SU(1,1)/\{\pm I\})^2$ be an irreducible torsion-free discrete subgroup. It follows from the Borel Density Lemma [B] that the projection of Γ into each of the direct factors $\mathrm{Aut}(\Delta)$ has a dense image. Let X be the irreducible quotient manifold Δ^2/Γ and H be the Hilbert space of square-integrable conjugate-holomorphic (0, 1)-forms on the first factor of Δ^2. Here the inner product is the natural one given by $\langle\overline{\alpha}, \overline{\beta}\rangle = \overline{i\int_\Delta \alpha\wedge\overline{\beta}}$ for square-integrable holomorphic 1-forms α, β on Δ. We have a unitary representation $\Theta : \mathrm{Aut}(\Delta) \to U(H)$, the group of unitary transformations of H, defined by $\Theta(\mu)(\overline{\alpha}) = \mu^*\overline{\alpha}$. Θ is a unitary representation since the natural inner product $\langle\cdot,\cdot\rangle$ on H is invariant under automorphisms. Let $\Phi : \Gamma \mapsto U(H)$ be defined by

$$\Phi(\gamma)(\overline{\alpha}) = \Theta(\gamma_1)(\overline{\alpha}) = \gamma_1^*\overline{\alpha}$$

for any $\gamma = (\gamma_1, \gamma_2) \in \Gamma \subset (\mathrm{Aut}(\Delta))^2$ and for any $\overline{\alpha} \in H$. Recall that the Bergman kernel form $K(z, \overline{w})dz \wedge d\overline{w}$ is invariant under automorphisms. For $z \in \Delta$ fixed and $K_z(\overline{w}) = K(z, \overline{w})$, $K_z(\overline{w})d\overline{w}$ is conjugate holomorphic and square-integrable. $K(z, \overline{w})dz \wedge d\overline{w}$ may thus be regarded as an $\mathrm{Aut}(\Delta)$-invariant closed H-valued holomorphic 1-form $\tilde{\omega}_0$ on Δ, where $\mathrm{Aut}(\Delta)$ acts on H via Θ. The pull-back to Δ^2 via the canonical projection then defines a Γ-invariant closed H-valued holomorphic 1-form $\tilde{\omega}$ on Δ^2, so that it descends to an E_Φ-valued closed holomorphic 1-form ω on $X = \Delta^2/\Gamma$, where as usual $E_\Phi \to X$ denotes the locally constant bundle of Hilbert spaces arising from $\Phi : \Gamma \to U(H)$. The canonical holomorphic foliation $\mathcal{F}$ defined by $\omega \in \Gamma(X, \Omega(E_\Phi))$ is everywhere defined and corresponds on the universal cover $\tilde{X} = \Delta^2$ to the canonical foliation $\tilde{\mathcal{F}}$ whose leaves are of the form $\{z_1\} \times \Delta$. Since $\Gamma \subset (\mathrm{Aut}(\Delta))^2$ is irreducible by the Borel Density Lemma every leaf of $\mathcal{F}$ on X is dense so that ω does not lead to a holomorphic fibration over a compact Riemann surface. We note here that we may take Γ such that the projection of Γ into each factor $\mathrm{Aut}(\Delta)$ does not contain any elliptic element, so that $\Phi : \Gamma \to U(H)$ is torsion-free, and the hypothesis of [(1.2), Theorem 1] is satisfied except for the finite-dimensionality of the representation space.

(4.2) We examine the analogous Lie group G (as in [(1.3), Proposition 1]) and the analogous plurisubharmonic function $\tilde{\varphi} : \tilde{X} = \Delta^2 \mapsto \mathbb{R}$ as defined in (2.2) and (2.3). We remark first of all that the representation $\Phi : \Gamma \to U(H)$ is irreducible. In particular there are no non-zero Γ-invariant vectors in H.

Let $\{\overline{\alpha}_1, \ldots, \overline{\alpha}_k, \ldots\}$ be an orthonormal basis of H. The Bergman kernel form $K(z, \overline{w})$ on Δ is given by $K(z, \overline{w}) = i\sum_{k=1}^{\infty} \alpha_k(z) \wedge \overline{\alpha_k(w)}$. Take $e_k = \overline{\alpha}_k \in H$; $k = 1, 2, \ldots$; to be basis vectors of H, so that the closed holomorphic 1-form $\tilde{\omega}$ on $\tilde{X} = \Delta^2$ with Euclidean coordinates (z_1, z_2), can be written $i\sum_{k=1}^{\infty}(\pi_1^*\alpha_k)(z_1, z_2)e_k$; where $\pi_1 : \Delta^2 \mapsto \Delta$ is the canonical projection onto the first factor. Let $f_k : \Delta \mapsto \mathbb{C}$ be holomorphic functions defined by $f_k(0) = 0$ and $\partial f_k \equiv \alpha_k$. Define $F_k(z_1, z_2) =$

$f_k(z_1)$. Then, $F = \sum_{k=1}^{\infty} F_k e_k : \tilde{X} = \Delta^2 \to H$ is the primitive of $\tilde{\omega}$ such that $F(0) = 0$. We have for $\tilde{\varphi}(z_1, z_2) = \sum_{k=1}^{\infty} |F_k(z_1, z_2)|^2$

$$i\partial\overline{\partial}\tilde{\varphi} = i\sum_{k=1}^{\infty} \partial F_k \wedge \partial\overline{F}_k = \pi_1^*\Big(i\sum_{k=1}^{\infty} \partial f_k \wedge \partial\overline{f}_k\Big)$$

$$= \pi_1^*\Big(i\sum_{k=1}^{\infty} \alpha_k(z) \wedge \overline{\alpha_k(z)}\Big) = \pi_1^*(K(z,\overline{z})dz \wedge d\overline{z})$$

is therefore up to a constant factor the pull-back of the Kähler form of the Bergman metric ds^2 on Δ. In the terminology of §3, $i\partial\overline{\partial}\tilde{\varphi}$ is the underlying semi-Kähler form $\tilde{\Omega}$ of the lifted semi-Kähler structure $(\tilde{X}, \tilde{\Omega}, \tilde{\mathcal{F}})$. Thus, $F = \pi_1^* f$, $f : \Delta \to H$, $f(z) = \sum_{k=1}^{\infty} f_k e_k$, is up to a constant multiple an isometric immersion of (Δ, ds^2) into the Hilbert space H.

The isometric immersion $f : \Delta \to H$ is in fact a proper embedding. To see this it suffices to verify that f is an embedding for some choice of orthonormal basis of H. We can choose $\alpha_k = \sqrt{\frac{k}{2\pi}} z^{k-1} dz$; $k = 1, 2, 3, \ldots$. Then $f_k = \frac{1}{\sqrt{2\pi k}} z^k$. Define

$$\psi(z) = \sum_{k=1}^{\infty} |f_k(z)|^2 = \frac{1}{2\pi} \sum_{k=1}^{\infty} \frac{|z|^{2k}}{k} = \frac{-1}{2\pi} \log(1 - |z|^2) \text{ on } \Delta\,.$$

Since ψ is an exhaustion function, $f : \Delta \mapsto H$ is a proper holomorphic embedding. We note that ψ is up to a constant factor a potential of ds^2 on Δ. Furthermore, $\tilde{\varphi}(z_1, z_2) = \psi(z_1)$.

Consider the semi-direct product $U(H) \ltimes H$, the group of affine isometries of H. The representation $\Theta : \mathrm{Aut}(\Delta) \to U(H)$, via $f : \Delta \mapsto H$, defines a representation $\hat{\Theta} : \mathrm{Aut}(\Delta) \mapsto U(H) \ltimes H$, given by $\hat{\Theta}(\mu)(\overline{\alpha}) = \Theta(\mu)(\overline{\alpha}) + c_\mu$, where

$$c_\mu = \sum_{k=1}^{\infty} f_k(\mu(0)) e_k \in H$$

Regard H as an $\mathrm{Aut}(\Delta)$-representation space via Θ. Then, $\{c_\mu : \mu \in \mathrm{Aut}(\Delta)\}$ defines a smooth 1-cocycle c in $Z^1(\mathrm{Aut}(\Delta), H)$ which is not a 1-coboundary, otherwise $\hat{\Theta}(\mu)$ would have a common fixed point for all $\mu \in \mathrm{Aut}(\Delta)$ and $\sum_{k=1}^{\infty} |f_k(z)|^2$ would be bounded on Δ. The existence of c is possible since $\mathrm{Aut}(\Delta) = SU(1,1)/\{\pm I\}$ violates Property (T) of Kazhdan.

The group of unitary representation $U(H)$, with the strong topology can be given the structure of a topological group. $\hat{\Theta} : \mathrm{Aut}(\Delta) \mapsto U(H) \ltimes H$ embeds $\mathrm{Aut}(\Delta)$ continuously as a subgroup G of the topological semi-direct product $U(H) \ltimes H$. G is a closed subgroup of affine isometries since $f : \Delta \mapsto H$ is a proper embedding.

Recall the representation $\Phi : \pi_1(X) \to U(H)$ for $X = \Delta^2/\Gamma$ irreducible. As the semi-Kähler structure $(\tilde{X}, \tilde{\Omega}, \tilde{\mathcal{F}})$ is not factorizable, by the same procedure as in the case of finite-dimensional representations, we obtain in this example by the Borel Density Lemma a pseudogroup of affine isometries S of H, which is contained in G. The conclusion of [(1.4), Proposition 1] remains valid for G, i.e. $\pi_1(X)$ normalizes G. As explained this does not lead to a contradiction since $G \cong SU(1,1)/\{\pm I\}$ is non-compact. Nonetheless, as $\pi_1(X)$ normalizes G, this leads to a representation Ξ of $\pi_1(X)$ on the Lie algebra $\mathbf{g}$ of G. In this case $\Xi(\pi_1(X))$ is Zariski-dense in the group G' of inner automorphisms of $\mathbf{g}$.

For the record we summarise the relevant features of the example of irreducible quotients $X = \Delta^2/\Gamma$ of the bidisk, $\Gamma \subset (\mathrm{Aut}(\Delta))^2$, in the following proposition.

Proposition 3. *There exists a compact Kähler manifold X with universal cover $\tilde{X}$, a Hilbert space H, and an irreducible unitary representation $\Phi : \pi_1(X) \to U(H)$, such that, for the corresponding locally constant bundle $E_\Phi \mapsto X$ of Hilbert spaces;*

(a) *there exists a non-zero $\omega \in \Gamma(X, \Omega(E_\Phi))$ with lifting $\tilde{\omega}$ to $\tilde{X}$ and an associated holomorphic map $F : \tilde{X} \mapsto H$, $dF = \tilde{\omega}$;*

(b) *the corresponding semi-Kähler structure $(X, \Omega, \mathcal{F})$ is not factorizable;*

(c) *there is a finite-dimensional noncompact simple Lie group G of affine isometries of H such that for the representation $\hat{\Phi} : \pi_1(X) \mapsto U(H) \ltimes H$ derived from $\Phi : \pi_1(X) \to U(H)$ via $F : \tilde{X} \to H$,*

$$\hat{\nu}^{-1} G \hat{\nu} = G \quad \textit{for any} \quad \hat{\nu} = \hat{\Phi}(\gamma)\,, \ \gamma \in \pi_1(X)\,.$$

As suggested by the example, for the analogue of [(1.3), Proposition 1], if we allow infinite-dimensional unitary representations, we can obtain a Lie group G of affine isometries of H normalized by $\pi_1(X)$, such that G is (I) compact and semisimple, (II) abelian or (III) noncompact and semisimple.

The new feature in the case of infinite-dimensional representations $\Phi : \pi_1(X) \to U(H)$ is the possibility that G is noncompact and semisimple. Under the assumption that $\Phi : \pi_1(X) \to U(H)$ is irreducible, one can show that in Case (III) there is always a Zariski-dense representation of $\pi_1(X)$ into some semisimple Lie group of the noncompact type. In other words, if we cannot prove an analogue of Castelnuovo–de Franchis, we will be led to harmonic maps, for which there are known theorems of fibrations over varieties of the general type, as in Mok [M1]. This is then the starting point of fibration theorems of compact Kähler manifolds over varieties of the general type, for instance in the case when the fundamental group is of subexponential growth, where by Mok [M2] there is always some non-trivial twisted holomorphic 1-forms arising from a possibly infinite-dimensional unitary representation. Details of the fibration theorem will appear in a forthcoming article.

References

[B] Borel, A., Density properties for central subgroups of semisimple groups without compact components, Ann. of Math. 72 (1980), 179–188.

[C] Catanese, F., Moduli and classification of irregular Kähler manifolds (and algebraic varieties) with Albanese general type fibrations, Invent. Math. 104 (1991), 263–289.

[K] Kaup, B., Über offene analytische Äquivalenzrelationen auf komplexen Räumen, Math. Ann. 183 (1969), 6–16.

[M1] Mok, N., Factorization of semisimple representation of Kähler groups, Invent. Math. 110 (1992), 557–614.

[M2] Mok, N., Semi-Kähler structures and algebraic dimensions of compact Kähler manifolds, in: Geometry and Global Analysis, Report of the First MSJ International Research Institute (1993), Tohoku University, Sendai, Japan, 1993.

[M3] Mok, N., Harmonic forms with values in locally constant Hilbert bundles, Proceedings of the conference in honor of J.-P. Kahane (Orsay 1993), Journal of Fourier Analysis and Applications (1995), Special Issue, 433–455.

[M4] Mok, N., Fibering compact Kähler manifolds over projective-algebraic varieties of the general type, Proceedings of the International Congress of Mathematicians (ICM) (ed. S. D. Chatterji), Zürich 1994, Birkhäuser 1995.

[R] Ran, Z., On subvarieties of Abelian varieties, Invent. Math. 62 (1981), 459–479.

[Z] Zimmer, R. J., Ergodic Theory and Semisimple Groups, Monographs Math. 81, Birkhäuser, Boston–Basel–Stuttgart 1984.

Poisson and symplectic geometry on loop spaces of smooth manifolds

*Oleg Mokhov**

Abstract. We study the Poisson and symplectic field structures of differential geometric type on loop spaces of smooth manifolds. These structures generate natural Hamiltonian representations for many important nonlinear systems of mathematical physics and field theory such as nonlinear sigma models with torsion, degenerate Lagrangian systems of field theory, systems of hydrodynamic type, N-component systems of Heisenberg magnet type, the Monge-Ampère equations, the equations of associativity in two-dimensional topological field theories (the Witten–Dijkgraaf–E. Verlinde–H. Verlinde system) and others. In particular, in this paper symplectic field structures corresponding to degenerate Lagrangian field systems in the case when Lagrangian function depends on higher derivatives of field variables $u^i(x, t)$ with respect to t and x are also presented.

1991 Mathematics Subject Classification: 58F05.

1. Poisson structures on loop spaces of manifolds

First of all, we recall very briefly some important notions of classical finite-dimensional Poisson geometry. Let an N-dimensional smooth manifold M with local coordinates $\{u^1, \dots, u^N\}$ be given. A classic Poisson bivector $\omega^{ij}(u)$ on M is by definition a skew-symmetric (2,0)-tensor on M ($\omega^{ij}(u) = -\omega^{ji}(u)$) that satisfies the well-known relation (the Jacobi identity):

$$\frac{\partial \omega^{ij}}{\partial u^s}\omega^{sk} + \frac{\partial \omega^{jk}}{\partial u^s}\omega^{si} + \frac{\partial \omega^{ki}}{\partial u^s}\omega^{sj} = 0. \tag{1.1}$$

Such a bivector gives a Poisson bracket on the space of smooth functions on M:

$$\{f(u), g(u)\} = \omega^{ij}(u)\frac{\partial f}{\partial u^i}\frac{\partial g}{\partial u^j} \tag{1.2}$$

* This work was partially supported by the Russian Foundation for Fundamental Research (Grant No. 96-01-01623), the International Science Foundation (Grant No. RKR300) and the INTAS (Grant No. 93-0166).

(or $\{u^i, u^j\} = \omega^{ij}(u)$) which is skew-symmetric

$$\{f(u), g(u)\} = -\{g(u), f(u)\}$$

and satisfies the Jacobi identity

$$\{f(u), \{g(u), h(u)\}\} + \{g(u), \{h(u), f(u)\}\} + \{h(u), \{f(u), g(u)\}\} = 0.$$

Generally speaking, nondegeneracy of the bivector ω^{ij} is not assumed. Recall that a smooth finite-dimensional manifold equipped by a Poisson bivector is called *a Poisson manifold.*

In this paper we study some natural infinite-dimensional generalizations of the Poisson bivectors and the corresponding Poisson brackets.

Let us consider the loop space of the manifold M, *i.e.*, the space $L(M)$ of all smooth parametrized mappings $\gamma : S^1 \to M$, $\gamma(x) = \{u^i(x),\ x \in S^1\}$. The classic Poisson bivector $\omega^{ij}(u)$ gives so-called *ultra-local Poisson bracket* on the loop space $L(M)$

$$\{u^i(x), u^j(y)\} = \omega^{ij}(u(x))\delta(x-y), \tag{1.3}$$

or, in other words, we have a Poisson bracket on the space of functionals on the loop space $L(M)$

$$\{F, G\} = \int_{S^1} \frac{\delta F}{\delta u^i(x)} \omega^{ij}(u(x)) \frac{\delta G}{\delta u^j(x)}\, dx \tag{1.4}$$

where F and G are arbitrary functionals on $L(M)$. In the infinite-dimensional case of $L(M)$ we can consider a natural generalization of the Poisson bracket (1.4). In fact, generally speaking, in this case a Poisson bivector ω^{ij} can depend also on derivatives of the fields $u^k(x)$: $\omega^{ij}(u, u_x, u_{xx}, \ldots)$.

Let us describe now all Poisson brackets on $L(M)$ of the following form (see also [2, 3, 61]):

$$\{F, G\} = \int_{S^1} \frac{\delta F}{\delta u^i(x)} \omega^{ij}(u, u_x, u_{xx}, \ldots) \frac{\delta G}{\delta u^j(x)}\, dx, \qquad \det(\omega^{ij}) \neq 0. \tag{1.5}$$

Remark 1.1. Note that in this paper we only consider tensors depending on a finite number of derivatives of the fields $u^k(x)$.

Remark 1.2. The nondegenerate Poisson brackets of the form (1.5) were considered in the papers [2, 3, 61]. The statement in [61] that all nondegenerate Poisson brackets of the form (1.5) are ultra-local, *i.e.*, in other words, $\omega^{ij}(u)$ does not depend on derivatives of the field variables $u^i(x)$, is erroneous (see [2, 3]). In [3] was obtained the classification of all the Poisson brackets (1.5) with respect to the Lie transformation.

Definition 1.1. A bivector $\omega^{ij}(u, u_x, u_{xx}, \ldots)$ is called *a generalized Poisson bivector on the loop space $L(M)$* if the tensor $\omega^{ij}(u, u_x, u_{xx}, \ldots)$ gives a Poisson bracket (1.5) on the space of functionals on $L(M)$.

Theorem 1.1. *A nondegenerate tensor $\omega^{ij}(u, u_x, u_{xx}, \ldots)$, $\det(\omega^{ij}) \neq 0$, on the loop space $L(M)$ is a generalized Poisson bivector on $L(M)$ if and only if there is a closed 2-form $\Omega_{ij}(u)$ on M ($d\Omega = 0$) and a closed 3-form $T_{ijk}(u)$ on M ($dT = 0$) such that $\det(T_{ijk}(u)u_x^k + \Omega_{ij}(u)) \neq 0$ and*

$$\omega^{ij}(T_{jks}(u)u_x^s + \Omega_{jk}(u)) = \delta_k^i. \tag{1.6}$$

Remark 1.3. Thus, it follows from (1.6) that nondegenerate generalized Poisson bivectors ω^{ij} on $L(M)$ can depend only on the first derivatives of the field variables $u^i(x)$.

Remark 1.4. If the manifold M is symplectic and the 2-form $\Omega_{ij}(u)$ is a symplectic form on M (*i.e.* $\det(\Omega_{ij}) \neq 0$) then for any closed 3-form $T_{ijk}(u)$ on M we always have always $\det(T_{ijk}(u)u_x^k + \Omega_{ij}(u)) \neq 0$ and formula (1.6) gives explicitly a generalized Poisson bivector ω^{ij}. We note here also that for any closed 3-form T_{ijk} on M, $\det(T_{ijk}(u)u_x^k) \equiv 0$.

Theorem 1.1 describes translation invariant Poisson brackets but we can also consider Poisson bivectors $\omega^{ij}(x, u(x), u_x, u_{xx}, \ldots)$ depending explicitly on the independent space variable x.

Theorem 1.2. *A tensor $\omega^{ij}(x, u, u_x, u_{xx}, \ldots)$, $\det(\omega^{ij}) \neq 0$, gives a Poisson bracket*

$$\{F, G\} = \int_{S^1} \frac{\delta F}{\delta u^i(x)} \omega^{ij}(x, u, u_x, u_{xx}, \ldots) \frac{\delta G}{\delta u^j(x)} dx$$

on the space of functionals on $L(M)$ if and only if there is a one-parameter family of 2-forms $\Omega_{ij}(z, u)$ on the manifold M and a closed 3-form $R_{ijk}(u)$ on M such that

$$\det[(\int_0^x (d\Omega)_{ijk}(z, u(x))dz + R_{ijk}(u))u_x^k + \Omega_{ij}(x, u(x))] \neq 0,$$

$$\omega^{is}[(\int_0^x (d\Omega)_{sjk}(z, u(x))dz + R_{sjk}(u))u_x^k + \Omega_{sj}(x, u(x))] = \delta_j^i, \tag{1.7}$$

where

$$(d\Omega)_{ijk}(z, u(x)) = \frac{\partial \Omega_{ij}(z, u(x))}{\partial u^k} + \frac{\partial \Omega_{ki}(z, u(x))}{\partial u^j} + \frac{\partial \Omega_{jk}(z, u(x))}{\partial u^i},$$

(note that there are no closure conditions for the 2-forms Ω_{ij} here in contrast with Theorem 1.1).

Remark 1.5. It is a much more complicated and unsolved problem to classify all degenerate generalized Poisson bivectors $\omega^{ij}(x, u, u_x, u_{xx}, \ldots)$ with respect to their dependence on derivatives of the fields $u^i(x)$. For example, it is interesting to describe all degenerate generalized Poisson bivectors of the simplest type

$$(\omega^{ij}(x, u, u_x, u_{xx}, \ldots)) = \begin{pmatrix} 0 & \omega^1 & \omega^2 \\ -\omega^1 & 0 & \omega^3 \\ -\omega^2 & -\omega^3 & 0 \end{pmatrix}.$$

In order to prove Theorems 1.1 and 1.2 we shall consider differential geometric objects which are inverse to nondegenerate Poisson bivectors ω^{ij}. As is well known, in the finite-dimensional case we shall get symplectic forms $\omega_{ij}(u)$ on the manifold M: $\omega_{ij}(u)\omega^{jk}(u) = \delta_i^k$, $(d\omega)_{ijk} = 0$. In the next section we shall consider the corresponding generalizations of symplectic structures in the infinite-dimensional case of the loop space $L(M)$.

2. Symplectic structures on loop spaces of manifolds

The tangent space of the loop space $L(M)$ at its point (a loop) γ consists of all the smooth vector fields ξ^i, $1 \leq i \leq N$, defined along the loop γ. We shall denote this space here by $T_\gamma L(M)$ (we note that $\xi^i(\gamma(x)) \in T_{\gamma(x)}M$, $\forall x \in S^1$, where $T_{\gamma(x)}M$ is the tangent space of the manifold M in the point $\gamma(x)$).

Consider a skew-symmetric bilinear form $\omega(\xi, \eta)$ on the loop space $L(M)$:

$$\omega(\xi, \eta) = \int_{S^1} \xi^i A_{ij} \eta^j \, dx, \tag{2.1}$$

$\omega(\xi, \eta) = -\omega(\eta, \xi)$, $\xi, \eta \in T_\gamma L(M)$, $\xi = (\xi^1, \ldots, \xi^N)$, $\eta = (\eta^1, \ldots, \eta^N)$, $A = (A_{ij})$ is a skew-symmetric operator defined by the loop γ

$$A[\gamma] : T_\gamma L(M) \to T_\gamma^* L(M).$$

Introduce a differential of a 2-form $\omega(\xi, \eta)$ (it is a natural infinite-dimensional generalization of the usual differential in the Lie algebra of vector fields on a manifold):

$$(d\omega)(\xi, \eta, \zeta) = \sum_{(\xi,\eta,\zeta)} \left\{ \int_{S^1} \xi^i \frac{\delta\omega(\eta, \zeta)}{\delta u^i} dx + \omega(\xi, [\eta, \zeta]) \right\}, \tag{2.2}$$

where $\xi, \eta, \zeta \in T_\gamma L(M)$,

$$[\eta, \zeta]^j = \eta^i_{(k)} \frac{\partial \zeta^j}{\partial u^i_{(k)}} - \zeta^i_{(k)} \frac{\partial \eta^j}{\partial u^i_{(k)}}$$

is the commutator of the vector fields η and ζ, $f_{(k)} = d^k f / dx^k$, and the sign $\sum_{(\xi,\eta,\zeta)}$ means that the sum is taken with respect to all cyclic permutations of the elements (ξ, η, ζ).

If $d\omega = 0$, *i.e.* the 2-form (2.1) is closed, then the operator A_{ij} is called a symplectic operator (ω is called a presymplectic form or presymplectic structure).

Remark 2.1. In the infinite-dimensional case a presymplectic structure is often said to be symplectic even if the corresponding 2-form ω is degenerate on $L(M)$. We shall adhere to this terminology here (see [7] for references on symplectic operators and general theory of local Hamiltonian and symplectic structures and their connections with integrability of nonlinear partial differential equations).

Lemma 2.1. *A nondegenerate tensor $\omega^{ij}(x, u, u_x, u_{xx}, \ldots)$, $\det \omega^{ij} \neq 0$, is a generalized Poisson bivector on $L(M)$ if and only if the tensor $\omega_{ij}(x, u, u_x, u_{xx}, \ldots)$, where $\omega^{is}\omega_{sj} = \delta^i_j$, is a symplectic operator.*

It was shown by the author in [34] that it is sufficient to verify the Jacobi identity only for all linear functionals of the form

$$F = \int_{S^1} f_i(x) u^i(x)\, dx,$$

where $f_i(x)$, $i = 1, \ldots, N$, are arbitrary functions. Consider vector fields ξ of the form $\xi^i = \omega^{ij} f_j(x)$. It is easy to check that the condition $(d\omega)(\xi, \eta, \zeta) = 0$ for a symplectic structure $\omega_{ij}(x, u, u_x, u_{xx}, \ldots)$ is equivalent to the Jacobi identity on linear functionals for the bracket

$$\{u^i(x), u^j(y)\} = \omega^{ij}(x, u, u_x, u_{xx}, \ldots)\delta(x - y).$$

Consider now the symplectic operators $\omega_{ij}(u, u_x, u_{xx}, \ldots)$.

Lemma 2.2. *An operator $\omega_{ij}(u, u_x, u_{xx}, \ldots)$ is symplectic if and only if*

(1) *$\omega_{ij} = -\omega_{ji}$ (skew-symmetry);*

(2) *for any vector-functions $\eta(x) = (\eta^1(x), \ldots, \eta^N(x))$, $\zeta(x) = (\zeta^1(x), \ldots, \zeta^N(x))$ we have the following identity:*

$$(-1)^k \left(\frac{d}{dx}\right)^k \left(\frac{\partial \omega_{ij}}{\partial u^s_{(k)}} \eta^i(x)\zeta^j(x)\right) + \frac{\partial \omega_{si}}{\partial u^j_{(k)}} \eta^i(x)\zeta^j_{(k)}(x) - \frac{\partial \omega_{sj}}{\partial u^i_{(k)}} \eta^i_{(k)}(x)\zeta^j(x) \equiv 0 \tag{2.3}$$

(the closure condition for ω_{ij}).

Consider now $R = \max\{m \ :\ \text{there exist } i, j, s \text{ such that } \partial\omega_{ij}/\partial u^s_{(m)} \neq 0\}$.

If $R \geq 2$ then the coefficient of the term $\eta^i_x(x)\zeta^j_{(R-1)}(x)$ in (2.3) is equal to $(-1)^R R \partial\omega_{ij}/\partial u^s_{(R)}$, *i.e.* $R \leq 1$ and the form ω_{ij} depends only on $u^k(x)$ and the first derivatives $u^k_x(x)$.

Now it is easy to show from the identity (2.3) that ω_{ij} must be quasilinear with respect to the first derivatives $u^k_x(x)$, *i.e.*

$$\omega_{ij}(u, u_x) = T_{ijk}(u) u^k_x + \Omega_{ij}(u) \tag{2.4}$$

Note that all symplectic operators of this type were described by the author in [30] where differential geometric homogeneous symplectic operators of the first order

$$A_{ij} = g_{ij}\frac{d}{dx} + b_{ijk}(u)u_x^k$$

were studied in detail (after the degeneration of g_{ij} we get a complete description of quasilinear symplectic structures $\omega_{ij}(u, u_x)$).

It turns out that the symplectic operators ω_{ij} of the form (2.4) give Hamiltonian (or symplectic) structures of some special Lagrangian systems in classical field theory (a certain degeneration of nonlinear sigma models).

Theorem 2.1. *Lagrangian systems $\delta S/\delta u = 0$ given by the action*

$$S = \int (a_{ij}(u)u_x^i u_t^j + b_i(u)u_t^i + U(x, t, u, u_x, u_{xx}, \ldots))\, dx\, dt \tag{2.5}$$

where $b_i(u)$ and $U(x, t, u, u_x, u_{xx}, \ldots)$ are arbitrary functions, and $a_{ij}(u)$ are skew-symmetric functions ($a_{ij} = -a_{ji}$), are Hamiltonian systems with respect to the corresponding symplectic structures of the form (2.4).

Conversely, any Hamiltonian system of the form

$$(T_{ijk}(u)u_x^k + \Omega_{ij}(u))u_t^j = \frac{\delta U}{\delta u^i}$$

is a Lagrangian system with respect to the action of the type (2.5).

In the next section we shall consider the symplectic geometry corresponding to the general nonlinear sigma models with torsion.

3. Nonlinear sigma models with torsion and symplectic geometry on loop spaces of manifolds

Consider two-dimensional systems generated by general nonlinear sigma model actions of the form

$$S = \int (\frac{1}{2}r_{ij}(u)u_x^i u_t^j + U(u))\, dx\, dt \tag{3.1}$$

where $r_{ij}(u)$ is an arbitrary (0,2)-tensor and $U(u)$ is an arbitrary function on M. As was shown by the author in [30], the corresponding Lagrangian system $\delta S/\delta u = 0$ always has the symplectic representation

$$A_{ij}u_t^j = \frac{\delta U}{\delta u} \tag{3.2}$$

where

$$A_{ij} = \frac{1}{2}(r_{ij} + r_{ji})\frac{d}{dx} + \frac{1}{2}\left(\frac{\partial r_{ki}}{\partial u^j} + \frac{\partial r_{ij}}{\partial u^k} - \frac{\partial r_{kj}}{\partial u^i}\right)u_x^k \tag{3.3}$$

is a symplectic operator.

If $\det(r_{ij} + r_{ji}) \neq 0$ (in this case we have a Riemannian or pseudo-Riemannian manifold M with the metric $g_{ij} = \frac{1}{2}(r_{ij} + r_{ji})$) then the symplectic operator (3.3) gives a symplectic form $\omega(\xi, \eta)$ on $L(M)$ that can be expressed as

$$\omega(\xi, \eta) = \int_\gamma \langle \xi, \nabla_{\dot\gamma} \eta \rangle \tag{3.4}$$

where $\langle \xi, \eta \rangle = \frac{1}{2}(r_{ij} + r_{ji})\xi^i\eta^j$ and $\nabla_{\dot\gamma}$ is the covariant derivation along γ generated by some differential geometric connection $\Gamma^i_{jk}(u)$ on M. Let us describe all symplectic forms of the form (3.4) on $L(M)$.

Theorem 3.1 ([30]). *Let a Riemannian or pseudo-Riemannian manifold (M, g_{ij}) be given. A differential geometric connection $\Gamma^i_{jk}(u)$ on M defines a symplectic form* (3.4), *where $\langle \xi, \eta \rangle = g_{ij}\xi^i\eta^j$, if and only if*

(1) *the connection $\Gamma^i_{jk}(u)$ is compatible with the metric g_{ij}, i.e.*

$$\nabla_k g_{ij} \equiv \frac{\partial g_{ij}}{\partial u^k} - g_{is}\Gamma^s_{jk} - g_{js}\Gamma^s_{ik} = 0;$$

(2) *the torsion*

$$T_{ijk} = g_{is}T^s_{jk}, \; T^i_{jk} \equiv \Gamma^i_{jk} - \Gamma^i_{kj},$$

is a closed 3-form on M.

Any closed 3-form on the (pseudo)-Riemannian manifold (M, g_{ij}) gives the only differential geometric connection $\Gamma^i_{jk}(u)$ compatible with the metric $g_{ij}(u)$, with torsion tensor determined explicitly by this closed 3-form and correspondingly any metric $g_{ij}(u)$ and any closed 3-form on M generate the symplectic form (3.4) on $L(M)$.

The corresponding symplectic operator has the form

$$M_{ij} = g_{ij}(u)\frac{d}{dx} + g_{is}(u)\Gamma^s_{jk}(u)u^k_x \tag{3.5}$$

where locally we have

$$\Gamma^i_{jk}(u) = \frac{1}{2}g^{is}(u)\left(\frac{\partial g_{sk}}{\partial u^j} + \frac{\partial g_{js}}{\partial u^k} - \frac{\partial g_{jk}}{\partial u^s} + T_{sjk}(u)\right), \tag{3.6}$$

$$T_{ijk}(u) = \frac{1}{2}\left(\frac{\partial f_{ij}}{\partial u^k} + \frac{\partial f_{jk}}{\partial u^i} + \frac{\partial f_{ki}}{\partial u^j}\right), \tag{3.7}$$

$$f_{ij}(u) = -f_{ji}(u). \tag{3.8}$$

Corollary. *On any Riemannian or pseudo-Riemannian manifold* (M, g_{ij}) *closed 2-forms of the form* (3.4) *on* $L(M)$ *are in one-to-one correspondence with closed 3-forms on* M.

An elementary conclusion is that on any two-dimensional Riemannian (or pseudo-Riemannian) manifold (M, g_{ij}) there is a single symplectic 2-form (3.4) on the loop space $L(M)$ and correspondingly a single symplectic operator (3.5) generated by the Levi-Civita connection. As for N-dimensional Riemannian (or pseudo-Riemannian) manifolds, the conclusion is that there always exist a symplectic 2-form of the form (3.4) and a symplectic operator (3.5), generated by the Levi-Civita connection.

It must be noted that the symplectic form (3.4) is always degenerate (*i.e.* in fact, it is only a presymplectic form on $L(M)$) and the null space of $\omega(\xi, \eta)$ is constituted by vector fields parallel along γ. In particular, the velocity vector field $\{u^i_x\}$ belongs to the null space of $\omega(\xi, \eta)$ if γ is a geodesic loop. Another subject is the interrelation between the symplectic forms under consideration and classical finite-dimensional symplectic structures on manifolds of geodesics. Recall that for N-dimensional Riemannian manifolds (M, g_{ij}) whose geodesics are periodic and of equal length, there exists a $(2N-2)$-dimensional symplectic manifold CM of geodesics. The symplectic structure on CM is determined by the curvature form of the S^1-connection in the principal bundle UM of unit tangent vectors of (M, g_{ij}), where the S^1-connection is generated by the canonical 1-form on T^*M (Reeb theorem, see [5, 54, 55, 62]). The tangent space $T_\gamma CM$ at the point γ is isomorphic to the space of normal Jacobi fields along the geodesic γ.

Theorem 3.2 ([30]). *The restriction of the symplectic form* (3.4) *generated on* $L(M)$ *by the Levi-Civita connection to the finite-dimensional subspace of normal Jacobi fields along the geodesics* γ *coincides with the symplectic Reeb form which is a closed nondegenerate 2-form on* CM *defining a symplectic structure.*

Theorem 3.3 ([30]). *Formulas* (3.5)–(3.8) *give a complete description of all symplectic operators of the following form*

$$A_{ij} = a_{ij}(u)\frac{d}{dx} + b_{ijk}(u)u^k_x, \quad \det a_{ij}(u) \neq 0 \tag{3.9}$$

(homogeneous differential operators of the first order with nondegenerate leading coefficient).

Remark 3.1. Hamiltonian operators of the form (3.9) were classified by Dubrovin and Novikov [16] (see also [15, 18]). In this case, the higher coefficient a^{ij} must be a flat metric on M and $b^{ij}_k(u)$ are generated by this flat metric. These Hamiltonian operators play a very important role in the theory of integrability of the diagonalizable systems of hydrodynamic type [59, 60] (see also [15, 18]).

Thus, any nonlinear sigma model (3.1) has a natural Hamiltonian (symplectic) representation (3.2) given by a symplectic operator (3.3). If a nonlinear sigma model system has also another Hamiltonian representation compatible with (3.2) then this nonlinear sigma model system will be integrable by the Lenart–Magri scheme ([27]). Thus, the problem of finding of Hamiltonian operators compatible with the symplectic operator (3.3) is very important and was formulated by the author in 1989 (see [30, 33, 39–41]). For some known two-component integrable nonlinear sigma models without torsion (r_{ij} is a symmetric tensor in (3.1)) such second Hamiltonian representations compatible with (3.2), (3.3) were found in [28].

4. Homogeneous symplectic forms on loop spaces of manifolds

Consider the matrix entries of the operator $A^{[m]}$ of the form

$$A^{[m]}_{ij} = a^{[m]}_{ij}(u)\frac{d^m}{dx^m} + b^{[m]}_{ijk}(u)u^k_x\frac{d^{m-1}}{dx^{m-1}} + (c^{[m]}_{ijkl}(u)u^k_x u^l_x + d^{[m]}_{ijk}(u)u^k_{xx})\frac{d^{m-2}}{dx^{m-2}} + \cdots, \tag{4.1}$$

with all the terms in (4.1) being degree m homogeneous with respect to the natural grading

$$\deg(fg) = \deg f + \deg g,\quad \deg f(u(x)) = \deg u(x) = 0,\quad \deg \frac{d^k u}{dx^k} = \deg \frac{d^k}{dx^k} = k.$$

Symplectic operators of the form (4.1) and the corresponding symplectic forms on the loop space $L(M)$

$$\omega(\xi, \eta) = \int_{S^1} \xi^i (A^{[m]}_{ij}\eta^j)\, dx \tag{4.2}$$

were introduced and studied by the author in [31, 39, 40].

It can be proved easily that for symplectic form (4.2) the coefficients of the operator (4.1) are transformed under changes of coordinates $u^i = u^i(v)$ on the N-dimensional manifold M as differential geometric objects. Under the assumption of nondegeneracy of the higher coefficient $a^{[m]}_{ij}(u)$ the conditions reflecting symplecticity of the operator (4.1) are in their essence some natural differential-geometric restrictions imposed on connections defined on the manifold M equipped with the metric $a^{[m]}_{ij}$ (symmetric for any odd $m = 2k + 1$, or skew-symmetric for any even $m = 2k$) and the coefficients can be expressed in an invariant form via the metric, the curvatures and the torsions of the connections. It allows us to obtain an invariant geometric description of the symplectic operators (4.1).

Definition 4.1. Symplectic operators of the form (4.1) are called *symplectic operators of differential geometric type*.

Remark 4.1. Poisson structures of the form (4.1) were introduced by Dubrovin and Novikov [17] as a natural generalization of Poisson structures of hydrodynamic type ($m = 1$). The case $m = 2$ for Poisson structures of the form (4.1) was studied in [52, 53] and later in [12].

For $m = 0$ the symplecticity conditions are

$$a^{[0]}_{ij}(u) = -a^{[0]}_{ji}(u), \quad \sum_{(i,j,k)} \frac{\partial a^{[0]}_{ij}}{\partial u^k} = 0,$$

where the sum is taken with respect to all cyclic permutations of i, j, k. It means that

$$\Omega_0 = a^{[0]}_{ij}(u) du^i \wedge du^j$$

is an arbitrary closed 2-form on M. If $\det a^{[0]}_{ij} \neq 0$, we have a symplectic structure Ω_0 on M and a symplectic structure

$$\omega(\xi, \eta) = \int_{S^1} \Omega_0(\xi, \eta)\, dx$$

on the loop space $L(M)$.

The case $m = 1$ corresponds exactly to symplectic forms (3.4). It is possible also to get a complete description of the matrix symplectic operators (4.1) for $m = 2$ [31]. In this case we deal with the operators of the form

$$A^{[2]}_{ij} = a^{[2]}_{ij}(u) \frac{d^2}{dx^2} + b^{[2]}_{ijk}(u) u^k_x \frac{d}{dx} + c^{[2]}_{ijkl}(u) u^k_x u^l_x + d^{[2]}_{ijk}(u) u^k_{xx}. \tag{4.3}$$

Theorem 4.1 ([31]). *Under the assumption* $\det a^{[2]}_{ij}(u) \neq 0$, *symplectic operators of the form* (4.3) *exist if and only if M is an even-dimensional manifold with an almost symplectic structure.*

We remind the reader that a manifold with an almost symplectic structure is an even-dimensional M endowed with a nondegenerate skew-symmetric metric $g_{ij}(u)$.

Let $a^{[2]}_{ij}(u) = g_{ij}(u)$. Consider an arbitrary symplectic operator of the form (4.2) and introduce coefficients $\Gamma^i_{jk}(u)$ by the relation

$$b^{[2]}_{ijk}(u) = 2 g_{is}(u) \Gamma^s_{jk}(u).$$

Lemma 4.2 ([31]). *The coefficients $\Gamma^i_{jk}(u)$ define a symplectic connection on the almost symplectic manifold (M, g_{ij}), i.e., a differential geometric connection compatible with the almost symplectic structure of the manifold:*

$$\nabla_k g_{ij} \equiv \frac{\partial g_{ij}}{\partial u^k} - g_{is}\Gamma^s_{jk} - g_{js}\Gamma^s_{ik} = 0.$$

It turns out that any symplectic connection $\Gamma^i_{jk}(u)$ on a manifold with an almost symplectic structure (M, g_{ij}) generates a unique symplectic operator (4.3).

Theorem 4.2 ([31]). *If (M, g_{ij}) is a manifold with an almost symplectic structure then there exists a one-to-one correspondence between the symplectic connections Γ^i_{jk} on (M, g_{ij}) and closed 2-forms of the form (4.2), $m = 2$, on the space $L(M)$ of smooth loops on M. This correspondence is expressed by the formula*

$$\omega(\xi, \eta) = \int_{S^1} \Big\{ \langle \xi, \nabla^2_\nu \eta \rangle + \frac{1}{2}\langle \nu, R(\xi, \eta)\nu \rangle + \langle \nu, \sum_{(\eta,\xi,\nu)} R(\eta, \xi)\nu \rangle$$

$$+ \langle T(\nu, \xi), T(\nu, \eta) \rangle + \langle T(\xi, \eta), \nabla_\nu \nu \rangle \Big\}\, dx, \tag{4.4}$$

where $\langle \xi, \eta \rangle = g_{ij}\xi^i\eta^j$, the sum $\sum_{(\eta,\xi,\nu)}$ is taken with respect to all cyclic permutations of the elements (η, ξ, ν), ν is the velocity vector of the loop $\gamma(x)$, i.e. $\nu^i = u^i_x$, ∇_ν is the covariant derivation along γ, $[T(\xi, \eta)]^i = T^i_{kl}\xi^k\eta^l$, $T^i_{jk} \equiv \Gamma^i_{jk} - \Gamma^i_{kj}$ is the torsion tensor of the connection and $[R(\xi, \eta)\zeta]^i = R^i_{jkl}\xi^k\eta^l\zeta^j$,

$$R^i_{jkl} = -\frac{\partial \Gamma^i_{jl}}{\partial u^k} + \frac{\partial \Gamma^i_{jk}}{\partial u^l} - \Gamma^i_{pk}\Gamma^p_{jl} + \Gamma^i_{pl}\Gamma^p_{jk},$$

is the curvature tensor of the connection.

The formula (4.4) gives the general form of closed 2-forms of the form (4.2), $m = 2$, and also describes all the symplectic operators of the form (4.3).

On symplectic manifolds there is a special class of symmetric symplectic connections, satisfying the condition $T^i_{jk} = 0$. Such connections exist on (M, g_{ij}) if and only if the manifold (M, g_{ij}) is in fact a symplectic one, *i.e.*, $(dg)_{ijk} = 0$.

Corollary 4.1 ([31]). *If a symplectic connection $\Gamma^i_{jk}(u)$ is a symmetric one then the closed 2-form* (4.8) *has the presentation*

$$\omega(\xi, \eta) = \int_{S^1} \Big\{ \langle \xi, \nabla^2_\nu \eta \rangle + \frac{1}{2}\langle \nu, R(\xi, \eta)\nu \rangle \Big\}\, dx. \tag{4.5}$$

This formula describes a most natural class of homogeneous symplectic forms of the second order on loop spaces of symplectic manifolds.

It is worth being noted that the class of symplectic connections on manifolds with almost symplectic structures is rich enough. Namely, any such a manifold has infinitely many distinct symplectic connections and correspondingly infinitely many distinct symplectic forms (4.4) on its loop space. Similarly, any symplectic manifold has infinitely many symmetric symplectic connections and correspondingly symplectic forms of the form (4.5) on its loop space.

5. Lagrangian systems and symplectic structures on loop spaces

We shall consider in this section two-dimensional Lagrangian systems generated by the actions of the form (see also [35])

$$S = \int \left(g_i(x, u, u_x, \ldots)u^i_t + h(x, u, u_x, \ldots)\right) dx\, dt \tag{5.1}$$

where g_i, h are arbitrary functions depending on the independent space variable x, some N fields $u^i(x, t)$ and their derivatives $u^i_{(n)} = \partial^n u^i/\partial x^n$ with respect to x.

The corresponding Lagrangian system $\delta S/\delta u^i = 0$ has the following form:

$$M_{ij} u^j_t = \frac{\delta H}{\delta u^i} \tag{5.2}$$

where

$$M_{ij} = \frac{\partial g_i}{\partial u^j_{(n)}} \frac{d^n}{dx^n} - (-1)^n \frac{d^n}{dx^n} \circ \frac{\partial g_j}{\partial u^i_{(n)}} \tag{5.3}$$

and $H = \int h\, dx$.

Lemma 5.1. *The differential operator M_{ij} is a symplectic operator and the bilinear form*

$$\omega(\xi, \eta) = \int_{S^1} \xi^i M_{ij} \eta^j\, dx, \tag{5.4}$$

where $\xi, \eta \in T_\gamma L(M)$, is a symplectic form on the loop space $L(M)$.

In other words, the inverse operator $K^{ij} = (M^{-1})^{ij}$, such that $K^{ij} M_{jk} = \delta^i_k$, always gives a nonlocal Poisson bracket

$$\{u^i(x), u^j(y)\} = K^{ij}[u(x)]\delta(x - y) \tag{5.5}$$

and hence we always have the following nonlocal Hamiltonian representation for Lagrangian systems generated by the actions (5.1):

$$u^i_t = K^{ij} \frac{\delta H}{\delta u^j} \equiv \{u^i(x), H\}. \tag{5.6}$$

The corresponding symplectic representation for these Lagrangian systems has the form

$$\omega(\delta u, u_t) = \delta H \tag{5.7}$$

where the relation (5.7) is valid for arbitrary variations $\delta u^i(x)$ of the fields $u^i(x)$, and H is a functional on the loop space $L(M)$.

Some special actions of the form (5.1) generate important nonlinear equations of mathematical physics and the field theory such as nonlinear sigma models (see above Section 3), Monge–Ampère equations, some systems of hydrodynamic type, the Krichever–Novikov equation and so on, and the corresponding symplectic representations are very useful and effective for investigations of integrability and the procedures of averaging. Of course, it is easy to generalize this construction for arbitrary degenerate Lagrangians quasilinear with respect to higher t-derivatives of the fields $u^i(x,t)$.

Let's consider some examples.

Example 5.1. The Korteweg–de Vries equation (KdV). Consider the following simple action of the type (5.1):

$$S = \int \big(\frac{1}{2}u_x u_t - \frac{1}{2}u_{xx}^2 - u_x^3\big)\, dx\, dt.$$

The symplectic representation (5.2)–(5.3) for the corresponding Lagrangian equation $\delta S/\delta u = 0$ has the form

$$\frac{d}{dx}(u_t) = -\frac{\delta H}{\delta u}, \quad H = \int \big(u_x^3 + \frac{1}{2}u_{xx}^2\big)\, dx.$$

After the field transformation $v(x) = u_x(x)$ we have obtained the usual Hamiltonian representation of the KdV equation:

$$v_t = \frac{d}{dx}\frac{\delta H}{\delta v}, \quad H = \int \big(v^3 + \frac{1}{2}v_x^2\big)\, dx, \quad v_t = 6vv_x - v_{xxx}.$$

Remark 5.1. Note that if $v^i(x) = u_x^i(x)$ then for any functional H the relation

$$\frac{\delta H}{\delta u^i} = -\frac{d}{dx}\frac{\delta H}{\delta v^i}$$

is valid for the variation derivatives.

Example 5.2 [32]. The Krichever–Novikov equation (KN). Consider the special action of the form (5.1)

$$S = \int \left(\frac{1}{2}\frac{u_t}{u_x} - \frac{1}{2}\frac{u_{xx}^2}{u_x^2} - \frac{1}{3}\frac{R(u)}{u_x^2}\right) dx\, dt, \tag{5.8}$$

where $R(u) = a_3u^3 + a_2u^2 + a_1u + a_0$ is an arbitrary polynomial of degree three, $a_i = \text{const}$, $i = 0, 1, 2, 3$. In this case the Lagrangian equation $\delta S/\delta u = 0$ has the symplectic representation (5.2)–(5.3) with the symplectic operator

$$M = -\frac{1}{u_x}\frac{d}{dx} \circ \frac{1}{u_x}, \tag{5.9}$$

$$\big(\frac{1}{u_x}\frac{d}{dx}\frac{1}{u_x}\big)u_t = \frac{\delta H}{\delta u}, \tag{5.10}$$

$$H = \int \big(\frac{1}{2}\frac{u_{xx}^2}{u_x^2} + \frac{1}{3}\frac{R(u)}{u_x^2}\big)\, dx. \tag{5.11}$$

The formulas (5.10)–(5.11) give the well-known symplectic (and Hamiltonian) representation ([56]) for the Krichever–Novikov equation ([25])

$$u_t = u_{xxx} - \frac{3}{2}\frac{u_{xx}^2}{u_x} + \frac{R(u)}{u_x}, \tag{5.12}$$

which is a completely integrable equation with an infinite collection of conservation laws, a representation of zero curvature with parameter on an elliptic curve and so on, and plays an important role in the contemporary theory of integrable nonlinear systems of mathematical physics.

We recall that the equation (5.12) appeared for the first time in [25] (in a slightly different form, see also [26, 49]) as the deformation equation (by virtue of the Kadomtsev–Petviashvili (KP) equation) for the Tyurin parameters corresponding to a general solution of the commutation equation of ordinary differential operators for a spectral curve of genus $g = 1$ and fixed rank $l = 2$ of the vector bundle of common eigenfunctions over the spectral curve (we note that the corresponding integrable Boussinesq type system of deformation of Tyurin parameters by virtue of the KP equation for general solution of the commutation equation of ordinary differential operators of genus $g = 1$ and rank $l = 3$ is obtained in the author's works [37, 38, 34]). Correspondingly any solution of the KN equation determines a solution of the KP equation by an explicit formula [25, 26].

The KN equation appeared as the first nonformal example of a system whose Hamiltonian structure is defined by a nontrivial differential symplectic operator. The action (5.8) gives a Lagrangian representation for the KN equation ([32]) and explains the natural origin of the symplectic operator (5.9).

In [7–9] symplectic structures of the degenerate KN equation ($R(u) \equiv 0$) are considered (as it is known ([58]), in this case and always, when the polynomial $R(u)$ has multiple roots, the KN equation reduces to the KdV equation by differential substitution). The degenerate KN equation has the form

$$u_t = u_x S[u], \tag{5.13}$$

where

$$S[u] = \frac{u_{xxx}}{u_x} - \frac{3}{2}\frac{u_{xx}^2}{u_x^2}$$

is the Schwarzian of u, and has a compatible pair of differential symplectic operators found by Dorfman ([7–9]) and a number of other interesting properties studied by various authors (see [63], for example). For multicomponent analogs of the degenerate KN equation (5.13) multisymplectic representation is found by Fordy and Antonowicz [1, 21]. In [11] a (2+1)-dimensional version of the degenerate KN equation (5.13) was studied by Dorfman and Nijhoff.

It is interesting that, as was discovered by the author and Ferapontov ([45], see also [19, 36]), the symplectic operator (5.9) and its multicomponent generalization

$$M_{ij} = \frac{1}{K}\frac{1}{u^i_x}\frac{d}{dx} \circ \frac{1}{u^j_x}, \tag{5.14}$$

$K = \text{const}$, arises naturally in the Dubrovin–Novikov theory of Hamiltonian systems of hydrodynamic type developed in [15–18]. The Hamiltonian structure of these systems introduced and studied by Dubrovin and Novikov in [16] is generated by metrics of zero curvature (all these Hamiltonian systems of hydrodynamic type have a natural Lagrangian representation of the type considered above, (5.1)–(5.3). It was shown in [45] that with the help of the operator (5.14) one can adequately extend the theory of Hamiltonian systems of hydrodynamic type to metrics of arbitrary constant Riemannian curvature K. In particular, such Hamiltonian diagonalizable systems of hydrodynamic type will be integrable (they are semi-Hamiltonian in the sense of Tsarev [59–60]). All systems of hydrodynamic type derived by the averaging of well-known integrable equations of the soliton theory (KdV, Nonlinear Schrödinger, Sine-Gordon and so on) have Hamiltonian structures generated by metrics of non-zero constant Riemanian curvature (along with structures of Dubrovin–Novikov type) (see [51]).

The corresponding symplectic theory can be developed for arbitrary Lagrangian systems. As example, we shall consider more general case of degenerate Lagrangian systems given by the actions of the form

$$S = \int (g_i(x, u, u_x, u_t, \dots, u_{x^s t^r}, \dots, u_{x^n t^m}) u^i_{t^{m+1}} + h(x, u, u_x, u_t, \dots, u_{x^s t^r}, \dots, u_{x^k t^m}))\, dx\, dt, \tag{5.15}$$

where $u_{x^s t^r} = \partial^{s+r} u / \partial x^s \partial t^r$, $r \le m$. In other words, the Lagrangian function is linear with respect to higher t-derivatives of the field variables $u^i(x, t)$.

Introducing the new variables $q^i_s(x, t) = u^i_{t^s}$, $0 \le s \le m$, we have the following relations for them: $(q^i_s)_t = q^i_{s+1}$, $0 \le s \le m - 1$.

Obviously, our Lagrangian system $\delta S / \delta u^i = 0$ is equivalent to the Lagrangian system $\delta \hat{S} / \delta q^i_s = 0$, $\delta \hat{S} / \delta \lambda^i_k = 0$, $0 \le k \le m - 1$, where

$$\hat{S} = \int (\hat{g}_i(x, \dots, (q^r_s)_{x^k}, \dots)(q^i_m)_t + q^i_s(\lambda^i_s)_t + \lambda^i_s q^i_{s+1} + \hat{h}(x, \dots, (q^r_s)_{x^k}, \dots))\, dx\, dt. \tag{5.16}$$

So we have the action $\hat{S}$ of the form (5.1) and the corresponding symplectic representation (5.2)–(5.7) for our Lagrangian system $\delta S / \delta u^i = 0$.

6. The equations of the associativity in 2D topological field theory and nondiagonalizable integrable systems of hydrodynamic type

In this section we shall the consider so-called nonlinear partial differential equations of the associativity in 2D topological field theories (see [6, 13, 14, 64]) and give their description as integrable nondiagonalizable weakly nonlinear systems of hydrodynamic type. For systems of this type the corresponding general differential geometric theory of integrability connected with hydrodynamic type Poisson structures on loop spaces of manifolds can be developed (general theory of local and nonlocal hydrodynamic type Poisson structures on loop spaces of pseudo-Riemannian manifolds is created in [15–19, 29, 36, 45, 46]).

Following Dubrovin [13], we recall very briefly the basic mathematical concepts connected with the Witten–Dijkgraaf–E. Verlinde–H. Verlinde (WDVV) system arising originally in two-dimensional topological field theories [6, 64] and its relations with the Dubrovin type equations of associativity.

Consider a function $F(t)$, $t = (t^1, \dots, t^N)$ such that the following three conditions are satisfied for its third derivatives that we denote

$$c_{\alpha\beta\gamma}(t) = \frac{\partial^3 F(t)}{\partial t^\alpha \partial t^\beta \partial t^\gamma} :$$

1) normalization, *i.e.*,

$$\eta_{\alpha\beta} = c_{1\alpha\beta}(t)$$

is a constant nondegenerate matrix;

2) associativity, *i.e.*, the functions

$$c^\gamma_{\alpha\beta}(t) = \eta^{\gamma\epsilon} c_{\epsilon\alpha\beta}(t)$$

for any t define the structure of an associative algebra A_t in the N-dimensional space with a basis $e_1, \dots, e_N$:

$$e_\alpha \cdot e_\beta = c^\gamma_{\alpha\beta}(t) e_\gamma .$$

3) $F(t)$ must be quasihomogeneous function of its variables:

$$F(c^{d_1} t^1, \dots, c^{d_N} t^N) = c^{d_F} F(t^1, \dots, t^N)$$

for any nonzero c and for some numbers $d_1, \dots, d_N, d_F$.

The resulting system of equations for $F(t)$ is called the Witten–Dijkgraaf–E.Verlinde–H.Verlinde (WDVV) system [6, 64] (see also [13, 14]). It was shown by Dubrovin [13] that solutions of the WDVV system can be reduced by a linear change of coordinates to two special types:

(1) in the physically most important case

$$F(t) = \frac{1}{2}(t^1)^2 t^N + \frac{1}{2} t^1 \sum_{\alpha=2}^{N-1} t^\alpha t^{N-\alpha+1} + f(t^2, \ldots, t^N) \tag{6.1}$$

for some function $f(t^2, \ldots, t^N)$;

(2) in a certain special case

$$F(t) = \frac{c}{6}(t^1)^3 + \frac{1}{2} t^1 \sum_{\alpha=1}^{N-1} t^\alpha t^{N-\alpha+1} + f(t^2, \ldots, t^N) \tag{6.2}$$

for a nonzero constant c.

If $N = 3$ (the first nontrivial case for the condition of associativity in the algebra A_t) then for solutions of the first type (6.1) of the WDVV system the associativity condition in the algebra A_t gives the following nonlinear partial differential equation for the function $f(x, y)$, where $x = t^2$, $y = t^3$, [13]:

$$f_{xxy}^2 = f_{yyy} + f_{xxx} f_{xyy}. \tag{6.3}$$

Let us introduce new variables a, b, c, d such that

$$a = f_{xxx}, \quad b = f_{xxy}, \quad c = f_{xyy}, \quad d = f_{yyy}. \tag{6.4}$$

The compatibility conditions have the form:

$$\begin{cases} a_y = b_x, \\ b_y = c_x, \\ c_y = d_x. \end{cases} \tag{6.5}$$

Besides, we have the relation

$$d = b^2 - ac \tag{6.6}$$

from (6.3).

So, the equation (6.3) is equivalent to the following homogeneous system of hydrodynamic type

$$\begin{pmatrix} a \\ b \\ c \end{pmatrix}_y = \begin{pmatrix} 0 & 1 & 0 \\ 0 & 0 & 1 \\ -c & 2b & -a \end{pmatrix} \begin{pmatrix} a \\ b \\ c \end{pmatrix}_x. \tag{6.7}$$

The system (6.7) is a nondiagonalizable weakly nonlinear homogeneous system of hydrodynamic type. In fact, the integrability of the system (6.7) follows from Dubrovin's results [14] but for the representation (6.7) it can be proved directly by the usual Hamiltonian and differential geometric methods for hydrodynamic type systems.

Theorem 6.1 ([44]). *Equation* (6.3) *is equivalent to the integrable nondiagonalizable weakly nonlinear homogeneous system of hydrodynamic type* (6.7).

As is recently shown by the author and Ferapontov [65] the system (6.7) is Hamiltonian with respect to a hydrodynamic Poisson structure on the loop space of some flat Riemannian manifold.

We recall very briefly the basic notions and results on Poisson structures of hydrodynamic type that we shall need here.

Consider the following action of type (5.1)

$$S = \int \left(\frac{1}{2} g_{ij} u^i_x u^j_t - h(u_x) \right) dx\, dt, \tag{6.8}$$

where (g_{ij}) is a constant nondegenerate symmetric tensor and $h(v)$ is an arbitrary function.

The corresponding Lagrangian system $\delta S / \delta u^i = 0$ has the symplectic representation (5.2)–(5.3) of the form

$$g_{ij} \frac{d}{dx} (u^j_t) = -\frac{\delta H}{\delta u^i}, \qquad H = \int h(u_x)\, dx. \tag{6.9}$$

After the field transformation $v^i(x) = u^i_x(x)$ (see Remark 5.1) we obtain a Hamiltonian system of hydrodynamic type in flat coordinates

$$v^i_t = g^{ij} \frac{d}{dx} \frac{\delta H}{\delta v^j}, \qquad H = \int h(v)\, dx. \tag{6.10}$$

In other words, the action (6.8) gives the general class of Hamiltonian systems of hydrodynamic type (written in flat coordinates $(v^1, \dots, v^N)$), introduced and studied by Dubrovin and Novikov [15–18].

Consider the one-dimensional systems of hydrodynamic type, *i.e.*, the one-dimensional evolution quasilinear systems of the first order

$$u^i_t = v^i_j(u) u^j_x, \tag{6.11}$$

where $v^i_j(u)$ is an arbitrary $N \times N$ matrix function of $u = (u^1, \dots, u^N)$, $u^i = u^i(x, t)$, $i = 1, \dots, N$.

The Hamiltonian systems of hydrodynamic type considered by Dubrovin and Novikov in [15–18] have the form

$$u^i_t = \{u^i, H\}, \tag{6.12}$$

where H is a functional of hydrodynamic type, *i.e.*,

$$H = \int h(u)\, dx, \tag{6.13}$$

and the Poisson bracket has the form

$$\{u^i(x), u^j(y)\} = g^{ij}(u(x)) \delta_x(x-y) + b^{ij}_k(u(x)) u^k_x \delta(x-y) \tag{6.14}$$

(the Poisson bracket of Dubrovin–Novikov type [16]). It was shown in [16] that if $\det [g^{ij}(u)] \neq 0$ then the expression (6.14) gives a Poisson bracket if and only if

(1) $g^{ij}(u)$ is a metric of zero Riemannian curvature (*i.e.*, simply a flat metric);

(2) $b_k^{ij}(u) = -g^{is}(u)\Gamma_{sk}^j(u)$, where $\Gamma_{sk}^j(u)$ are the coefficients of the differential geometric connection generated by the metric g^{ij}, *i.e.*, the only symmetric connection compatible with the metric (the Levi-Civita connection).

Thus, there always exist local variables $v^i = v^i(u)$ in which the Poisson bracket (6.14) is simply constant:

$$\{v^i(x), v^j(y)\} = e^{ij}\delta_x(x-y), \tag{6.15}$$

where $e^{ij} = 0, \pm 1$; $i, j = 1, \dots, N$. Correspondingly, it is easy to give a Lagrangian description (6.8) of these systems. The considered Hamiltonian systems of hydrodynamic type have the form

$$u_t^i = [\nabla^i \nabla_j h(u)]\, u_x^j, \tag{6.16}$$

where ∇ is the covariant derivative generated by a zero curvature metric.

Note that the Poisson brackets (6.14) define natural differential geometric Poisson brackets on loop spaces of arbitrary flat Riemannian manifolds, just as symplectic structures of differential geometric type do on loop spaces of arbitrary Riemannian manifolds (see Sections 1–4).

The system (6.7) is generated by the following hydrodynamic type Poisson structure on the loop space of the flat Riemannian manifold [65]:

$$M = \begin{pmatrix} -\frac{3}{2} & \frac{1}{2}a & b \\ \frac{1}{2}a & b & \frac{3}{2}c \\ b & \frac{3}{2}c & 2(b^2-ac) \end{pmatrix} \frac{d}{dx} + \begin{pmatrix} 0 & \frac{1}{2}a_x & b_x \\ 0 & \frac{1}{2}b_x & c_x \\ 0 & \frac{1}{2}c_x & (b^2-ac)_x \end{pmatrix}, \tag{6.17}$$

and the Hamiltonian

$$H = \int c\,dx. \tag{6.18}$$

The flat coordinates (v^1, v^2, v^3) are connected to the initial variables (a, b, c) by the Viète formulas

$$a = v^1 + v^2 + v^3, \quad b = -\frac{1}{2}(v^1v^2 + v^2v^3 + v^3v^1), \quad c = v^1v^2v^3. \tag{6.19}$$

In the flat coordinates the system (6.7) has the following Hamiltonian representation of the form (6.10):

$$\begin{pmatrix} v^1 \\ v^2 \\ v^3 \end{pmatrix}_t = \frac{1}{2}\begin{pmatrix} v^2v^3 - v^1v^2 - v^1v^3 \\ v^1v^3 - v^2v^1 - v^2v^3 \\ v^1v^2 - v^3v^1 - v^3v^2 \end{pmatrix}_x = \frac{1}{2}\begin{pmatrix} 1 & -1 & -1 \\ -1 & 1 & -1 \\ -1 & -1 & 1 \end{pmatrix} \frac{d}{dx} \begin{pmatrix} \partial h/\partial v^1 \\ \partial h/\partial v^2 \\ \partial h/\partial v^3 \end{pmatrix}, \tag{6.20}$$

where $h = c = v^1 v^2 v^3$. Hence, the system under consideration is Hamiltonian with the Hamiltonian operator

$$M = \frac{1}{2}\begin{pmatrix} 1 & -1 & -1 \\ -1 & 1 & -1 \\ -1 & -1 & 1 \end{pmatrix} \frac{d}{dx} \tag{6.21}$$

and the Hamiltonian $H = \int c\,dx = \int v^1 v^2 v^3\,dx$. The density of momentum and Casimirs of the corresponding Poisson bracket have the following form: $2b = -v^1 v^2 - v^2 v^3 - v^3 v^1$ (the density of momentum); v^1, v^2, v^3 (Casimirs).

Similarly, for $N = 3$ and for the second type special solutions (6.2) of the WDVV system the associativity condition for the algebra A_t gives the following Dubrovin equation for the function $f(x, y)$ [13]:

$$f_{xxx} f_{yyy} - f_{xxy} f_{xyy} = 1. \tag{6.22}$$

After introducing new variables a, b, c, d (6.4) from (6.22) we obtain the relation

$$d = \frac{1 + bc}{a}. \tag{6.23}$$

The compatibility conditions (6.5) and the relation (6.23) generate the following homogeneous system of hydrodynamic type

$$\begin{pmatrix} a \\ b \\ c \end{pmatrix}_y = \begin{pmatrix} 0 & 1 & 0 \\ 0 & 0 & 1 \\ -\frac{(1+bc)}{a^2} & \frac{c}{a} & \frac{b}{a} \end{pmatrix} \begin{pmatrix} a \\ b \\ c \end{pmatrix}_x . \tag{6.24}$$

Theorem 6.2 ([44]). *Equation* (6.22) *is equivalent to the integrable nondiagonalizable weakly nonlinear homogeneous system of hydrodynamic type* (6.24).

Now consider the more general case of the equations of the associativity when the normalization relations $\eta_{\alpha\beta} = c_{1\alpha\beta}(t)$ are not assumed (this normalization condition means that the element e_1 is the unit for all the associative algebras A_t), *i.e.*, in this case $\eta_{\alpha\beta}$ is an arbitrary constant nondegenerate matrix and it is possible that the associative algebras A_t have no units.

The equations of associativity have the following form

$$\frac{\partial^3 F}{\partial t^i \partial t^j \partial t^s} \eta^{sr} \frac{\partial^3 F}{\partial t^r \partial t^k \partial t^l} = \frac{\partial^3 F}{\partial t^i \partial t^l \partial t^s} \eta^{sr} \frac{\partial^3 F}{\partial t^r \partial t^j \partial t^k} \tag{6.25}$$

and they are nontrivial even for $N = 2$. Really, for $N = 2$ we have the following equation of associativity

$$F_{t^1 t^1 t^1} F_{t^2 t^2 t^2} = F_{t^1 t^1 t^2} F_{t^1 t^2 t^2}, \tag{6.26}$$

if

$$(\eta_{ij}) = \begin{pmatrix} 0 & 1 \\ 1 & 0 \end{pmatrix},$$

and the equation of associativity

$$F_{t^1t^1t^1}F_{t^1t^2t^2} + F_{t^1t^1t^2}F_{t^2t^2t^2} = (F_{t^1t^1t^2})^2 + (F_{t^1t^2t^2})^2, \tag{6.27}$$

if

$$(\eta_{ij}) = \begin{pmatrix} 1 & 0 \\ 0 & 1 \end{pmatrix}.$$

These equations were considered by Witten in [64]. After introducing the new variables a, b, c, d by analogy with (6.4) these equations transform to the following homogeneous systems of hydrodynamic type:

$$\begin{pmatrix} a \\ b \\ c \end{pmatrix}_y = \begin{pmatrix} 0 & 1 & 0 \\ 0 & 0 & 1 \\ -bc/a^2 & c/a & b/a \end{pmatrix} \begin{pmatrix} a \\ b \\ c \end{pmatrix}_x \tag{6.28}$$

and

$$\begin{pmatrix} a \\ b \\ c \end{pmatrix}_y = \begin{pmatrix} 0 & 1 & 0 \\ 0 & 0 & 1 \\ -c/b & (1 - c^2/b^2 + ac/b^2) & (2c/b - a/b) \end{pmatrix} \begin{pmatrix} a \\ b \\ c \end{pmatrix}_x, \tag{6.29}$$

respectively.

Theorem 6.3 ([44]). *Both systems* (6.28) *and* (6.29) *are integrable nondiagonalizable weakly nonlinear homogeneous systems of hydrodynamic type.*

References

[1] Antonowicz, M., Fordy, A. P., Multicomponent Schwarzian KdV hierachies, Rep. Math. Phys. 32 (1993), 223–233.

[2] Astashov, A. M., Normal forms of Hamiltonian operators in field theory, Dokl. Akad. Nauk SSSR 270 (5) (1983) = Soviet Math. Dokl. 27 (3) (1983).

[3] —, Vinogradov, A. M., On the structure of Hamiltonian operators in field theory, J. Geom. Phys. 3 (2) (1986), 263–287.

[4] Barouch, E., Fokas, A. S., Papageorgiou, V. G., J. J. Math. Phys. 29 (12) (1988), 2628–2633.

[5] Besse, A., Manifolds with closed geodesics, Mir, Moscow 1981.

[6] Dijkgraaf, R., Verlinde, H., Verlinde, E., Topological strings in $d < 1$, Nuclear Phys. B 352 (1991), 59–86.

[7] Dorfman, I. Ya., Dirac structures and integrability of nonlinear evolution equations, Wiley, England 1993.

[8] —, Krichever–Novikov equation and local symplectic structures, Dokl. Akad. Nauk SSSR 302 (4) (1988), 792–795.

[9] —, Dirac structures of integrable evolution equations, Phys. Letters A 125 (5) (1987), 240–246.

[10] —, Mokhov, O. I., Local symplectic operators and structures related to them, J. Math. Phys. 32 (12) (1991), 3288–3296.

[11] —, Nijhoff, F. W., On a (2+1)-dimensional version of the Krichever–Novikov equation, Phys. Lett. A 157 (1991), 107–112.

[12] Doyle, P. W., Differential geometric Poisson bivectors in one space variable, J. Math. Phys. 34 (4) (1993), 1314–1338.

[13] Dubrovin, B. A., Geometry of 2D topological field theories, Preprint SISSA–89/94, 1994, 204 p.

[14] —, Integrable systems in topological field theory, Nuclear Phys. B 379 (1992), 627–689.

[15] —, Novikov, S. P., Hydrodynamics of soliton lattices, Sov. Sci. Rev. Sect. C Math. Phys. Rev. 9 (1993), part 4, 1–136.

[16] —,—, Hamiltonian formalism of one-dimensional systems of hydrodynamic type and the Bogolyubov–Whitham averaging method, Dokl. Akad. Nauk SSSR 270 (4) (1983) 781–785 = Soviet Math.Dokl. 27 (1983), 665–669.

[17] —, —, On Poisson brackets of hydrodynamic type, Dokl. Akad. Nauk SSSR 279 (2) (1984), 294–297 = Soviet Math. Dokl. 30 (1984), 651–654.

[18] —, —, Hydrodynamics of weakly deformed soliton lattices. Differential geometry and Hamiltonian theory, Uspekhi Mat. Nauk 44 (6) (1989), 29–98 = Russian Math. Surveys 44 (6) (1989) 35–124).

[19] Ferapontov, E. V., Differential geometry of nonlocal Hamiltonian operators of hydrodynamic type, Funktsional. Anal. i Prilozhen. 25 (3) (1991), 37–49.

[20] Fokas, A. S., Fuchssteiner, B., On the structure of symplectic operators and hereditary symmetries, Lettere al Nuovo Cimento 28 (8) (1980), 299–303.

[21] Fordy, A. P., in: Nonlinear Evolution Equations and Dynamical Systems, Proc. of the 8th Internat. Workshop (NEEDS'92), 6–17 July, 1992, Dubna, Russia (V. G. Makhan'kov, ed.), World Scientific, Singapore 1993.

[22] Fuchssteiner, B., Fokas, A. S., Symplectic structures, their Bäcklund transformations and hereditary symmetries, Physica D 4 (1981), 47–66.

[23] Gelfand, I. M., Dorfman, I. Ya., Hamiltonian operators and algebraic structures related to them, Funktsional. Anal. i Prilozhen. 13 (4) (1979), 13–30.

[24] —, —, Schouten bracket and Hamiltonian operators, Funktsional. Anal. i Prilozhen. 14 (3) (1980), 71–74.

[25] Krichever, I. M., Novikov, S. P., Holomorphic bundles and nonlinear equations. Finite-zone solutions of rank 2, Dokl. Akad. Nauk SSSR 247 (1) (1979), 33–37.

[26] —,—, Holomorphic bundles over algebraic curves and nonlinear equations, Uspekhi Mat. Nauk 35 (6) (1980), 47–68.

[27] Magri, F., A simple model of the integrable Hamiltonian equation, J. Math. Phys. 19 (5) (1978), 1156–1162.

[28] Meshkov, A. G., Hamiltonian and recursion operators for two-dimensional scalar fields, Phys. Lett. 170 A (6) (1992), 405–408.

[29] Mokhov, O. I., Poisson brackets of Dubrovin–Novikov type (DN-brackets), Funktsional. Anal. i Prilozhen. 22 (4) (1988), 92–93 = Functional Anal. Appl. 22 (1988), 336–338.

[30] —, Symplectic forms on loop space and Riemannian geometry, Funktsional. Anal. i Prilozhen. 24 (3) (1990), 86–87 = Functional Anal. Appl. 24 (1990).

[31] —, Homogeneous second order symplectic structures on loop spaces and symplectic connections, Funktsional. Anal. i Prilozhen. 25 (2) (1991), 65–67 = Functional Anal. Appl. 25 (1991).

[32] —, Canonical Hamiltonian representation of the Krichever–Novikov equation, Mat. Zametki 50 (3) (1991), 87–96 = Math. Notes 50 (1991), 939–945.

[33] —, Two-dimensional nonlinear sigma models and symplectic geometry on loop spaces of (pseudo) Riemannian manifolds, Nonlinear Evolution Equations and Dynamical Systems, in: Proceedings of the 8th Internat. Workshop (NEEDS'92), 6–17 July, 1992, Dubna, Russia (V. G. Makhan'kov, ed.), World Scientific, Singapore 1993, 444–456; hep-th/9301048.

[34] —, Geometry of commuting differential operators of rank 3 and Hamiltonian flows, Ph.D. Thesis, Moscow State University, Moscow, Russia, 1984.

[35] —, Symplectic structures on loop spaces of smooth manifolds and Lagrangian systems of the field theory, Abstracts of the Internat. Geom. Colloquium, May 10–14, 1993, Moscow, Russia, 38–39.

[36] —, Hamiltonian systems of hydrodynamic type and constant curvature metrics, Phys. Lett. A 166 (3–4) (1992), 215–216.

[37] —, Commuting ordinary differential operators of rank 3 corresponding to an elliptic curve, Uspekhi Mat. Nauk 37 (4) (1982), 169–170.

[38] —, Commuting differential operators of rank 3 and nonlinear equations, Izv. Akad. Nauk SSSR Ser. Mat. 53 (6) (1989), 1291–1315.

[39] —, Symplectic geometry on loop spaces of smooth manifolds and nonlinear systems, Internat. Workshop "Theory of Nonlinear Waves", September 1991, Kaliningrad University, Kaliningrad, Russia.

[40] —, Symplectic forms on loop spaces of Riemannian manifolds, Internat. Conference "Differential equations and related problems" in honour of 90-th anniversary of I. G. Petrovsky (1901–1973), May 1991, Moscow State University, Moscow, Russia.

[41] —, Two-dimensional σ-models in the field theory: symplectic approach, Abstracts of the 9th Workshop "Modern Group Analysis. Methods and Applications", June 1992, Nizhniy Novgorod, Russia.

[42] —, On Hamiltonian structure of evolution with respect to the space variable x for the Korteweg-de Vries equation, Uspekhi Mat. Nauk 45 (1) (1990), 181–182.

[43] —, Killing–Poisson bivectors on Riemannian manifolds and integrable systems, Abstracts of Internat. Congress of Mathematicians, 3–11 August 1994, Zürich, Switzerland, 50.

[44] —, Differential equations of associativity in 2D topological field theories and geometry of nondiagonalizable systems of hydrodynamic type, Abstracts of Internat. Conference on Integrable Systems "Nonlinearity and Integrability: from Mathematics to Physics", February 21–24, 1995, Montpellier, France .

[45] —, Ferapontov, E. V., On the nonlocal Hamiltonian hydrodynamic type operators connected with constant curvature metrics, Uspekhi Mat. Nauk 45 (3) (1990), 191–192.

[46] —,—, Hamiltonian pairs associated with skew-symmetric Killing tensors on spaces of constant curvature, Funktsional. Anal. i Prilozhen. 28 (2) (1994), 60–63.

[47] —, Nutku, Y., Bianchi transformation between the real hyperbolic Monge-Ampère equation and the Born–Infeld equation, Lett. Math. Phys. 32 (2) (1994), 121–123.

[48] Novikov, S. P., Andrejewski Lectures, Berlin, November–December 1993, Sfb 288 Preprint No. 117, 42 p.

[49] —, Two-dimensional Schrödinger operators in periodic fields, Itogi Nauki i Tekhn., Seriya Sovremennye Problemy Matematiki 23 (1983), 3–32.

[50] Nutku, Y., Sarıoğlu, Ö., An integrable family of Monge–Ampère equations and their multi-Hamiltonian structure, Phys. Lett. A 173 (3) (1993), 270–274.

[51] Pavlov, M. V., Multihamiltonian structures of the Whitham equations, Dokl. Akad. Nauk 338 (2) (1994), 165–167

[52] Potemin, G.V., On Poisson brackets of differential-geometric type, Dokl. Akad. Nauk SSSR 286 (1) (1986), 39–42 = Soviet Math. Dokl. 33 (1986), 30–33.

[53] —, Some problems of differential geometry and algebraic geometry in the theory of solitons, Ph.D. Thesis, Moscow State University, Moscow, Russia. 1991.

[54] Reeb, G., Comptes Rendus 229 (20) (1949), 969–971.

[55] —, Bull. Cl. Sciences, Acad. Roy. Belg. 5 Série, 36 (4) (1950), 324–329.

[56] Sidorenko, Yu. N., Zap. Nauchn. Sem. LOMI 161 (7) (1987), 76–87.

[56] Sokolov, V. V., On Hamiltonian property of the Krichever–Novikov equation, Dokl. Akad. Nauk SSSR 272 (1) (1984), 48–50.

[57] Svinolupov, S. I., Sokolov, V. V., Evolution equations with nontrivial conservation laws, Funktsional. Anal. i Prilozhen. 16 (4) (1982), 86–87.

[58] —,—, Yamilov, R. I., Bäcklund transformations for integrable evolution equations, Dokl. Akad. Nauk SSSR, 271 (4 (1983), 802–805.

[59] Tsarev, S.P., On Poisson brackets and one-dimensional Hamiltonian systems of hydrodynamic type, Dokl. Akad. Nauk SSSR 282 (3) (1985), 534–537 = Soviet Math. Dokl. 31 (1985), 488–491.

[60] —, Geometry of Hamiltonian systems of hydrodynamic type. The generalized hodograph method, Izvestiya Akad. Nauk SSSR, Ser. Mat. 54 (5) (1990), 1048–1068 = Math. USSR Izvestiya 37 (2) (1991), 397–419.

[61] Vinogradov, A. M., Hamiltonian structures in field theory. Dokl. Akad. Nauk SSSR 241 (1) (1978), 18–21 = Soviet Math. Dokl. 19 (4) (1978).

[62] Weinstein, A., J. Differential Geom. 9 (4) (1974), 513–517.

[63] Wilson, G., On the quasi-hamiltonian formalism of the KdV equation, Phys. Lett. A 132 (8–9) (1988), 445–450.

[64] Witten, E., On the structure of the topological phase of two-dimensional gravity, Nuclear Phys. B 340 (1990), 281–332.

[65] Mokhov, O. I., Ferapontov, E. V., Equations of associativity in two-dimensional topological field theory as integrable Hamiltonian nondiagonalizable systems of hydrodynamic type, to be published in Funktsional. Anal. i Prilozhen. 30 (2) (1996) = Funct. Anal. Appl. 30 (1996); hep-th/9505180.

An algebraic geometry view of a model quantum field theory on a curve

A. K. Raina

Abstract. We survey an algebraic geometry approach that we have developed to a certain model quantum field theory on a compact Riemann surface, which first arose in string theory. Our aim has been to use algebraic geometry to formulate it in a way which is not only mathematically rigorous, but which brings out the physics as well as the geometry of the model. This is achieved through an algebro-geometric formulation of the *analyticity constraints* implicit in the physicist's description of the model by the so-called *operator product expansion* (OPE). Algebraic geometry serves not only to formulate the model, but also provides powerful computational tools. As a consequence we are able to show that these analyticity axioms are sufficient to determine all the *field correlation functions* (needed by the physicist to describe the system), which we show to be sections (in general meromorphic) of line bundles on cartesian products of the Riemann surface. In the process we obtain a rigorous, unified proof of celebrated identities of Cauchy, Frobenius and Fay, which are hereby seen to express the *fermionic* nature of the quantum system. The *current correlation functions*, formally defined by coalescing arguments in the field correlation functions, are obtained using *ringed spaces with nilpotent elements*. This provides an example of a solution using *global geometry* to a problem of *normal-ordering* a product of fields. The technique of using ringed spaces can be extended to give a global geometric formulation of the *Sugawara* construction of the energy-momentum tensor as a normal-ordered product of currents. We thereby arrive at a new algebro-geometric construction of a *projective connection*. Algebraic geometry proofs of classical formulas in the theory of functions on a Riemann surface are also obtained in the course of this study.

1. Algebraic geometry formulation

The procedure that we shall follow here repeatedly will be to first introduce the heuristic expressions from quantum field theory that we wish to interpret in terms of algebraic geometry. The mathematician may regard them as symbolic expressions yet to be defined. We shall then formulate those that we shall work with in terms of algebraic geometry. Thus the sytem that we shall consider consists of a pair of "quantum fields" b and c on a compact, connected Riemann surface M of arbitrary genus $g \geq 0$. The quotes indicate that we do not attempt to define a quantum field, which is a very singular object and extremely difficult to deal with mathematically with any rigour. We shall always deal, instead, with certain well-behaved functionals of the fields, called the *correlation functions* of the system. These are, in fact, what the physicist would like to compute.

We denote the general correlation function $C(m,n)$ by the *symbolic* expression

$$C(m,n) = \langle b(Q_1)\dots b(Q_m)c(P_1)\dots c(P_n)\rangle \tag{1.1}$$

where $Q_1, \dots, Q_m, P_1, \dots, P_n$ are points on M. In physics, such a correlation function should give the expectation value of finding m particles of the field b and n of the field c at $Q_1, \dots, Q_m$ and $P_1, \dots, P_n$ respectively, in their ground state. It is difficult, however, to sustain a serious physical interpretation of the correlation functions in view of the physical artificiality of the model, although these fields do arise naturally in string theory as Faddeev–Popov ghost fields. Their rather pedestrian origin, however, does not give any idea of the remarkable interest of this model from the viewpoint of mathematical physics. Our aim here is to develop a single coherent viewpoint of this system and accordingly we make only occasional mention of some of the many interesting papers that have been written about the system.

We shall regard the correlation functions $C(m,n)$ as functionals of the fields, complete knowledge of which specifies the model. More precisely, the $C(m,n)$ will be seen to be - not necessarily holomorphic- *sections of holomorphic line bundles.* The analyticity properties of these sections are constrained by the *operator product expansion* (*OPE*) of the b and c fields, which is given by the heuristic relation [1]

$$b(z)c(w) = \frac{I}{z-w} + \text{holomorphic terms,} \tag{1.2}$$

where I is the identity operator and this relation is supposed to make sense only inside one of the $C(m,n)$.

We begin the development of our algebraic geometry formulation of the model by associating a holomorphic line bundle α on M to the "quantum field" c and a holomorphic line bundle β on M to b. Thus $C(m,n)$ is a meromorphic section of the line bundle

$$p_1^*(\beta) \otimes \dots \otimes p_m^*(\beta) \otimes p_{m+1}^*(\alpha) \otimes \dots \otimes p_{m+n}^*(\alpha) \tag{1.3}$$

on $M^{m+n} \equiv M_1 \times \dots \times M_{m+n}$, the cartesian product of $m+n$ copies of M, where $p_i : M^{m+n} \to M_i$ is the i-th canonical projection. The *OPE* (1.2) will now be interpreted as saying that the only singularities of $C(m,n)$ are simple poles when a Q and a P argument coincide. *We must emphasise that this global principle in no way restricts the possible singularities that can arise in practice.* All it means is that we require that *any singularity, other than the ones of physical origin coming from the OPE (1.2), must be forced on us by rigorous mathematical analysis.* (See the discussion of the spin $(1-J)-J$ system below). While our principle has a certain resemblance to the "principle of maximal analyticity" in vogue in physics many years ago, it is more compelling here, since our model belongs to a class of models essentially defined through their *OPE*'s. It should be noted, however, that our principle extends the meaning of the *OPE*, a property of the model on the *complex plane* [1], to the *compact Riemann surface M*.

Our first conclusion from this is that the "one point functions" $C(1,0) = \langle b \rangle$, $C(0,1) = \langle c \rangle$ are simply holomorphic sections of β and α respectively. A physicist who was studying this system starting from a lagrangian would call these one point functions the *zero modes* of the system and would want to eliminate them. Starting from the *OPE* (1.2) as we are doing, however, there is no reason *up to now* to introduce constraints on the line bundles α and β so as to eliminate such sections.

The first nontrivial case is that of the "two point function" $C(1,1) = \langle b(Q)c(P) \rangle$. In view of the *OPE* (1.2) we are led to the following definition:

Definition 1.1. $\langle bc \rangle$ is a meromorphic section of the line bundle $p_1^*(\beta) \otimes p_2^*(\alpha)$ whose only singularity is a simple pole along the diagonal Δ of $M \times M$.

This means that a non-zero two point function exists if and only if the map i in the exact sequence

$$O \to H^0(M \times M, p_1^*(\beta) \otimes p_2^*(\alpha)) \xrightarrow{i} H^0(M \times M, p_1^*(\beta) \otimes p_2^*(\alpha) \otimes \mathcal{O}(\Delta)) \quad (1.4)$$

is not an isomorphism.

So far the line bundles α and β have been completely arbitrary. We cannot, of course, expect to get anything interesting without some restriction, but it is important to make these restrictions as weak as possible. The optimum condition turns out to be to simply bound the sum of the *degrees* of the two line bundles by $2g - 2$. We then have the following elegant characterisation of the $b - c$ system (see [2] for the proof):

Theorem 1.2. *Let* $\deg(\alpha) + \deg(\beta) \leq 2g - 2$. *Then a non-zero two point function* $\langle bc \rangle$ *exists if and only if:*

(i) $\beta \otimes \alpha = K \equiv$ *the holomorphic cotangent bundle of M,*

(ii) $\deg(\alpha) = g - 1 = \deg(\beta)$,

(iii) *neither α nor β have any holomorphic sections.*

If these conditions hold then the two point function not only exists, but it is also unique (after normalisation).

Thus we see that simply requiring the existence of a non-zero two point function imposes very stringent conditions on the line bundles α and β, even though the degree condition we imposed is very much weaker than the conditions one would be led to impose in a field-theoretic approach. In fact *we can use Theorem 1.2 to understand the conventional formulation of the model.* Thus condition (i) is precisely the condition that the integrand of the standard action for the $b - c$ system, viz. $S \sim \int_M b\bar{\partial}c$, is indeed a volume form, as it must be for the integration over M to make sense. A field associated with the line bundle $K^{\otimes\lambda}$, the λ-th tensor power of the holomorphic cotangent bundle K, will be said to have "conformal spin λ" and so condition (ii) means that we are in the case when b and c are fields of conformal spin $1/2$, or rather a "twisted" version of it since we do not require $\beta = \alpha = \sqrt{K}$. Finally, condition (iii) says that zero modes must be absent. However, we now see that this is necessary in order to have a two point function whose singularity structure is determined by the *OPE* (1.2), rather than because some undefined functional integral will otherwise give trouble, as is usually argued!

We seem to be excluding from consideration the (conformal) spin $(1 - J) - J$ version of the $b - c$ system (J is a positive integer or half-integer) which is usually considered in the literature. That is, however, not the case. For indeed let us take $\deg(\alpha) = 2J(g - 1)$. The Riemann–Roch theorem tells us that α has holomorphic sections ("zero modes") if $J \geq 1$. Theorem 1.2 asserts that in that case *the two point function $\langle bc \rangle$ must have extra singularities not coming from the OPE* (1.2) *and these extra singularities must be such that we obtain a new $b - c$ system which does satisfy the conditions of the theorem.* One way is to introduce points $x_1, \ldots, x_I$, where $I = (2J - 1)(g - 1)$ and let D denote the divisor $x_1 + \cdots + x_I$. Define $\tilde{\alpha} \equiv \alpha \otimes \mathcal{O}(-D)$, $\tilde{\beta} \equiv \beta \otimes \mathcal{O}(D)$. Then for $\tilde{\alpha}$ and $\tilde{\beta}$ we have the required properties $\deg(\tilde{\alpha}) = g - 1 = \deg(\tilde{\beta})$ and $\tilde{\alpha} \otimes \tilde{\beta} = \alpha \otimes \beta = K$. Condition (iii) of Theorem 1.2 will also be satisfied for a generic choice of the points $\{x_i, i = 1, \ldots, I\}$. This is *effectively* how physicists handle the spin $(1 - J) - J$ case of the $b - c$ system. For complete details we refer to [3].

Another interesting consequence of Theorem 1.2 is that it provides a proof of one of the folk theorems of the physics literature, viz. a kind of "charge conservation theorem" for "spin fields" (more generally, for "twist fields"). We are given pairs of points and rational numbers $\{x_i, \mu_i \mid 1 \leq i \leq N_+\}$, $\{y_j, -\nu_j \mid 1 \leq j \leq N_-\}$, where the x_i, y_j are points on M. The μ_i, ν_j are positive rational numbers which satisfy the constraint $\sum_{i=1}^{N_+} \mu_i - \sum_{j=1}^{N_-} \nu_j = \ell$ (ℓ is a positive or negative integer called the "total twist") and which describe the monodromy of the b and c fields near the corresponding points:

$$\begin{aligned} b(z) &\sim (z - x_i)^{-\mu_i} & c(z) &\sim (z - x_i)^{\mu_i} \\ &\sim (z - y_j)^{\nu_j} & &\sim (z - y_j)^{-\nu_j} \end{aligned} \tag{1.5}$$

Then with the help of techniques from algebraic geometry we can reduce the problem of studying the $b - c$ system in the presence of such a "twist structure" to the generalised system of Theorem 1.2 on a finite cyclic covering $\tilde{M} \to M$, defined by a positive divisor D of M and a line bundle $\mathcal{L}$ such that $\mathcal{L}^{\otimes d} = \mathcal{O}_M(D)$, where d is the degree of the cyclic covering. *Theorem* 1.2 *then implies that the total twist ℓ must be zero.* In the case of spin fields this immediately implies that for a nonzero two point function we must have as many spin fields with a positive square-root behaviour as with negative, a well known folk theorem [4]. For details of the construction and proofs we refer to [2].

2. Field correlation functions

In the previous section we saw that algebraic geometry helped us to achieve a rather detailed qualitative understanding of the $b - c$ system from its *OPE* (1.2). In fact, algebraic geometry enables us to do much more and we will need no more input from physics (apart from the question of statistics). We shall from now on assume that the line bundles α and β satisfy the conditions of Theorem 1.2, *i.e.* that $\alpha \in \mathrm{Pic}^{g-1}(M)$ and has no holomorphic sections and that $\beta = K \otimes \alpha^{-1}$. We shall not give any further discussion of the spin $(1-J)-J$ system, for which we refer to [3]. We can now obtain an explicit expression for the two point function $\langle bc \rangle$ with the help of the following lemma:

Lemma 2.1. *Let* $\mathcal{M}_\zeta(1,1) \equiv p_1^*(K \otimes \zeta^{-1}) \otimes p_2^*(\zeta) \otimes \mathcal{O}(\Delta)$, *where* $\deg(\zeta) = g-1$. *Then:*

(i) *if* $g = 0$, $\mathcal{M}_\zeta(1,1)$ *is the trivial line bundle on* $M \times M$,

(ii) *if* $g \geq 1$, $\mathcal{M}_\zeta(1,1) = \pi_\zeta^*(\mathcal{O}(\Theta))$ *where* $\pi_\zeta : M \times M \to \mathrm{Pic}^{g-1}(M)$ *is given by* $(Q, P) \mapsto \mathcal{O}(Q - P) \otimes \zeta$. *Here* Θ *denotes the canonical theta divisor in* $\mathrm{Pic}^{g-1}(M)$. *(This becomes a translate of the usual theta divisor by the Riemann constant once a marking is chosen on* M, *which defines a Riemann matrix in canonical form).*

We also need the concept of the "prime form" $E(Q, P)$ for which we have found it convenient to introduce a new *algebro-geometric* definition:

Definition 2.2. We define the *prime form* to be the image of the canonical element $1 \in \mathcal{O}_{M\times M}$ in the exact sequence:

$$0 \longrightarrow \mathcal{O}_{M\times M} \overset{1\mapsto E(Q,P)}{\longrightarrow} \mathcal{O}_{M\times M}(\Delta) \tag{2.1}$$

This definition of the prime form $E(Q, P)$ can be related to the usual function-theoretic definitions in various genera to be found in the books of Fay [7] and Mumford [8] with the help of Lemma 2.1. Then by once again using Lemma 2.1, we can obtain the two point function $\langle bc \rangle$ explicitly and for $g \geq 1$ it coincides with the *Szegö kernel for a compact Riemann surface*, which was introduced by Hawley and Schiffer [5]. We refer to [2] and [6] for details of our approach.

The question of determining the higher point field correlation functions of the system is, of course, not meaningful until we have specified the *statistics* of the system. In [2] we have analysed the possible statistics of the system from an axiomatic analysis of the *OPE* (1.2). We shall not go into that here but merely report the conclusion that the usual Fermi/Bose dichotomy holds (if we weaken our requirements then some more exotic possibilities do exist [2]). Of the two cases the fermionic one turns out to be more interesting and we shall confine our attention to that case, although the bosonic case requires only a simple modification.

Our claim that the fermionic case is more interesting than the bosonic one only holds if we implement the condition of fermionic statistics in a special way, viz. by adding to the *OPE* (1.2) the following *OPE*'s for two b or two c fields:

$$b(z)b(w) \sim O(z-w), \qquad c(z)c(w) \sim O(z-w) \tag{2.2}$$

whose meaning is that the correlation function $C(m,n)$ should vanish when the arguments of two b or two c fields coincide. We can now write down simple axioms for all the field correlation functions $C(m,n)$:

Axioms 2.3.
Each field correlation function $C(m,n) = \langle b(Q_1)\dots b(Q_m)c(P_1)\dots c(P_n)\rangle$ is a meromorphic section of the holomorphic line bundle

$$\mathcal{F}_\alpha(m,n) \equiv p_1^*(K\otimes\alpha^{-1})\otimes\cdots\otimes p_m^*(K\otimes\alpha^{-1})\otimes p_{m+1}^*(\alpha)\otimes\cdots\otimes p_{m+n}^*(\alpha) \tag{2.3}$$

on M^{m+n} having:

($\mathcal{A}1$) a simple zero for $Q_i = Q_j$ or $P_i = P_j$,

($\mathcal{A}2$) a simple pole for $Q_i = P_j$,

($\mathcal{A}3$) no singularities other than those required by the second axiom.

Axioms ($\mathcal{A}1$) and ($\mathcal{A}2$) define *divisors* (formal sums with integral coefficients of codimension 1 subvarieties) of M^{m+n} which we respectively denote by $D_z(m,n)$ and $D_p(m,n)$ and the total divisor is $D(m,n) = D_z(m,n) - D_p(m,n)$, where we follow the usual convention of putting a plus sign for zeros and a minus sign for poles (see [9] for an introduction to this concept for physicists). Then by ($\mathcal{A}3$) we conclude that $C(m,n)$ defines a *holomorphic* section of the line bundle

$$\mathcal{M}_\alpha(m,n) = \mathcal{F}_\alpha(m,n)\otimes\mathcal{O}(-D(m,n)) \tag{2.4}$$

Thus, $C(m,n)$ defines an element of $H^0(M^{m+n},\mathcal{M}_\alpha(m,n))$ and we can pose a precise mathematical question, viz. what is the dimension of the space of holomorphic sections of $\mathcal{M}_\alpha(m,n)$? The answer is given by the following theorem (see [6] for the proof):

Theorem 2.4.

(i) *If* $m \neq n$, $\dim H^0(M^{m+n},\mathcal{M}_\alpha(m,n)) = 0$.

(ii) *If* $m = n$, $\dim H^0(M^{2n},\mathcal{M}_\alpha(n,n)) = 1$.

This theorem has significant implications for the physics of the $b-c$ system, for part (i) implies that $C(m,n) = 0$ for $m \neq n$, which is in perfect agreement with the physicist's argument that this must happen due to the requirement of *charge conservation*. This latter argument would normally be based on the lagrangian of the system, but here we see that the *OPE*'s serve equally well. Part (ii) of the theorem is even more remarkable, as it implies the validity of the *Wick representation*

$$C(n,n) = \det(\langle b(Q_i)c(P_j)\rangle)\Big|_{i,j=1}^{n} \tag{2.5}$$

since it is clear that the determinant of two point functions on the right-hand side of (2.5) satisfies all our axioms. This result shows that indeed our method of introducing the condition of fermionic statistics through the *OPE*'s (2.2) was correct. Moreover, it also leads to—rigorous proofs of—interesting identities once we can write down the correlation functions explicitly.

For this all we need is the following lemma and our previous observations on the prime form:

Lemma 2.5.

(i) *If* $g = 0$, *then* $\mathcal{M}_\alpha(n,n)$ *is the trivial line bundle;*

(ii) *for* $g \geq 1$, $\mathcal{M}_\alpha(n,n) = \pi_\alpha^{n*}(\mathcal{O}(\Theta))$ *where* $\pi_\alpha^n : M^{2n} \to \mathrm{Pic}^{g-1}(M)$ *is given by* $(Q_1, \ldots, Q_n, P_1, \ldots, P_n) \mapsto \mathcal{O}(\sum_1^n (Q_i - P_i)) \otimes \alpha$.

With the help of this lemma we can write down the $2n$-point function $C(n,n)$ directly as a product of the unique section of $\mathcal{M}_\alpha(n,n)$ (1 for $g = 0$ and a theta function for $g \geq 1$) and the canonical meromorphic section of $\mathcal{O}(D(n,n))$ (a ratio of products of prime forms).

On the other hand, by the Wick representation (2.5), *which we have proved*, it can also be expressed as a determinant of two point functions. In this way we obtain, with complete mathematical rigour, interesting identities. These identities are usually referred to as the *bosonization identities* in the physics literature because of the way they first appeared in physics, viz. they provided a proof that two methods of obtaining the correlation functions of the $b - c$ system were consistent [10].

Our analysis, on the other hand, traces their physics origin to the *OPE*'s of the system and is more powerful in that it provides a *proof* of the identities instead of using them to show consistency. Of course, the bosonization viewpoint can also be made rigorous in the Grassmannian formulation[11]. The identities take different forms in different genera and were first obtained by the mathematicians whose names are attached to them:

(i) *Cauchy's bialternant identity* ($g = 0$):

$$\frac{\prod_{1 \leq i < j \leq n} (Q_i - Q_j)(P_j - P_i)}{\prod_{1 \leq i, j \leq n} (Q_i - P_j)} = \det \left(\frac{1}{(Q_i - P_j)} \right) \Big|_{i,j=1}^{n} \tag{2.6}$$

(ii) *Frobenius' identity* ($g = 1$):

$$\frac{\sigma(\alpha + \sum_1^n (Q_i - P_i))}{\sigma(\alpha)} \frac{\prod_{1 \leq i < j \leq n} \sigma(Q_i - Q_j)\sigma(P_j - P_i)}{\prod_{1 \leq i, j \leq n} \sigma(Q_i - P_j)} = \det \left(\frac{\sigma(\alpha + Q_i - P_j)}{\sigma(\alpha)\sigma(Q_i - P_j)} \right) \Big|_{i,j=1}^{n} \tag{2.7}$$

(iii) *Fay's identity* ($g \geq 2$):

$$\frac{\theta[\alpha](\sum_1^n(Q_i - P_j))}{\theta[\alpha](0)} \quad \frac{\prod_{1\leq i<j\leq n} E(Q_i,Q_j)E(P_j,P_i)}{\prod_{1\leq i,j\leq n} E(Q_i,P_j)}$$

$$= \det\left(\frac{\theta[\alpha](Q_i-P_j)}{\theta[\alpha](0)E(Q_i,P_j)}\right)\Bigg|_{i,j=1}^{n} \tag{2.8}$$

Note that in (2.7) and (2.8) the conditions that $\sigma(\alpha) \neq 0$ and $\theta[\alpha](0) \neq 0$ hold if and only if α has no holomorphic sections, which is ($\mathcal{A}3$) of our axioms. The case $n = 2$ of (2.8) is usually known in the mathematics literature as the *trisecant identity* for geometrical reasons into which we shall not go here [8]. Detailed proofs of the results of this section can be found in [6] and [2]. An exposition of our proof for mathematicians, in which the connections with physics have been eliminated, has also appeared in the treatise [12].

Before we conclude our discussion of the field correlation functions of the $b - c$ system, let us expand on a remark in [2] concerning a variant of the system discussed above, which sometimes appears in the literature (see the paper of the Verlindes in [10]). The only difference is that the defining line bundle α is now taken to be an *odd theta characteristic*. More generally, we take $\deg(\alpha) = g - 1$, $h^0(M, \alpha) = 1$, *i.e.* α is a smooth point of the canonical theta divisor Θ in $\mathrm{Pic}^{g-1}(M)$. Of course, Theorem 1.2 says that this system has no two point function in the sense of Definition 1.1, but in the physics context what is of interest is the question of the existence of higher point functions satisfying Axioms 2.3. By a careful analysis of the proof [6] of part (ii) of Theorem 2.4 we obtain:

Theorem 2.6. $\dim H^0(M^{2n}, \mathcal{M}_\alpha(n,n)) = 1 \quad (n = 1, 2, \ldots) \iff$ *either* $\alpha \in \mathrm{Pic}^{g-1}(M) - \Theta$ *or* α *is a smooth point of* Θ.

With this theorem not only can we show that the system has $2n$−point functions (for $n \geq 2$) uniquely determined by Axioms 2.3, but with this we can also give a *direct proof* of a corollary to Fay's identity (2.8), which Fay [7] obtains by a limiting argument from (2.8) (see the equation following eqn. 43 on p. 33 of [7]). Thus we see that this case of the $b - c$ system is also covered by our approach.

3. Currents

We shall now describe some recent work [13] (details will appear in [14]) on an algebraic geometry approach to the current correlation functions. The heuristic definition of the current $j(z)$ is through point-splitting and subtracting the leading singularity:

$$j(z) = \lim_{Q\to P=z} b(Q)c(P) - \frac{1}{Q - P} \tag{3.1}$$

This definition does not lend itself in any obvious way to a geometric formulation, but we shall show (see [13]) that in fact the modern Grothendieck formulation of algebraic geometry provides us with the necessary concepts to achieve this. To understand the problem, let us first consider the one point function $\langle j(z)\rangle$. According to the heuristic definition (3.1), $\langle j(z)\rangle$ should be identified with the coefficient of $(Q-P)$ in an expansion of $\langle b(Q)c(P)\rangle E(Q,P)-1$ about the diagonal Δ of $M\times M$. This definition suggests that $\langle j(z)\rangle$ is a holomorphic one form on M, but this procedure does not offer a global geometric definition. It involves subtracting sections of (at least for $g\geq 1$) two different line bundles, viz. $\mathcal{M}_\alpha(1,1)$ and the trivial line bundle on $M\times M$, and performing a Taylor series expansion.

Our solution to this problem is based on the observation that we do not require $\mathcal{M}_\alpha(1,1)$ to be trivialisable on the whole of $M\times M$, which in any case is not true for $g\geq 1$, but only on the *first infinitesimal neighbourhood of* Δ *in* $M\times M$, which is defined through the concept of a *ringed space*. Since ringed spaces are not very familiar to physicists, let us first consider the variety Δ itself as a ringed space. It is defined as a pair $(\Delta,\mathcal{O}_\Delta)$ consisting of the topological space Δ and its structure sheaf of holomorphic functions $\mathcal{O}_\Delta$, which is the quotient of the sheaf of holomorphic functions on $M\times M$ modulo those vanishing on Δ. The restriction of $\mathcal{O}_\Delta$ to an affine open set $\mathcal{U}$ is of the form $k[x,y]/\Im(x-y)$, where $k[x,y]$ is the polynomial ring in the variables x and y and $\Im(x-y)$ is the ideal in $k[x,y]$ generated by $(x-y)$. This quotient is of the form of a polynomial ring in *one* variable $k[t]$, where $t=x+y$, which is as it should be for it to be the structure sheaf of the one dimensional variety Δ, which is simply a copy of M.

The first infinitesimal neighbourhood 2Δ of Δ consists of the pair $(\Delta,\mathcal{O}_{2\Delta})$, where Δ is, as before, the topological space, but with a new structure sheaf $\mathcal{O}_{2\Delta}$. This latter is the quotient of the sheaf of holomorphic functions on $M\times M$ modulo those with a *double zero* on Δ. The restriction of this to an affine open set $\mathcal{U}$ is of the form $k[x,y]/\Im(x-y)^2$, where $\Im(x-y)^2$ is the square of the ideal generated by $\Im(x-y)$. This quotient is of the form $k[t]\oplus k[t]dt$, where $t=x+y$ and $dt=x-y$ so that $(dt)^2=0$, *i.e. the ringed space* $(\Delta,\mathcal{O}_{2\Delta})$ *contains nilpotents.* The global geometric way of describing this is that

$$p_{1*}(\mathcal{O}_{2\Delta}) \;=\; \mathcal{O}_M\oplus K_M, \tag{3.2}$$

which says that the direct image of the structure sheaf of $\mathcal{O}_{2\Delta}$ to the first factor of $M\times M$ is the direct sum of the trivial line bundle and the cotangent bundle of M (or rather their associated sheaves). The important point is that the decomposition (3.2) is *canonical*. This means that we can in a natural way find the component of an element of the l.h.s. of (3.2) in each factor on the r.h.s.

4. Field-current correlation functions

We shall now see how the concepts introduced in the previous section enable us to compute not merely $\langle j(z)\rangle$, but also the general field-current correlation function $\langle b(Q_1)\dots b(Q_{n-1})c(P_1)\dots c(P_{n-1})j(z)\rangle$. By (3.1), this field-current correlation function is the coefficient of $(Q_n - P_n)$ in an expansion of $C(n,n)E(Q_n, P_n)$ about the diagonal $Q_n = P_n = z$ of $M_n \times M_{2n}$. This suggests that the field-current correlation function is a meromorphic one form for fixed $\{Q_i, P_i, 1 \le i \le n-1\}$. As explained in the last section, our proposal is to study $C(n,n)E(Q_n, P_n)$ on *the first infinitesimal neighbourhood* 2Δ of Δ in $M_n \times M_{2n}$ and to use the *canonical global splitting* of (3.2) above to determine this meromorphic one form.

Now $C(n,n)E(Q_n, P_n)$ is a meromorphic section of $\mathcal{F}_\alpha(n,n) \otimes \mathcal{O}(D_{n,2n})$, where D_{ij} denotes the divisor of M^{2n} defined by the diagonal of $M_i \times M_j$. Since the only relevant variables of $C(n,n)E(Q_n, P_n)$ are Q_n and P_n, we can restrict $\mathcal{F}_\alpha(n,n) \otimes \mathcal{O}(D_{n,2n})$ to general position on each M_i for $i \ne n, 2n$ to get the line bundle $\mathcal{L} = \mathcal{M}_\gamma(1,1) \otimes \mathcal{G}$ on $M_n \times M_{2n}$, where $\mathcal{G} = p_n^*(\mathcal{O}(D)) \otimes p_{2n}^*(\mathcal{O}(-D))$, $\gamma = \alpha \otimes \mathcal{O}(D)$ and $D = \sum_{i=1}^{n-1}(Q_i - P_i)$. As explained in the last section, our approach will only make sense if $\mathcal{L}$ is trivialisable on the ringed space 2Δ. While it is simple to see that $\mathcal{L}$ is trivialisable on Δ, this is a very subtle question on 2Δ, but nevertheless true:

Proposition 4.1. *The line bundles $\mathcal{M}_\gamma(1,1)$ and $\mathcal{G}$ are both trivialisable on 2Δ and thus $\mathcal{L}$ is so as well.*

We can now use the canonical splitting (Equation 3.2) to compute the image of $C(n,n)E(Q_n, P_n)$ in (3.2), where for $g \ge 1$ we take the classical trivialisation of $\mathcal{M}_\gamma(1,1)$ by Riemann's theta function $\theta[\alpha](\sum_1^{n-1}(Q_i - P_i) + \vec{u})/\theta[\alpha](\sum_1^{n-1}(Q_i - P_i))$. Here α in square brackets denotes theta characteristics determined by α (after having chosen a marking on M, a symplectic homology basis and the dual basis of holomorphic one forms w_i, $1 \le i \le g$), $\vec{u}$ is in $\mathbb{C}^g$ and the sums are Abel sums. We also need to determine what happens to the natural meromorphic section of $\mathcal{G}$, which is a product of prime forms.

Lemma 4.2. *Let 1_D denote the the natural section of $\mathcal{O}(D)$ and now consider $p_n^*(1_D) \otimes p_{2n}^*(1_{-D})$, which can be written in the notation of prime forms as $(\prod_1^{n-1} E(Q_n, Q_i)E(P_n, P_i))/(\prod_1^{n-1} E(Q_n, P_i)E(P_n, Q_i))$. Then the restriction of this meromorphic section to 2Δ takes, in the canonical decomposition (3.2) above, the form $1 + w_D$, where w_D is a meromorphic one form on Δ (identified with M_n) having simple poles at $Q_1, \dots, Q_{n-1}$ with residue $+1$ and at $P_1, \dots, P_{n-1}$ with residue -1. The signs of these residues are fixed by choosing an isomorphism between K_Δ and $\mathcal{O}(-\Delta)|\Delta$ which, when $g \ge 1$, is compatible with the Abel map.*

Lemma 4.2 is an algebro-geometric form of a well known formula for the prime form: $w_{a-b}(z) = d_z \ln(E(z,a)/E(z,b))$. Note that there is no canonical trivialisation of $\mathcal{G}$ on 2Δ. The effect of choosing a different trivialisation is to simply change w_{a-b}

by the addition of a *holomorphic* one form, which does not matter since w_{a-b} was only defined up to the addition of such a one form. When we want to write the formulas for the correlation function in function-theoretic form, the correct choice will be made by the chosen marking of M and the representation of the prime form as a function on $\mathbf{U}\times\mathbf{U}$, where $\mathbf{U}$ is the universal covering space of M.

With these two results it is easy to obtain the field-current correlation function:

Theorem 4.3. *The field-current correlation function is given by:*

$$\langle b(Q_1)\dots b(Q_{n-1})c(P_1)\dots c(P_{n-1})j(z)\rangle = C(n-1,n-1)$$

$$\times\left(w_D(z)+\sum_{j=1}^{g} w_j(z)\,\partial\ln\theta[\alpha]\Big(\sum_{i=1}^{n-1}(Q_i-P_i)+\vec{u}\Big)/\partial u^j\Big|_{\vec{u}=0}\right) \quad (4.1)$$

where the second term is absent for $g=0$.

Alternatively, we can compute the field-current correlation function starting from the determinantal form of the field correlation function. For this we need an algebro-geometric analogue of a simple formula given by Mumford (see part (a) of the lemma on page 3.225 of [8]) for the prime form, viz. $d_z(E(z,a)/E(z,b))|_{z=a} = 1/E(a,b)$. From the proof it is clear that Mumford actually obtains $d_zE(z,a)|_{z=a} = 1$ from which this formula follows. Our algebro-geometric analogue of the latter is that we can find the image of the canonical section "1" of the trivial line bundle $\mathcal{O}_{M\times M}$ on $M\times M$ in $\mathcal{O}_{2\Delta}(\Delta)$ by traversing the following commutative diagram in two different ways:

$$\begin{array}{ccc} \mathcal{O}_{M\times M}|\Delta & \hookrightarrow & \mathcal{O}_{2\Delta}(\Delta) \\ \uparrow & & \uparrow \\ \mathcal{O}_{M\times M} & \hookrightarrow & \mathcal{O}_{M\times M}(\Delta). \end{array} \quad (4.2)$$

As a result of this calculation we obtain the following alternative form for the field-current correlation function:

Theorem 4.4. *From the Wick representation we obtain,*

$$\langle b(Q_1)\dots b(Q_{n-1})c(P_1)\dots c(P_{n-1})j(z)\rangle = C(n-1,n-1)\langle j(z)\rangle \quad (4.3)$$

$$-\sum_{k=1}^{n-1}\langle b(Q_1)\dots b(Q_{n-1})c(P_1)\dots c(P_k=z)\dots c(P_{n-1})\rangle\langle b(z)c(P_k)\rangle$$

The two expressions (4.1) and (4.3) for the field-current correlation functions give us an identity, which is a simple generalisation of the "first corollary to Fay's identity" given by Mumford [8], valid in all genera. We refer to [14] for details.

5. The two point function of currents

The two point function of currents $\langle j(z_1)j(z_2)\rangle$ is the most important current correlation function from the point of view of physics and so it is very important to see whether our techniques generalise to this case.

From the *OPE* (1.2) we see that $\langle j(z_1)j(z_2)\rangle$ is given in terms of the 4-point function $C(2,2)$ by the following heuristic double limit:

$$\begin{aligned}\langle j(z_1)j(z_2)\rangle &= \{C(2,2)E(Q_1,P_1)E(Q_2,P_2) \\ &-1-\langle j(z_1)\rangle - \langle j(z_2)\rangle\}\Big|_{Q_i\to P_i=z_i\,(i=1,2)}\end{aligned} \tag{5.1}$$

It may look very unlikely that we can make mathematical sense out of (5.1), since we have to make sense first of the sum of 1, $\langle j(z_1)\rangle$ and $\langle j(z_2)\rangle$ and then of subtracting it from the first term in (5.1)!

We expect $\langle j(z_1)j(z_2)\rangle$ to be a (meromorphic) one form in each variable, *i.e.* that it is a meromorphic section of the *canonical bundle* $\omega_{M\times M}\equiv p_1^*(K)\otimes p_2^*(K)$ of $M\times M$. Now $C(2,2)E(Q_1,P_1)E(Q_2,P_2)$, which appears in (5.1), is a meromorphic section of the line bundle

$$\mathcal{R}\equiv\mathcal{F}_\alpha(2,2)\otimes\mathcal{O}(D_{13}+D_{24})=\mathcal{M}_\alpha(2,2)\otimes\mathcal{A}, \tag{5.2}$$

where

$$\mathcal{A}\equiv\mathcal{O}(D_{12}+D_{34}-D_{14}-D_{23}) \tag{5.3}$$

The natural generalisation of the procedure of the previous section is to consider the restriction of the line bundle $\mathcal{R}$ to the sub-*scheme* $Z\equiv 2\Delta_{13}\times 2\Delta_{24}$ of M^4. Our basic tool for studying this restriction is the following beautiful exact sequence

$$O\longrightarrow\omega_Y\longrightarrow\mathcal{O}_Z\longrightarrow\mathcal{O}_{2Y}\longrightarrow O \tag{5.4}$$

where $Y\equiv\Delta_{13}\times\Delta_{24}$.

The exact sequence (5.4) does not seem to exist in the mathematics literature and its only *raison d'être* appears to be to enable us to make sense of the heuristic formula (5.1), for which it has just the right properties. For if pr_{12} denotes the projection $M^4\to M_1\times M_2$ and we take the direct image of (5.4) by it, we get a new exact sequence

$$O\longrightarrow\omega_{M_1\times M_2}\longrightarrow pr_{12*}(\mathcal{O}_Z)\longrightarrow pr_{12*}(\mathcal{O}_{2Y})\longrightarrow O \tag{5.5}$$

The remarkable feature of (5.5) is that $pr_{12*}(\mathcal{O}_{2Y})$ is a rank 3 vector bundle on $M_1\times M_2$ which is *canonically* the direct sum of three line bundles:

$$pr_{12*}(\mathcal{O}_{2Y})=\mathcal{O}_{M_1\times M_2}\oplus K_{M_1}\oplus K_{M_2} \tag{5.6}$$

Note that this splitting enables us to make sense out of the formal sum $1+\langle j(z_1)\rangle+\langle j(z_2)\rangle$ in (5.1). In addition, $pr_{12*}(\mathcal{O}_Z)$ is a rank 4 vector bundle on $M_1\times M_2$

which is *canonically* the direct sum of four line bundles:

$$pr_{12*}(\mathcal{O}_Z) = \mathcal{O}_{M_1 \times M_2} \oplus K_{M_1} \oplus K_{M_2} \oplus \omega_{M_1 \times M_2} \tag{5.7}$$

Now all we need is to know whether the line bundle $\mathcal{R}$ is trivialisable on Z.

Proposition 5.1. *The line bundles $\mathcal{M}_\alpha(2,2)$ and $\mathcal{A}$ are trivialisable on Z and hence so is $\mathcal{R}$.*

In order to be able to actually compute $\langle j(z_1) j(z_2)\rangle$ we need to know what happens to $C(2,2)E(Q_1, P_1)E(Q_2, P_2)$ on restriction to Z. It is the product of the unique section of $\mathcal{M}_\alpha(2,2)$, which is easy to handle by a simple generalisation of the procedure for the case of the field-current correlation function, and the canonical meromorphic section of the line bundle $\mathcal{A}$, defined in (5.3), which we shall denote as $1_{\mathcal{A}}$. The answer in this case can be expressed in terms of what is sometimes called the *Bergmann kernel* ω_B [5] or, more appropriately, the *generalised Weierstrass* $\wp$ *function*. This concept was introduced for a compact Riemann surface in function-theoretic form by Hawley and Schiffer [5]. This way is not suitable for us and so we introduce it through an algebro-geometric definition: it is a symmetric meromorphic section of $\omega_{M\times M}$, defined by a holomorphic section of $\omega_{M\times M}(2\Delta)$ which is 1 on restriction to the diagonal Δ of $M \times M$.

Lemma 5.2. *Let $1_{\mathcal{A}}$ denote the canonical meromorphic section of the line bundle $\mathcal{A}$ defined in* (5.3), *which is $\{E(Q_1, Q_2)E(P_2, P_1)\}/\{E(Q_1, P_2)E(Q_2, P_1)\}$ in the notation of prime forms. Then its restriction to Z has the following decomposition in the canonical decomposition of eqn.* (5.7)*:*

$$1_{\mathcal{A}}|Z = 1 + \omega_B(z_1, z_2) \tag{5.8}$$

Since there is no canonical trivialisation of $\mathcal{A}$ on Z, $\omega_B(z_1, z_2)$ is defined only up to the addition of a holomorphic bidifferential. However, a definite one is automatically fixed when the appropriate choices have been made, just as for $w_D(z)$ in Lemma 4.2.

If we think about the meaning of restricting $1_{\mathcal{A}}$ to Z we easily realise that Lemma 5.2 is simply an algebro-geometric proof of the following well known formula $\omega_B(x, y) = \partial^2 \ln E(x, y)/\partial x \partial y$ (see [7] and [8]). When $g = 1$ this gives the following well known formula linking two functions of Weierstrass: $\wp(z) = -d^2 \ln \sigma(z)/dz^2$. The link between Lemma 5.2 and this equation is the elementary formula

$$\lim_{\substack{\epsilon\to 0\\ \delta\to 0}} \left(\frac{f(x+\epsilon, y+\delta) f(x, y)}{f(x+\epsilon, y) f(x, y+\delta)} - 1 \right) = \frac{\partial^2}{\partial x \partial y} \ln f(x, y) dx\, dy. \tag{5.9}$$

Lemma 5.2 gives a rather remarkable geometric interpretation of this formula for ω_B, for if we write down $1_{\mathcal{A}}$ on an affine subset of the Riemann sphere (case when $g = 0$), *i.e.* on the complex plane, we see that it is simply the *anharmonic ratio* of four points on the complex plane. *Thus $1_{\mathcal{A}}$ is a natural generalisation of the notion of the anharmonic ratio (or cross ratio) to a compact Riemann surface of arbitrary* *genus*. Thus the formula for the Weierstrass function $\wp(z)$ comes by coalescing two pairs of arguments in the generalised anharmonic ratio in the case of genus $g = 1$. This viewpoint accords an unexpected fundamental role to this well known formula as well as a new insight. In Lemma 5.2 we wrote down this generalised anharmonic ratio in terms of prime forms. By the dictionary we have earlier established for the prime form in [2] and [6], we find that this combination of prime forms coincides with a function-theoretic definition of a generalised anharmonic ratio proposed recently by Gunning [15].

We can now compute the image of $C(2,2)E(Q_1, P_1)E(Q_2, P_2)|Z$ in (5.7) to obtain:

Theorem 5.3. *The two point function of currents is given by*

$$\langle j(z_1)j(z_2)\rangle = \omega_B(z_1, z_2) + \sum_{i,j=1}^{g} w_i(z_1)w_j(z_2)\frac{\partial^2\theta[\alpha](\vec{u}+\vec{v})}{\partial u_i \partial v_j \theta[\alpha](0)}\Big|_{\vec{u}=\vec{v}=0} \tag{5.10}$$

where the second term on the right-hand side is absent if $g = 0$.

Theorem 5.4. *From the Wick representation we obtain*

$$\langle j(z_1)j(z_2)\rangle = \langle j(z_1)\rangle\langle j(z_2)\rangle - \langle b(z_1)c(z_2)\rangle\langle b(z_2)c(z_1)\rangle. \tag{5.11}$$

Equating these two expressions for $\langle j(z_1)j(z_2)\rangle$ gives us only a tautology for $g = 0$, but for $g \geq 1$ it gives the "second corollary to the trisecant identity" [8]:

Theorem 5.5. *The following identity holds for all genera* $g \geq 1$,

$$\omega_B(z_1, z_2) + \sum_{i,j=1}^{g} w_i(z_1)w_j(z_2)\frac{\partial^2 \ln\theta[\alpha](0)}{\partial u_i \partial v_j} = \frac{\theta[\alpha](z_1 - z_2)\theta[\alpha](z_2 - z_1)}{(\theta[\alpha](0))^2(E(z_1, z_2))^2}. \tag{5.12}$$

It is important to note that our results for the field-current correlation functions (eqns.(4.1) and (4.3)) as well as for the two point function of the currents (eqns.(5.10) and (5.11)) are perfectly consistent with the standard field-current and current-current operator product expansions as found in [1] (with an adjustment for a difference of notation):

$$b(w)j(z) \sim \frac{b(w)}{z-w}, \quad c(w)j(z) \sim -\frac{c(w)}{z-w} \tag{5.13}$$

$$j(z)j(w) \sim \frac{1}{(z-w)^2}. \tag{5.14}$$

Details of the proofs of the results of this section will appear in a subsequent paper.

6. n-point current correlation functions

We shall briefly summarise the procedure by which we can extend the methods of the previous section so as to calculate the n-point function of the currents $\langle j(z_1)\dots j(z_n)\rangle$. Details will be given in a subsequent paper. The heuristic double limit (5.1) is now replaced by the multiple limit

$$\langle j(z_1)\dots j(z_n)\rangle = \lim_{Q_i\to P_i=z_i(1\le i\le n)} C(n,n)\prod_{i=1}^{n} E(Q_i,P_i)$$
$$-1-\sum_{i=1}^{n}\langle j(z_i)\rangle-\dots-\sum_{1\le i_1<\dots<i_{n-1}\le n}\langle j(z_{i_1})\dots j(z_{i_{n-1}})\rangle \tag{6.1}$$

Our starting point is now the line bundle $\mathcal{F}_\alpha(n,n)\otimes\mathcal{O}(\sum_{i=1}^{n} D_{i,n+i})$ of which the first term on the right-hand side of (6.1) is a meromorphic section. This line bundle can be shown to be trivialisable on restriction to the product Z_n of the first order neighbourhoods $2\Delta_{i,n+i}$ of the diagonals Δ_{ij} of $M_i\times M_j$, over $1\le i\le n$. Instead of the exact sequence (5.4) it turns out that we now have an *array* of $n-1$ short exact sequences. We define the canonical projection $pr_n : M^{2n}\to M^n$ and, as before, taking the direct image of $\mathcal{O}_{Z_n}$ by this map enables us to make sense of the sum of lower order current correlation functions appearing in (6.1). We have then to compute the image of the first term on the r.h.s. of (6.1) in $pr_{n*}\mathcal{O}_{Z_n}$ and to pick out, in the canonical decomposition of the latter, the component giving a (meromorphic) section of the canonical bundle of M^n. This is a rather complicated expression (except when $g=0$), which we do not write down, but it can be regarded as coming from the non-determinantal side of the expressions we had obtained for $C(n,n)$ in Section 2. The computation is much simpler when one starts with the determinantal expression for $C(n,n)$. In this case the n-point function of currents is given by the determinant of the $n\times n$ matrix $\mathbf{A}\equiv(a_{ij})$, where $a_{ij}=\langle b(z_i)c(z_j)\rangle$ if $i\ne j$ and $a_{ii}=\langle j(z_i)\rangle$. Equating these two expressions for the n-point function of currents leads to interesting identities in all genera:

Theorem 6.1.

(a) *In the case of genus $g=0$, where $\langle j(z_1)\dots j(z_n)\rangle=0$ if n is odd, let $z_1,\dots,z_{2m}(m\ge 2)$ be arbitrary complex numbers and define the antisymmetric matrix $\mathbf{A}\equiv\{a_{ij}\}$ by $a_{ij}=1/(z_i-z_j)$ for $i\ne j$, 0 otherwise. Then*

$$\det(\mathbf{A})=\sum\prod_{\gamma}1/(z_{i_\gamma}-z_{j_\gamma})^2, \tag{6.2}$$

where (i_γ, j_γ) is an element of a partition of $1,\dots,2n$ into non-intersecting pairs with $i_\gamma<j_\gamma$, the product being over distinct pairs in a given partition and the sum over different partitions.

(b) *Let $g\ge 1$ $(n\ge 3)$ and let D_z denote the vector field $\sum_{i=1}^{g} w_i(z)\partial/\partial u_i$, where,* *as before, the w_i are a basis of holomorphic 1-forms. Then*

$$D_{z_1} \dots D_{z_n} \ln \theta[\alpha](0) = (-1)^{(n-1)}\{\sum \prod S_\alpha(z_{i_\gamma}, z_{j_\gamma})\} \tag{6.3}$$

where each term of the sum is labelled by an irreducible n-cycle (such as $(12)(23)\dots(n-1\ n)(n\ 1)$*) and the sum is over all such cycles.* $S_\alpha(z,w)$ *is the Szegö kernel, as usual.*

We were unable to find the identity (6.2) in the literature, though, of course, it would be very surprising if it were new. A direct elementary proof of it was obtained by Prof. Don Zagier. The second identity (6.1) can be found in the book of Fay [7] when α is an even theta characteristic (see eqn. (40) on p. 27) and $n = 4$.

There is a well known procedure by which one sees that the *OPE* (5.14)

$$j(z)j(w) \sim \frac{1}{(z-w)^2}$$

is equivalent to the infinite oscillator algebra

$$\{c, j_n \mid [j_m, j_n] = mc\delta_{m+n,0}, [j_n, c] = 0\}.$$

It is then natural to regard the set $\{\langle j(z_1)\dots j(z_n)\rangle | n = 1, 2\dots\}$ as a *realisation* of the oscillator algebra with $c = 1$ on an arbitrary compact Riemann surface.

7. Energy-momentum tensor

In this section we shall summarise results of [14] on the only aspect of the system that we have not discussed as yet, viz. the *energy-momentum tensor*. We shall find that once again our ideas of trivialisation play a role, but that this time we have to consider *second order neighbourhoods*. Formally, the energy momentum tensor is the normal ordered product of the currents: $T(z) = (1/2) : jj :$. It is to be calculated heuristically by the following "point splitting formula":

$$T(z) = \lim_{z_1 \to z_2 = z} \frac{1}{2}\left(\langle j(z_1)j(z_2)\rangle - \frac{1}{(z_1 - z_2)^2}\right) \tag{7.1}$$

This means that the one point function $\langle T(z)\rangle$ should be obtained by expanding $\langle j(z_1)j(z_2)\rangle(E(z_1,z_2))^2 - 1$ about the diagonal and looking at the coefficient of $(z_1 - z_2)^2$, which is of a higher order than in the corresponding situation for currents. We are thus led to consider the holomorphic line bundle $\omega_{M\times M}(2\Delta)$ restricted to the *second order neighbourhood* of the diagonal, which we shall denote as 3Δ; it represents the ringed space $(\Delta, \mathcal{O}_{3\Delta})$ where $\mathcal{O}_{3\Delta} = \mathcal{O}_{M\times M}/\mathcal{O}(-3\Delta)$. The exact sequence that we have to consider is the following canonical exact sequence:

$$O \longrightarrow K_\Delta^2 \longrightarrow \mathcal{O}_{3\Delta} \longrightarrow \mathcal{O}_{2\Delta} \longrightarrow O \tag{7.2}$$

We then have the following proposition:

Proposition 7.1. *The line bundle $\omega_{M \times M}(2\Delta)$ has a canonical trivialisation on 2Δ and is trivialisable on 3Δ (in fact on $n\Delta$).*

We shall normalise the trivialisation by requiring that the section which defines it should restrict to $1/2$ on Δ.

Definition 7.2. The *one-point function* $\langle T(z) \rangle$ of the energy-momentum tensor is defined to be a trivialisation of $\omega(2\Delta)$ on 3Δ such that its restriction to 2Δ coincides with the canonical trivialisation, taking the value $1/2$ on Δ.

It should be noted that the exact sequence (7.2) implies that two such $\langle T(z) \rangle$ differ by a holomorphic quadratic differential.

Theorem 7.3. *The above definition of $\langle T(z) \rangle$ is equivalent to an assignment of quadratic differentials to each open set in a complex analytic coordinate covering $\mathcal{U}$ of Δ, which is simply a copy of M, such that on overlaps they transform as quadratic differentials plus an inhomogeneous term, viz.* $(1/12)\times$ *the schwarzian of the coordinate transformation.*

According to a standard mathematical terminology (see [16]), this implies that $12\langle T(z) \rangle$ is a *projective connection.* On the other hand, according to the definition of Belavin, Polyakov and Zamolodchikov [17], this implies that *the system has central charge $c = 1$. This is an algebro-geometric counterpart on a compact Riemann surface of any genus of a well known result in the theory of the Virasoro algebra* (see case (i) of Remark 4.2 on page 46 of [18]. Since we have shown that the spin $(1 - J) - J$ version of this system is a disguised version of the (twisted) conformal spin $1/2$ case, it is not surprising that our methods can be extended to the other cases of the anomaly formula.

It should also be noted that we have obtained a rather elegant geometric interpretation of both $\langle T(z) \rangle$ as well as of the concept of a projective connection, viz. that they are just an infinitesimal version of our generalised anharmonic ratio! This may be regarded, perhaps, as a concrete form of the characterisation of Deligne that "Intuitivement, se donner une connection projective ... permet de définir le birapport (=rapport anharmonique) de 4 points infiniment voisins ..." [19].

8. Concluding remarks

We have seen how every detail of the structure of the $b - c$ system is a consequence of its *OPE* (1.2) and statistics. Algebraic geometry provides a rigorous mathematical language for describing the system and one which is as physically natural as the language of Hilbert space is for Quantum Mechanics. Further developments are under investigation.

It is a pleasure to thank the Max Planck Institut für Mathematik, Bonn, and its director Prof. F. Hirzebruch for hospitality during the preparation of this paper. Discussions with Prof. Werner Nahm and Prof. Don Zagier were most useful and are gratefully acknowledged.

References

[1] Friedan, D., Martinec, E., Shenker, S. Conformal invariance, supersymmetry, and string theory, Nuclear Phys. B 271 (1986), 93–165.

[2] Raina, A. K., An algebraic geometry study of the $b - c$ system with arbitrary twist fields and arbitrary statistics, Commun. Math. Phys. 140 (1991), 373–397.

[3] Raina, A. K., Analyticity and chiral fermions on a Riemann surface, Helv. Phys. Acta 63 (1990), 694–704.

[4] Atick, J. J., Sen, A., Spin field correlators on an arbitrary genus Riemann surface and non-renormalization theorems in string theories, Phys. Lett. B 186 (1987), 339–346.

[5] Hawley, N. S., Schiffer, M., Half-order differentials on Riemann surfaces, Acta Math. 115 (1966), 199–236.

[6] Raina, A. K., Fay's trisecant identity and conformal field theory, Commun. Math. Phys. 122 (1989), 625–641; Fay's trisecant identity and Wick's theorem: an algebraic geometry viewpoint, Exposition. Math. 8 (1990), 227–245.

[7] Fay, J., Theta functions on Riemann surfaces, Lecture Notes in Math. 352, Springer-Verlag, 1973.

[8] Mumford, D., Tata lectures on Theta. II, Springer-Verlag, 1984.

[9] Raina, A.K., Chiral fermions on a compact Riemann surface and the trisecant identity, in: Proc. 3rd. Regional Conf. on Math. Phys. (F. Hussain, A. Qadir, ed.), World Scientific, Singapore 1990.

[10] Eguchi, T., Ooguri, H., Chiral bosonization on a Riemann surface, Phys. Lett. B 187 (1987), 127–134; Verlinde, E., Verlinde, H., Chiral bosonization, determinants and the string partition function, Nuclear Phys. B 288 (1987), 357–396; Alvarez-Gaumé, L., Bost, J.-B., Moore, G., Nelson, P., Vafa, C., Bosonization on higher genus Riemann surfaces, Commun. Math. Phys. 112 (1987), 503–552.

[11] Alvarez-Gaumé, L., Gomez, C., Reina, C., New methods in string theory, in: Superstrings '87 (L. Alvarez-Gaumé, ed.), World Scientific, Singapore 1988; Kawamoto, N., Namikawa, Y., Tsuchiya, A., Yamada, Y., Geometric realisation of conformal field theory on Riemann surfaces, Commun. Math. Phys. 116 (1988), 247–308.

[12] Lange, H., Birkenhake, C., Complex abelian varieties, Springer-Verlag, 1992.

[13] Raina, A. K., An algebraic geometry view of currents in a model quantum field theory on a curve. C. R. Acad. Sci. Paris Sér. I 318 (1994), 851–856.

[14] Biswas, I., Raina, A. K., Projective structures on a Riemann surface, Internat. Math. Res. Notices 15 (1996), 753–768.

[15] Farkas, H. M., The trisecant formula and hyperelliptic surfaces, Contemp. Math. 136 (1992), 161–169.

[16] Tyurin, A. N., On periods of quadratic differentials, Russian Math. Surveys 33 (6) (1978), 169–221.

[17] Belavin, A. A., Polyakov, A. M., Zamolodchikov, A.B., Infinite conformal symmetry in two-dimensional quantum field theory, Nuclear Phys. B 241 (1984), 333–380.

[18] Kac, V. G., Raina, A. K., Highest weight representations of infinite dimensional Lie algebras, World Scientific, Singapore 1987.

[19] Deligne, P., Equations Différentielles à Points Singuliers Réguliers, Lecture Notes in Math. 163, Springer-Verlag, 1970.

Deforming convex hypersurfaces by the linear combination of the mean curvature and the square root of the scalar curvature

Weimin Sheng

Abstract. We study the deformation of convex hypersurfaces in R^{n+1} by the linear combination of the mean curvature and the square root of the scalar curvature. We show that the hypersurfaces remain convex in a finite time interval and converge to a round point in the sense of Huisken provided that the initial hypersurface satisfies a certain pinching condition. This result includes the results of both Huisken and Chow.

1991 Mathematics Subject Classification: Primary 58G11; Secondary 53A10.

1. Introduction

Let M_0 be a smooth, closed, strictly convex hypersurface in a Euclidean $(n+1)$-space R^{n+1}. Suppose that M_0 is given by a smooth embedding $F_0 : S^n \to R^{n+1}$. We consider the initial value problem

$$\begin{aligned} &\frac{\partial}{\partial t} F = -k(x,t)\nu(x,t), \\ &F(\cdot,0) = F_0, \end{aligned} \tag{1}$$

where $k(\cdot,t)$ is a suitable curvature function of the hypersurface M_t parametrized by $F(\cdot,t) : S^n \to R^{n+1}$, and $\nu(\cdot,t)$ is the outward unit normal vector field to M_t.

In [5], Huisken considered the case $k(\cdot,t) = H(\cdot,t)$, where H is the mean curvature, and showed that in this case the initial value problem (1) has a unique smooth solution for some maximum time interval [0, T), and as $t \to T$, the hypersurfaces M_t converge to a round point, in a sense. The corresponding one-dimensional result was proved by Gage and Hamilton [3].

Using different methods, Tso [10] considered the case $k(\cdot,t) = K(\cdot,t)$, where K is the Gauss–Kronecker curvature, and proved that (1) has a unique solution for a maximal time interval [0,T) and M_t converges to a point as $t \to T$.

Combining the methods of Huisken and Tso, Chow [1], [2] was able to prove analogous results for the case $k(\cdot,t) = K(\cdot,t)^{1/n}$ and $k(\cdot,t) = R(\cdot,t)^{1/2}$, where K

and R are the Gauss-ronecker and scalar curvatures respectively. In the case $k(\cdot, t) = R(\cdot, t)^{1/2}$, his results were not as complete as those of Huisken for the mean curvature case in that the initial hypersurface M_0 had to satisfy a suitable pinching condition to obtain the existence of a solution as well as the convergence to a round point.

In this paper we would like to consider the case $k(\cdot, t) = aH(\cdot, t) + bR(\cdot, t)^{1/2}$. We obtain the following

Theorem 1.1. *Let M_0 be a smooth, closed and strictly convex hypersurface in R^{n+1} $(n \geq 2)$ parametrized by a smooth embedding $F_0 : S^n \to R^{n+1}$. Then the initial value problem* (1) *for the case $k(\cdot, t) = aH(\cdot, t) + bR(\cdot, t)^{1/2}$ has a unique solution on a maximum time interval $[0, T)$ and the M's converge to a round point as $t \to T$ provided that the initial hypersurface M_0 satisfies the following additional assumption: for all points $p \in M_0$, $\frac{R}{H^2} > C(n) > 0$, where the constant $C(n) = \frac{n-2}{n-1}$ is chosen so that the inequality will imply M_0 is strictly convex. Here a and b are nonnegative constants which do not vanish simultaneously, H and R are the mean curvature and the scalar curvature of M_t respectively.*

Our theorem includes both [5] for the case where $a = 1, b = 0$ and [2] for the case where $a = 0, b = 1$. The case $k(\cdot, t) = aH(\cdot, t) + bK(\cdot, t)^{1/n}$ was studied in [8] with analogous results being obtained.

I would especially like to thank professors Zhengguo Bai and Yibing Shen for their help and encouragement.

2. Short time existence

If $M^n \subset R^{n+1}$ is a hypersurface, the mean curvature is given by $H = g^{ij}h_{ij}$ and the scalar curvature is given by $R = H^2 - |A|^2$, where A is the second fundamental form $h_{ij} = \langle \partial^2 F/\partial x^i \partial x^j, \nu\rangle$ and $g_{ij} = \langle \partial F/\partial x^i, \partial F/\partial x^j\rangle$ is the induced metric on M from R^{n+1} in the coordinate system $\{x^i\}$. In this section we prove short time existence for the equation

$$\begin{aligned} \frac{\partial F}{\partial t} &= -(aH(x,t) + bR(x,t)^{1/2}) \cdot \nu(x,t), \\ F(x,0) &= F_0(x), \quad x \in S^n \end{aligned} \tag{2}$$

with strictly convex initial data. Equation (2) is only weakly parabolic due to its invariance under the diffeomorphism group of S^n. Short time existence will follow from Hamilton's existence theorem for evolution equations with an integrability condition [4].

Theorem 2.1. *If M_0 is strictly convex, then equation* (2) *has a unique smooth solution $F(\cdot, t)$ for a short time.*

In this paper, we use the same notation as in [5] and we always sum over repeated indices from 1 to n. Let $\Theta = (Hg^{kl} - h^{kl})\nabla_k\nabla_l$. Then Θ is elliptic provided M is strictly convex. We remark that the second order partial differential operator $L = a\Delta + bR^{-1/2}\Theta$ plays the same role for (2) as does the Laplace–Beltrami operator Δ in [5]. If X and Y are 1-forms and $m = m_{ij}dx^i \otimes dx^j$ is a symmetric 2-tensor, then $\langle X, Y\rangle_m = m^{ij}X_iY_j$, where $m^{ij} = m_{kl}g^{ik}g^{jl}$. If A_{ikl} is a covariant 3-tensor, then

$$|A_{ikl}|^2_{h,g} = h^{ij}g^{km}g^{ln}A_{ikl}A_{jmn}.$$

3. Evolution of the metric and curvature

The evolution of the metric g and second fundamental form A are given by:

Lemma 3.1. *We have*

$$\frac{\partial}{\partial t}g_{ij} = -2(aH + bR^{1/2})h_{ij}, \tag{3}$$

$$\frac{\partial}{\partial t}\nu = \nabla(aH + bR^{1/2}), \tag{4}$$

$$\begin{aligned}\frac{\partial}{\partial t}h_{ij} = {} & (a\Delta + bR^{-\frac{1}{2}}\Theta)h_{ij} + bR^{-\frac{1}{2}}(\nabla_i H\nabla_j H - \nabla_i h_{kl}\nabla_j h_{pq}g^{kp}g^{lq}) \\ & - \frac{b}{4}R^{-\frac{3}{2}}\nabla_i R\nabla_j R - 2(aH + bR^{\frac{1}{2}})h_{ik}g^{kl}h_{lj} + a|A|^2h_{ij} \\ & + bR^{-\frac{1}{2}}(H|A|^2 - C)h_{ij},\end{aligned} \tag{5}$$

or equivalently,

$$\begin{aligned}\frac{\partial}{\partial t}h_{ij} = {} & a\Delta + bR^{-\frac{1}{2}}\Theta h_{ij} - bR^{-\frac{1}{2}}\Bigg[\frac{H^4}{4R}\nabla_i\left(\frac{|A|^2}{H^2}\right)\nabla_j\left(\frac{|A|^2}{H^2}\right) \\ & + \frac{1}{H^2}\big(H\nabla_i h_{kl} - \nabla_i H\cdot h_{kl}\big)\big(H\nabla_j h_{pq} - \nabla_j H\cdot h_{pq}\big)g^{kp}g^{lq}\Bigg] \\ & + a|A|^2h_{ij} + bR^{-\frac{1}{2}}(H|A|^2 - C)h_{ij} - 2(aH + bR^{\frac{1}{2}})h_{ik}g^{kl}h_{lj},\end{aligned} \tag{6}$$

where $C = h_{ij}h_{kl}h_{pq}g^{iq}g^{jk}g^{lp}$.

Proof. The lemma follows from a straightforward computation as in [5] and [2]. □

We now compute the evolution equations for H and $|A|^2$.

Lemma 3.2. *We have the following evolution equations:*

$$\frac{\partial}{\partial t}H = (a\Delta + bR^{-\frac{1}{2}}\Theta)H - bR^{-\frac{1}{2}}\left[\frac{1}{H^2}|H\nabla_i h_{kl} - \nabla_i H \cdot h_{kl}|^2 + \frac{H^4}{4R}\left|\nabla\left(\frac{|A|^2}{H^2}\right)\right|^2\right] + a|A|^2H + bR^{-\frac{1}{2}}(H|A|^2 - C)H, \tag{7}$$

$$\begin{aligned}\frac{\partial}{\partial t}H^2 &= (a\Delta + bR^{-\frac{1}{2}}\Theta)H^2 - 2a|\nabla H|^2 - 2bR^{-\frac{1}{2}}|\nabla H|^2_{Hg-h} \\ &\quad - 2bR^{-\frac{1}{2}}H\left[\frac{1}{H^2}|H\nabla_i h_{kl} - \nabla_i H \cdot h_{kl}|^2 + \frac{H^4}{4R}\left|\nabla\left(\frac{|A|^2}{H^2}\right)\right|^2\right] \\ &\quad + 2a|A|^2H^2 + 2bR^{-\frac{1}{2}}(H|A|^2 - C)H^2,\end{aligned} \tag{8}$$

$$\begin{aligned}\frac{\partial}{\partial t}|A|^2 &= (a\Delta + bR^{-\frac{1}{2}}\Theta)|A|^2 - 2a|\nabla A|^2 - 2bR^{-\frac{1}{2}}|\nabla A|^2_{Hg-h} \\ &\quad - 2bR^{-\frac{1}{2}}\left[\frac{1}{H^2}|H\nabla_i h_{kl} - \nabla_i H \cdot h_{kl}|^2_{h,g} + \frac{H^4}{4R}\left|\nabla\left(\frac{|A|^2}{H^2}\right)\right|^2_h\right] \\ &\quad + 2a|A|^4 + 2bR^{-\frac{1}{2}}(H|A|^2 - C)|A|^2,\end{aligned} \tag{9}$$

$$\begin{aligned}\frac{\partial}{\partial t}(|A|^2 - \frac{1}{n}H^2) &= (a\Delta + bR^{-\frac{1}{2}}\Theta)(|A|^2 - \frac{1}{n}H^2) - 2a(|\nabla A|^2 - \frac{1}{n}|\nabla H|^2) \\ &\quad - 2bR^{-\frac{1}{2}}(|\nabla A|^2_{Hg-h} - \frac{1}{n}|\nabla H|^2_{Hg-h}) \\ &\quad - \frac{b}{2}R^{-\frac{3}{2}}H^4\left|\nabla\left(\frac{|A|^2}{H^2}\right)\right|^2_{\mathring{h}} \\ &\quad - 2bR^{-\frac{1}{2}}H^{-2}|H\nabla_i h_{kl} - \nabla_i H \cdot h_{kl}|^2_{\mathring{h}} + 2a|A|^2(|A|^2 - \frac{1}{n}H^2) \\ &\quad + 2bR^{-\frac{1}{2}}(H|A|^2 - C)(|A|^2 - \frac{1}{n}H^2),\end{aligned} \tag{10}$$

where $\mathring{h}^{ij} = h^{ij} - \frac{1}{n}Hg^{ij}$.

Proof. Equations (7), (8), (9) and (10) follow from (3) and (6) by direct computation. □

4. Preserving convexity

We first derive the evolution equation for $\frac{|A|^2}{H^2}$.

Theorem 4.1. *We have the evolution equation*

$$\frac{\partial}{\partial t}\Big(\frac{|A|^2}{H^2}\Big) = (a\Delta + bR^{-\frac{1}{2}}\Theta)\left(\frac{|A|^2}{H^2}\right) + \frac{2a}{H}\left\langle \nabla\left(\frac{|A|^2}{H^2}\right), \nabla H\right\rangle \tag{11}$$
$$- \frac{2a}{H^4}|H\nabla_i h_{kl} - \nabla_i H \cdot h_{kl}|^2 + \frac{2b}{n}R^{-\frac{1}{2}}\left\langle \nabla\left(\frac{|A|^2}{H^2}\right), \nabla H\right\rangle_{Hg-h}$$
$$- \frac{2bR^{\frac{1}{2}}}{H^5}|H\nabla_i h_{kl} - \nabla_i H \cdot h_{kl}|^2 + \frac{b}{2}R^{-\frac{3}{2}}H\left|\nabla\left(\frac{|A|^2}{H^2}\right)\right|^2_{|A|^2 g - H\cdot h}.$$

Proof. Equation (11) follows directly from (8) and (9). □

By applying the maximum principle to Theorem 4.1, we have

Corollary 4.2. *If for any real number C, $\frac{|A|^2}{H^2} \le C$ at $t = 0$, then it remains so for $0 \le t < T$.* □

Corollary 4.3. *For $n = 2$, if for any real number $\epsilon > 0$, $h_{ij} \ge \epsilon H g_{ij}$ at $t = 0$, then it remains so for $0 \le t < T$. For $n \ge 3$, if $\frac{|A|^2}{H^2} \le C < C(n)$ at $t = 0$, where $C(n) = \frac{1}{n-1}$ is chosen small enough to ensure convexity, then there exist an $\epsilon > 0$ depending only on C such that $h_{ij} \ge \epsilon H g_{ij}$ for all time.* □

Lemma 4.4. *We have*

$$\frac{\partial}{\partial t}R = (a\Delta + bR^{-\frac{1}{2}}\Theta)R - 2a|\nabla H|^2 + 2a|\nabla A|^2$$
$$- \frac{1}{2}bHR^{-\frac{3}{2}}|\nabla R|^2 + \frac{1}{2}bR^{-\frac{3}{2}}|\nabla R|^2_h + 2a|A|^2 R \tag{12}$$
$$+ 2b(H|A|^2 - C)R^{\frac{1}{2}}.$$

Proof. By (5), we can derive the evolution equations of H^2 and $|A|^2$, then (12) follows directly from the Gauss equation $R = H^2 - |A|^2$. □

The equality

$$\langle \nabla R, \nabla H\rangle = 2H|\nabla H|^2 - \langle \nabla |A|^2, \nabla H\rangle$$

implies

$$|\nabla H|^2 \le \frac{1}{H}\langle \nabla R, \nabla H\rangle + |\nabla A|^2. \tag{13}$$

Combining Lemma 4.4 and (13), we have

$$\begin{aligned}\frac{\partial}{\partial t}R \geq (a\Delta + bR^{-\frac{1}{2}}\Theta)R - \frac{2a}{H}\langle\nabla R, \nabla H\rangle \\ - \frac{1}{2}bR^{-\frac{3}{2}}|\nabla R|^2_{Hg-h} + 2a|A|^2R + 2b(H|A|^2 - C)R^{\frac{1}{2}}.\end{aligned} \tag{14}$$

By the maximum principle to (14), if $R \geq C$ at $t = 0$, then it remains so. The inequality

$$(H|A|^2 - C) \geq \frac{1}{3}H(H^2 - |A|^2) = \frac{1}{3}HR$$

implies

$$\begin{aligned}\frac{\partial}{\partial t}R \geq (a\Delta + bR^{-\frac{1}{2}}\Theta)R - \frac{2a}{H}\langle\nabla R, \nabla H\rangle \\ - \frac{1}{2}bR^{-\frac{3}{2}}|\nabla R|^2_{Hg-h} + \left(\frac{2a}{n} + \frac{2b}{3}\right)R^2.\end{aligned}$$

We consider the ODE

$$\frac{\partial}{\partial t}\phi = \left(\frac{2a}{n} + \frac{2b}{3}\right)\phi^2, \qquad \phi(0) = R_{\min}(0) = \min{}_{M_0} R > 0.$$

It is easy to obtain

$$\phi(t) = \frac{1}{R_{\min}(0)^{-1} - \left(\frac{2a}{n} + \frac{2b}{3}\right)t},$$

$\phi(t) \to +\infty$ as $t \to \left(\left(\frac{2a}{n} + \frac{2b}{3}\right)^{-1} R_{\min}(0)^{-1}\right)^{-}$. We have from the maximum principle $R \geq \phi(t)$. Then the maximum time $T \leq \left(\frac{2a}{n} + \frac{2b}{3}\right)^{-1} R_{\min}(0)^{-1} < +\infty$. We thus obtain

Theorem 4.5. *Under the assumption of Corollary* 4.3, *the maximum time* T *that the Eq.* (2) *has a solution on the interval* $0 \leq t < T$ *is finite.*

5. Pinching of principal curvature

In this section, we shall show that the principal curvatures approach each other, at least in the sense of the following

Theorem 5.1. *If there exists an* $\epsilon > 0$ *such that* $h_{ij} \geq \epsilon H g_{ij}$ *for all time, then there exist constants* $\sigma > 0$ *and* $C_0 < \infty$ *depending only on the initial hypersurface* M_0

and the constants a and b such that

$$|A|^2 - \frac{1}{n}H^2 \leq C_0 H^{2-\sigma}$$

holds for all $0 \leq t < T$.

Therefore the conclusion of Theorem 5.1 holds under the assumptions of Theorem 1.1. The remainder of this section consists of the proof of Theorem 5.1.

We define $f_\sigma = (|A|^2 - \frac{1}{n}H^2)/H^{2-\sigma}$ and $f = f_0$. Then Theorem 4.1 implies

Lemma 5.2. *We have the evolution equation*

$$\begin{aligned}\frac{\partial}{\partial t} f &= (a\Delta + bR^{-\frac{1}{2}}\Theta) f + \frac{2a}{H}\langle \nabla f, \nabla H\rangle - \frac{2a}{H^4}|H\nabla_i h_{kl} - \nabla_i H \cdot h_{kl}|^2 \\ &+ \frac{2b}{H}R^{-\frac{1}{2}}\langle \nabla f, \nabla H\rangle_{Hg-h} - \frac{2b}{H^5}R^{\frac{1}{2}}|H\nabla_i h_{kl} - \nabla_i H \cdot h_{kl}|^2 \\ &+ \frac{b}{2}R^{-\frac{3}{2}}H|\nabla f|^2_{|A|^2 g - H\cdot h}.\end{aligned}$$

□

Lemma 5.3. *If there exists an* $\epsilon > 0$ *such that* $h_{ij} \geq \epsilon H g_{ij}$ *for all time, then there exist constants* $\sigma_1 > 0$, $C_1 > 0$ *and* $C_2 < +\infty$ *depending only on a, b and* M_0 *such that for* $\sigma < \sigma_1$,

$$\begin{aligned}\frac{\partial}{\partial t} f_\sigma &\leq (a\Delta + bR^{-\frac{1}{2}}\Theta) f_\sigma - C_1 H^{\sigma-2}|\nabla H|^2 + C_2 H^{-\sigma}|\nabla f_\sigma|^2 \\ &+ a\sigma f_\sigma |A|^2 + b\sigma R^{-\frac{1}{2}} f_\sigma (H|A|^2 - C).\end{aligned}$$

Proof. Combining Lemma 5.2 and (7), we have

$$\begin{aligned}\frac{\partial}{\partial t} f_\sigma &= H^\sigma (a\Delta + bR^{-\frac{1}{2}}\Theta) f + 2aH^{\sigma-1}\langle \nabla f, \nabla H\rangle \\ &- 2H^{\sigma-4}(a + bR^{\frac{1}{2}}H^{-1})|H\nabla_i h_{kl} - \nabla_i H \cdot h_{kl}|^2 \\ &+ 2bR^{-\frac{1}{2}}H^{\sigma-1}\langle \nabla f, \nabla H\rangle_{Hg-h} + \frac{b}{2}R^{-\frac{3}{2}}H^{1+\sigma}|\nabla f|^2_{|A|^2 g - Hh} \\ &+ \sigma H^{\sigma-1} f (a\Delta + bR^{-\frac{1}{2}}\Theta) H - b\sigma H^{\sigma-1} f R^{-\frac{1}{2}}(H^{-2}|H\nabla_i h_{kl} - \nabla_i H \cdot h_{kl}|^2 \\ &+ \frac{1}{4}H^4 R^{-1}|\nabla f|^2) + a\sigma H^\sigma |A|^2 f + b\sigma H^\sigma R^{-\frac{1}{2}} f (H|A|^2 - C).\end{aligned}$$

The equalities

$$\begin{aligned} R^{-\frac{1}{2}}\Theta f_\sigma &= R^{-\frac{1}{2}}H^\sigma\Theta f + \sigma(\sigma-1)H^{\sigma-2}fR^{-\frac{1}{2}}|\nabla H|^2_{Hg-h} \\ &\quad + \sigma R^{-\frac{1}{2}}H^{\sigma-1}f\Theta H + 2\sigma H^{\sigma-1}R^{-\frac{1}{2}}\langle\nabla H, \nabla f\rangle_{Hg-h}, \\ \Delta f_\sigma &= \sigma H^{\sigma-1}f\Delta H + \sigma(\sigma-1)H^{\sigma-2}f|\nabla H|^2 \\ &\quad + H^\sigma\Delta f + 2\sigma H^{\sigma-1}\langle\nabla H, \nabla f\rangle, \\ |\nabla f|^2 &= H^{2\sigma}|\nabla f_\sigma|^2 - 2\sigma H^{-1-2\sigma}f_\sigma\langle\nabla H, \nabla f_\sigma\rangle + \sigma^2H^{-2-2\sigma}f_\sigma^2|\nabla H|^2, \\ \langle\nabla H, \nabla f\rangle &= H^{-\sigma}\langle\nabla H, \nabla f_\sigma\rangle - \sigma H^{-\sigma-1}f_\sigma|\nabla H|^2, \end{aligned}$$

imply

$$\begin{aligned} \frac{\partial}{\partial t}f_\sigma &= (a\Delta + bR^{-\frac{1}{2}}\Theta)f_\sigma + \frac{b}{2}R^{-\frac{3}{2}}H^{1-\sigma}|\nabla f_\sigma|^2_{|A|^2g-Hh} \\ &\quad - \frac{b\sigma}{4}R^{-\frac{3}{2}}f_\sigma H^{3-2\sigma}|\nabla f_\sigma|^2 - b\sigma R^{-\frac{3}{2}}H^{-\sigma}f_\sigma\langle\nabla H, \nabla f_\sigma\rangle_{|A|^2g-Hh} \\ &\quad + \frac{b\sigma^2}{2}R^{-\frac{3}{2}}f_\sigma^2H^{2-2\sigma}\langle\nabla H, \nabla f_\sigma\rangle + 2a(1-\sigma)H^{-1}\langle\nabla H, \nabla f_\sigma\rangle \\ &\quad + 2b(1-\sigma)H^{-1}R^{-\frac{1}{2}}\langle\nabla H, \nabla f_\sigma\rangle_{Hg-h} \\ &\quad - [2H^{\sigma-4}(a + bR^{\frac{1}{2}}H^{-1}) + b\sigma H^{\sigma-3}fR^{-\frac{1}{2}}]|H\nabla_ih_{kl} - \nabla_iH\cdot h_{kl}|^2 \\ &\quad + \frac{b}{2}\sigma^2R^{-\frac{3}{2}}H^{-1-\sigma}f_\sigma^2|\nabla H|^2_{|A|^2g-Hh} - 3b\sigma(1-\sigma)H^{\sigma-2}fR^{-\frac{1}{2}}|\nabla H|^2_{Hg-h} \\ &\quad - 3a(1-\sigma)\sigma H^{-2}f_\sigma|\nabla H|^2 + a\sigma|A|^2f_\sigma + b\sigma R^{-\frac{1}{2}}(H|A|^2 - C)f_\sigma. \end{aligned} \tag{15}$$

The coefficient of $|\nabla f_\sigma|^2$ in (15)

$$\frac{b}{2}H^{-\sigma}R^{-\frac{1}{2}}H\left(\frac{|A|^2}{R}g - \frac{Hh}{R} - \frac{\sigma H^2}{2R}fg\right) \le C_3H^{-\sigma}g$$

where the constant $C_3 < +\infty$. The coefficient of $\langle\nabla f_\sigma, \nabla H\rangle$ in (15)

$$R^{-\frac{1}{2}}\left(-b\sigma f\cdot\frac{|A|^2}{R}g + b\sigma f\frac{Hh}{R} + \frac{b\sigma^2}{2}f^2\frac{H^2}{R}g + 2a(1-\sigma)\frac{R^{\frac{1}{2}}}{H}g \right.$$
$$\left. + 2b(1-\sigma)g - 2b(1-\sigma)\frac{h}{H}\right)$$

is between 0 and a positive constant provided σ is sufficiently small. By Lemma 2.3 of [5],

$$|H\nabla_ih_{kl} - \nabla_iH\cdot h_{kl}|^2 \ge \frac{1}{2}\epsilon^2H^2|\nabla H|^2,$$

thus

$$[2H^{\sigma-4}(a+bR^{\frac{1}{2}}H^{-1})+b\sigma H^{\sigma-3}fR^{-\frac{1}{2}}]\,|H\nabla_i h_{kl}-\nabla_i H\cdot h_{kl}|^2 \geq \big[(a+bR^{\frac{1}{2}}H^{-1})+\frac{b\sigma}{2}HR^{-\frac{1}{2}}f\big]\epsilon^2 H^{\sigma-2}|\nabla H|^2.$$

Therefore applying the inequality $\langle p,q\rangle \leq \frac{\eta}{2}|p|^2+\frac{1}{2\eta}|q|^2$, $\eta>0$ with $p=H^{\frac{\sigma-1}{2}}\nabla H$ and $q=H^{\frac{1-\sigma}{2}}\nabla f_\sigma$ to (15), we obtain Lemma 5.3 for sufficiently small σ. □

The evolution of the L^p-norms of f_σ is given by

Lemma 5.4. *If there exists an $\epsilon>0$ such that $h_{ij}\geq \epsilon Hg_{ij}$ for all time, then there exist constants $0<C_5,C_6,C_7,C_8<+\infty$ depending only on a, b and M_0 such that for $p\geq C_5$,*

$$\frac{\partial}{\partial t}\int_M f_\sigma^p d\mu \leq \int_M[-C_6p^2f_\sigma^{p-2}|\nabla f_\sigma|^2 - C_7pH^{\sigma-2}f_\sigma^{p-1}|\nabla H|^2+C_8p\sigma H^2 f_\sigma^p].$$

Proof. By Lemma 5.3 and following equations

$$\Delta f_\sigma^p = p(p-1)f_\sigma^{p-2}|\nabla f_\sigma|^2+pf_\sigma^{p-1}\Delta f_\sigma,$$
$$\Theta f_\sigma^p = p(p-1)f_\sigma^{p-2}|\nabla f_\sigma|^2_{Hg-h}+pf_\sigma^{p-1}\Theta f_\sigma,$$

we have

$$\begin{aligned}\frac{\partial}{\partial t}f_\sigma^p \leq{}& (a\Delta+bR^{-\frac{1}{2}}\Theta)f_\sigma^p - ap(p-1)f_\sigma^{p-2}|\nabla f_\sigma|^2\\ &- bp(p-1)f_\sigma^{p-2}R^{-\frac{1}{2}}|\nabla f_\sigma|^2_{Hg-h}-pC_1f_\sigma^{p-1}H^{\sigma-2}|\nabla H|^2\\ &+pC_2H^{-\sigma}f_\sigma^{p-1}|\nabla f_\sigma|^2+a\sigma pf_\sigma^p|A|^2+pb\sigma R^{-\frac{1}{2}}f_\sigma^p(H|A|^2-C)\end{aligned}$$

and hence

$$\begin{aligned}\frac{\partial}{\partial t}\int f_\sigma^p d\mu \leq \int_M &[bR^{-\frac{1}{2}}\Theta f_\sigma^p - ap(p-1)f_\sigma^{p-2}|\nabla f_\sigma|^2\\ &- bp(p-1)f_\sigma^{p-2}R^{-\frac{1}{2}}|\nabla f_\sigma|^2_{Hg-h}-C_1pf_\sigma p-1H^{\sigma-2}|\nabla H|^2\\ &+C_2pf_\sigma^{p-1}H^{-\sigma}|\nabla f_\sigma|^2+a\sigma pf_\sigma^p|A|^2+pb\sigma R^{-\frac{1}{2}}f_\sigma^p(H|A|^2-C)].\end{aligned} \tag{16}$$

Now

$$\int_M R^{-\frac{1}{2}}\Theta f_\sigma^p = -\frac{p}{2}\int_M R^{-\frac{3}{2}}H^{2-\sigma}f_\sigma^{p-1}|\nabla f_\sigma|^2_{Hg-h} + \frac{\sigma p}{2}\int_M R^{-\frac{3}{2}}H^{1-\sigma}f_\sigma^p\langle\nabla H,\nabla f_\sigma\rangle_{Hg-h} + p\int_M R^{-\frac{1}{2}}H^{-1}f_\sigma^{p-1}\langle\nabla H,\nabla f_\sigma\rangle_{Hg-h}.$$

Clearly $f_\sigma \leq H^\sigma$. Therefore

$$\int_M R^{-\frac{1}{2}}y\Theta f_\sigma^p \leq C_4 p\int_M R^{-\frac{1}{2}}f_\sigma^{p-1}|\nabla H||\nabla f_\sigma|, \tag{17}$$

where $C_4 < +\infty$ is a constant depending only on a, b and M_0. Applying the inequalities $\langle a,b\rangle \leq \frac{\eta}{2}|a|^2 + \frac{1}{2\eta}|b|^2$, $\eta > 0$ and $R \leq H^2$ to (16) and (17), we have

$$\frac{\partial}{\partial t}\int_M f_\sigma^p d\mu \leq \int_M [-C_6 p^2 f_\sigma^{p-2}|\nabla f_\sigma|^2 - C_7 pH^{\sigma-2}f_\sigma^{p-1}|\nabla H|^2 + C_8 p\sigma H^2 f_\sigma^p],$$

where $0 < C_5, C_6, C_7, C_8 < +\infty$ depending only on a, b and M_0 provided $p \geq C_5$. This proves Lemma 5.4. □

By Lemma 5.4 of Chow [1], we can easily obtain suitable L^p bounds on f_σ provided p is sufficiently large and σ is sufficiently small (see [8] for details). Now by using the Sobolev inequality for submanifolds of R^n in [7] and an iteration type argument, we may continue exactly as in [5] to obtain a supremum bound on f_σ for σ sufficiently small. This completes the proof of Theorem 5.1. □

6. Long time existence

In this section, following the method of K. Tso [10] we prove long existence for Eq. (2) by using the support function of M. To define the support function $\bar{h}$ we parametrize M by the inverse of the Gauss map. That is, $F : S^n \to M^n$ is defined by $\nu(F(x)) = x$ for all $x \in S^n$. F is well-defined since M is strictly convex. The support function is defined by $\bar{h}(\nu) = \langle F(\nu), \nu\rangle$, $\nu \in S^n$. Then the Eq. (2) is equivalent to

$$\frac{\partial}{\partial t}\bar{h}(\nu,t) = -(aH + bR^{\frac{1}{2}}), \quad \nu \in S^n. \tag{18}$$

We extend $\bar{h}$ to R^{n+1} so that it is homogeneous of degree 1. That is, $\bar{h}(\nu,t) = |x|\bar{h}(x/|x|)$ for $x \in R^{n+1}$. $\bar{h}$ is uniquely determined by its restrictions to hyperplanes tangent to S^n. As a typical case, we consider the hyperplane $\{x \in R^{n+1} : x_{n+1} = -1\}$. Define $u : R^n \to R$ by $u(x) = \bar{h}(x,-1)$, $x \in R^n$. We refomulate Eq. (2) as an equation in u by computing the second fundamental form and metric in terms of u.

By [2], we have

$$h_{ij} = \frac{1}{\sqrt{1+|x|^2}} u_{ij}, \quad g_{ij} = u_{ik} u_{jl} (\delta^{kl} + x^k x^l), \tag{19}$$

where we denote $u_{ij} = \dfrac{\partial^2 u}{\partial x^i \partial x^j}$ and $u_t = \dfrac{\partial u}{\partial t}$. Then

$$\begin{aligned} H &= \frac{1}{\sqrt{1+|x|^2}} (\delta_{kl} - \frac{x_k x_l}{1+|x|^2} (u^{-1})^{kl}), \\ |A|^2 &= \frac{1}{(1+|x|^2)^3} (\delta_{ij} - \frac{x_i x_j}{1+|x|^2})(\delta_{kl} - \frac{x_k x_l}{1+|x|^2})(u^{-1})^{ik} (u^{-1})^{jl}, \\ R &= \frac{1}{(1+|x|^2)^3} (\delta_{ij} - \frac{x_i x_j}{1+|x|^2})(\delta_{kl} - \frac{x_k x_l}{1+|x|^2})((u^{-1})^{ij} (u^{-1})^{kl} \\ &\quad - (u^{-1})^{ik} (u^{-1})^{jl}), \end{aligned} \tag{20}$$

and Eq. (2) becomes

$$\begin{aligned} u_t &= (1+|x|^2)^{1/2} (aH + bR^{1/2}) \\ &= -(a(\delta_{kl} - \frac{x_k x_l}{1+|x|^2})(u^{-1})^{kl} + b(1+|x|^2)^{-1} (\delta_{ij} - \frac{x_i x_j}{1+|x|^2})^{\frac{1}{2}} \\ &\quad \cdot (\delta_{kl} - \frac{x_k x_l}{1+|x|^2})^{\frac{1}{2}} ((u^{-1})^{ij} (u^{-1})^{kl} - (u^{-1})^{ik} (u^{-1})^{jl})^{\frac{1}{2}}). \end{aligned} \tag{21}$$

As is usual for parabolic equations, the Hölder norms $C^{k,\alpha}(S^n \times [0,t])$ are defined with time derivatives counting as two space derivatives. We then have

Lemma 6.1. *If $\bar{h} \in C^2(S^n \times [0,t])$ is a smooth solution of* (18) *and $\bar{h} \geq r > 0$, then there exists a constant $C < +\infty$ depending only on r, a, b and M_0 such that*

$$\|\bar{h}\|_{C^2(S^n \times [0,t])} \leq C.$$

The proof of Lemma 6.1 follows by trivial modifications of the proofs in [2] and [8]. □

The $C^{2,\alpha}$ estimate of $\bar{h}$ now follows from a theorem of Krylov-Safonov [6] and an adaptation of an argument due to Trudinger [9] (again see [10] for details). The C^∞ estimate for $\bar{h}$ now follows from standard arguments. By applying the previous estimates to the normalized equation, we may show C^∞-convergence to a hypersurface $\widetilde{M}_T$ for the normalized equation.

We now show how C^∞-convergence to a convex hypersurface $\widetilde{M}_T$ and the estimate $f = (|A|^2 - (1/n)H^2)/H^2 \leq CH^\sigma$ for some $\sigma > 0$ and $C < +\infty$ implies Theorem 1.1. Clearly there is some point p on $\widetilde{M}_T$ with $K(p) > 0$. By continuity there is a neighborhood U of p with $K > 0$ in U. Then for the unnormalized equation, $H(x)$ approaches infinity for every $x \in U$. By the invariance of f under scaling, $f(x) = 0$

for every $x \in U$. Therefore U is totally umbilical and hence K is a constant in U. Clearly, K must then be a constant in $\widetilde{M}_T$. Hence $\widetilde{M}_T$ is a standard sphere. This completes the proof of Theorem 1.1.

References

[1] B. Chow, Deforming convex hypersurfaces by the n-th root of the Gaussian curvature, J. Differential Geom. 33 (1985), 117–138.

[2] —–, Deforming convex hypersurfaces by the square root of the scalar curvature, Invent. Math. 87 (1987), 63–82.

[3] M. Gage, R. S. Hamilton, The shrinking of convex plane curves by the heat equation, J. Differential Geom. 23 (1986), 69–96.

[4] R. S. Hamilton, Three-manifolds with positive Ricci curvature, J. Differential. Geom. 17 (1982), 255–306.

[5] G. Huisken, Flow by mean curvature of convex surfaces into sphere, J. Differential Geom. 20 (1984), 237–266.

[6] N. V. Krylov, M. V. Safonov A certain property of solutions of parabolic equations with measurable coefficients, Math. USSR. Izv. 16 (1981), 151–164.

[7] J. H. Michael, L. M. Simon, Sobolev and mean value inequalities on generalized submanifolds of R^n, Comm. Pure Appl. Math. 26 (1973), 316–379.

[8] Y. B. Shen, W. M. Sheng, Deforming convex hypersurfaces by the linear combination of the mean curvature and n-th root of the Gauss–Kronecker curvature, to appear in Acta Math. Sci. 17 (1997).

[9] N. S. Trudinger, Elliptic equations in nondivergence form, Proc. Miniconference on PDE, Canberra 1981, 1–6.

[10] K. Tso, Deforming a hypersurface by its Gauss–Kronecker curvature, Comm. Pure Appl. Math. 38 (1985), 867–882.

Asymptotics for exterior solutions of quasilinear elliptic equations

*Leon Simon**

Abstract. We consider asymptotic behaviour of solutions of quasilinear elliptic equations which are entire (defined and smooth on all of $\mathbb{R}^n$) or exterior (defined and smooth over all of $\mathbb{R}^n \setminus K$ where K is compact). There are a number of new results, but we also take this opportunity to give a brief survey of the known results about exterior and entire solutions, beginning with the classical work of Bernstein.

Introduction

We here consider exterior and entire solutions of elliptic equations of the form

$$\sum_{i,j=1}^{n} a_{ij} D_i D_j u = 0 \qquad \ddagger$$

where (a_{ij}) is a positive definite matrix. The solution u is always assumed to be C^2, and is said to be *exterior* if ‡ holds outside some bounded subset of $\mathbb{R}^n$, and *entire* if ‡ holds everywhere on $\mathbb{R}^n$. An exterior solution u is said to have *polynomial growth* if $\exists$ constants c, γ such that $|u(x)| \le c|x|^\gamma$ for $|x| \ge 1$; it is said to have *linear growth* if this holds with $\gamma = 1$, and *sub-linear growth* if $|x|^{-1}u(x) \to 0$ as $|x| \to \infty$.

The coefficients a_{ij} in ‡ may depend on the variables $x = (x^1, \ldots, x^n)$ (i.e. the independent variables) and also on the dependent variables u, Du (gradient of u), and, in some cases, on D^2u. We shall mainly concentrate on entire solutions, only occasionally mentioning results for exterior solutions. There are quite a few new results concerning exterior and entire solutions of ‡ discussed here, but we also take the present paper as an opportunity to give a brief survey of the principal known results, and to draw attention to some interesting questions which, as far as the present author is aware, are still open.

---

* This paper is a revision and updating of the (unpublished) research report [SL9]. The research for this paper was partially supported by NSF grants DMS–9504456 and DMS–9207704 at Stanford University.

1. The linear case

Here we assume $a_{ij} = a_{ij}(x)$ are locally Lipschitz in $\mathbb{R}^n$. In this case there are many entire solutions. Indeed Lax in [LP] demonstrated quite generally that if the equation ‡ has the unique continuation property (this is guaranteed by the local Lipschitzness of the coefficients—see e.g. [H], [GL]) and if u is any C^2 solution of ‡ on any bounded domain $\Omega \subset \mathbb{R}^n$, then there exists a sequence $\{u_k\}$ of *entire* solutions which converge to u uniformly on compact subsets of Ω. Thus there always exists a very large class of entire solutions in the linear case. However one should keep in mind that all these entire solutions may have very restricted behaviour with respect to their asymptotics at infinity.

For example in the case when $a_{ij}(x) \equiv \delta_{ij} + x^i x^j$ the equation (‡) is just $\Delta u + r^2 \dfrac{\partial^2 u}{\partial r^2} = 0$ and it is possible to prove that all solutions $u \in C^2(\mathbb{R}^n)$ are asymptotically "conic" on approach to infinity in the at least in the weak sense that for each eigenfunction φ_j of $-\Delta_{S^{n-1}}$ there are constants $C_1^{(j)}$, $C_2^{(j)}$ such that $\langle u(r), \varphi_j \rangle_{L^2(S^{n-1})} \sim C_1^{(j)} + C_2^{(j)} r$ as $r \uparrow \infty$. One can check this by making an expansion $u(r\omega) = \sum_j \psi_j(r)\varphi_j(\omega)$, where φ_j are the complete orthonormal set of eigenfunctions for the Laplacian operator on S^{n-1}. Then since $-\Delta\varphi_j = \lambda_j \varphi_j$ with $\lambda_j \geq 0$, we can check that ψ_j satisfies the second order ODE $r^2(1 + r^2)\psi_j'' + (n - 1)r\psi_j' - \lambda_j \psi_j = 0$. Now, as one checks by changing variable $s = r^{-1}$ and computing series solutions of the resultant ODE near $s = 0$, all solutions of this ODE behave asymptotically as $C_{j,1} + C_{j,2}r + E_j(r)$ as $r = \infty$, with $|E_j(r)| \leq C_{j,3}/r$ for some constants $C_{j,1}$, $C_{j,2}$, $C_{j,3}$, whence the result claimed above.

Of course if we further make asymptotic assumptions about the behaviour of the coefficients (e.g. that

$$a_{ij}(x) \to a_{ij}^0 \text{ as } |x| \to \infty,$$

where (a_{ij}^0) is a constant positive definite matrix, and where the convergence means $\sup_{|x|=\rho} |a_{ij}(x) - a_{ij}^0| \to 0$ as $\rho \to \infty$), then the above-mentioned result of Lax combined with standard growth estimates implies that there are many entire solutions which grow faster than $|x|^\gamma$ for any given $\gamma > 0$. For the convenience of the reader we describe the argument for this: First note that by an orthogonal transformation of coordinates we can assume that $a_{ij}^0 = \delta_{ij}$, and for $\rho > 0$ we let $a_{ij}^\rho(x) = a_{ij}(\rho x)$, so that $a_{ij}^\rho \to \delta_{ij}$ uniformly on $\mathbb{R}^n \setminus B_\sigma(0)$ as $\rho \to \infty$ for each $\sigma > 0$. Now consider the solution $u = u^\rho$ of Dirichlet problem $L^\rho u = 0$ on B_1, and $u|\partial B_1 = v|\partial B_1$, where $L^\rho = \sum_{ij} a_{ij}^\rho D_i D_j$ and where v is any homogeneous degree k harmonic polynomial on $\mathbb{R}^n$. Using elliptic L^2 estimates it is easily checked that $u^\rho \to v$ in L^2 on B_1 and by the Lax result mentioned above we then have entire solutions $\widetilde{u}^\rho$ of the equation $L^\rho u = 0$ with $\widetilde{u}^\rho \to v$ on B_1.

Now for $\sigma < \rho$ let $\|u\|_{\rho,\sigma} \equiv \rho^{-n/2}\|u\|_{L^2(B_\rho \setminus B_\sigma)}$, and take q to be any positive real number different than all the exponents $(n-2)/2 \pm q_j$. We claim there are

$R = R(n, q) \geq 2$ and $\epsilon = \epsilon(n, q) > 0$ such that if $\rho_0 > 1$ is large enough to ensure that $\sup_{|x|\geq\rho_0} |a_{ij}(x) - \delta_{ij}| < \epsilon$, and if u is a solution of ‡ on $\mathbb{R}^n \backslash B_{\rho_0}$, then we have the implication

(i) $\|u\|_{\rho R^2, \rho R} \geq R^q \|u\|_{\rho R, \rho} \Rightarrow \|u\|_{\rho R^3, \rho R^2} \geq R^q \|u\|_{\rho R^2, \rho R}$.

The number ϵ is independent of ρ, hence the above inequality can be indefinitely iterated (with $\rho = \rho_0 R^j$, $j = 1, 2, \ldots$) to show that in fact $\|u\|_{\rho_0 R^j, \rho_0 R^{j-1}} \geq CR^{jq}$ for all j, which of course guarantees that u satisfies the lower growth bound $\liminf_{R\to\infty} R^{-q}\|u\|_{L^2(B_R)} > 0$, provided only that we have the "starting hypothesis" $\|u\|_{\rho_0 R^2, \rho_0 R} \geq R^q \|u\|_{\rho_0 R, \rho_0}$. Of course for a given q the starting hypothesis is satisfied by the functions $\widetilde{u}(x)$ obtained by rescaling the $\widetilde{u}^{\rho_0}$ constructed above, according to $\widetilde{u}(x) = \widetilde{u}^{\rho_0}(\rho_0^{-1}x)$, provided we choose $k > q$ to begin with, and provided ρ_0 is sufficiently large (depending on the constant $\epsilon = \epsilon(n, q)$ above); this works because $\widetilde{u}^{\rho_0}$ was chosen to satisfy $\widetilde{u}^{\rho_0} \to v$ in B_1, which gives $\rho_0^{-n}\|\widetilde{u} - \rho_0^{-k}v\|^2_{L^2(B_{\rho_0})} \to 0$ as $\rho_0 \to \infty$. So then we have the starting hypotheses for example with $R^{-3}\rho_0$ in place of ρ_0 ($R \geq 2$ arbitrary), provided we choose ρ_0 large enough depending on R.

One proves (i) by first checking that it suffices to prove the same fact about functions u which are *harmonic* on the annular region $B_{\rho R^3} \backslash B_\rho$. (Cf. the simple indirect proof described in [SL10, pp. 249–250]; note that the argument in [SL10] involves a translation of the t variable to reduce the parameter τ in [SL10] to zero, and also that to apply the argument of [SL10, p. 249] to the present context we need to make the change of variable $t = \log r$; then we are in the setting of [SL10] with $\Sigma = S^{n-1}$.)

Having reduced the proof of (i) to the case when u is actually harmonic on $B_{\rho R^3} \backslash B_\rho$, we next note that for any u which is harmonic on the annulus $B_{\rho R^3} \backslash B_\rho$ it is standard that we can write an expansion in terms of a complete orthonormal set $\varphi_0 (\equiv \text{const.}), \varphi_1(\omega), \varphi_2(\omega), \ldots$ of eigenfunctions of the Laplacian on S^{n-1}; thus for $n \geq 3$

(ii) $u(r\omega) = r^{-(n-2)/2} \sum_{j=0}^{\infty} (a_j r^{q_j} + b_j r^{-q_j})\varphi_j$,

for $n \geq 3$, where $q_j = \sqrt{((n-2)/2)^2 + \lambda_j}$ and λ_j is the eigenvalue corresponding to φ_j, so that $-\Delta_{S^{n-1}}\varphi_j = \lambda_j \varphi_j$ and $\lambda_0 = 0$, $\varphi_0 = $ constant. (Actually the eigenvalues are given as $\lambda_j = k_j(k_j + n - 2)$, where $k_j \leq k_{j+1}$ are non-negative integers, $k_0 = 0$, $k_1 = 1$, and each positive integer appears with finite multiplicity (greater than one) in the sequence $k_1, k_2, \ldots$; furthermore, each $r^{k_j}\varphi_j$ is a homogeneous harmonic polynomial of degree k_j. (Because all eigenfunctions of Laplacian on S^{n-1} arise as the restriction of homogeneous harmonic polynomials on $\mathbb{R}^n$.) In case $n = 2$ we have a similar expression but with a logarithmic entry in the first term:

(ii)$'$ $u(r\omega) = a_0 + b_0 \log r + \sum_{j=1}^{\infty} (a_j r^{q_j} + b_j r^{-q_j})\varphi_j$,

Now the norm $\|u\|^2_{\rho R,\rho}$ introduced above is given by

$$\|u\|^2_{\rho R,\rho} = (\rho R)^{-n} \int_\rho^{\rho R} \|u(r)\|^2_{L^2(S^{n-1})} r^{n-1} dr,$$

and hence by orthonormality of the φ is given explicitly for $n \geq 3$ by

$$(\rho R)^{-n} \int_\rho^{\rho R} \sum_{j=0}^{\infty} (a_j r^{q_j} + b_j r^{-q_j})^2 \, r dr,$$

while for $n = 2$ the first term in the sum is $\int_\rho^{\rho R} (a_0 + b_0 \log r)^2 \, r dr$. We claim that then in fact, for $R \geq R_0$,

$$\|u\|^2_{\rho,\rho R} \approx \sum_{\{0 \leq q_j \leq 1\}} \left(a_j^2 (\rho R)^{2\gamma_j^+} + b_j^2 (\rho R)^{2\gamma_j^-}\right) + \sum_{\{q_j > 1\}} \left(a_j^2 (\rho R)^{2\gamma_j^+} + R^{-n} b_j^2 \rho^{-2\gamma_j^-}\right),$$

where, here and subsequently, we use the notation that $\|u\|^2_{\rho,\rho R} \approx E$ (with E the expression on the right of the above inequality) means that there is a fixed constant C, depending only on n, such that

$$C^{-1} E \leq \|u\|^2_{\rho,\rho R} \leq C(1 + \log^2(\rho R) + \log^2 R)\, E.$$

To see this we need the following elementary calculus lemma (with $\delta = q_j$), the proof of which is left to the reader:

Lemma. *If a, b are real, $\rho > 0$ and $\theta \in (0, 1)$, then*

$$C^{-1}(a^2(\rho R)^{2\delta+2} + b^2(\rho R)^{-2\delta+2}) \leq \int_\rho^{\rho R} (ar^\delta + br^{-\delta})^2 \, r dr$$
$$\leq C(a^2(\rho R)^{2\delta+2} + b^2(\rho R)^{-2\delta+2}), \qquad \text{if } \theta < \delta < 1 - \theta$$
$$C^{-1}\delta^{-1}(a^2(\rho R)^{2\delta+2} + b^2\rho^{-2\delta+2}) \leq \int_\rho^{\rho R} (ar^\delta + br^{-\delta})^2 \, r dr$$
$$\leq C\delta^{-1}(a^2(\rho R)^{2\delta+2} + b^2\rho^{-2\delta+2}), \qquad \text{if } \delta \geq 1 + \theta$$
$$C^{-1}(R\rho)^2(a^2 + b^2) \leq \int_\rho^{\rho R} (a + b\log r)^2 \, r dr \leq C(R\rho)^2(a^2 + b^2(\log^2 R + \log^2 \rho))$$
$$C^{-1}(a^2(\rho R)^4 + b^2 \log R) \leq \int_\rho^{\rho R} (ar + br^{-1})^2 \, r dr \leq C(a^2(\rho R)^4 + b^2 \log R),$$

for $R \geq R_0$, where in each case $C \geq 1$ and $R_0 \geq 2$ are constants depending only on n, and also on θ in the first two parts, and not depending on R, ρ or a, b.

Using the above expession for $\|u\|_{\rho R,\rho}$ it is now elementary to check (keeping in mind that, by a suitable rescaling, we may reduce to the case $\rho = 1$) the growth property (i) holds as claimed.

Thus, subject to the above condition that $a_{ij}(x) \to a^0_{ij} (= \delta_{ij})$ as $|x| \to \infty$, for any given k we can indeed find entire solutions with growth faster than $\|x\|^k$ in the L^2-sense that $R^{-n/2}\|u\|_{L^2(B_R \setminus B_{R/2})} \geq R^k$ for all sufficiently large R. Presumably in this case there are also solutions which grow faster than any power of $|x|$ in some directions, although the author is not aware of general results of this type.

Another question of interest in the linear case is whether or not, subject to the above restriction that $a_{ij} \to a^0_{ij}$ as $|x| \to \infty$, one can find solutions asymptotic to any given solution of the constant coefficient equation $a^0_{ij} D_i D_j u = 0$. (After a linear transformation of coordinates, this is the same as asking whether or not there exist solutions asymptotic to any given homogeneous harmonic polynomial, assuming that the coefficients $a_{ij}(x)$ converge to δ_{ij} as $|x| \to \infty$.)

In the non-linear case (i.e. the case when a_{ij} can depend on x, u, Du, and even D^2u), the situation is more complicated; in certain cases there may be no entire solutions at all (as is the case when $n = 2$ and ‡ is the minimal surface equation—see the discussion below). We divide our discussion of the non-linear case into the two cases $n = 2$, $n \geq 3$; as the reader will see, far more is known about questions of asymptotics in case $n = 2$.

2. The non-linear case, $n = 2$

A striking theorem of very general type is the following one due to Bernstein [B]:

Theorem 1. *In case $n = 2$ the only entire solutions of ‡ which have sub-linear growth are the constant functions.*

What is particularly striking about this theorem is its generality; the coefficient matrix (a_{ij}) need only be positive definite, no continuity restrictions are imposed. In fact Bernstein actually proved the more general "geometric theorem" that if G is the graph of any C^2 function of sublinear growth, and if the Gauss curvature K of G satisfies $K \leq 0$, then $K \equiv 0$. This geometric theorem evidently implies Theorem 1 for the following well-known reason:

At a given point x_0, after a linear change of variable the Hessian D^2u of u is diagonalized—$(D^2u) = \text{diag}\,(u_{11}, u_{22})$—and the equation can be written in the form

$$\alpha u_{11} + \beta u_{22} = 0,$$

where α, β are positive. By squaring and dividing out by $\alpha\beta$ this gives $-2u_{11}u_{22} = \frac{\alpha}{\beta}u_{11}^2 + \frac{\beta}{\alpha}u_{22}^2 \geq \min\{\frac{\alpha}{\beta}, \frac{\beta}{\alpha}\}(u_{11}^2 + u_{22}^2)$. In terms of the original coordinates, this gives

$$-(u_{11}u_{22} - u_{12}^2) \geq \gamma(u_{11}^2 + u_{22}^2 + 2u_{12}^2)$$

where $\gamma = \frac{1}{2}\min\{\frac{\alpha}{\beta}, \frac{\beta}{\alpha}\} \geq \frac{1}{2}\frac{\lambda_1}{\lambda_2}$, with λ_1, λ_2 the minimum and maximum eigenvalues of the positive definite coefficient matrix (a_{ij}). Since $K = \frac{u_{11}u_{22} - u_{12}^2}{(1 + |Du|^2)^2}$, this shows

$$-K \geq \gamma(1 + |Du|^2)^{-2}(u_{11}^2 + u_{22}^2 + 2u_{12}^2),$$

so that $K \leq 0$ and equality holds if and only if Hessian $u = 0$, so that the alternative $K \equiv 0$ in Bernstein's general theorem in this case leads to the conclusion that u is linear + const. (Hence we get Theorem 1 immediately by virtue of the boundedness of u.)

Bernstein's original proof of his general geometric theorem actually had a gap in the final stage of the argument (related to connectivity properties of open subsets of $\mathbb{R}^2$); an appropriate connectivity property was formulated and proved by E. Hopf [HE1], who then reworked Bernstein's entire proof in [HE2]. E. Mickle [ME] also published a revision of Bernstein's proof.

Notice that there are counterexamples to Theorem 1 if u is allowed to have linear growth—indeed there are graphs G with $K < 0$ everywhere and $|u(x)| \leq c|x|$ for $|x| \geq 1$. (For example, $u(x, y) = x \arctan y$ is such an example; more generally $u(x, y) = x\varphi(y)$ with φ smooth, positive, bounded and strictly increasing—thanks to Bruce Solomon for pointing out these examples.) Such couterexamples also give counterexamples to Theorem 1 in case the hypothesis of sublinear growth is replace by linear growth, because $K < 0$ implies that u does indeed satisfy an equation of the form ‡ with postive definite coefficient matrix (a_{ij}).

Bernstein was able to apply Theorem 1 to deduce his famous result (also referred to as Bernstein's Theorem) that

Theorem 2. *Other than the affine functions (i.e. linear + constant) there are no entire solutions of the minimal surface equation*

$$\Delta u - \sum_{i,j=1}^{2} \frac{D_i u D_j u}{1 + |Du|^2} D_i D_j u = 0$$

(i.e. the equation ‡ in case $a_{ij} = \delta_{ij} - \frac{D_i u D_j u}{1 + |Du|^2}$*).*

This immediately follows from Theorem 1 once we notice that $\varphi = \arctan D_1 u$ satisfies the elliptic equation

$$\sum_{i,j=1}^{2} \left(\delta_{ij} - \frac{D_i u D_j u}{1 + |Du|^2}\right) D_i D_j \varphi = 0,$$

because then Theorem 1 gives $\varphi \equiv$ const, hence $D_1 u \equiv$ const. Similarly $D_2 u \equiv$ const, and hence u is affine as required. The fact that $\arctan D_1 u$ and $\arctan D_2 u$ satisfy the above equation is rather unpleasant to check by direct calculation, but one can easily see it by a geometric argument: The equation for φ simply says that φ, thought of as a function on the graph G, is harmonic relative to the intrinsic Laplacian operator on G. (See e.g. [OR] for discussion of this.) On the other hand the Gauss map of the graph of a solution of the MSE is conformal as is the stereographic projection S of $S^2\backslash\{(0,0,-1)\}$ into $\mathbb{R}^2$, and hence also the composition of S and the isometry of S^2 obtained by permuting the 2nd and 3rd coordinate functions (i.e. the map $(y^1, y^2, y^3) \mapsto S(y^1, y^3, y^2)) : S^2\backslash\{(0,-1,0)\} \to \mathbb{R}^2$). Thus, with respect to suitable conformal coordinates for the graph of u, the function $-(1+|Du|^2)^{-1/2}u_x + i(1+|Du|^2)^{-1/2}$ is holomorphic, so that arctan of the ratio of the real and imaginary parts (i.e. arctan u_x) is harmonic in these coordinates, hence harmonic on the graph of u, as required.

An interesting problem is to try to classify the class of quasilinear equations for which such a "Bernstein property" (i.e. the result "entire $\Rightarrow$ affine") holds. As shown in [SL1] (for an alternative exposition of the proof, see [GT, Chapter 16]), this class includes any "equation of minimal surface type" (the class of equations introduced by Finn [FR]); i.e. equations of the form ‡ for which there is a constant $c > 0$ with

$$c^{-1}\sum_{i,j=1}^{2}\Big(\delta_{ij} - \frac{D_iuD_ju}{1+|Du|^2}\Big)\xi^i\xi^j \le \sum_{i,j=1}^{2} a_{ij}\xi^i\xi^j \le c\sum_{i,j=1}^{2}\Big(\delta_{ij} - \frac{D_iuD_ju}{1+|Du|^2}\Big)\xi^i\xi^j$$

for all $x,\ \xi \in \mathbb{R}^2$.

Here we want to add a new class of equations for which the Bernstein property holds: we consider the class of equations of variational type; that is, equations of the form ‡ with

$$a_{ij} = \frac{\partial^2 F}{\partial p_i \partial p_j}(Du) \qquad ‡‡$$

where $F = F(p),\ p \in \mathbb{R}^2$, is a given C^2 function with $\dfrac{\partial^2 F}{\partial p_i \partial p_j}(p)$ positive definite for each $p \in \mathbb{R}^2$. For the present discussion we make two additional assumptions concerning F (although these will not be uniformly assumed subsequently):

(i) $$\left|\frac{\partial F}{\partial p_j}(p)\right| \le c \ \ \forall\ p \in \mathbb{R}^2 \ \ j = 1,2$$

(ii) $$\begin{cases} \displaystyle\sum_{i,j=1}^{2} p_i p_j \frac{\partial^2 F}{\partial p_i \partial p_j}(p) \le c|p|^{-\epsilon} \ \forall\ p \in \mathbb{R}^2,\ |p| \ge 1 \\ \displaystyle\sum_{i,j=1}^{2} \xi_i \xi_j \frac{\partial^2 F}{\partial p_i \partial p_j}(p) \le c|p|^{-\epsilon}|\xi|^2 \ \forall\ p,\xi \in \mathbb{R}^2,\ |p| \ge 1, \end{cases}$$

where $\epsilon > 0$, $c > 0$ are fixed constants. We note that the minimal surface equation is included in this class—in fact it corresponds to the choice $F(p) = \sqrt{1+|p|^2}$, and (i),

(ii) are satisfied with $\epsilon = 1$. On the other hand, there are many examples of equations *not* of minimal surface type for which (i), (ii) hold. For example, if $F(p)$ is given (at least for $|p|$ large) by

$$F(p) = (1 + |p|^m)^{1/m}$$

with $m > 1$. In this case we have (for $|p|$ sufficiently large)

$$\sum_{i,j=1}^{2} p_i p_j \frac{\partial^2 F}{\partial p_i \partial p_j}(p) = \frac{d^2}{dt^2} F(tp)|_{t=1} \equiv \frac{d^2}{dt^2}(1 + t^m |p|^m)^{1/m}|_{t=1}$$
$$= (m-1)\frac{|p|^m}{1 + |p|^m}(1 + |p|^m)^{-(m-1)/m}$$

and

$$\left|\frac{\partial^2 F}{\partial p_i \partial p_j}(p)\right| \leq c(1 + |p|)^{-1},$$

so that (i), (ii) hold with $\epsilon = \min\{1, m-1\}$.

We claim

Theorem 3. *If* (i), (ii) *hold for some* $\epsilon > 0$, $c > 0$, *then any entire solution of* ‡, ‡‡ *is affine.*

Remarks. (1) Notice that the equation

$$\frac{u_{xx}}{1 + u_x^2} + \frac{u_{yy}}{1 + u_y^2} = 0$$

has the non-linear solution $u(x, y) = xy$, and has the form (a) with

$$F(p_1, p_2) = p_1 \arctan p_1 + p_2 \arctan p_2 - \log\sqrt{(1 + p_1^2)(1 + p_2^2)}$$

and (i), (ii) hold with $\epsilon = 0$; thus Theorem 3 is false if the condition $\epsilon > 0$ is replaced by $\epsilon \geq 0$ in (i), (ii).

(2) An example of an equation with the Bernstein property which does not fit into any of the above classes is given by G. Aronsson in [AG]. Aronsson shows that the equation

$$u_x^2 u_{xx} + 2u_x u_y u_{xy} + u_y^2 u_{yy} = 0$$

has the Bernstein property (i.e. entire $\Rightarrow$ affine). Note that this example technically does not fit into the class being considered here because the coefficient matrix $(u_i u_j)$ is only semi-positive definite. There are closely related elliptic equations for which existence of non-linear solutions is open; for example, it is not clear if the equation

$$(1 + u_x^2)u_{xx} + 2u_x u_y u_{xy} + (1 + u_y^2)u_{yy} = 0$$

has the Bernstein property.

Proof of Theorem 3. The proof involves quite standard techniques. The equation for u has the divergence form

$$\sum_{i=1}^{2} D_i\left(\frac{\partial F}{\partial p_i}(Du)\right) = 0,$$

which can be written in the weak form

$$\int_{\mathbb{R}^n} \sum_{i=1}^{2} \frac{\partial F}{\partial p_i}(Du) D_i \zeta = 0 \ \forall \zeta \in C_c^\infty(\mathbb{R}^n). \tag{1}$$

In this equation we first replace ζ by $\gamma(u)\zeta$, with $\zeta \in C_c^\infty(\mathbb{R}^n)$ non-negative and

$$\gamma(t) = \begin{cases} 1, & t > \rho \\ (2\rho)^{-1}(t+\rho), & -\rho \le t \le \rho \\ 0, & t < -\rho\,. \end{cases}$$

This gives

$$\int_{-\rho<u<\rho} \zeta\, Du \cdot \frac{\partial F}{\partial p}(Du) = -2\rho \int_{-\rho<u} \gamma(u) D\zeta \cdot \frac{\partial F}{\partial p}(Du). \tag{2}$$

We can of course assume without loss of generality that $D_p F(0) = 0$ (otherwise replace $F(p)$ by $F(p) - p \cdot F_p(0)$); it then follows from the positivity of $\left(\frac{\partial^2 F}{\partial p_i \partial p_j}\right)$ that

$$p \cdot \frac{\partial F}{\partial p}(p) \ge c|p|, \quad |p| \ge 1,$$

for some fixed constant $c > 0$. In view of (i) we then conclude from (2) above

$$\int_{-\rho<u<\rho} \zeta \sqrt{1+|Du|^2} \le c\rho^3 \sup |D\zeta|. \tag{3}$$

We now select ζ to be identically 1 in the ball B_ρ, zero outside $B_{2\rho}$ and $\sup|D\zeta| \le 2/\rho$. (3) then gives

$$\int_{-\rho<u<\rho,\, |x|\le\rho} \sqrt{1+|Du|^2} \le c\rho^2 \ \forall \rho \in \mathbb{R}. \tag{4}$$

Since $\sqrt{1+|Du|^2}dx$ is the surface measure for $G = \operatorname{graph} u$, this in particular implies that

$$|G \cap \{X \in \mathbb{R}^3 : |X| < \rho\}| \le c\rho^2,$$

where $|\ |$ denotes surface area (i.e. 2-dimensional Hausdorff measure). For later purposes we note that the identity $2\int_{|X|}^{R} \rho^{-3}\, d\rho = |X|^{-2} - R^{-2}$ together with Fubini's theorem and the above area bound, implies directly that

$$\int_{G\cap\{\rho_1<|X|<\rho_2\}} \frac{1}{|X|^2}\,d\sigma \le c\big(1+\log\frac{\rho_2}{\rho_1}\big) \quad \text{for } 0<\rho_1<\rho_2<\infty, \tag{5}$$

where $d\sigma$ denotes surface measure on G and where c is a fixed constant independent of ρ_1 and ρ_2. Notice that in deriving this inequality we only used (i) and the positivity of $\left(\frac{\partial^2 F}{\partial p_i \partial p_j}\right)$; no use has yet been made of the condition (ii).

To complete the proof, we return to the equation (1); this time we replace ζ by the expression

$$\sum_{\ell=1}^{2} D_\ell\left(\zeta^2 \frac{D_\ell u}{v^\alpha}\right),$$

where $v = \sqrt{1+|Du|^2}$ and α is a constant $\in (0,1)$ to be chosen, and then integrate by parts with repect to x^ℓ. This yields

$$\int \zeta^2 \frac{A^{ij}}{v^\alpha}\left(\delta_{k\ell} - \alpha\frac{D_\ell u D_k u}{1+|Du|^2}\right)(D_i D_\ell u)(D_j D_k u) = -2\int \zeta \frac{A^{ij}}{v^\alpha}(D_\ell u)(D_j D_\ell u) D_i\zeta, \tag{6}$$

where $A^{ij} = \dfrac{\partial^2 F}{\partial p_i \partial p_j}(Du)$. Then (A^{ij}) is a positive matrix and also the minimum eigenvalue of the matrix $(\delta_{\ell k} - \alpha\frac{D_\ell u D_k u}{1+|Du|^2})$ is $> 1-\alpha$. Thus (6) yields (after using the Cauchy–Schwartz inequality)

$$\begin{aligned}(1-\alpha)\int \zeta^2 \frac{\sum_{i,j,\ell}^2 A^{ij}(D_i D_\ell u)(D_j D_\ell u)}{v^\alpha} &\le 2\int \left|\frac{\zeta A^{ij}(D_j D_\ell u)(D_\ell u)(v D_i\zeta)}{v^{\alpha+1}}\right| \\ &\le 2\left(\int \frac{\zeta^2 A^{ij}(D_j D_\ell u)(D_\ell u)(D_j D_k u)(D_k u)}{v^{\alpha+2}}\right)^{1/2}\left(\int \frac{v^2 A^{ij} D_i\zeta\, D_j\zeta}{v^\alpha}\right)^{1/2}.\end{aligned}$$

Thus with $T = \int \zeta^2 \frac{A^{ij}}{v^\alpha}(D_i D_\ell u)(D_j D_\ell u)$, we get

$$(1-\alpha)T \le 2T^{1/2}\left(\int v^2 \frac{A^{ij}}{v^\alpha} D_i\zeta\, D_j\zeta\right)^{1/2},$$

and hence

$$T \le c_\alpha \int v^2 \frac{A^{ij}}{v^\alpha} D_i\zeta\, D_j\zeta. \tag{7}$$

Now we use the "logarithmic cut-off trick": let $r = r(x) = \sqrt{|x|^2 + u(x)^2}$ and choose $\zeta = \mu(r)$ in (7), where μ has the form

$$\mu(r) = \begin{cases} 0, & r > \rho \\ \dfrac{\log(r/\sqrt{\rho})}{\log(1/\sqrt{\rho})}, & \sqrt{\rho} < r < \rho \\ 1, & 0 < r < \sqrt{\rho}, \end{cases}$$

where $\rho > 1$ is arbitrary. Now notice that $|\mu'(r)| \leq \dfrac{1}{r \log \sqrt{\rho}}$, and hence this choice of ζ in (7) (together with the conditions (i) and (ii)) yields

$$\int_{r<\sqrt{\rho}} \frac{A^{ij}(D_i D_\ell u)(D_j D_\ell u)}{v^\alpha} \leq \frac{c}{(\log \rho)^2} \int_{\sqrt{\rho}<r<\rho} \frac{v^{-\epsilon-\alpha+2}}{r^2}.$$

Since $\epsilon > 0$ we can choose $\alpha \in (0, 1)$ such that $\epsilon + \alpha \geq 1$, whereupon

$$\begin{aligned} \int_{r<\sqrt{\rho}} \frac{A^{ij}(D_i D_\ell u)(D_j D_\ell u)}{v^\alpha} &\leq \frac{c}{(\log \rho)^2} \int_{\sqrt{\rho}<r<\rho} \frac{v}{r^2}\, dx \\ &\leq \frac{c}{\log \rho} \qquad \text{by (5).} \end{aligned}$$

Thus, letting $\rho \uparrow \infty$, we conclude $D_i D_j u \equiv 0$; i.e., that u is affine as required.

Actually if we drop the assumption (ii) there is still a good deal that can be said about entire solutions of ‡, ‡‡:

Theorem 4. *If $n = 2$ and* (i) *holds, and if u is an entire solution of the equation ‡, ‡‡ then either*

(1) *u is affine, or*

(2) *$|Du(x)| \to \infty$ as $|x| \to \infty$ (in the uniform sense that $\inf_{|x|=\rho} |Du(x)| \to \infty$ as $\rho \to \infty$).*

We prove this below; first though we want to point out the following corollary of Theorem 4:

Corollary. *If the hypotheses are as in Theorem 4, and if in addition u has linear growth (i.e., $\exists\, c > 0$ such that $|u(x)| \leq c|x|$ $\forall\, |x| \geq 1$), then u is affine.*

Proof of Corollary. We proved (using (i)) in the first part of the proof of Theorem 3 that

$$\int_{-\rho<u<\rho,\ |x|\leq\rho} \sqrt{1 + |Du|^2} \leq c\rho^2 \ \forall \rho \in \mathbb{R}.$$

With $u - t$ in place of u, we thus have

$$\int_{-\rho<u-t<\rho,\,|x|\le\rho} \sqrt{1+|Du|^2} \le c\rho^2 \ \forall \rho > 0.$$

Selecting $t = j\rho$ and summing $j = -1 - [m_\rho/\rho]$ up to $j = 1 + [m_\rho/\rho]$, where $m_\rho = \sup |u|$, we get

$$\begin{aligned}\int_{|x|<\rho} \sqrt{1+|Du|^2} &\le c(1+\frac{m_\rho}{\rho})\rho^2 \\ &\le c\rho^2,\end{aligned}$$

with c independent of ρ. In particular this shows that the alternative (2) of Theorem 4 cannot occur, hence the alternative (1) must hold, so the corollary is established.

Proof of Theorem 4. First, since it only used assumption (i), we still have at our disposal the inequality

$$|G \cap \{X \in \mathbb{R}^3 : |X| < \rho\}| \le c\rho^2, \tag{1}$$

which we proved at the beginning of the proof of Theorem 3 .

Also, since $K \le 0$ by the inequality $-K \ge \gamma(1+|Du|^2)^{-2}(u_{11}^2 + u_{22}^2 + 2u_{12}^2)$ (see the discussion at the beginning of §2), we have by [SS]

$$\begin{aligned}\int_{G\cap\{|X|<\rho\}} -K &\le c\rho^{-2}|G \cap \{|X| < 2\rho\}| \\ &\le c \ \text{ by (1)},\end{aligned}$$

so that

$$\int_G -K < \infty. \tag{2}$$

On the other hand, since the ratio between the maximum and minimum eigenvalues of the matrix $\left(\dfrac{\partial^2 F}{\partial p_i \partial p_j}\right)$ is bounded by a constant c_Λ for $|p| \le \Lambda$ (for arbitrary $\Lambda > 0$), we deduce, using (2) and the inequality $-K \ge \gamma(1+|Du|^2)^{-2}(u_{11}^2 + u_{22}^2 + 2u_{12}^2)$ from §2), that

$$\int_{\{x:|Du(x)|<L)\}} |D^2u|^2 \le c(L) \ \forall \ L > 0.$$

Using this in combination with the Schwartz inequality, we conclude that for each $\epsilon > 0$ and each $L > 0$

$$\int_{\{x:\rho/2<|x|<\rho,\,|Du(x)|<L)\}} |D^2u| \le \epsilon\rho \ \ \forall \rho > \rho_0,$$

where $\rho_0 = \rho_0(\epsilon, L)$. Thus, since

$$\int_{\{\rho/2\le|x|\le\rho\}} g = \int_{\rho/2}^{\rho}\Big(\int_{\{|x|=\sigma\}} g\Big)\, d\sigma \ge \operatorname*{ess\,inf}_{\sigma\in(\rho/2,\,\rho)}\Big(\int_{|x|=\sigma} g\Big)\rho/2,$$

we deduce that for each $\rho > \rho_0(\epsilon, L)$, there exist (many) $\sigma \in (\rho/2, \rho)$ such that

$$\int_{\{x\,:\,|x|=\sigma,\ |Du(x)|<L\}} |D^2u| \le 2\epsilon,$$

so that in particular

$$\int_{\{|x|=\sigma\}} |Dv_L|\, ds \le 2\epsilon,$$

where $v_L = \max\{|Du|, L\}$.

By 1-dimensional integration we thus have

$$|v_L(x) - v_L(y)| \le 2\epsilon \ \ \forall\, x, y \in \{|x| = \sigma\}. \tag{3}$$

On the other hand it is standard that, for equations of the form ‡ for $n = 2$, the quantity $|Du|$ (hence v_L) satisfies a maximum principle. Let $L \ge 2$ be arbitrary and select $\epsilon = \frac{1}{4}$. Then (3) $\Rightarrow$ *either* $|Du| < L + 1$ everywhere on $\{|x| < \sigma\}$ (and hence $|Du| < L + 1$ everywhere on $\{|x| < \sigma\}$ by the maximum principle), *or* $|Du| \ge L$ everywhere on $\{|x| = \sigma\}$.

In case of the latter alternative we repeat the argument with 4ρ in place of ρ; this gives (for $\rho \ge \rho_0$ with ρ_0 depending on L), that there also exists $\sigma' \in (2\rho, 4\rho)$ such that

either $|Du| < L + 1$ on $\{|x| < \sigma'\}$ *or* $|Du| \ge L$ everywhere on $\{|x| = \sigma'\}$.

Thus we conclude that (for each ρ sufficiently large, depending only on L), either

(i) $|Du| < L + 1$ on $\{x : |x| < \rho/2\}$

or

(ii) $|Du| \ge L$ everywhere on $\{x : |x| = \sigma'\} \cup \{x : |x| = \sigma\}$

for some $\sigma \in (\rho/2, \rho)$ and some $\sigma' \in (2\rho, 4\rho)$. The latter alternative means that the Gauss map $X = (x, u(x)) \in G \mapsto \dfrac{(-Du(x), 1)}{\sqrt{1 + |Du(x)|^2}} \in S^2$ maps the boundary of the region $U \equiv G \cap \{X : \sigma < |x| < \sigma'\}$ into the region $\{\xi = (\xi^1, \xi^2, \xi^3) : 0 < \xi^3 < 1/L\}$. Since K (which is the Jacobian of the Gauss map) has constant sign, this would imply that the Gauss image of U is contained in the region $\{0 < \xi^3 < 1/L\}$, and hence that the Gauss image of $G \cap \{X : \rho < |x| < 2\rho\}$ is contained in the same region, and hence that $|Du| \ge L$ in $G \cap \{X : \rho < |x| < 2\rho\}$.

In view of the fact that the above discussion is valid for arbitrary L and arbitrary $\rho > \rho(L)$, we thus finally conclude that either $|Du|$ is bounded, or else $|Du| \to \infty$ as $|x| \to \infty$, as required. In case $|Du|$ is bounded, since the equation ‡ has the divergence-form $\sum_i D_i F_{p_i}(Du) = 0$, we can differentiate with respect to x^ℓ ($\ell \in \{1, \dots, n\}$ arbitrary), thus showing that $v = D_\ell u$ satisfies the equation

$$\sum_{i,j=1}^{n} D_i(a_{ij} D_j v) = 0,$$

where $a_{ij}(x) = F_{p_i p_j}(Du)$. But by the De Giorgi Nash estimate (see [GT] for an exposition of this) we know that if v is a weak solution on a ball B_R of such a divergence form equation $\sum_{i,j=1}^{n} D_i(a_{ij} D_j v) = 0$, with Dv of class L^2 and with $\sum_{i,j=1}^{n} a_{ij}\xi_i\xi_j \geq \lambda|\xi|^2$ and $|a_{ij}| \leq \Lambda$ on B_R, where λ, Λ are fixed positive constants, then v is Hölder continuous and there is an estimate of the form

$$R^\alpha [v]_{\alpha;\, B_R} \leq C \sup_{B_R} |v|,$$

where $C > 0$ and $\alpha \in (0, 1)$ are fixed constants depending only on n, λ and Λ. Thus applying this with $D_\ell u$ in place of v, where u is our solution of the MSE, we deduce

$$R^\alpha \, [Du]_{\alpha; B_{R/2}} \leq C \sup_{B_R} |Du|,$$

with C depending on upper and lower bounds for the maximum and minimum eigenvalues of the coefficient matrix a_{ij}. Thus actually (since the coefficient matrix depends continuously on Du and $|Du|$ is bounded) we conclude that

$$R^\alpha \, [Du]_{\alpha; B_{R/2}} \leq C,$$

with C a fixed constant independent of R, so letting $R \uparrow \infty$ we conclude that $Du \equiv 0$ as claimed.

Remark. Note that no result of the type of Theorem 4 can hold for Laplace's equation. (For example $u(x, y) = \sin x \cosh y$ is an entire solution of $\Delta u = 0$, with $Du(x_k) = 0$, where $x_k = ((k + 1/2)\pi, 0)$, $k = 0, \pm 1, \dots$.)

Concerning *exterior* (as distinct from entire) solutions, we note that, by straightforward modifications of the argument used in the proof of Theorem 4 above, we can prove

Theorem 5. *If F is as in Theorem 4, and if u is an exterior solution of ‡, ‡‡, then either $|Du|$ is bounded and $\lim_{|x|\to\infty} Du(x)$ exists, or $|Du(x)| \to \infty$ as $|x| \to \infty$ (in the sense that $\inf_{|x|=\rho} |Du(x)| \to \infty$ as $\rho \to \infty$).*

We leave the necessary modifications to the reader.

Under suitable restrictions it is possible to prove that the second alternative here cannot occur, so that in such cases one can assert that *all* exterior solutions are such that Du is bounded and has a limit at ∞. The classical example of this phenomenon is the result of L. Bers [BL], who demonstrated it for the minimal surface equation (i.e. $F(p) = \sqrt{1+|p|^2}$). This result was extended to all the cases when F is such that ‡, ‡‡ is an equation of minimal surface type in [SL2]. (For discussion of higher dimensional analogues, see the next section.)

3. The non-linear case, $n \geq 3$

In the case of dimension $n \geq 3$, the situation concerning entire and exterior solutions of ‡ is far less well-understood than in the 2-dimensional case. From the early 1950's, attention focussed on the minimal surface equation (i.e. the equation $\sum_{i,j=1}^{n}\left(\delta_{ij} - \frac{D_i u D_j u}{1+|Du|^2}\right) D_i D_j u = 0$) as a key example. The prime question was whether or not Theorem 2 generalized to dimension $n \geq 3$. There were many proofs of Theorem 2 (some very elementary) in case $n = 2$, but up to 1960 all seemed to have no potential for generalization to higher dimensions. In 1960 W. Fleming [FW] gave a proof which, for the first time, seemed to leave some hope of such a generalization. Fleming in fact proved a connection between the Bernstein problem and the study of area minimizing surfaces (not necessarily graphs) in Euclidean space. Specifically, he was able to prove that (for any dimension $n \geq 2$) the Bernstein property (entire ⇒ affine) holds provided there are no non-planar area minimizing hypercones in $\mathbb{R}^{n+1}$. (For a discussion of the terminology, see e.g. [G] or [SL3].) Since he was able to prove that no such non-trivial hypercones existed in $\mathbb{R}^3$, he could give a new proof of the Bernstein theorem in dimension 2, but his most important contribution was to suggest the new approach to the higher-dimensional problem referred to above.

Next De Giorgi [DG] made an important improvement in Fleming's method, by showing that in fact, to prove the Bernstein theorem for solutions of the MSE defined on $\mathbb{R}^n$, it sufficed to check that no non-trivial hypercones existed in $\mathbb{R}^n$ (rather than in $\mathbb{R}^{n+1}$), thus in particular proving the Bernstein property in the case $n = 3$. De Giorgi's observation amounted to the fact that a non-affine minimal graph in $\mathbb{R}^{n+1}$ has a tangent cone at ∞ of the cylindrical form $\mathbb{C} \times \mathbb{R}$, with $\mathbb{C}$ a non-planar area minimizing hypercone in $\mathbb{R}^n$, hence his improvement of Fleming's result. For an elegant treatment of this argument see Giusti [G, Chapter 17] (see also Massari & Miranda [MM]); Giusti's discussion shows that if u is a non-affine entire solution in $\mathbb{R}^n$ then for any sequence $\rho_j \downarrow 0$ there is a subsequence $\rho_{j'}$ such that the boundary of the set $\{x \in \mathbb{R}^n : \rho_{j'} u(\rho_{j'}^{-1} x) \to \infty\}$ is a non-planar area minimizing hypercone $\mathbb{C}$ in $\mathbb{R}^n$, which gives a contradiction if no such area minimizing hypercones $\mathbb{C}$ exist in $\mathbb{R}^n$. Actually it can be shown that the whole "sub-graph" $\{(x, y) \in \mathbb{R}^{n+1} : y < \rho_{j'} u(\rho_{j'}^{-1} x)\}$ approaches the region on one side of the cylindrical cone $\mathbb{C} \times \mathbb{R}$, but this requires more effort, using work of Bombieri & Giusti [BG] as in Miranda [M] and Simon [SL5, 6] and in any case it is not needed for the present application to the Bernstein problem.

Ultimately, the Fleming/De Giorgi approach was completely justified when Almgren [A] (for $n = 4$) and J. Simons [SJ] (for $n \leq 7$) were able to prove the appropriate non-existence of area-minimizing hypercones in $\mathbb{R}^n$ (so that the Bernstein theorem was established for $n \leq 7$), and Bombieri, De Giorgi & Giusti [BDG] were able to prove that the result is false in dimension $n \geq 8$. Thus the Bernstein property holds for entire solutions of the MSE on $\mathbb{R}^n$ for $n \leq 7$, but there are counterexamples (i.e. non-affine entire solutions) in all dimensions $n \geq 8$. (And hence there are also non-trivial area minimizing hypercones in $\mathbb{R}^n$, $n \geq 8$ by De Giorgi's 1965 result.)

After the first non-affine examples of entire solutions of the minimal surface equation were shown to exist (in dimension $n \geq 8$), it was natural to try to obtain more information about the nature (especially with respect to qualitative behaviour) and number of such solutions. The work in [BDG] showed that there is at least one distinct non-affine solution of the minimal surface equation in $\mathbb{R}^n$ for each even $n \geq 8$; of course we here ignore the additional solutions obtained by homotheties and rigid motions of graph u, and by the procedure of "adding redundant variables" (i.e. getting a solution $\tilde{u}$ in $\mathbb{R}^{n+k}$ from a solution u in $\mathbb{R}^n$ by defining $\tilde{u}(x^1, \ldots, x^{n+k}) = u(x^1, \ldots, x^n)$).

Recently many other non-affine solutions have been shown to exist in [SL6]. The examples shown to exist in [BDG] and [SL6] all have polynomial growth, and it is interesting to speculate whether or not *all* entire solutions of the minimal surface equation have this property. For some partial results in this direction the reader is referred to the discussion in [SL6]. (See in particular Theorem 5 of [SL6].) Another interesting open question is whether or not there exists a solution which is actually a polynomial in the variables $x^1, \ldots, x^n$.

While on the subject of growth of solutions of the minimal surface equation, we should mention recent work on *lower* growth bounds for such solutions. Caffarelli, Nirenberg, and Spruck [CNS] first proved that if u is a non-affine solution of the minimal surface equation, then $|Du(x_k)| \geq c|x_k|^{1/2}$ for some fixed constant $c > 0$ and some sequence $\{x_k\}$ with $|x_k| \to \infty$. Ecker & Huisken [EH] improved this by showing that the same holds with $|x_k|$ in place of $|x_k|^{1/2}$ on the right. (Working independently Nitsche [N] proved that the same holds with $|x_k|^{\mu}$ in place of $|x_k|^{1/2}$ for any $\mu < 1$.) Finally it has been claimed in [SL7] that the result holds with $|x_k|^{\beta_n - \epsilon}$ in place of $|x_k|^{1/2}$ for any $\epsilon > 0$ (with $c = c(\epsilon)$), where $\beta_n = \dfrac{n-3}{2} - \sqrt{\left(\dfrac{n-3}{2}\right)^2 - (n-2)}$ (> 1); it is shown in [SL7] that much can be said about the x_k for which the above lower bound holds. Notice that this lower bound is best possible for its type, because the non-affine solutions constructed in [BDG] (and some of the examples in [SL6]) attain the lower bound mentioned above.

Concerning *exterior* solutions of the minimal surface equation in the case $n \geq 3$, Ber's Theorem (Theorem 5 above) has a generalization to $\mathbb{R}^n$, $n \leq 7$; this is proved in [SL4]. In fact in [SL4] a more general result, valid in all dimensions $n \geq 3$, is proved. Specifically:

Theorem 6. *If $n \geq 2$ and if u is a C^2 solution of the minimal surface equation on $\mathbb{R}^n \sim \Omega$ (Ω bounded open), then either $|Du|$ is bounded and $\lim_{|x|\to\infty} Du(x)$ exists, or else the tangent cone of* graph u *at ∞ has the form of a cylinder $C \times \mathbb{R}$ ($\subset \mathbb{R}^{n+1}$), with C an area minimizing hypercone in $\mathbb{R}^n$ with* sing $C \neq \emptyset$.

For a discussion of the notion of tangent cone at ∞, we refer the reader to [SL4]. Notice that the required generalization of Theorem 5 to $n \leq 7$ follows trivially from this general result by virtue of the fact (mentioned in our discussion of the Bernstein Theorem) that there exist no non-trivial hypercones in $\mathbb{R}^n$ in case $n \leq 7$.

For equations of the form ‡ other than the minimal surface equation not so much is known about when the Bernstein Theorem holds and whether or not one can expect reasonable qualitative results about growth properties of non-linear solutions (in case they exist). We mention the result of [SL4], where it is shown that the Bernstein Theorem generalizes to dimension $n = 3$ in case of equations of the form ‡, ‡‡ with $F(p) \equiv G(p, -1)$, where G is a constant coefficient parametric functional. In this setting, it is shown in [SL4] that the Bernstein Theorem generalizes to all dimensions $n \leq 7$ in case the parametric functional G is sufficiently close to the area functional in the appropriate sense.

It would be of particular interest to know if there are any equations other than those fitting into the special classes mentioned above for which the Bernstein Theorem holds.

We conclude this discussion by mentioning some results in the direction of Theorem 4 and its corollary. For example, by virtue of the De Giorgi–Nash theory, we know that any non-affine solution of an equation of the form ‡, ‡‡ must have unbounded gradient. (See for example the discussion in [GT].) On the other hand there are a large class of equations of the form ‡, ‡‡ for which any entire solution with linear growth automatically has bounded gradient (and hence must be affine in view of the general result just mentioned). We refer the reader to e.g. [SL8] for a discussion of a class of such equations. In this context it would be of interest to know if there are more general classes for which the result "linear growth and entire ⇒ affine."

References

[A] F. Almgren, Some interior regularity theorems for minimal surfaces and an extension of Bernstein's theorem, Ann. of Math. 84 (1966), 277–292

[AG] G. Aronsson, On the partial differential equation $u_x^2 u_{xx} + 2u_x u_y u_{xy} + u_y^2 u_{yy} = 0$, Ark. Mat. 7 (1968), 395–425.

[B] S. Bernstein, Über ein geometrisches Theorem und seine Anwendung auf die partielle Differentialgleichungen vom elliptischen Typus, Math. Z. 26 (1927), 151–158.

[BL] L. Bers, On isolated singularities of minimal surfaces, Ann. of Math. 53 (1951), 417–491.

[BDG] E. Bombieri, E. De Giorgi, E. Giusti, Minimal cones and the Bernstein problem, Invent. Math. 7 (1969), 243–268.

[BG] E. Bombieri & E. Giusti, Harnack's inequality for elliptic differential equations on minimal surfaces. Invent. Math. 15 (1972), 24–46.

[CNS] L. Caffarelli, L. Nirenberg, J. Spruck, On a form of Bernstein's Theorem, Analyse mathématique et applications 55–56 Gauthier-Villars, Paris 1990.

[DG] E. De Giorgi, Una estensione del teorema di Bernstein, Ann. Scuola Norm. Sup. Pisa 19 (1965), 79–85.

[EH] K. Ecker & G. Huisken, A Bernstein result for minimal graphs of controlled growth, J. Differential Geom. 31 (1990), 397–400.

[FH] H. Federer, Geometric Measure Theory, Springer-Verlag, Berlin–Heidelberg–New York 1969.

[FR] R. Finn, On equations of minimal surface type, Ann. of Math. 60 (1954), 397–416.

[FW] W. Fleming, On the oriented Plateau problem, Rend. Circ. Mat. Palermo 11 (1962), 1–22.

[GL] N. Garofalo & F.-H. Lin, Monotonicity properties of variational integrals, A_p-weights, and unique continuation, Indiana Univ. Math J. 35 (1986), 245–268.

[GT] D. Gilbarg & N. Trudinger, Elliptic Partial Differential Equations of Second Order, Springer-Verlag 1983.

[G] E. Giusti, Minimal Surfaces and Functions of Bounded Variation, Birkhäuser, Boston 1983.

[HS] R. Hardt & L. Simon, Nodal sets for solutions of elliptic equations, J. Differential Geom. 30 (1989), 643–688.

[HE1] Hopf E., A theorem on the accessibility of boundary parts of an open point set. Proc. Amer. Math. Soc. 1 (1950), 76–79.

[HE2] Hopf E., On S. Bernstein's theorem on surfaces $z(x, y)$ of non-positive curvature. Proc. Amer. Math. Soc. 1 (1950), 80–85.

[H] L. Hörmander, Uniqueness theorems for 2nd-order elliptic operators, Comm. Partial Differential Equations 8 (1983), 21–64.

[LP] P. Lax, A stability theorem for solutions of abstract differential equations and its application to the study of local behavior of some elliptic equations, Comm. Pure Appl. Math. 9 (1956), 747–766.

[MM] U. Massari & M. Miranda, Minimal surfaces of codimension one, Mathematics Studies # 91, North Holland, Amsterdam–New York 1984.

[M] M. Miranda, Superficie minime illimitate, Ann. Scuola Norm. Sup. Pisa, Ser. IV 4 (1977), 313–322.

[ME] E. Mickle, A remark on a theorem of Serge Bernstein, Proc. Amer. Math. Soc. 1 (1950), 86–89.

[N] J. C. C. Nitsche, Lectures on Minimal Surfaces, Vol. 1, Cambridge University Press, 1989.

[OR] R. Osserman, A Survey of Minimal Surfaces, Dover, 1986.

[SS] R. Schoen & L. Simon, Regularity of simply connected surfaces with quasi-conformal Gauss map, Seminar on Minimal Submanifolds, Annals of Math. Stud. 103 (1983), 127–146.

[SJ] J. Simons, Minimal varieties in Riemannian manifolds, Ann. of Math. 88 (1968), 62–105.

[SL1] L. Simon, A Hölder estimate for quasiconformal mappings between surfaces in Euclidean space, with applications to graphs having quasiconformal Gauss map, Acta Math. 139 (1977), 19–51.

[SL2] —, Equations of mean curvature type in 2 independent variables, Pacific J. Math. 69 (1977), 245–268.

[SL3] —, Lectures on Geometric Measure Theory, Centre for Math. Analysis, Australian Nat. Univ. 1983.

[SL4] —, On some extensions of Bernstein's theorem, Math. Z. 154 (1977), 265–273.

[SL5] —, Asymptotic behaviour of minimal graphs over exterior domains, Ann. H. Poincaré 4 (1987), 231–242.

[SL6] —, Entire solutions of the minimal surface equation, J. Differential Geom. 30 (1989), 643–688.

[SL7] —, Lower growth estimates for solutions of the minimal surface equation, in preparation.

[SL8] —, Interior gradient bounds for non-uniformly elliptic equations, Indiana Univ. Math. J. 25 (1976), 821–855.

[SL9] —, Growth properties for exterior solutions of quasilinear elliptic equations, Research report, CMA (Canberra) 1989.

[SL10] —, Isolated singularities of extrema of geometric variational problems, in: Lecture Notes in Mathematics 1161, Springer-Verlag 1985, 206–277.

External geometry of p-minimal surfaces

Vladimir G. Tkachev *

Abstract. A surface $\mathcal{M}$ is said to be p-minimal if one of the coordinate functions is p-harmonic in the inner metric. We show that in the two dimensional case the Gaussian map of such surfaces is quasiconformal. In the case when the surface is a tube we study the geometrical structure of such surfaces. In particular, we establish the second order differential inequality for the form of the sections of $\mathcal{M}$ which generalizes the known ones in the theory of minimal surfaces.

1. Introduction

1.1. Let $\mathcal{M} = (M; x)$ be a surface given by a C^2-immersion $x : M \to \mathbb{R}^{n+1}$ of an n-dimensional orientable noncompact manifold M.

Definition 1. A surface $\mathcal{M}$ is said to be *minimal* if its mean curvature vector $H(m) \equiv 0$.

The well-known property of minimal immersions in Euclidean space is the harmonicity of their coordinate functions. Moreover, if *one* coordinate function of an immersion is harmonic then *all* coordinates satisfy this property and the immersion is minimal. On the other hand, for $n = 2$ this condition yields the fact that the Gauss map of such surfaces is conformal [14].

The natural question arises: what happens if we replace the requirement of harmonicity by p-harmonicity?

Definition 2. For a fixed $p > 1$ a surface $\mathcal{M}$ is said to be *p-minimal* if one of the coordinate functions is p-harmonic with respect to the inner metric of $\mathcal{M}$. In other words, there exists a direction $e \in \mathbb{R}^n$ such that

$$\Delta_p f \equiv \operatorname{div} |\nabla f|^{p-2} \nabla f = 0, \tag{1.1}$$

where $f(m) = \langle x(m), e \rangle$ and ∇ is the covariant derivative on M.

One can easily show that p-harmonicity of one coordinate function can't be extended to the others provided $p \neq 2$. This means that $\mathbb{R}^{n+1}$ is equipped with a specified

*The paper was supported by RFRF, project 93-011-176

direction e. This kind of asymmetry is typical for the Minkowski spaces $\mathbb{R}_1^{n+1}$ with time-axis Oe. Another example is the tubular minimal surfaces (see the definition below) which are Euclidean analogues of the relative strings in nuclear physics.

We should also mention that equation (1.1) is of great importance in the nonlinear potential theory [2] and the elliptic type PDE's [7], [4]. On the other hand, it is closely linked to quasiregular mappings (see [3] for detailed information).

This paper is devoted to two sides of the p-minimal surface theory only. The first part of our paper concerns the basic facts of p-minimal surface theory itself. In particar, we show that the Gauss map of a two-dimensional p-minimal surface is $K(p)$-quasiconformal. In Section 3 we apply a new method to study the shape of p-minimal tubes. In particular, we establish the estimates for the sizes of sections of such surfaces which give us information about the evolution of a tube in the time-direction.

It is the aim of this paper to demonstrate the properties which are common to the tubular minimal and p-minimal surfaces. We only touch on the nonparametric case for p-minimal surfaces and don't mention the properly immersed non-tubular p-minimal surfaces at all. The general theory of p-minimal surfaces and some examples will be given in a paper in preparation.

2. Preliminary properties of p-minimal surfaces

2.1. We have noticed above that the case $p = 2$ corresponds to minimal surfaces. To clarify the geometrical meaning of (1.1) for arbitrary p we denote by $k_e(m)$ the curvature of $\mathcal{M}$ in the e-direction (*i.e.* the sectional curvature of the 2-plane spanned by e and the unit normal ν to $\mathcal{M}$ at a point m).

Proposition 1. *Let m be a noncritical point, i.e., $e \wedge \nu(m) \neq 0$. Then*

$$H(m) = -(p-2)k_e(m). \tag{2.2}$$

Proof. Really, let $\overline{\nabla}$ and ∇ denote the standard covariant derivatives in $\mathbb{R}^{n+1}$ and $\mathcal{M}$ respectively. Then

$$\nabla f(m) = \left(\overline{\nabla}\langle x(m), e\rangle\right)^{\top} = e^{\top},$$

where $e^{\top}$ is the projection of e onto the tangent space to the surface $\mathcal{M}$ at a point m. It follows from the assumptions of the proposition that $|e^{\top}| \neq 0$ or, equivalently, $|\nabla f(m)| \neq 0$. Thus, for any tangent vector X we obtain

$$\nabla_X |\nabla f| = \nabla_X |e^{\top}| = \frac{\langle \overline{\nabla}_X e^{\top}, e^{\top}\rangle}{|e^{\top}|} = \frac{\langle \overline{\nabla}_X(-e^{\perp}), e^{\top}\rangle}{|e^{\top}|} = \langle e, \nu\rangle \langle A(X), \frac{e^{\top}}{|e^{\top}|}\rangle.$$

Here A is the Weingarten map of $\mathcal{M}$ and $e^{\perp}$ is the projection of e onto the normal space to $\mathcal{M}$. By virtue of the symmetry of A we conclude that

$$\nabla|\nabla f| = \langle e, \nu\rangle A(\tau), \tag{2.3}$$

where $\tau = e^{\top}/|e^{\top}|$ is well defined at m. After substituting (2.3) into (1.1) we have

$$\begin{aligned}\Delta_p f &= \operatorname{div}(|\nabla f|^{p-2}\nabla f) = (p-2)|\nabla f|^{p-3}\langle \nabla f, \nabla(|\nabla f|)\rangle + |\nabla f|^{p-2}\Delta f \\ &= |\nabla f|^{p-4}\left(|e^{\top}|^2\Delta f + \langle e, \nu\rangle (p-2)\langle A(e^{\top}), e^{\top}\rangle\right).\end{aligned} \tag{2.4}$$

Finally, the definition of $k_e(m)$ together with the well known connection between the mean curvature $H(m)$ and the inner Laplacian [6]: $\Delta f(m) = H(m)\langle e, \nu\rangle$ yield from (2.4)

$$\Delta_p f = |\nabla f|^{p-4}\langle e, \nu\rangle |e^{\top}|^2\big(H(m) + (p-2)k_e(m)\big), \tag{2.5}$$

everywhere in the regular part $M_0 \equiv \{m \in M : |e^{\top}(m)| \neq 0\}$.

We now assume that equality (2.2) doesn't hold at some noncritical point $m_1 \in M_0$. Then in view of (2.5) and (1.1), $\langle e, \nu(m_1)\rangle \equiv 0$, and by continuity of the expression in parentheses in (2.5), the last identity holds everywhere in some neighbourhood $\Omega(m_1)$. Thus, in $\Omega(m_1) \cap M_0$ the coordinate function $f(m) = \langle e, x\rangle$ is constant and, it follows that $A \equiv 0$ in $\Omega(m_1)$. But this conclusion trivially yields (2.2), which contradicts our assumption. □

The following assertion is an immediate consequence of the Meusnier theorem.

Corollary 1. *The mean curvature H of a p-minimal surface $\mathcal{M}$ and the mean curvature h of the section $\Sigma(\tau)$ are related by*

$$h(m) = -\frac{p-1}{\omega}k_e(m) = \frac{p-1}{p-2}\,\frac{H(m)}{\omega} \tag{2.6}$$

where $\omega = \langle \nu_m, e\rangle$.

We use further the auxiliary assertion which clarifies the local structure of a p-minimal surface near a critical point. We notice that this property has no analogue in minimal surface theory.

Lemma 1. *Let $\mathcal{M}$ be a p-minimal surface given as a graph of a C^2-function $f(x)$ defined in a domain $G \subset \mathbb{R}^n$. Let $x_0 \in G$ be a critical point of $f(x)$, i.e. $\overline{\nabla} f(x_0) = 0$. Then the Hessian $\overline{\nabla}^2 f$ is degenerate. In other words, x_0 is a planar point.*

Proof. To prove this assertion we rewrite (2.2) in a more suitable way. In the local coordinates we have the following formulas for the mean curvature $H(m)$ and the Laplace-Beltrami operator Δ respectively:

$$\begin{aligned} H(m) &= \frac{1}{g^{3/2}}\sum_{i,j=1}^{n}(g\delta_{ij} - \overline{\nabla}_i f\overline{\nabla}_j f)\overline{\nabla}_{ij}^2 f, \\ \Delta u &= \frac{1}{\sqrt{g}}\sum_{i,j=1}^{n}\overline{\nabla}_j(g^{ij}\sqrt{g}\,\overline{\nabla}_j u),\end{aligned} \tag{2.7}$$

where $\overline{\nabla}_i$ denotes the covariant derivative along the coordinate vector e_i, g^{ij} is the inverse matrix to the metric tensor $g_{ij} = \delta_{ij} + \overline{\nabla}_i f \overline{\nabla}_j f$ and $g = \det \|g_{ij}\|$. Hence, we obtain from (1.1) and (2.4)

$$g|\overline{\nabla} f|^2 \operatorname{tr} \overline{\nabla}^2 f + \sum_{l,s=1}^{n} (p - 2 - |\overline{\nabla} f|^2) \overline{\nabla}_l f \overline{\nabla}_s f \overline{\nabla}^2_{ls} f = 0, \tag{2.8}$$

Here $\overline{\nabla}^2 f$ is the Hessian of $f(x)$ and the trace $\operatorname{tr} \overline{\nabla}^2 f$ is equal to the euclidean Laplace operator in $\mathbb{R}^n$. Write $a_{ij} = \overline{\nabla}^2_{ij}(x_0)$ and $A = ||a_{ij}||$. Then for an appropriate choice of $\varepsilon > 0$ and every vector $y \in \mathbb{R}^n$ such that $|y| < \varepsilon$ we have

$$\overline{\nabla}_k f(x_0 + y) = \sum_{i=1}^{n} a_{ki} y_i + o(|y|),$$

and

$$|\overline{\nabla} f(x_0 + y)|^2 = O(|y|^2).$$

Substituting these relations in (2.8), we arrive at

$$\sum_{k,l,s=1}^{n} \sum_{i,j=1}^{n} (a_{ki} a_{kj} \operatorname{tr} A + (p-2) a_{li} a_{sj} a_{ls}) y_i y_j = o(|y|^2).$$

Taking into account the validity of the last equality for all sufficiently small $y \in \mathbb{R}^n$ we obtain a matrix equation

$$A^2 (I \operatorname{tr} A + (p-2)A) = 0, \tag{2.9}$$

where I is the unit matrix. By virtue of the symmetry of the Hessian A we can choose an orthonormal basis of $\mathbb{R}^n$ consisting of the eigenvectors of A. Namely, A takes a diaganal form $\lambda_i \delta_{ij}$ and from (2.9) we have for $i : 1 \le i \le n$,

$$\lambda_i \big(\lambda_i (p-2) + \operatorname{tr} A\big) = 0.$$

We see from the last identity that all non-zero eigenvalues λ_i must be equal to $-(p-2)^{-1} \operatorname{tr} A$. Let $\lambda_1, \dots, \lambda_k$ be all such numbers. Then after summing we obtain

$$\operatorname{tr} A = \sum_{i=1}^{k} \lambda_i = -\frac{k}{p-2} \operatorname{tr} A. \tag{2.10}$$

On the other hand $\operatorname{tr} A = k\lambda_1 \neq 0$. It follows from (2.10) that $p = 2 - k$, where $k \geq 1$ is a positive integer. But this contradicts $p > 1$ and hence, all λ_i are zero. Now the theorem follows from the standard properties of symmetric matrices. □

2.2. Given a surface $\mathcal{M}$ in $\mathbb{R}^3$ we denote by $\gamma(m) : \mathcal{M} \to S^2$ the standard Gauss map. A result of Gauss states that, if the surface is minimal, that map is conformal. Here we extend this property to p-minimal surfaces. First we recall

Definition 3 ([1], [2]). A map $F : M_1 \to M_2$ of two smooth Riemannian manifolds M_1, M_2 is called a *quasiconformal map* if the Jacobian $\det d_x F$ doesn't change sign on M_1 and for almost every $x \in M_1$,

$$\max |d_x F(E)| \leq K_m \min |d_x F(E)| \tag{2.11}$$

where min and max are given over all unit tangent vectors E of $T_x M_1$. The number $K = \max_{m \in M_1} K_m$ is called the *distortion coefficient* of F.

Theorem 1. *Let $\mathcal{M}$ be a two dimensional p-minimal surface in $\mathbb{R}^3$. Then the Gauss map is a $K(p)$-quasiconformal map with distortion coefficient*

$$K(p) = \max\{p - 1;\ 1/(p-1)\}. \tag{2.12}$$

Proof. We notice that the tangent spaces $T_m M$ to M and $T_{\gamma(m)} S^2$ to the unit sphere S^2 can be regarded as canonical isomorphic ones. Really, we identify the vector $A(E)$ with $d\gamma_m(E)$, where $d\gamma_m$ is the differential of the Gauss map at m. We specify a point $m \in M$ and choose an orthonormal basis E_1, E_2 of the tangent space $T_m M$ which diagonalizes A, *i.e.*

$$A(E_i) = \lambda_i E_i,$$

where λ_1, λ_2 are the principal curvatures of $\mathcal{M}$ at m. Without loss of generality we can arrange that $|e^T(m)| \neq 0$. Really Lemma 1 yields that the homomorphism A is identically zero and (2.11) is trivial.

Let us denote $\tau = e^T/|e^T|$. Then for some angle $\psi \in [0;\ 2\pi]$,

$$\tau = E_1 \cos\psi + E_2 \sin\psi,$$

and by the Meusnier theorem we have

$$\langle A\tau, \tau\rangle = \lambda_1 \cos^2\psi + \lambda_2 \sin^2\psi = -\frac{1}{p-2}(\lambda_1 + \lambda_2).$$

Hence

$$\lambda_1 = -\lambda_2 \frac{1 + (p-2)\sin^2\psi}{1 + (p-2)\cos^2\psi}.$$

It is a direct consequence of the last identity that the Jacobian $\det(d_m \gamma) = \lambda_1 \lambda_2$ must be negative. Standard facts of quadratic form theory allows us to conclude that the distortion coefficient of γ at a point m is less or equal to

$$K_m = \max_{\psi} \left\{q;\ \frac{1}{q}\right\}, \qquad q = \frac{1 + (p-2)\sin^2\psi}{1 + (p-2)\cos^2\psi}.$$

Then varying ψ we obtain the required maximum of K_m. □

L. Simon in [16] established that every entire two dimensional nonparametric surface with quasiconformal Gauss map must be a plane. As a consequence of this result we obtain a version of the well-known Bernstein theorem.

Corollary 2. *Let $\mathcal{M}$ be an entire p-minimal graph in $\mathbb{R}^3$. Then $\mathcal{M}$ is a plane.*

Remark. It follows from [10] and [15] that every minimal n-dimensional graph $\mathcal{M}$ in $\mathbb{R}^{n+1}$ has parabolic conformal type. In other words, every compact set on $\mathcal{M}$ has zero n-capacity. In these papers we applied the quasiconformal mapping theory to minimal surfaces. The methods used there allows us to conclude that a similar property holds for p-minimal graphs also if $p \geq n$. These facts together with Corollary 2 make the following very plausible:

Conjecture. Let $\mathcal{M}$ be an entire p-minimal graph given over the whole of $\mathbb{R}^n$. If $p \geq n$ then $\mathcal{M}$ is a hyperplane.

3. Tubular p-minimal hypersurfaces

3.1. In this section we deal with tubular type p-minimal surfaces. This class of surfaces in the two dimensional case was investigated by J. C. C. Nitsche [13] and have been studied by V. M. Miklyukov [9] in the high dimensional situations.

Definition 4. We say that a surface $\mathcal{M}$ is a *tube* with the projection interval $\tau(\mathcal{M}) \subset Ox_{n+1}$, if

(1) for any $\tau \in \tau(\mathcal{M})$ the sections $\Sigma_\tau = x(M) \cap \Pi_\tau$ by hyperplanes $\Pi_\tau = \{x \in \mathbb{R}_1^{n+1} : x_{n+1} = \tau\}$ are not empty compact sets;

(2) for $\tau', \tau'' \in \tau(\mathcal{M})$ any part of $\mathcal{M}$ situated between two different $\Pi_{\tau'}$ and $\Pi_{\tau''}$ is a compact set.

If $\tau(\mathcal{M})$ is an infinite interval we call the surface an *infinite tube*. Otherwise, we call a length of $\tau(\mathcal{M})$ the *life-time* of $\mathcal{M}$.

Let

$$\rho(\tau) = \max_{m \in \Sigma(\tau)} \left(\sum_{i=1}^{n} x_i^2 \right)^{1/2}.$$

It follows from the results of [12], [11], [5], that every n-dimensional minimal tube of arbitrary codimension satisfies the following differential inequality

$$\rho(\tau)\rho''(\tau) \geq (n-1)(1+\rho'(\tau)^2), \tag{3.13}$$

which is crucial for the theory of minimal tubes. As a consequence every minimal tube for $n \geq 3$ is contained in a slab between two parallel planes. Hence, there are no many-dimensional infinite minimal tubes. In contrast, the two dimensional case essentially differs from the high dimensional one: there are tubes of finite life-time as well as infinite tubes. Moreover, we show in [19] that the life-time in the first case is derived by the angle between the full-flow vector of a minimal tube and the time-axis.

Lemma 2. *Let V be a convex compact set in $\mathbb{R}^n$ and W be a compact set such that $W \setminus V \neq \emptyset$. Then there exists a closed ball $B \subset \mathbb{R}^n$ such that*

$$W \subset B \tag{3.14}$$

and

$$\partial B \cap (W \setminus V) \neq \emptyset. \tag{3.15}$$

Proof. The distance function $f(x) = \operatorname{dist}(x, V)$ is continuous on $\mathbb{R}^n$. It follows from the conditions of the lemma that this function attains the maximum value on W at some point $a \in W$ and $d = f(a) > 0$. On the other hand, by virtue of the convexity of V there exists a unique point $b \in \partial V$ such that $f(a) = \|b - a\|$.

Choose a new coordinate system of $\mathbb{R}^n$ with the origin at a: let the first coordinate vector be

$$e_1 = \frac{b-a}{d},$$

and choose the others, $e_2, \ldots, e_n$ so that we have an orthonormal system. Then, the hyperplane given by $x_1 = d$ is one of support to V at a. It follows from the triangle inequality that W is contained in a halfspace $\{x_1 \geq 0\}$ and V in $\{x_1 \geq d\}$.

Given positive h and R we specify an open ball

$$B(R, h) = \{x \in \mathbb{R}^n : (x_1 + R)^2 + x_2^2 + \cdots + x_n^2 < (R + h)^2\}.$$

By our choice and the compactness of V, given a positive ϵ there exists $R > 0$ such that V is contained in a ball $B(R, \epsilon)$.

Suppose $\epsilon = d/2$ and R_0 is the corresponding radius. Then the definition of d yields that $a \notin B(R_0, d/2)$, however the greater ball $B(R_0, 3d/2)$ contains V as well as W. Let δ_0 be the minimum over all $\delta \in (0; d)$ such that

$$W \subset \overline{B(R_0, d/2 + \delta)}.$$

Then $a \in \partial B$, where $B = \overline{B(R_0, d/2 + \delta_0)}$ and $V \cap B = \emptyset$. □

Corollary 3 (Maximum Principle). *Let $\mathcal{M} = (M, x)$ be an immersed compact p-minimal hypersurface in $\mathbb{R}^{n+1}$ with nonempty boundary ∂M. Then*

$$\operatorname{conv} x(\partial M) = \operatorname{conv} x(M), \tag{3.16}$$

where conv E *is the convex hull of E.*

Proof. Let us denote $\Omega = \operatorname{conv} x(\partial M)$ and assume that (3.16) fails. Then this implies $x(M) \setminus \Omega \neq \emptyset$. By Lemma 2 we can find a closed ball B such that $x(M) \subset B$ and there exists a point $m \in \operatorname{int} M$, $x(m) \in \partial B$. We choose a neighbourhood $\mathcal{O}$ of m such that the restriction of x on $\mathcal{O}$ is an embedding. Further arguments will be local and we can arrange that $\mathcal{M} = x(\mathcal{O})$ without loss of generality.

Because of the choice of B, the tangent spaces to $\mathcal{M}$ and ∂B at $x(m)$ coincide. Moreover, $\mathcal{M} \subset B$ and the standard comparison principle for touching surfaces gives the following inequality

$$\lambda_i \geq \frac{1}{R}, \tag{3.17}$$

where λ_i are the principal curvatures of $\mathcal{M}$ at m with respect to the inward normal of ∂B and R is the radius of B.

We now turn to identity (2.2). By the definition of $k_e(m)$ there exists a system of positive numbers $\alpha_i \leq 1$ such that

$$\sum_{i=1}^{n} \alpha_i = 1 \qquad \text{and} \qquad \mathrm{k_e(m)} = \sum_{\mathrm{i=1}}^{\mathrm{n}} \alpha_{\mathrm{i}} \lambda_{\mathrm{i}}.$$

It follows from these relations, (3.17) and (2.2), that

$$0 = \sum_{i=1}^{n} \lambda_i (1 + (p-2)\alpha_i) \geq \frac{n+p-2}{R} > 0.$$

This contradiction proves Corollary 3. □

3.2. Further, we use the Minkowski operations. Namely, given $A, B \subset \mathbb{R}^n$ the notations $A \oplus B$ and λA denote the sets $\{x = a + b : a \in A, b \in B\}$ and $\{x = \lambda a : a \in A\}$.

Definition 5. A family of convex sets $\{\Omega(\tau) : \tau \in [\alpha, \beta]\}$ is called [8] *convex* if for arbitrary $\tau_1 < \tau_2$ from interval $[\alpha; \beta]$ and a nonnegative $t \leq 1$ one has

$$\Omega\left(\tau_1 t + \tau_2(1-t)\right) \subset t\Omega(\tau_1) \oplus \bar{t}\Omega(\tau_2).$$

Let $\mathcal{M}$ be an n-dimensional p-minimal tube in $\mathbb{R}^{n+1}$. Let us denote by $\Omega(\tau)$ the projection of the convex hull of the section $\Sigma(\tau)$ onto the hyperplane $\Pi_0 = \{x_{n+1} = 0\}$. Then

$$\operatorname{conv} \Sigma(\tau) = \tau e_{n+1} \oplus \Omega(\tau).$$

Theorem 2. *The family* $\{\Omega(\tau) : \tau \in \tau(\mathcal{M})\}$ *is convex.*

Proof. We specify $\tau_1 < \tau_2$ from the interval $\tau(\mathcal{M})$ and $t \in [0; 1]$. Let H be the slab $\{x : x_{n+1} \in (\tau_1; \tau_2)\}$ and $M' = x^{-1}(H \cap x(M))$. Then Corollary 3 gives

$$V \equiv \operatorname{conv}(\Sigma(\tau_1) \cup \Sigma(\tau_2)) = \operatorname{conv} x(M').$$

Let $\tau_0 = t\tau_1 + \bar{t}\tau_2$. Then $\Sigma(\tau_0) \subset V$ and by the definition of the convex hull we conclude that $\operatorname{conv} \Sigma(\tau_0) \subset V$.

We choose an arbitrary $z \in \Omega(\tau_0)$. Then $y = z + \tau_0 e_{n+1} \in \Pi(\tau_0) \cap V$, and there exist $y_i \in \operatorname{conv} \Sigma(\tau_i)$ and $\lambda \in [0; 1]$ such that

$$y = \lambda y_1 + \bar{\lambda} y_2. \tag{3.18}$$

Then the decomposition of $y_i = z_i + \tau_i e_{n+1}$ for certain $z_i \in \Omega(\tau_i)$ and (3.18) gives

$$z = \lambda z_1 + \bar{\lambda} z_2, \qquad \tau_0 = \lambda \tau_1 + \bar{\lambda} \tau_2.$$

Hence, $\lambda = t$ and it follows that $z \in t\Omega(\tau_1) \oplus \bar{t}\Omega(\tau_2)$ as required. □

The following assertion gives a sample of applications of the last result.

Corollary 4. *Let $R(\tau)$ be the radius of the least ball which contains $\Sigma(\tau)$* (such a ball is said to be circumscribed near $\Sigma(\tau)$). *Then $R(\tau)$ is a convex function.*

Proof. We denote by $B(\tau)$ the projection onto Π_0 of the ball circumscribed near $\Sigma(\tau)$. Then, by virtue of convexity of $B(\tau)$ we have $B(\tau) \supset \Omega(\tau)$, and Theorem 2 yields for arbitrary $t \in [0; 1]$:

$$\Omega(\tau_0) \subset t\Omega(\tau_1) \oplus \bar{t}\Omega(\tau_2) \subset tB(\tau_1) \oplus \bar{t}B(\tau_2) = B_0,$$

where $\tau_0 = \tau_1 t + \tau_2(1-t)$. By definition, $R(\tau_0) \le R_0$, where R_0 is the radius of B_0. On the other hand, $R_0 = tR(\tau_1) + \bar{t}R(\tau_2)$ and we obtain the required inequality

$$R(\tau_1 t + \tau_2(1-t)) \le R(\tau_1) + \bar{t}R(\tau_2).$$

□

3.3. Now we study the structure of $\Sigma(\tau)$ more completely. This requires further delicate information not only about $R(\tau)$ but about the curve of the centers of the balls $B(\tau)$ as well. Let us denote by $\xi(\tau)$ the center of $B(\tau)$. We recall without proof the well known extremal property of $B(\tau)$ (see [8], Theorem 7.5).

Lemma 3. *Let E be a closed subset of $\mathbb{R}^n$ and $B(E)$ the ball circumscribed near E with the center ξ. Then for all unit vectors $y \in \mathbb{R}^n$ there exists $b \in \partial E \cap \partial B(E)$ such that*

$$\langle b - \xi, y \rangle \ge 0. \tag{3.19}$$

We denote by

$$\sigma(E) = \min_{y \in S^{n-1}} \max_{b \in \partial B \cap E} \frac{\langle b - \xi, y \rangle}{R},$$

where B is the circumscribed ball near a compact set E, R is the radius and ξ is the center of B. It follows from (3.19) that $0 \le \sigma(E) \le 1$. Moreover, one easily shows that $\sigma(E) = 0$ if and only if the intersection of the boundary sphere $S = \partial B$ with F lies in some equatorial hemisphere of S.

Theorem 3. *Let $\mathcal{M}$ be a p-minimal tube in $\mathbb{R}^{n+1}$ such that*

$$\sigma(\Sigma(\tau)) \geq \epsilon > 0, \qquad \forall \tau \in \tau(\mathcal{M}). \tag{3.20}$$

Then $\xi(\tau)$ is a δ-convex curve of τ. In other words, any coordinate function $\xi_k(\tau)$ admits the composition

$$\xi_k(\tau) = \varphi_k(\tau) - \psi_k(\tau),$$

with $\varphi_k(\tau)$, $\psi_k(\tau)$ being convex functions.

Proof. We consider τ_1, τ_2 from $\tau(\mathcal{M})$ and $t \in [0; 1]$. Let us denote by $B(\tau_i) = B_i(\xi(\tau_i), R_i)$ the corresponding balls circumscribed near $\Sigma(\tau_i)$. As above we have for $\tau_0 = t\tau_1 + \bar{t}\tau_2$

$$\Omega(\tau_0) \subset tB(\tau_1) \oplus \bar{t}B(\tau_2).$$

By Lemma 3 we can find $y \in \partial B(\tau_0) \cap \Sigma(\tau_0)$ such that

$$\langle y - \xi(\tau_0), \xi(\tau_0) - \xi_0 \rangle \geq \epsilon |y - \xi(\tau_0)| \cdot |\xi(\tau_0) - \xi_0|,$$

where $\xi_0 = t\xi(\tau_1) + \bar{t}\xi(\tau_2)$. Hence,

$$\begin{aligned} |y - \xi_0|^2 &= \left|(y - \xi(\tau_0)) + (\xi(\tau_0) - \xi_0)\right|^2 \\ &\geq |y - \xi(\tau_0)|^2 + |\xi(\tau_0) - \xi_0|^2 + 2\epsilon |y - \xi(\tau_0)| \cdot |\xi(\tau_0) - \xi_0|, \end{aligned}$$

and taking into account that $|y - \xi(\tau_0)| = R(\tau_0)$ and $|y - \xi_0| \leq R_0$ we obtain

$$|\xi(\tau_0) - \xi_0|^2 + 2\epsilon |y - \xi(\tau_0)| \cdot |\xi(\tau_0) - \xi_0| + (R^2(\tau_0) - R_0^2) \leq 0,$$

and as a consequence,

$$|\xi(\tau_0) - \xi_0| \leq \frac{R_0^2 - R^2(\tau_0)}{R(\tau_0)\epsilon + \sqrt{R_0^2 - R^2(\tau_0)(1 - \epsilon^2)}}. \tag{3.21}$$

By Corollary 4 we have $R_0 \geq R(\tau_0)$ and from (3.21),

$$|\xi(\tau_0) - \xi_0| \leq \frac{R_0^2 - R^2(\tau_0)}{\epsilon(R(\tau_0) + R_0)} = \frac{1}{\epsilon}(R_0 - R(\tau_0)). \tag{3.22}$$

We consider the coordinate function $\xi_k(\tau) = \langle \xi(\tau), e_k \rangle$. Then (3.22) yields

$$t\xi_k(\tau_1) + \bar{t}\xi_k(\tau_2) - \xi_k(\tau_0) \leq \frac{1}{\epsilon}\big(tR(\tau_1) + \bar{t}R(\tau_2) - R(\tau_0)\big).$$

This inequality means that the difference $\psi(\tau) = \epsilon^{-1}R(\tau) - \xi_k(\tau)$ is convex. Therefore, by Corollary 4 we obtain the required decomposition of $\xi_k(\tau)$ into the difference of two convex functions

$$\xi_k(\tau) = \frac{1}{\epsilon}R(\tau) - \psi(\tau),$$

and the lemma is proved. □

Theorem 4. *Let $\mathcal{M}$ be a p-minimal surface with assumption* (3.20) *and* $\beta = (n-1)/(p-1)$. *Then $R(\tau)$ and $\xi(\tau)$ satisfy the differential inequality*

$$R(\tau)R''(\tau) \geq \beta(1 + R'(\tau)^2) + |\xi'(\tau)|^2 \min\{\beta; 1\} \tag{3.23}$$

almost everywhere in $\tau(\mathcal{M})$.

Proof. Convexity of a function provides existence a.e. of the second differential (see [8] or [2], Theorem 5.3). It follows from Corollary 4, Theorem 3 that $R(\tau)$ as well as $\xi_k(\tau)$ have the second differentials almost everywhere in $\tau(\mathcal{M})$. We denote by $\tau'(\mathcal{M})$ the set of full measure where the second differentials of $R(\tau)$ and $\xi_k(\tau)$, $1 \leq k \leq n+1$ do exist.

Let S^{n-1} be the unit sphere in $\Pi_0 \sim \mathbb{R}^n$ endowed with the standard metric. We consider the hypersurface $\mathcal{M}_0$ given by

$$w(\theta, \tau) = \xi(\tau) + R(\tau)\theta + \tau e_{n+1} : S^{n-1} \times \mathbb{R} \to \mathbb{R}^{n+1}$$

where $\theta \in S^{n-1}$. We have shown in [18] that for such a surface the curvature $k_{e,\mathcal{M}_0}$ in the e-direction can be calculated from

$$k_{e,\mathcal{M}_0}(\theta, \tau) = \frac{\omega^3}{R(\tau)} \Big[R(\tau)R''(\tau) + R(\tau)\langle \xi''(\tau), \theta\rangle + \langle \xi'(\tau), \theta\rangle^2 - |\xi'|^2 \Big] \tag{3.24}$$

where

$$\omega^2 = \langle \nu_m, e\rangle^2 = \frac{1}{1 + \big(R'(\tau) + \langle \theta, \xi'(\tau)\rangle\big)^2}.$$

By the definition of functions $R(\tau)$ and $\xi(\tau)$ we conclude that the surface $\mathcal{M}$ is contained in $\mathcal{M}_0$ in the sense that $\Sigma(\tau)$ is a subset of $\Pi(\tau) \cap \mathcal{M}_0$ for all $\tau \in \tau(\mathcal{M})$.

Let us consider an arbitrary $\tau \in \tau'(\mathcal{M})$ and $E = \Omega(\tau) \cap \partial B(\tau)$. The surfaces $\mathcal{M}$ and $\mathcal{M}_0$ have the common outward normal ν_m at $m = y \oplus \tau e_{n+1}$ for every $y \in E$ (we mean by *outward* the normal which is directed out from the inside of $B(\tau)$). Let $\mathcal{O}$ be the neighbourhood of m where $x(\cdot)$ is an embedding. It is a consequence of the definition of $\mathcal{M}_0$ that $\nu_m \wedge e_{n+1} \neq 0$. We denote by $\gamma(\tau)$ and $\gamma_0(\tau)$ the sections of $x(M)$ and $\mathcal{M}_0$ by the two-plane spanned by ν_m and e_{n+1}. Then the comparison principle for touching surfaces yields

$$k_{e,\mathcal{M}}(m) \leq k_{e,\mathcal{M}_0}(m).$$

We write $h(m)$ and $h_0(m)$ for the mean curvatures at m of the sections $\Sigma(\tau)$ and $\Pi(\tau) \cap \mathcal{M}_0 = \xi(\tau) \oplus \tau e_{n+1} \oplus B(\tau)$ with respect to their common outward normal. Then by the comparison principle we arrive at the inequality

$$h(m) \leq h_0(m) \equiv -\frac{n-1}{R(\tau)},$$

and after (2.6)

$$-\frac{p-1}{\omega} k_e(m) \leq -\frac{n-1}{R(\tau)}.$$

By (3.24) we obtain after simplification

$$R(\tau)R''(\tau) - \beta(1 + R'(\tau)^2) \geq (\beta - 1)\langle \xi'(\tau), \theta\rangle^2 + |\xi'|^2 + \langle \theta, y\rangle, \qquad (3.25)$$

where $y = 2\beta R'(\tau)\xi'(\tau) - R(\tau\xi''(\tau))$. Thus, Lemma 3 applied to the vector y provides $b \in E$ such that $\langle b - \xi(\tau), y\rangle \geq 0$. We take

$$\theta_0 = \frac{b - \xi(\tau)}{R(\tau)}$$

and it follows from (3.25) that

$$R(\tau)R''(\tau) - \beta(1 + R'(\tau)^2) \geq (\beta - 1)\langle \xi'(\tau), \theta_0\rangle^2 + |\xi'(\tau)|^2 \geq |\xi'(\tau)|^2 \min\{\beta; 1\},$$

and the theorem is proved completely. □

Remark. Finally, we notice that the quantity $R(\tau)$ measures the size of the section $\Sigma(\tau)$ instead of the distance of this section from the time-axis in the previous inequalities (3.13). Moreover, in the base case $p = 2$ the established inequality (3.23) is stronger than (3.13).

On the other hand, δ-convex functions belong to the class $\overline{W}^2_{1,\mathrm{loc}}(\tau(\mathcal{M}))$; that is, they have a second-order generalized derivative that is a measure (see [2], Chapter 2, §4.10, Corollary). This allows us to proceed to the integration of (3.23) to completion in the standard way [12]:

Corollary 5. *Let $\mathcal{M}$ be a p-minimal tube,* $\dim \mathcal{M} = n > p > 1$. *Then it has finite life-time* length $\tau(\mathcal{M})$. *Moreover,*

$$\operatorname{length} \tau(\mathcal{M}) \leq 2c_\beta r(\mathcal{M}), \quad \beta = \frac{n-1}{p-1}$$

where

$$r(\mathcal{M}) \equiv \min_{\tau \in \tau(\mathcal{M})} R(\tau) > 0$$

and

$$c_\beta = \int_0^{+\infty} \frac{dt}{(1 + t^{2\beta})^{1/2}}.$$

We notice that the previous inequality is precise for the p-minimal surfaces with rotational symmetry.

References

[1] L. Ahlfors, Lectures on quasiconformal mappings, Toronto–New York–London: Van Nostrand Math. Studies, 1966.

[2] V. M. Goldstein and Yu. G. Reshetnyak, Introduction to the theory of functions with generalized derivatives and quasiconformal mappings, Nauka, Moscow 1983.

[3] J. Heinonen, T. Kilpelainen and O. Martio, Nonlinear potential theory of degenerate elliptic equations, Oxford Univ. Press, London 1993.

[4] J. Heinonen, T. Kilpelainen and J. Rossi, The growth of A-subharmonic functions and quasiregular mappings along asymptotic paths, Indiana Univ. Math. J. 38 (1989), 581–601.

[5] V. A. Klyachin, Estimate of spread for minimal surfaces of arbitrary codimension, Sibirsk. Mat. Zh. 33 (5) (1992), 201–207.

[6] S. Kobayashi and K. Nomizu, Foundations of differential geometry, Vol. 2, Interscience, 1969.

[7] P. Lindqvist, On the definition and properties of p-superharmonic functions, J. Reine Angew. Math. 365 (1986), 67–75.

[8] K. Leichtweiss, Konvexe Mengen, Springer-Verlag, 1980.

[9] V. M. Miklyukov, On some properties of tubular minimal surfaces in R^n, Dokl. Akad. Nauk SSSR 247 (1979), No. 3, 549–552; English transl. in Soviet Math. Dokl. 20 (1979).

[10] V. M. Miklyukov and V. G. Tkachev, On the structure in the large of externally complete minimal surfaces in $\mathbb{R}^3$, Soviet Math. (Iz. VUZ) 31 (1987), 30–36.

[11] V. M. Miklyukov and V. G. Tkachev, Some properties of tubular minimal surfaces of arbitrary codimension, Mat. Sb. 180 (9) (1989), 1278–1295; English transl. in Math. USSR Sb. 68 (1) (1991), 133–150.

[12] V. M. Miklyukov and V. D. Vedenyapin, Extrinsic dimension of tubular minimal hypersurfaces, Mat. Sb. 131 (1986), 240–250; English transl. in Math. USSR Sb. 59 (1988).

[13] J. C. C. Nitsche, A uniqueness theorem of Bernstein's type for minimal surfaces in cylindrical coordinates, J. Math. Mech. 11 (3) (1962), 293–302.

[14] R. Osserman, A survey of minimal surfaces, Dover Publications, New York 1987.

[15] N. O. Pyleva and V. G. Tkachev, Projective volume of minimal graphs, in preparation.

[16] L. Simon, A Hölder estimate for quasiconformal mappings between surfaces in Euclidean space, with application to graphs, having quasiconformal Gauss map, Acta Math. 139 (1977), 19–51.

[17] L. Simon, Equation of mean curvature type in two independent variables, Pacif. J. Math. 69 (1977), 245–268.

[18] V. G. Tkachev, The external estimates of clasp function for elliptic hypersurfaces, Preprint 2031-B92, deposited at VINITI, 1992, 17 pp. (in Russian).

[19] V. G. Tkachev, Minimal tubes and coefficients of holomorphic functions in an annulus, Bull. Soc. Sci. Lett. Łódź Sér. Rech. Déform. XLV (1995), 19–26.

Geometry of torus quasi bundles over curves

*Kenji Ueno**

Abstract. In the present note, we shall give an overview of recent developments on a torus quasi bundle over a compact Riemann surface. We shall show that the notion of logarithmic transformation plays an important role to study geometrically a torus quasi bundle.

1991 Mathematics Subject Classification: 32J18, 32J20, 32J15, 14D99

1. Introduction

In the following a curve means a compact Riemann surface and a torus means a complex torus. A torus quasi bundle over a curve is a generalization of a notion of a principal torus bundle over a compact Riemann surface whose structure group is a complex torus T. It is also a complex analytic generalization of the notion of Seifert fibre spaces. Torus quasi bundles often give interesting examples of non-Kähler manifolds.

Principal torus bundles over compact complex manifolds are studied by Höfer [Ho]. Our starting point is an observation of Höfer on the classifying cohomology of isomorphism classes of principal torus bundles. Let $T = \mathbb{C}^n/\Lambda$ be an n-dimensional complex torus and $\mathcal{O}_M(T)$ be a sheaf of local holomorphic mappings from a compact complex manifold M to the torus T. Then, there is a one-to-one correspondence between the set of isomorphism classes of principal T-bundles over the complex manifold M and the first cohomology group $H^1(M, \mathcal{O}_M(T))$. Höfer defined a homomorphism

$$\Phi : \Lambda \otimes_{\mathbb{Z}} H^1(M, \mathcal{O}_M^*) \to H^1(M, \mathcal{O}_M(T)),$$

which is surjective if M is a curve. He did not give geometric meaning to this homomorphism Φ. In the present notes we shall show that for a curve C the homomorphism Φ has a geometric meaning. Namely, each element ξ of $\Lambda \otimes_{\mathbb{Z}} H^1(M, \mathcal{O}_M^*)$ gives the data of logarithmic transformations and if we apply the logarithmic transformations to the trivial principal torus bundle $C \times T \to C$ we obtain the principal T-bundle

*Partially supported by Grant in Aid for Scientific Research on Priority Areas 231 "Infinite Analysis".

corresponding to the cohomology class $\Phi(\xi)$. If a base manifold is of dimension bigger than or equal to 2, it is not easy to define logarithmic transformations. (See Fujimoto [Fu] and Ueno [Ue]). It is an interesting open problem to understand the above homomorphism by means of geometry by generalizing the notion of logarithmic transformations. Fortunately in case of curves logarithmic transformation is a good tool to construct a torus quasi bundle from the trivial principal torus bundle.

In the present note we shall mainly discuss a torus quasi bundle from the view point of logarithmic transformations. The notion of logarithmic transformation was introduced by Kodaira [Ko1] to study elliptic surfaces. It is a complex analytic surgery which makes a new complex manifold from a torus bundle over a curve. A new manifold is either a principal torus bundle or a torus quasi bundle. For example, from a trivial principal T-bundle $C \times T \to C$ over a curve C where T is a complex torus, by applying logarithmic transformations we obtain all principal torus bundles over the curve C.

Let us briefly explain the contents of the present note. In §2 we shall explain basic facts on principal torus bundles over a curve. Analytic elliptic surfaces which are principal bundles of elliptic curves will be discussed also.

In §3 we shall explain logarithmic transformations. In this section we shall only consider the ones which produce no multiple fibres.

In §4 we shall introduce more general logarithmic transformations so that by applying logarithmic transformations we shall obtain a torus quasi bundle from a principal torus bundle. Also we shall show that any torus quasi bundle over a curve C whose general fibre is a torus T is a finite ramified covering of a principal torus bundle over the same curve C whose structure group is a quotient torus T_0 of the torus T. This gives a new approach to studying torus quasi bundles. They have been studied by taking a ramified covering of the base curve to obtain a principal torus bundle.

In §5 we shall study the Picard group and the Neron–Severi group of a torus quasi bundle. We shall give a review of the results in [BU].

Finally in §6 we shall give a new proof of the projectivity criterion of torus quasi bundles due to Kawazoe [Ka]. Here, we shall fully use the fact that a torus quasi bundle over the curve C is a finite ramified covering of a principal torus bundle over the curve C constructed in §4.

2. Principal torus bundle

First let us consider a principal torus bundle over a curve C. Let T be a complex torus. The torus T can be obtained by a quotient manifold $\mathbb{C}^n/\Lambda$, where Λ is a lattice of rank $2n$ in $\mathbb{C}^n$. We have an exact sequence

$$0 \to \Lambda \to \mathbb{C}^n \to T \to 0.$$

We let $\mathcal{O}_C(T)$ be a sheaf of local holomorphic mappings from the curve C to the torus T. Then, we have an exact sequence of sheaves

$$0 \to \Lambda \to \mathcal{O}_C^n \to \mathcal{O}_C(T) \to 0,$$

and from it we have an exact sequence of cohomology groups

$$\to H^1(C, \mathcal{O}_C^n) \to H^1(C, \mathcal{O}_C(T)) \xrightarrow{c} H^2(C, \Lambda) \to . \tag{1}$$

Note that there is a one-to-one correspondence between the set of isomorphism classes of principal T-bundles over the curve C and the first cohomology group $H^1(C, \mathcal{O}_C(T))$. Hence to each principal T-bundle $\pi : M \to C$ we can associate a cohomology class $\eta \in H^1(C, \mathcal{O}_C(T))$. Then, by (1) we have a cohomology class $c(\eta) \in H^2(C, \Lambda)$, which is called the *characteristic class* of the principal torus bundle $\pi : M \to C$. The following proposition is an easy consequence of Blanchard's theorem ([Bl]).

Proposition 2.1. *A principal T-bundle $\pi : M \to C$ is Kähler if and only if $c(\eta)$ is of finite order in $H^2(C, \Lambda)$, where $\eta \in H^1(C, \mathcal{O}_C(T))$ is the cohomology class corresponding to the isomorphism class of the torus bundle $\pi : M \to C$.*

Now assume that our torus T is projective, that is T is an abelian variety. Then the following theorem is a generalization of a special case of Kodaira's theorem ([Ko1], Theorem 11.5). For a proof see, for example, Kawazoe[Ka].

Proposition 2.2. *If the torus T is an abelian variety, a torus bundle $\pi : M \to C$ is projective if and only if the cohomology class $\eta \in H^1(C, \mathcal{O}_C(T))$ corresponding to $\pi : M \to C$ is of finite order.*

Let us give some examples of principal torus bundles. For simplicity let us consider a one-dimensional complex torus, that is an elliptic curve E. An elliptic curve is obtained as a quotient manifold $\mathbb{C}/\Lambda$ where the lattice Λ is normalized in such a way that it is generated by 1 and $\tau \in H$ over the integer ring $\mathbb{Z}$, where H is the upper half plane. Put

$$q = \exp(2\pi i \tau).$$

Then, the complex number q operates on $\mathbb{C}^* = \mathbb{C} - \{0\}$ by multiplication and generates a cyclic group $q^{\mathbb{Z}}$. By $\mathbb{C}^*/\langle q \rangle$ we denote the quotient space $\mathbb{C}^*/q^{\mathbb{Z}}$. Then the quotient space is isomorphic to the original elliptic curve E.

Let $L \to C$ be a complex line bundle over the curve C. Then, the cyclic group $q^{\mathbb{Z}}$ operates on each fibre of

$$L^* = L \setminus \{\text{zero section}\},$$

hence, on L^*, and the quotient manifold

$$S_q(L) = L^*/q^{\mathbb{Z}}$$

has a structure of a principal bundle over C whose structure group is the elliptic curve E. We call such a principal bundle an elliptic principal bundle. If the line bundle L is trivial, the quotient manifold is the product $C \times E$. The exact sequence

$$0 \to q^{\mathbb{Z}} \to \mathcal{O}_C^* \to \mathcal{O}_C(E) \to 0$$

induces a natural homomorphism

$$\psi : H^1(C, \mathcal{O}_C^*) \to H^1(C, \mathcal{O}_C(E)). \tag{2}$$

The isomorphism class of the line bundle L is given by a cohomology class $\xi \in H^1(C, \mathcal{O}_C^*)$. Then, the cohomology class corresponding to the isomorphism class of $S_q(L)$ is given by the image $\psi(\xi)$ of ξ by the natural homomorphism (2).

Also a natural inclusion $\mathbb{Z} \hookrightarrow \Lambda$ induces a natural injection

$$H^2(C, \mathbb{Z}) \to H^2(C, \Lambda). \tag{3}$$

Then, the first Chern class $c_1(\xi)$ of the line bundle L can be regarded as an element of $H^2(C, \Lambda)$ by the inclusion (3). It is easily shown that this cohomology class is the characteristic class of the elliptic bundle $S_q(L)$.

Proposition 2.3.

1) *$S_q(L)$ and $S_q(L^{-1})$ are isomorphic as elliptic principal bundles.*
2) *The elliptic bundle $S_q(L)$ is Kähler if and only if* $\deg L = 0$. *In this case the first Betti number $b_1(S_q(L))$ of the elliptic bundle is $2g(C) + 2$ where $g(C)$ is the genus of the base curve C.*
3) *If* $\deg L \neq 0$, *then $S_q(L)$ is non-Kähler and $b_1(S_q(L)) = 2g(C) + 1$.*

Note that in case $C = P^1$ and $\deg L = \pm 1$ the elliptic principal bundle is a primary Hopf surface. Thus, $S_q(L)$ is a generalization of a Hopf surface.

Now let us come back to a general complex torus $T = \mathbb{C}^2/\Lambda$ and a principal T-bundle.

An exact sequence

$$0 \longrightarrow \mathbb{Z} \longrightarrow \mathbb{C} \xrightarrow{\exp(2\pi\sqrt{-1}*)} \mathbb{C}^* \longrightarrow 0$$

and the inclusion $\Lambda \hookrightarrow \mathbb{C}^n$ induce the following commutative diagram.

$$\begin{array}{ccccccccc} 0 & \longrightarrow & \Lambda & \longrightarrow & \Lambda \otimes_{\mathbb{Z}} \mathbb{C} & \xrightarrow{\mathrm{id} \otimes \exp(2\pi\sqrt{-1}*)} & \Lambda \otimes_{\mathbb{Z}} \mathbb{C}^* & \longrightarrow & 0 \\ & & \| & & \downarrow & & \downarrow & & \\ 0 & \longrightarrow & \Lambda & \longrightarrow & \mathbb{C}^n & \longrightarrow & T & \longrightarrow & 0 \end{array} \tag{4}$$

Then, the homomorphism $\Lambda \otimes_{\mathbb{Z}} \mathbb{C}^* \to T$ induces a natural group homomorphism

$$\Phi : \Lambda \otimes_{\mathbb{Z}} H^1(C, \mathcal{O}^*) = \Lambda \otimes_{\mathbb{Z}} \mathrm{Pic}(C) \to H^1(C, \mathcal{O}_C(T)) \tag{5}$$

where we regard $H^1(C, \mathcal{O}^*)$ as a $\mathbb{Z}$-module. The following proposition is due to Höfer ([Ho], Proposition 7.1 and Corollary 7.2).

Proposition 2.4. *The natural mapping* (5) *is surjective and the following diagram is commutative:*

$$\begin{array}{ccc} \Lambda \otimes_{\mathbb{Z}} H^1(C, \mathcal{O}^*) & \to & H^1(C, \mathcal{O}_C(T)) \\ \downarrow \mathrm{id} \otimes c_1 & & \downarrow c \\ \Lambda \otimes_{\mathbb{Z}} H^2(C, \mathbb{Z}) & \to & H^2(C, \Lambda). \end{array}$$

Proof. The commutative diagram of the sheaves associated with diagram (4) gives the following commutative diagram of exact sequences of cohomology groups:

$$\begin{array}{ccc} 0 & & 0 \\ \downarrow & & \downarrow \\ H^1(C, \Lambda) & = & H^1(C, \Lambda) \\ \downarrow & & \downarrow \\ \Lambda \otimes_{\mathbb{Z}} H^1(C, \mathcal{O}_C) & \xrightarrow{\varphi} & \mathbb{C}^n \otimes_{\mathbb{C}} H^1(C, \mathcal{O}_C) \\ \downarrow & & \downarrow \\ \Lambda \otimes_{\mathbb{Z}} H^1(C, \mathcal{O}_C^*) & \xrightarrow{\Phi} & H^1(C, \mathcal{O}_C(T)) \\ \downarrow & & \downarrow \\ H^2(C, \Lambda) & = & H^2(C, \Lambda) \\ \downarrow & & \downarrow \\ 0 & & 0. \end{array}$$

Since the homomorphism φ is surjective, the homomorphism Φ is also surjective by the above commutative diagram. □

Note that Höfer claimed that Φ is always injective, but this is not true. In the proof of the above proposition, if Φ is isomorphic, then φ is also isomorphic. But this is not the case, since we have $\dim_{\mathbb{C}} \Lambda \otimes_{\mathbb{Z}} H^1(C, \mathcal{O}_C) = 2gn$ where g is the genus of the base curve C while we have $\dim_{\mathbb{C}} \mathbb{C}^n \otimes_{\mathbb{C}} H^1(C, \mathcal{O}_C) = gn$, hence the homomorphism φ is not isomorphic.

3. Logarithmic transformation

In this section we fix a curve C and a complex torus $T = \mathbb{C}^n/\Lambda$. Choose points P_1, $P_2, \dots, P_\ell$ on C and put $C' = C - \{P_1, P_2, \dots, P_\ell\}$. Let

$$D_j = \{t_j \in \mathbb{C} \mid |t_j| < \epsilon\}$$

be a coordinate neighbourhood of the point P_j. To each point P_j we associate an element $\vec{a}_j \in \Lambda$. The mapping

$$\begin{array}{llll} \ell_{\vec{a}_j} & : & D_j^* \times T & \to & D_j \times T \\ & & [t_j, [\zeta]] & \mapsto & (t_j, [\zeta - \frac{\vec{a}_j}{2\pi\sqrt{-1}} \log t_j]) \end{array} \tag{6}$$

is an isomorphism. Note that the mapping $\ell_{\vec{a}_j}$ cannot be extended to that of $D_j \times T$. We can patch together $C' \times T$ and $D_j \times T$'s by means of the mappings $\ell_{\vec{a}_i}$'s and obtain a new complex manifold. It is denoted by

$$L_{P_1}(\vec{a}_1, 1)L_{P_2}(\vec{a}_2, 1)\dots L_{P_\ell}(\vec{a}_\ell, 1)(C \times T)$$

and is called a complex manifold obtained from $C \times T$ by means of logarithmic transformation. Note that there is a natural holomorphic mapping

$$\tilde{\pi} : X = L_{P_1}(\vec{a}_1, 1)L_{P_2}(\vec{a}_2, 1)\dots L_{P_\ell}(\vec{a}_\ell, 1)(C \times T) \to C$$

and $\tilde{\pi} : X \to C$ gives a structure of a principal T-bundle.

Remark 3.1.

1) The above construction depends on the choice of a local coordinate with center P_j. But it is easy to see that different choices of local coordinates give principal T-bundles isomorphic to X.
2) In the above definition we only use local triviality of a torus bundle. Therefore, the logarithmic transformation can be defined for a fibre space $\pi : M \to C$ over a curve C if any general fibre of π is a complex torus T and if in neighbourhoods of the points P_j's the fibre space is complex analytically trivial. In this case we also use the notation

$$L_{P_1}(\vec{a}_1, 1)L_{P_2}(\vec{a}_2, 1)\dots L_{P_\ell}(\vec{a}_\ell, 1)(M).$$

Now Proposition 2.4 implies the following theorem.

Theorem 3.2. *Let*

$$\pi : X = L_{P_1}(\vec{a}_1, 1)L_{P_2}(\vec{a}_2, 1)\dots L_{P_\ell}(\vec{a}_\ell, 1)(C \times T) \to C$$

be a principal T-bundle obtained from the trivial principal T-bundle by means of logarithmic transformation. Then, the cohomology class corresponding to the iso-

morphism class of the principal T-bundle $\pi : X \to C$ is the image of

$$\sum_{j=1}^{\ell} \vec{a}_j \otimes [P_j] \in \Lambda \otimes_{\mathbf{Z}} \mathrm{Pic}(C)$$

in $H^1(C, \mathcal{O}_C(T))$, where for a point P on the curve C the symbol $[P]$ means the complex line bundle determined by the point P. The characteristic class of $\pi : X \to C$ is given by

$$\sum_{j=1}^{\ell} \vec{a}_j$$

if we identify $H^2(C, \Lambda)$ with Λ. Conversely, for any element

$$\Xi = \sum_{i=1}^{m} \vec{b}_i \otimes [Q_i] \in \Lambda \otimes_{\mathbf{Z}} \mathrm{Pic}(C)$$

the isomorphism class of the principal T-bundle

$$L_{Q_1}(\vec{b}_1, 1) L_{Q_2}(\vec{b}_2, 1) \dots L_{Q_m}(\vec{b}_m, 1)(C \times T)$$

corresponds to the cohomology class $\Phi(\Xi) \in H^1(C, \mathcal{O}_C(T))$. Hence, any principal T-bundle is obtained form the trivial T-bundle by means of logarithmic transformations.

By the above theorem and Proposition 2.1 we obtain the following corollary.

Corollary 3.3. *The principal T-bundle*

$$\pi : X = L_{P_1}(\vec{a}_1, 1) L_{P_2}(\vec{a}_2, 1) \dots L_{P_\ell}(\vec{a}_\ell, 1)(C \times T) \to C$$

is Kähler if and only if

$$\sum_{j=1}^{\ell} \vec{a}_j = \vec{0}.$$

4. Torus quasi bundle

Let $T = \mathbb{C}^n/\Lambda$ be an n-dimensional torus. By a quasi T-bundle $\pi : X \to C$ over a curve C we mean that π is a principal T-bundle over $C - \{P_1, P_2, \dots, P_\ell\}$, the fiber $\pi^{-1}(P_j)$ over the point P_j is of the form $m_j T_j$ as a divisor with $m_j \geq 2$ and T_j a torus, and that by taking a suitable finite covering $\mu : \widetilde{C} \to C$ branched along points on C including all the points, P_j, $j = 1, \dots, \ell$, the normalization of the pull back $X \times_C \widetilde{C}$ over $\widetilde{C}$ is a principal T-bundle over the curve $\widetilde{C}$. The fiber $m_j T_j$ is called a multiple fiber with multiplicity m_j.

To construct such a quasi T-bundle we first generalize the notion of logarithmic transformations.

Choose points $P_1, P_2, \dots, P_\ell$ on C and put $C' = C - \{P_1, P_2, \dots, P_\ell\}$. For each point P_j fix a positive integer m_j. We let $\vec{a}_j$ be an element of $\frac{1}{m_j}\Lambda$ such that the order of the point $[\vec{a}_j]$ of the torus T corresponding to $\vec{a}_j$ is precisely m_j. Let

$$D_j = \{t_j \in \mathbb{C} \mid |t_j| < \epsilon\}$$

be a coordinate neighbourhood of the point P_j and put

$$\widehat{D}_j = \{s_j \in \mathbb{C} \mid |s_j| < \epsilon^{1/m_j}\}.$$

By the mapping

$$\begin{aligned} g \ : \ \widehat{D}_j &\to D_j \\ s_j &\mapsto s_j^{m_j}, \end{aligned}$$

$\widehat{D}_j$ is an m_j-sheeted ramified covering of D_j. A holomorphic mapping $g : \widehat{D}_j \times T \to \widehat{D}_j \times T$ defined by

$$g_j : (s_j, [\zeta]) \mapsto (e_{m_j} s_j,\ [\zeta + \vec{a}_j]), \quad e_{m_j} = e^{2\pi\sqrt{-1}/m_j} \tag{7}$$

is an analytic automorphism of order m_j and generates the cyclic group $G_j = \langle g_j \rangle$ of order m_j. Since the automorphism g_j has no fixed points, the quotient $\widehat{D}_j \times T/G_j$ is a complex manifold. Let

$$\mu_j : \widehat{D}_j \times T \to \widehat{D}_j \times T/G_j$$

be the canonical quotient mapping. By $[s_j, [\zeta]]$ we denote the point of the quotient manifold $\widehat{D}_j \times T/G_j$ corresponding to a point $(s_j, [\zeta])$ of $\widehat{D}_j \times T$. We have a holomorphic mapping

$$\begin{aligned} \pi_j \ : \ \widehat{D}_j \times T/G_j &\to D_j \\ [s_j, [\zeta]] &\mapsto s_j^{m_j}. \end{aligned}$$

Over the punctured disk D_j^* the holomorphic mapping π_j gives a principal T-bundle but over the origin 0 the equation

$$\pi_j = 0$$

defines a divisor mT_j where $T_j = T/\langle [\vec{a}_j] \rangle$ is a torus obtained as the quotient by a finite subgroup generated by the point $[\vec{a}_j]$. It is easy to see that π_j gives a structure of quasi T-bundle over D_j.

The mapping

$$\begin{aligned} \ell_{\vec{a}_j} \ : \ \widehat{D}_j^* \times /G &\to D_j^* \times T \\ [s_j, [\zeta]] &\mapsto (s_j^m, [\zeta - \frac{m\vec{a}_j}{2\pi\sqrt{-1}} \log s_j]) \end{aligned} \tag{8}$$

is a well defined holomorphic mapping and is isomorphic. Therefore, we can patch together $\widehat{D}_j \times / G_j$, $i = 1, 2, \ldots, \ell$ and $C' \times T$ by the isomorphisms $\ell_{\vec{a}_j}$ to obtain a compact complex manifold X, which is denoted by

$$X = L_{P_1}(\vec{a}_1, m_1) L_{P_2}(\vec{a}_2, m_2) \ldots L_{P_\ell}(\vec{a}_\ell, m_\ell)(C \times T) \tag{9}$$

and is called the manifold obtained from $C \times T$ by means of logarithmic transforma tions. There is a natural holomorphic mapping $\pi : X \to C$ given by π_j on $\widehat{D}_j \times T / G_j$ and the projection to the first factor on $C' \times T$. The fiber space $\pi : X \to C$ is a principal T-bundle over C' and has multiple fibers with multiplicity m_j if $m_j \geq 2$ and is a quasi T-bundle over C.

Since locally any ramified covering of a curve is cyclic, the following theorem is an immediate consequence of the definition of logarithmic transformations and Theorem 3.2.

Proposition 4.1. *Every quasi T-bundle can be obtained from the trivial T-bundle $C \times T \to C$ by means of logarithmic transformations defined above.*

Let us consider the quasi T-bundle

$$\pi : X = L_{P_1}(\vec{a}_1, m_1) L_{P_2}(\vec{a}_2, m_2) \ldots L_{P_\ell}(\vec{a}_\ell, m_\ell)(C \times T) \to C.$$

We let Λ_0 be the lattice in $\mathbb{C}^n$ generated by Λ and all $\vec{a}_j$'s and put

$$T_0 = \mathbb{C}^n / \Lambda_0.$$

If we denote by H a finite subgroup of the torus T generated by the points $[\vec{a}_j] \in T$ corresponding to $\vec{a}_j$, then the torus T_0 can be written as T/H. Put

$$\pi_0 : Y = L_{P_1}(\vec{a}_1, 1) L_{P_2}(\vec{a}_2, 1) \ldots L_{P_\ell}(\vec{a}_\ell, 1)(C \times T_0) \to C.$$

Then, $\pi_0 : Y \to C$ is a principal T_0 bundle over the curve C. Note that by our construction there exists a surjective holomorphic mapping $f : X \to Y$ such that the following diagram is commutative.

$$\begin{array}{ccc} X & \xrightarrow{f} & Y \\ \pi \downarrow & & \downarrow \pi_0 \\ C & = & C \end{array}$$

Moreover, the holomorphic mapping is only ramified along the multiple fibers of $\pi : X \to C$. Many properties of the quasi T-bundle X can be deduced from those of the principal T_0-bundle Y. For example, we have the following proposition and the corollary.

Proposition 4.2. *The torus quasi bundle X is Kähler (resp. projective) if and only if the principal torus bundle Y is Kähler (resp. projective).*

Proof. Assume that Y is Kähler. Let ω be a Kähler form of Y. Note that $f : X \to Y$ is an abelian covering ramified along the support of T_i of the multiple fibers. Hence, the pull-back $f^*\omega$ is positive definite on $X \setminus \cup_{i=1}^{\ell} T_i$ and at each point of T_i it is positive semi-definite. Near the multiple fiber $m_i T_i$, X is isomorphic to $\widehat{D}_i \times T/G_i$. As a (1,1)-form

$$\frac{\sqrt{-1}}{2\pi}\partial\bar{\partial}\left(\sum_{\nu=1}^{n}|\zeta_\nu|^2 + |s_i|^2\right)$$

is G_i-invariant, it defines a Kähler form on $\widehat{D}_i \times T/G_i$. Let ρ_i be a non-negative C^∞-function in t satisfying

$$\rho_i(t) = \begin{cases} 1, & \text{if } |t| < \frac{1}{3}\epsilon^{2/m_i}; \\ 0, & \text{if } |t| \geq \frac{2}{3}\epsilon^{2/m_i}. \end{cases}$$

Then, a form

$$\omega_i = \frac{\sqrt{-1}}{2\pi}\partial\bar{\partial}\{\rho_i(|s_i|^2)(\sum_{\nu=1}^{n}|\zeta_\nu|^2 + |s_i|^2)\}$$

is positive definite on $\pi^{-1}(D_i(\frac{1}{3}\epsilon^{2/m_i}))$ and $\omega_i \equiv 0$ on $\pi^{-1}(D_i(2\epsilon^{2/m_i}/3))$, where we put $D_i(r) = \{s_i \mid |s_i| < r\}$. Hence, we may regard ω_i as a global (1,1)-form on X. Since, $f^*\omega$ is positive definite on $X \setminus \cup_{i=1}^{\ell} T_i$, and ω_i is positive definite in a neighbourhood of T_i and zero outside a certain neighbourhood of T_i, the form

$$\alpha f^*\omega + \sum_{i=1}^{\ell}\omega_i$$

is positive definite on X if we choose α sufficiently large. Hence, X is Kähler.

Conversely, assume that X is Kähler. Put

$$d = m_1 \cdot m_2 \ldots m_\ell, \quad m_0 = LCM\{m_1, m_2, \ldots, m_\ell\}.$$

We can always find a d-fold abelian covering $\sigma : \widetilde{C} \to C$ of the curve C branched at P_1, $P_2, \ldots, P_\ell$ and a point $P_0 \in C \setminus \{P_1, P_2, \ldots, P_\ell\}$ such that σ has d/m_i ramification points $\{P_i^{(m)}\}$, $m = 1, 2, \ldots, d/m_i$, $i = 0, 1, 2, \ldots, \ell$. Then, the normalization $\widetilde{X}$ of $X \times_C \widetilde{C}$ has a natural structure of a principal T-bundle over $\widetilde{C}$ and it is isomorphic to

$$\prod_{i=1}^{\ell}\prod_{m=1}^{d/m_i} L_{P_i^{(m)}}(m_i\vec{a}_i, 1)(\widetilde{B} \times T). \tag{10}$$

The natural holomorphic mapping $\widetilde{\sigma} : \widetilde{X} \to X$ is only branched along $\pi^{-1}(P_0)$. By a similar argument to that above we can show that $\widetilde{X}$ is Kähler if X is Kähler. Then, by Proposition 2.1, $\widetilde{X}$ is Kähler if and only if

$$\sum_{i=1}^{\ell}\sum_{m=1}^{d/m_i} m_i\vec{a}_i = \vec{0}.$$

The equality can be rewritten as

$$\sum_{i=1}^{\ell}\frac{d}{m_i}m_i\vec{a}_i = d\sum_{i=1}^{\ell}\vec{a}_i = \vec{0}.$$

Hence, by Corollary (3.3) Y is Kähler. Moreover, since f is a finite holomorphic mapping, X is projective if and only if Y is projective. □

Corollary 4.3. *The quasi torus bundle X defined by* (9) *is Kähler if and only if we have*

$$\sum_{j=1}^{\ell}\vec{a}_j = \vec{0}.$$

5. Picard group and Neron–Severi group of a torus quasi bundle

First let us consider the Picard group of a principal T-bundle

$$\pi : X = L_{P_1}(\vec{a}_1, 1)L_{P_2}(\vec{a}_2, 1)\dots L_{P_\ell}(\vec{a}_\ell, 1)(C \times T) \to C.$$

For any line bundle $L \in \mathrm{Pic}(T)$ we have the homomorphism

$$\varphi_L : T \to \mathrm{Pic}^0(T) = T^\vee,\ \varphi_L(x) = \text{ isom. class of } T_x^* L \otimes L^{-1}, \qquad (11)$$

where $T_x : T \to T$ is the translation by $x \in T$. The principal T-bundle $\pi : X \to C$ being fixed, we can associate to any line bundle $L \in \mathrm{Pic}(T)$ an element in $H^1(\mathcal{O}_C(T^\vee))$ in the following way: for a Čech 1-cocycle $\{\xi_{ij}\}$ defining our T-principal bundle, $\xi_{ij} : U_i \cap U_j \to T$, we put

$$\eta_{ij}^L := \varphi_L \circ \xi_{ij} : U_i \cap U_j \to T^\vee.$$

Then $\{\eta_{ij}^L\}$ is a Čech 1-cocycle (φ_L is a homomorphism) and defines a cohomology class in $H^1(C, \mathcal{O}_C(T^\vee))$, denoted by $\tilde{\varphi}_L(\xi)$.

Definition 5.1. Let $\xi \in H^1(\mathcal{O}_C(T))$ be fixed. For any $L \in \mathrm{Pic}(T)$ the $T^\vee$-principal bundle described by $\widetilde{\varphi}_L(\xi)$ is called the *associated $T^\vee$-bundle to L.*

Choose a general point $P \in C$ and fix it. In the following we identify the torus T with the fiber $\pi^{-1}(P)$. A proof of the following lemma can be found in [BU, Lemma 3].

Lemma 5.2. *Let L be a line bundle over the complex torus $T = \pi^{-1}(P)$. Then, the obstruction to extending L to a line bundle on the total space of the principal T-bundle $\pi : X \to C$ is the associated $T^\vee$-bundle to L, $\widetilde{\varphi}_L(\xi)$. That is, the line bundle L can be extended to X if and only if $\widetilde{\varphi}_L(\xi) = 0$.*

Note that in the proof of this lemma, $H^2(C, \mathcal{O}_C^*) = 0$ for a curve C plays the crucial role.

Now restricting a line bundle $\mathcal{L}$ on X to the fiber $\pi^{-1}(P)$, we have a natural group homomorphism

$$r : \mathrm{Pic}(X) \to \mathrm{Pic}(\pi^{-1}(P)) = \mathrm{Pic}(T). \tag{12}$$

Then $\mathrm{Ker}\, r$ consists of isomorphism classes of line bundles whose restriction to the fiber $\pi^{-1}(P)$ is trivial, hence the restriction to each fiber of π is a line bundle of degree 0 on the torus under identification of the torus with each fiber.

Let $\{U_j\}$ be an open covering of C with trivialization

$$\pi^{-1}(U_j) \simeq U_j \times T. \tag{13}$$

For each line bundle $\mathcal{L}$ belonging to $\mathrm{Ker}\, r$ there exists a holomorphic mapping

$$\varphi_j : U_j \to \mathrm{Pic}^0(T) = T^\vee$$

with

$$\mathcal{L}|_{\pi^{-1}(U_j)} \simeq (\varphi_j \times \mathrm{id}_T)^*(\mathcal{P}),$$

where $\mathcal{P}$ is the Poincaré bundle on $\mathrm{Pic}^0(T) \times T$. Since any line bundle of degree 0 on the torus is invariant by the translations by an element of T, on $U_j \cap U_k \neq \emptyset$ we have

$$\varphi_j = \varphi_k.$$

Hence, the line bundle $\mathcal{L}$ defines a holomorphic mapping

$$\varphi : C \to T^\vee. \tag{14}$$

Since the restriction $\mathcal{L}|_{\pi^{-1}(P)}$ is trivial, the holomorphic mapping (14) satisfies

$$\varphi(P) = [0]. \tag{15}$$

The line bundle $\mathcal{L}$ and the holomorphic mapping φ are related by

$$\mathcal{L} \simeq \pi^*(M) \otimes \varphi^*(\mathcal{P}),$$

where M is a line bundle on the curve C and $\varphi^*(\mathcal{P})$ is the line bundle on X whose restriction to $\pi^{-1}(U_j)$ is $(\varphi_j \times \mathrm{id}_T)^*(\mathcal{P})$. Using the fact that $H^2(C, \mathcal{O}_C^*) = 0$, we can patch together $(\varphi_j \times \mathrm{id}_T)^*(\mathcal{P})$'s to obtain $\varphi^*(\mathcal{P})$, since the line bundle of degree 0 on a torus is invariant under the translations. (For the details, see [BU, §3]).

Also note that there is a one to one correspondence between the set of holomorphic mappings (14) with property (15) and $\mathrm{Hom}(J(C), T^\vee)$.

Let us consider a group homomorphism

$$R : \mathrm{Pic}(X) \xrightarrow{r} \mathrm{Pic}(T) \xrightarrow{c} H^2(T, \mathbb{Z}). \tag{16}$$

The homomorphism R is essentially equivalent to a natural homomorphism

$$\mathrm{Pic}(X) \to \mathrm{Pic}(T)/\mathrm{Pic}^0(T)$$

induced by the homomorphism r. A line bundle $\mathcal{L}$ belonging to Ker R is the one whose restriction to each fiber of π is of degree 0. Note that each line bundle $L \in \mathrm{Pic}^0(T)$ can be extended to a line bundle $\mathcal{L}$ on X in such a way that its restriction to each fiber is isomorphic to L. (For the details, see [BU].) Hence, there is an isomorphism

$$\mathrm{Ker}\, R/\mathrm{Ker}\, r \simeq \mathrm{Pic}^0(T). \tag{17}$$

Define subgroups P_j of Pic(X) by

$$P_2 = \pi^* \mathrm{Pic}(B), \quad P_1 = \mathrm{Ker}\, r, \quad P_0 = \mathrm{Pic}(X). \tag{18}$$

Then, $\{P_\bullet\}$ defines a decreasing filtration of Pic(X). By the above consideration and the arguments in [BU, §2], we have the following theorem.

Theorem 5.3. *We have the following isomorphisms.*

$$P_1/P_2 \simeq \mathrm{Hom}(J(C), \mathrm{Pic}^0(T)) \tag{19}$$
$$P_0/P_1 \simeq \{ L \in \mathrm{Pic}(T) \mid \widehat{\varphi}_L(\xi) = 0 \} \tag{20}$$

where $\xi \in H^1(C, \mathcal{O}_C(T))$ is the cohomology class corresponding to the principal T-bundle $\pi : X \to C$ and $\widetilde{\varphi}_L(\xi) = 0$ is defined above.

The Neron–Severi group $NS(X)$ of a complex manifold X is the kernel of the homomorphism of the cohomology groups $\iota : H^2(X, \mathbb{Z}) \to H^2(X, \mathcal{O}_X)$. This is nothing but the image of $H^1(X, \mathcal{O}_X^*) \xrightarrow{c} H^2(X, \mathbb{Z})$, that is, the subgroup of $H^2(X, \mathbb{Z})$ generated by the Chern classes of line bundles on X. The following theorem is one of the main results of [BU]. We can prove it by using the above theorem.

Theorem 5.4. *Let $\pi : X \to C$ be a principal T-bundle over the curve C defined by a cohomology class $\xi \in H^1(C, \mathcal{O}_C(T))$ with $c(\xi) \neq 0$ (i.e. X is non-Kähler). Then we have an exact sequence of free abelian groups*

$$0 \to \mathrm{Hom}(J(C), T^\vee) \to NS(X)/\mathrm{Tors}\, NS(X) \to \widetilde{N} \to 0,$$

where $\widetilde{N}$ is the subgroup of the Neron–Severi group of the torus T defined by

$$\widetilde{N} = \{c_1(L) \in NS(T) \mid \widetilde{\varphi}_L(\xi) = 0\}.$$

6. Projectivity of a torus quasi bundle

In this section we shall give a different proof of Kawazoe's result ([Ka]) on projectivity of a torus quasi bundle.

First let us consider the relationship between $\mathrm{Pic}^0(C)$ and the Jacobian variety $J(C)$ of a curve C of genus g regarded as the Albanese variety. Let us choose a symplectic basis $\{\alpha_1, \dots, \alpha_g, \beta_1, \dots, \beta_g\}$ of the first homology group $H_1(C, \mathbb{Z})$ and a basis $\{\omega_1, \dots, \omega_g\}$ of holomorphic one-forms on C with the following properties.

$$\begin{aligned} &\alpha_i \cdot \alpha_j = 0, && \beta_i \cdot \beta_j = 0, \quad 1 \le i, j \le g \\ &\alpha_i \cdot \beta_j = \delta_{i,j} && 1 \le i, j \le g \\ &\int_{\alpha_i} \omega_j = \delta_{ij}, && \int_{\beta_i} \omega_j = \tau_{ij}, \quad 1 \le i, j \le g \end{aligned}$$

where we call the matrix $\tau = (\tau_{i,j})$ the period matrix of the curve C. Then, the Jacobian variety $J(C)$ is a g-dimensional complex torus $\mathbb{C}^g/\Omega$ where the lattice Ω is generated by row vectors of the $2g \times g$ matrix

$$\begin{pmatrix} I_g \\ \tau \end{pmatrix}.$$

Let us give an explicit isomorphism $p : \mathrm{Pic}^0(C) \to J(C)$. First let us consider two distinct points P, Q on the curve C. Then the line bundle $[P - Q]$ is an element of $\mathrm{Pic}^0(C)$ and $p([P - Q])$ is given by

$$[(\int_Q^P \omega_1, \int_Q^P \omega_2, \dots, \int_Q^P \omega_g)] \in J(C).$$

This mapping can be rewritten by using a meromorphic differential form. The classical theory of a compact Riemann surface tells us that there exists a unique meromorphic differential form $\omega_{P,Q}$ (differential of the third kind) such that $\omega_{P,Q}$ is holomorphic outside P and Q, $\omega_{P,Q}$ has a pole of order one at P and Q with

$$\mathrm{Res}_P(\omega_{P,Q}) = 1/2\pi\sqrt{-1}, \quad \mathrm{Res}_Q(\omega_{P,Q}) = -1/2\pi\sqrt{-1}$$

and that

$$\int_{\alpha_i} \omega_{P,Q} = 0 \quad 1 \le i \le g.$$

Moreover, we have

$$\int_{\beta_j} \omega_{P,Q} = \int_Q^P \omega_j, \quad 1 \le j \le g.$$

Hence we can write

$$p([P - Q]) = [(\int_{\beta_1} \omega_{P,Q}, \int_{\beta_2} \omega_{P,Q}, \dots, \int_{\beta_g} \omega_{P,Q})].$$

This expression can be easily generalized. Note that any element of $\mathrm{Pic}^0(C)$ is expressed by a line bundle $[D]$ of a divisor $D = \sum_{j=1}^{\ell} m_j P_j$ of degree 0. There exists a unique meromorphic one-form ω_D on C such that ω_D is holomorphic outside the P_j's, ω_D has a pole of order one at P_j with

$$\mathrm{Res}_{P_j}(\omega_D) = m_j/2\pi\sqrt{-1}$$

and that

$$\int_{\alpha_i} \omega_D = 0, \quad 1 \leq i \leq g.$$

Then, the image $p([D])$ is given by

$$p([D]) = [(\int_{\beta_1} \omega_D, \int_{\beta_2} \omega_D, \ldots, \int_{\beta_g} \omega_D)]. \tag{21}$$

Now let us consider projectivity of a principal T-bundle

$$\pi : X = L_{P_1}(\vec{a}_1, 1)L_{P_2}(\vec{a}_2, 1)\ldots L_{P_\ell}(\vec{a}_\ell, 1)(C \times T) \to C$$

over the curve C where we now assume that the complex torus T is algebraic, that is, an abelian variety. The cohomology class $\xi \in H^1(C, \mathcal{O}_C(T))$ is given by $\Phi(\sum_{j=1}^{\ell} \vec{a}_j \otimes [P_j])$ where Φ is given in (5). By Proposition 2.2 X is projective if and only if the cohomology class $\Phi(\sum_{j=1}^{\ell} \vec{a}_j \otimes [P_j])$ is of finite order in $H^1(C, \mathcal{O}_C(T))$.

We need the following lemma.

Lemma 6.1. *A cohomology class $\xi \in H^1(C, \mathcal{O}_C(T))$ is of finite order if and only if there exists an element $\sum_{j=1}^{\ell} \vec{a}_j \otimes [P_j] \in \Lambda \otimes \mathrm{Pic}(C)$ of finite order with $\Phi(\sum_{j=1}^{\ell} \vec{a}_j \otimes [P_j])$.*

A proof is obtained by using the diagram in the proof of Proposition 2.4. Now if $\sum_{j=1}^{\ell} \vec{a}_j \otimes [P_j]$ is of finite order, then we have

$$\sum_{j=1}^{\ell} \vec{a}_j \otimes [P_j] \in \Lambda \otimes_{\mathbb{Z}} \mathrm{Pic}^0(C), \quad \sum_{j=1}^{\ell} \vec{a}_j = \vec{0}.$$

We let $\{\vec{e}_1, \ldots, \vec{e}_{2n}\}$ be a $\mathbb{Z}$-basis of Λ. Then, we have

$$\vec{a}_j = \sum_{k=1}^{2n} m_{jk}\vec{e}_k.$$

The equality $\sum_{j=1}^{\ell} \vec{a}_j = \vec{0}$ implies

$$\sum_{j=1}^{\ell} m_{jk} = 0. \tag{22}$$

Since we have

$$\sum_{j=1}^{\ell} \vec{a}_j \otimes [P_j] = \sum_{k=1}^{2n} \vec{e}_k \otimes (\sum_{j=1}^{\ell} m_{jk}[P_j]),$$

$\sum_{j=1}^{\ell} \vec{a}_j \otimes [P_j]$ is of finite order in $\Lambda \otimes_{\mathbb{Z}} \mathrm{Pic}^0(C)$, if and only if $\sum_{j=1}^{\ell} m_{jk}[P_j]$ is of finite order in $\mathrm{Pic}^0(C)$ for all $k = 1, 2, \ldots, 2n$. By (22) there exists a unique meromorphic one-form η_k holomorphic outside P_j's such that η_k has a pole of order one at P_j with

$$\mathrm{Res}_{P_j}(\eta_k) = m_{jk}/2\pi\sqrt{-1}, \qquad \int_{\alpha_i} \eta_k = 0, \quad 1 \le i \le g.$$

Then, we have

$$p(\sum_{j=1}^{\ell} m_{jk}[P_j]) = [(\int_{\beta_1} \eta_k, \int_{\beta_2} \eta_k, \ldots, \int_{\beta_g} \eta_k)].$$

Thus, $\sum_{j=1}^{\ell} m_{jk}[P_j]$ is of finite order in $\mathrm{Pic}^0(C)$, if and only if there exist positive integers m and integers $n_1, \ldots, n_g, n_{k1}, \ldots n_{kg}$ with

$$\begin{aligned} m(\int_{\beta_1} \eta_k, \int_{\beta_2} \eta_k, \ldots, \int_{\beta_g} \eta_k) &= (n_1, n_2, \ldots, n_g) + \sum_{j=1}^{g} n_{kj}(\tau_{j1}, \tau_{j2}, \ldots, \tau_{jg}) \\ &= (n_1, n_2, \ldots, n_g) + \sum_{j=1}^{g} n_{kj}(\int_{\beta_1} \omega_j, \int_{\beta_2} \omega_j, \ldots, \int_{\beta_g} \omega_j) \end{aligned}$$

where we use the fact that $\tau_{kj} = \tau_{jk}$. Thus, if we put

$$\eta'_k = \eta_k - \sum_{j=1}^{g} \frac{n_{jk}}{m} \omega_j$$

then we have

$$m \int_{\gamma} \eta'_k \in \mathbb{Z}$$

for any element $\gamma \in H_1(C, \mathbb{Z})$. Hence, if $\sum_{j=1}^{\ell} \vec{a}_j \otimes [P_j]$ is of finite order, then for all k we can find meromorphic one-forms η'_k holomorphic outside P_j's such that they have poles of order one at P_j with

$$\mathrm{Res}_{P_j}(\eta'_k) = m_{jk}/2\pi\sqrt{-1}$$

and that there exists a positive integer m with

$$m \sum_{k=1}^{2n} (\int_{\gamma} \eta'_k) \vec{e}_k \in \Lambda.$$

Conversely, it is easy to show that if we can find meromorphic one-forms η'_k with above properties, then $\sum_{j=1}^{\ell} a_j \otimes [P_j]$ is of finite order. Regarding Λ as a subset of $\mathbb{C}^n$, we may write

$$\sum_{j=1}^{\ell} \vec{e}_k \otimes \eta'_k = (\xi_1, \xi_2, \ldots, \xi_n).$$

Hence, from the above argument and Proposition 4.2 we infer the following theorem ([Ka]).

Theorem 6.2. *Let*

$$\pi : X = L_{P_1}(\vec{a}_1, m_1)L_{P_2}(\vec{a}_2, m_2)\ldots L_{P_\ell}(\vec{a}_\ell, m_\ell)(C \times T) \to C$$

be a torus quasi bundle over the curve C. Assume that the complex torus $T = \mathbb{C}^n/\Lambda$ is an abelian variety. Then, X is projective if and only if there exist meromorphic one-forms ξ_k, $k = 1, 2, \ldots, n$ over C holomorphic outside P_j's, such that they have poles of order one at P_j with

$$\mathrm{Res}_{P_j}\, \xi_k = a_{jk}/2\pi\sqrt{-1}$$

and that

$$\int_\gamma \xi_k \in \mathbb{Q}\Lambda, \quad k = 1, 2, \ldots, n$$

for all $\gamma \in H_1(C, \mathbb{Z})$ where we write $\vec{a}_j = (a_{j1}, a_{j2}, \ldots, a_{jn})$.

References

[Bl] Blanchard, A., Espaces fibrés kählériens compacts, C.R. Acad. Sci. 238 (1954), 2281–2283.

[Bo] Borel, A., A spectral sequence for complex analytic bundles, Appendix II to [Hi].

[BU] Brînzănescu, V. & Ueno, K., Neron–Severi group of torus quasi bundles on curve, to appear in the Proceedings of Taniguchi conference.

[Fu] Fujimoto, Y., Logarithmic transformations on elliptic fiber spaces, J. Math. Kyoto Univ. 28 (1988), 91–110.

[Hi] Hirzebruch, F., Topological Methods in Algebraic Geometry, 3rd Edition, Springer-Verlag, Berlin–Heidelberg–New York 1966.

[Ho] Höfer, T., Remarks on torus principal bundles, J. Math. Kyoto Univ. 33 (1993), 227–259.

[Ka] Kawazoe, M., Projectivity of torus fibre spaces over curves and multiple fibres, Internat. J. Math. 7 (1996), 183–192.

[Ko1] Kodaira, K., On compact analytic surfaces III, Ann. of Math. 78 (1963), 1–40.

[Ko2] Kodaira, K., On the structure of compact analytic surfaces, I, Amer. J. Math. 86 (1964), 751–798.

[Ue] Ueno, K., Degeneration of elliptic surfaces and certain non-Kähler manifolds, Progr. Math. 39, Birkhäuser, Boston–Basel–Stuttgart, 1983, 545–566.

Stability of minimal graphs in products of surfaces

*Tom Y. H. Wan**

1. Introduction

Suppose M and N are two oriented complete (possibly noncompact) surfaces with strongly negative curvatures. We show that any entire minimal graph in the product 4-manifold $M \times N$ with the product metric is stable provided the curvatures of M and N satisfy a suitable inequality. Such a stability result for the case where the curvatures are equal to the same negative constant was used in the study of minimal Lagrangian submanifolds and minimal maps by Schoen [5] and Lee [3]. The stability result under this stronger curvature assumption follows from a theorem of Micallef and Wolfson [4] concerning minimal Lagrangian surfaces in a 4-dimensional Kähler-Einstein manifold.

In this note, we give a direct proof of this result. In fact, we will derive a second variational formula for minimal surfaces in a product manifold of (possibly more than two) surfaces in terms of second variations of the energies of the projections (which are harmonic maps into corresponding surfaces) and variations of their Hopf differentials. Then assuming a suitable inequality between the curvatures in the case of a product of two surfaces, the second variational formula gives us the required stability of the entire minimal graph.

While visiting Stanford University in the summer of 1994, the author was inspired to write this note by a discussion with Y. I. Lee aimed at extending Schoen's construction of the middle point in Teichmüller spaces [5] to a construction of center of mass. The author would like to thank R. Schoen and Y. I. Lee for helpful discussions.

*Partially supported by the Earmarked Grant, Hong Kong.

2. Harmonic maps between surfaces and minimal graphs

2.1. Harmonic maps between surfaces

Let us recall some basic facts about harmonic maps between surfaces. Let $\rho^2|dz|^2$ and $\sigma^2|du|^2$ be the metrics on M and N respectively, where z and u are local complex coordinates of M and N respectively. A map u from M to N is harmonic if and only if u satisfies

$$\frac{\partial^2 u}{\partial z \partial \bar{z}} + 2\frac{\partial \log \sigma(u(z))}{\partial u}\frac{\partial u}{\partial z}\frac{\partial u}{\partial \bar{z}} = 0.$$

The Hopf differential of u is defined by

$$\Phi = \phi(z)\, dz^2 = \sigma^2(u(z))\frac{\partial u}{\partial z}\frac{\partial \bar{u}}{\partial z}\, dz^2,$$

which is the (2,0)-part of the pull-back metric $u^*(\sigma^2|du|^2)$. If u is harmonic, then it is well-known that Φ is a holomorphic quadratic differential on M.

Let

$$\|\partial u\|^2(z) = \frac{\sigma^2(u(z))}{\rho^2(z)}\left|\frac{\partial u}{\partial z}\right|^2(z),$$

and

$$\|\bar{\partial} u\|^2(z) = \frac{\sigma^2(u(z))}{\rho^2(z)}\left|\frac{\partial u}{\partial \bar{z}}\right|^2(z),$$

be the ∂ and $\bar{\partial}$-energy densities of u respectively. Then the energy density and the Jacobian of u are given by

$$e(u) = \|\partial u\|^2 + \|\bar{\partial} u\|^2 \quad \text{and} \quad J(u) = \|\partial u\|^2 - \|\bar{\partial} u\|^2$$

respectively. Moreover, the pull-back metric can be expressed in terms of Φ and $\|\partial u\|^2$ or Φ and $\|\bar{\partial} u\|^2$. In fact,

$$\begin{aligned} u^*(\sigma^2|du|^2) &= \phi\, dz^2 + e(u)\rho^2|dz|^2 + \bar{\phi}\, d\bar{z}^2 \\ &= \phi\, dz^2 + (\|\partial u\|^2 + \|\Phi\|^2\|\partial u\|^{-2})\rho^2|dz|^2 + \bar{\phi}\, d\bar{z}^2 \\ &= \phi\, dz^2 + (\|\Phi\|^2\|\bar{\partial} u\|^{-2} + \|\bar{\partial} u\|^2)\rho^2|dz|^2 + \bar{\phi}\, d\bar{z}^2, \end{aligned}$$

where $\|\Phi\|(z) = \rho^{-2}(z)|\phi(z)|$.

2.2. Minimal graphs

Let us now discuss the relationship between minimal graphs and harmonic maps in two dimensions. Suppose we have a minimal surface in the product manifold $\prod_{j=1}^k \Sigma_j$ with the product metric, where Σ_j, $j = 1, \ldots, k$ are complete surfaces with metric g_j. Then the induced metric determines a complex structure on the minimal surface.

Let us denote the surface by Σ. Since harmonicity depends only on the conformal structure of Σ and an isometric immersion is minimal if and only if it is harmonic (see [2]), we conclude that the projections

$$u_j : \Sigma \to \Sigma_j$$

are harmonic maps and the map from Σ to $\prod_{j-1}^{k} \Sigma_j$ given by

$$F : z \mapsto (u_1(z), \ldots, u_k(z))$$

is conformal, *i.e.*,

$$\sum_{j=1}^{k} \Phi(u_j) \equiv 0,$$

where $\Phi(u_j)$ are Hopf differentials of the corresponding harmonic maps u_j.

Finally, if the minimal surface is in fact an entire graph over Σ_1, say, then the corresponding harmonic map $u_1 : \Sigma \to \Sigma_1$ is a diffeomorphism. We will need this fact to prove our main result.

3. The second variational formula

In this section, we derive a second variational formula for the area functional of a minimal surface in a product manifold $\prod_{j=1}^{k} \Sigma_j$ of surfaces. First of all, one can fix any conformal metric on the Riemann surface Σ detemined by the induced metric on the minimal graph as in the previous section. Suppose the conformal metric is given by $\rho^2(z)|dz|^2$ on Σ in local conformal coordinate z. Then we can determine the energy densities $e(u_j)$ and the norms $\|\Phi(u_j)\|$ of their Hopf differentials. (We use the same notation as in the previous section.) Let F be the conformal immersion determined by the minimal surface as in the previous section. Then, for any $\Omega \Subset \Sigma$, the area $A(F(\Omega))$ is given by

$$A(F(\Omega)) = \int_{\Omega} \Big[\big(\sum_j e(u_j)\big)^2 - 4 \, \| \sum_j \Phi(u_j) \|^2 \Big]^{\frac{1}{2}} dv_{\Sigma},$$

Now let $F_t : z \mapsto (u_1^t(z), \ldots, u_k^t(z))$ from Σ to $\prod_{j=1}^{k} \Sigma_j$ be a variation of the minimal immersion on Ω, such that $F_0 = F$. Then, as above, the area $A(F_t(\Omega))$ is given by

$$A_t = A(F_t(\Omega)) = \int_{\Omega} \left[\big(\sum_j e(u_j^t)\big)^2 - 4 \| \sum_j \Phi(u_j^t) \|^2 \right]^{\frac{1}{2}} dv_{\Sigma},$$

where $\Phi(u_j^t)$ is the (2,0)-part of the pull-back of the metric on Σ_j via u_j^t.

Then we have the following:

Proposition 1. *Let $\dot{A}$ and $\ddot{A}$ be the first and second variations of the area functional respectively; $\dot{E}_j$ and $\ddot{E}_j$ be the first and second variations of the energy functional of u_j^t respectively; and $\dot{\Phi}_j$ be the first variation of the* (2, 0)*-part of the pull-back via u_j^t. Then, we have the following equalities.*

$$\dot{A} = \sum_j \dot{E}_j, \tag{3.1}$$

$$\ddot{A} = \sum_j \ddot{E}_j - 4\int_\Omega \Big(\sum_j e(u_j)\Big)^{-1} \| \sum_j \dot{\Phi}_j \|^2 \, dv_\Sigma. \tag{3.2}$$

Proof. The proof is straightforward using the fact that $\sum_j \Phi(u_j) = 0$ $(u_j = u_j^0)$.

Equation (3.1) just confirms that $F = F_0$ is minimal if and only if u_j are harmonic under the condition that $\sum_j \Phi(u_j) = 0$ (which is equivalent to saying that F is conformal). Equation (3.2) relates the second variation of the minimal surface $F(\Sigma)$ in $\prod_{j=1}^k \Sigma_j$ to the second variations of the energy functional of the harmonic maps u_j and the first variations of their Hopf differentials.

4. The main result

In the case where we have an entire minimal graph Σ in $\Sigma_1 \times \Sigma_2$ over Σ_1, we have $k = 2$ in Proposition 1 and u_1 is a harmonic diffeomorphism. Then all possible variations can be realized as variations on u_2 only since a variation in u_1 can be regarded as a change of coordinates which leave the area functional invariant. Therefore, in order to calculate the second variation for Σ, we can use the equation (3.2) with $k = 2$ and assume that $u_1^t \equiv u_1$. That is

$$\ddot{A} = \ddot{E}_2 - 4\int_\Omega \frac{\|\dot{\Phi}_2\|^2}{e(u_1) + e(u_2)} \, dv_\Sigma. \tag{4.3}$$

Using this equation, we can prove the following:

Theorem 2. *Suppose M and N are two oriented complete (possibly noncompact) Riemannian surfaces. Let Σ be an entire minimal graph over M in the product manifold $M \times N$ with the product metric. If*

$$-K_M(\pi_M(z)) \le -K_N(\pi_N(z)) \quad \text{and} \quad -b^2 \le K_N \le -a^2 < 0$$

for some positive constants a and b, where K_M and K_N are the curvatures of M and N respectively; π_M and π_N are projections from Σ to M and N respectively. Then Σ is stable.

Proof. Let us use the same notation as before with $\Sigma_1 = M$ and $\Sigma_2 = N$. Then $u_1 = \pi_M$ and $u_2 = \pi_N$ are harmonic maps. Using the Bochner formula for the ∂- and $\bar{\partial}$-energy densities [6] and writing $\Phi = \Phi_1 = -\Phi_2$ (since $\Phi_1 + \Phi_2 = 0$), the functions $\log \|\partial u_i\|$, $\log \|\bar{\partial} u_i\|$, $i = 1, 2$, wherever they are defined, satisfy equations of the following form

$$\Delta_\Sigma w = h(z)(e^{2w} - \|\Phi\|^2 e^{-2w}) + K_\Sigma, \tag{4.4}$$

where $h(z) = -K_{\Sigma_i}(u_i)$ for $\log \|\partial u_i\|$ and $\log \|\bar{\partial} u_i\|$.

Note that the minimal submanifold is an entire graph over $M = \Sigma_1$. So u_1 is a diffeomorphism. We may also assume that it is orientation preserving. (Otherwise, we can consider the function $\log \|\bar{\partial} u_1\|$ instead of $\log \|\partial u_1\|$ in the following argument.) Then $J(u_1) = \|\partial u_1\|^2 - \|\bar{\partial} u_1\|^2 > 0$ implies that the function $\log \|\partial u_1\|$ is well-defined and the Bochner formula (4.4) for $\|\partial u_1\|^2$ is satisfied on the whole of Σ. Moreover, the conformal metric $\|\partial u_1\|^2 ds_\Sigma^2$ is complete. In fact, the pull-back of the metric on M via u_1 is dominated by $2\|\partial u_1\|^2 ds_\Sigma^2$. So the surjectivity of u_1 and the completeness of M imply the completeness of $\|\partial u_1\|^2 ds_\Sigma^2$.

The other three functions $\log \|\bar{\partial} u_1\|$, $\log \|\partial u_2\|$, and $\log \|\bar{\partial} u_2\|$ may not be well-defined on the whole surface. However, from [6], $\|\bar{\partial} u_1\|$, $\|\partial u_2\|$, and $\|\bar{\partial} u_2\|$ are either identically zero or have at most isolated zeroes of the same type as that of holomorphic functions. Moreover, they satisfy, in the distributional sense,

$$\Delta_\Sigma w \geq h(z)(e^{2w} - \|\Phi\|^2 e^{-2w}) + K_\Sigma \quad \text{on the whole surface } \Sigma, \tag{4.5}$$

where $h(z) = -K_{\Sigma_1}(u_1)$ if $w = \log \|\bar{\partial} u_1\|$ and $h(z) = -K_{\Sigma_2}(u_2)$ if $w = \log \|\partial u_2\|$ or $\log \|\bar{\partial} u_2\|$. Note that the singularities of w in these cases are of the form $c \log |z - z_o|$ for some $c > 0$. So $w(z) \to -\infty$ near singularities.

Now we are in a similar situation as that in Lemma 1.2 in [7]. Let w be one of the three functions (with isolated singularities) $\log \|\bar{\partial} u_1\|$, $\log \|\partial u_2\|$, or $\log \|\bar{\partial} u_2\|$ and let $\eta = w - \log \|\partial u_1\|$. Then, from the assumption on the curvatures, and from equations (4.4) and (4.5), we have, in the distributional sense, that

$$\|\partial u_1\|^{-2} \Delta_\Sigma \eta \geq -K_{\Sigma_2}(u_2)(e^{2\eta} - |\mu_1|^2 e^{-2\eta} - 1 + |\mu_1|^2),$$

where $|\mu_1|^2 = \|\Phi\| \|\partial u_1\|^{-2} \leq 1$. Notice that the singularities of η are also of the type $c \log |z - z_o|$ for $c > 0$. The generalized maximum principle of Cheng and Yau [1] also works in our situation. Therefore, as in the proof of Lemma 1.2 in [7], we conclude that $\|\eta\|_\infty \leq 0$, and hence $w(z) \leq \log \|\partial u_1\|(z)$ for all $z \in \Sigma$ (including the singularities which are trivially true). Note that even if $\|\bar{\partial} u_1\|$, $\|\partial u_2\|$, or $\|\bar{\partial} u_2\|$ vanish identically, the inequality is still satisfied trivially. If $w = \log \|\bar{\partial} u_1\|$, the inequality is just $J(u_1) > 0$. For the other two functions, we have

$$\|\partial u_2\|^2 \leq \|\partial u_1\|^2 \quad \text{and} \quad \|\bar{\partial} u_2\|^2 \leq \|\partial u_1\|^2.$$

Then, using the increasing properties of the function $e^{2x} + c^2 e^{-2x}$ on the interval

$(\frac{1}{2}\log c, \infty)$, we obtain the following inequality between the energy densities of u_1 and u_2:

$$e(u_2)(z) \leq e(u_1)(z).$$

Now, if $e(u_2)$ is not identically zero, *i.e.*, u_2 is not a constant map, then putting the above inequality into the second variational formula (4.3), we have

$$\ddot{A} \geq \ddot{E}_2 - 2\int_\Omega \frac{\|\dot{\Phi}_2\|^2}{e(u_2)}\, dv_\Sigma. \tag{4.6}$$

Let ζ be the variation of u_2, *i.e.*, $\zeta = \frac{d}{dt}\big|_{t=0} u_2^t$. Suppose the metric is $g_2 = \sigma^2(w)|dw|^2$ on Σ_2. Then, it is a standard fact that

$$\ddot{E}_2 = 2\int \sigma^2\left(\|\nabla_z^{u_2}\zeta\|^2 + \|\nabla_{\bar{z}}^{u_2}\zeta\|^2\right) dv_\Sigma - \int K_N\sigma^4\|\bar{\zeta}u_{2,z} + \zeta\overline{u_{2,z}}\|^2\, dv_\Sigma,$$

where

$$u_{2,z} = \frac{\partial u_2}{\partial z}, \quad \overline{u_{2,z}} = \overline{u_{2,\bar{z}}} = \overline{\frac{\partial u_2}{\partial z}},$$

$$\nabla_z^{u_2} = \frac{\partial}{\partial z} + 2(\log\sigma)_w u_{2,z}, \quad \text{and} \quad \nabla_{\bar{z}}^{u_2} = \frac{\partial}{\partial \bar{z}} + 2(\log\sigma)_w u_{2,\bar{z}}.$$

It is easy to show by the Schwarz inequality that

$$\int_\Omega \frac{\|\dot{\Phi}_2\|^2}{e(u_2)}\, dv_\Sigma \leq \int \sigma^2\left(\|\nabla_z^{u_2}\zeta\|^2 + \|\nabla_{\bar{z}}^{u_2}\zeta\|^2\right) dv_\Sigma.$$

Therefore, by combining all the above with (4.6), we have

$$\ddot{A} \geq -\int K_N\sigma^4\|\bar{\zeta}u_{2,z} + \zeta\overline{u_{2z}}\|^2\, dv_\Sigma \geq 0, \quad (\text{since } K_N < 0),$$

which implies that the minimal graph is stable.

Finally, if $e(u_2) \equiv 0$, then it is easy to see that the variation $\dot{\Phi}_2 \equiv 0$. Therefore, we have $\ddot{A} = \ddot{E}_2 \geq 0$ since $K_N < 0$. This completes the proof of the theorem.

Corollary 3. *Suppose M and N are two oriented complete (possibly noncompact) surfaces with constant negative curvature such that $-K_M \leq -K_N$. Then any entire minimal graph over M in the product manifold $M \times N$ with the product metric is stable.*

From the proof of Theorem 2, we see that we only need to assume that one of the projections, say u_1, is a diffeomorphic injection and the completeness of $\|\partial u_1\|^2\, ds_\Sigma^2$ if it is orientation preserving or the completeness of $\|\bar{\partial} u_1\|^2\, ds_\Sigma^2$ if it is orientation reversing.

References

[1] S. Y. Cheng & S. T. Yau, Differential Equations on Riemannian manifolds and their Geometric Applications, Comm. Pure Appl. Math. 28 (1975), 333–354.

[2] J. Eells & L. Lemaire Selected Topics in Harmonic Maps, Regional Conference Series in Mathematics 50, Amer. Math. Soc., Providence, RI, 1983.

[3] Y. I. Lee, Lagrangian Minimal Surfaces in Kähler-Einstein Surfaces of Negative Scalar Curvature, preprint 1994.

[4] M. Micallef & J. Wolfson, The Second Variation of Area of Minimal Surfaces in Four-Manifolds, Math. Ann. 295 (1993), 245–267.

[5] R. Schoen, The Role of Harmonic Mappings in Rigidity and Deformation problems, Complex Geometry (Osaka 1990), Lecture Notes in Pure and Appl. Math. 143, Dekker, New York 1993, 179–200.

[6] R. Schoen & S. T. Yau, On Univalent Harmonic Maps between Surfaces, Invent. Math. 44 (1978), 265–278.

[7] L.-F. Tam & T. Wan, Harmonic Diffeomorphisms between Cartan–Hadamard Surfaces with Prescribed Hopf Differentials, Comm. Anal. Geom. 2 (1994), 593–625.

Log Fano threefolds and quotients of $K3$ surfaces

D. -Q. Zhang

Abstract. Let X be a normal projective 3-fold over $\mathbb{C}$ with, at worst, isolated log terminal singularities, such that $c(K_X + H) \sim 0$ for a minimal positive integer c and a Cartier ample divisor H on X. We show that a general member of $|H|$ is a log Enriques surface and give a bound $c \leq 168$. We also try to classify quotients of $K3$ surfaces.

1991 Mathematics Subject Classification: 14J45, 14J28, 14E20

Introduction

Let X be a normal projective 3-fold defined over the complex number field $\mathbb{C}$ and with, at worst, log terminal singularities in the sense of Kawamata or Kollár [KMM, Ko], such that $c(K_X + H) \sim 0$ for a minimal positive integer c and an ample Cartier divisor H on X.

In his paper [R], Reid proved that a general member of $|H|$ is a $K3$ surface with, at worst, rational double point singularities under a stronger condition that X has only Gorenstein canonical singularities. On the other hand, Alexeev [A] shows, among other important theorems, that a general member of $|H|$ is reduced and has only simple quadruple point singularities in codimension one.

In the present note, we shall use a theorem of Prokhorov to prove Theorem 1.1.

Theorem 1.1. *Let X be a normal projective 3-fold over $\mathbb{C}$ with, at worst, isolated log terminal singularities such that $c(K_X + H) \sim 0$ for a minimal positive integer c and an ample Cartier divisor H on X. Then the following two assertions are true.*

(1) *A general member of $|H|$ is a log Enriques surface (see Remark 1.2 below).*

(2) *We have $c \leq 168$. Moreover, for every prime divisor p of c we have $p \leq 19$.*

Remark 1.2.

1) In [Z1], *a log Enriques surface V* is defined as a normal, projective and regular surface with only log terminal singularities (*i.e.*, quotient singularities because $\dim V = 2$) such that IK_V is a trivial Cartier divisor for some positive integer I. The smallest positive integer $I(V)$ leading to a linear equivalence $I(V)K_V \sim 0$, is called *the index of V*.

In [Z1], we proved that the index $I(V) \leq 66$ and that every prime divisor of $I(V)$ is bounded by 19 (which is the best possible bound). Blache [B] has improved the first bound to $I(V) \leq 21$.

The *canonical cover of* V is the Galois $(\mathbb{Z}/I(V)\mathbb{Z})$-cover $U \longrightarrow V$, defined by the linear equivalence $I(V)K_V \sim 0$. Clearly, $K_U \sim 0$ and hence either U is an abelian surface or U is a $K3$ surface with, at worst, rational double point singularities.

In [Z1], we classified log Enriques surfaces with smooth canonical cover. In [Z2], we showed that for every log Enriques surface S, there exists a new log Enriques surface S' of the same index and a crepant blowing-up $S' \to S$ such that the canonical cover of S' is a $K3$ surface with, at worst, Dynkin type A_1 rational double point singularities. Blache's recent paper [B] essentially deals with log Enriques surfaces also.

2) Without the assumption that the singular locus $\operatorname{Sing} X$ is isolated, one can still prove, using a theorem proved recently by Prokhorov [P], that a general member of $|H|$ is a log Enriques surface (Corollary 1.7). However, we do not know whether statement (2) of Theorem 1.1 remains true.

3) After the first version [Z3] was circulated, the author was kindly informed that Conte and Murre [CM] have reproved G. Fano's theorems on 3-folds with Enriques surfaces as hyperplane sections. They also gave a list of known examples of such 3-folds.

4) Under a much stronger condition that X has at worst terminal and cyclic-quotient singularities, Sano [S1, 2] proved that $c \leq 2$. It is conjectured, in general, that $c \leq 2$.

In Theorem 2.7 below, we describe quotients of $K3$ surfaces with, at worst, rational double point singularities. Theorem 2.7 is a corollary of Theorems 2.4, 2.5 and 2.6 in §2. Part (1) of Theorem 2.7 is described more precisely in either of Theorem 2.4 (3-3) and Theorem 2.6 (1); and for parts (2) and (3) of Theorem 2.7, we refer to Theorems 2.4 (3-2) and Theorem 2.6 (2) respectively.

Theorem 2.7. *Let T be a K3 surface with, at worst, rational double point singularities. Suppose that $\mathbb{Z}/\sigma\mathbb{Z}$ acts on T faithfully. Let $S = T/(\mathbb{Z}/\sigma\mathbb{Z})$. Then one of the following three cases occurs.*

(1) *The fundamental group of S^o, the smooth part of S, is finite. S^o and its universal covering of S^o are all rational surfaces.*

(2) *There exists a composition $S \to \overline{S}$ of contractions of extremal rays such that $\overline{S}$ is a log Enriques surface (see Remark 1.2 above).*

(3) *There exist a composition $S \to \overline{S}$ of contractions of extremal rays, a relatively minimal (smooth) ruled surface $\widehat{S}$ with $q(\widehat{S}) = 1, 2$, and a Galois covering $\widehat{S} \to \overline{S}$ which is unramified over $\overline{S} - \operatorname{Sing} \overline{S}$.*

This paper is organized as follows. In §1, we take an index one cover to reduce to the case of the Gorenstein Fano 3-fold and then apply Reid's result in [R] to prove Theorems 1.4 and 1.1. In §2, we apply the log minimal model programme [KMM] to S which tells us to consider three cases (cf. Theorem 2.4). Two cases of these cases are treated in Theorems 2.5 and 2.6. Blache's result in [B] (see also [Z1]) and the main theorem in [GZ1, 2] (see also [FKL]) are used in the proofs.

Remark 2.8. In a recent paper [OZ1], it is proved that, up to isomorphisms, there is exactly one log Enriques surface S (resp. S') whose canonical cover has a rational double point singularity of Dynkin type D_{19} (resp. A_{19}). These are nothing but Examples 6.11 and 3.2 in [Z1].

1. The proof of Theorem 1.1

The following lemma follows from the fact that X is simply connected and hence $\operatorname{Pic} X$ has no non-trivial torsion elements.

Lemma 1.3. *Let X be as in Theorem 1.1. Then the minimal positive integer c such that $c(K_X + H) \sim 0$ for an ample Cartier divisor H is equal to the Cartier index c' of X, i.e., the minimal positive integer c' such that $c'K_X$ is a Cartier index.*

We now prove Theorem 1.4, which, together with Corollary 1.7 below, implies Theorem 1.1.

Theorem 1.4. *Let X be a normal projective 3-fold over $\mathbb{C}$ with, at worst, isolated log terminal singularities such that $c(K_X + H) \sim 0$ for a minimal positive integer c and an ample Cartier divisor H on X. Then there exists a log Enriques surface S of index $I(S)$ on X such that the following properties are satisfied.*

(1) *S is $\mathbb{Q}$-linearly equivalent to H. Further, S is linearly equivalent to H if and only if S is a Cartier divisor.*
(2) *There exists a Galois $\mathbb{Z}/c\mathbb{Z}$-cover $\pi : T \to S$ from a $K3$ surface T with, at worst, rational double point singularities such that π is unramified over the smooth locus $S - \operatorname{Sing} S$ of S.*
(3) *We have $c \leq 8I(S) \leq 168$. Moreover, for every prime divisor p of c we have $p \leq 19$.*

Proof. Let X be as in Theorem 1.4. Then we have a linear equivalence $c(H+K_X) \sim 0$. Let

$$Y := \operatorname{Spec} \bigoplus_{i=0}^{c-1} \mathcal{O}_X(-i(H + K_X)).$$

Then Y is a normal 3-fold with, at worst, Gorenstein canonical singularities (cf. Lemma 1.3 and the proof of [Ka, Proposition 1.7]). Moreover, the natural quotient morphism $\pi : Y \to X$ is a Galois $\mathbb{Z}/c\mathbb{Z}$-cover unramified over $X - \mathrm{Sing}\, X$. Since $c(H + K_X) \sim 0$, we have $\pi^*(H + K_X) \sim 0$. Hence we have $-K_Y = -\pi^*(K_X) \sim \pi^* H$.

Let T be a general member of $|-K_Y| = |\pi^* H|$. By [R], T is a $K3$ surface with, at worst, rational double point singularities. Let $S := \pi(T)$. Then S is irreducible and we have $\sigma S = \pi_*(T) \sim cH$, where σ is the degree of the induced covering $\pi : T \to S'$. Hence $S \sim_{\mathbb{Q}} \frac{c}{\sigma} H$.

Since $\mathrm{Sing}\, X$ is isolated by assumption, the covering $\pi : T \to S$ is unramified outside a finite set. In particular, one has $K_S \equiv 0$.

On the other hand, $K_S \sim (K_X + S)_{|S} \sim_{\mathbb{Q}} (-H + \frac{c}{\sigma} H) = (-1 + c/\sigma) H_{|S}$. So $c = \sigma$, $S \sim_{\mathbb{Q}} H$ and $S = \pi_*(T)$. Thus S is the quotient space $T/(\mathbb{Z}/c\mathbb{Z})$, and since T is a normal surface, so is S.

Note that the Galois $\mathbb{Z}/c\mathbb{Z}$-cover $\pi : T \to S$ is unramified outside $\mathrm{Sing}\, S$ by the purity of branch locus. The second part of Theorem 1.4 (1) follows from the fact that X is simply connected and the remaining parts of the theorem follow from Proposition 2.1 in §2. □

Remark 1.5. We believe that S in Theorem 1.4 is a Cartier divisor and hence $S \in |H|$, though we are not able to prove it at the moment.

Recently Prokhorov [P] has proved the following without assuming that $\mathrm{Sing}\, X$ is isolated:

Theorem 1.6. *Let X be a normal projective 3-fold over $\mathbb{C}$ with, at worst, log terminal singularities, such that $c(K_X + H) \sim 0$ for a minimal positive integer c and an ample Cartier divisor H on X. Then a general member of $|H|$ is a normal surface with at worst log terminal singularities.*

Corollary 1.7. *Let X be as in Theorem 1.6. Then a general member S of $|H|$ is a log Enriques surface.*

Proof. By Theorem 1.6 and the adjunction formula, $K_S \sim_{\mathbb{Q}} 0$. So we only have to show that $H^1(S, \mathcal{O}_S) = 0$. This follows from the exact sequence :

$$0 = H^1(X, \mathcal{O}_X) \to H^1(S, \mathcal{O}_S) \to H^2(X, \mathcal{O}_X(-S)) = 0,$$

where $H^1(X, \mathcal{O}_X) = 0 = H^2(X, \mathcal{O}_X(-S))$ by the vanishing theorem [KMM, Th 1-2-5]. □

2. Quotients of $K3$ surfaces

In this section, we fix the following notations :

Let T be a $K3$ surface with, at worst, rational double point singularities. Suppose that $\mathbb{Z}/\sigma\mathbb{Z}$ acts faithfully on T. Let $S := T/(\mathbb{Z}/\sigma\mathbb{Z})$ and $\pi : T \to S$ be the quotient morphism.

Note that S has only quotient singularities, *i.e.* log terminal singularities, because $\dim S = 2$ (cf. [Ka, Cor. 1.9]), and hence S is $\mathbb{Q}$-factorial.

We first prove the following proposition.

Proposition 2.1. *We use the above notation and assumptions. Let $\langle g\rangle = \mathbb{Z}/\sigma\mathbb{Z}$ and let $\langle g^r\rangle \cong \mathbb{Z}/\frac{\sigma}{r}\mathbb{Z}$ be the maximal subgroup of $\langle g\rangle$ fixing a nowhere vanishing 2-form on T. Then the following four statements are true:*

(1) $\sigma/r \leq 8$;

(2) *The quotient space $T_1 := T/\langle g^r\rangle$ is a $K3$ surface with, at worst, rational double point singularities;*

(3) *We have $\varphi(r) \leq 21$, where φ is the Euler function; in particular, $r \leq 66$. For every prime divisor p of σ we have $p \leq 19$;*

(4) *$K_S \sim_{\mathbb{Q}} 0$ if and only if $T \to S$ is unramified over $S - \operatorname{Sing} S$. In this case, S is a log Enriques surface of index r with $r \leq 21$ and the canonical cover of S is T_1 (see Remark 1.2 in the introduction).*

Proof. (1) and (2) are from [Ni, §5]. By the definition of r, we obtain (1), (2) and (3) in the following claim, while (4) follows from (3) and the Hodge decomposition (see the claim in the proof of Lemma 2.3 in [Z1]).

Claim:

1) If t is a nowhere vanishing 2-form on T, then $t^g = \xi_r t$ for a primitive r^{th} root ξ_r of 1.

2) There is a natural $\langle\overline{g}\rangle = \langle g\rangle/\langle g^r\rangle \cong \mathbb{Z}/r\mathbb{Z}$ action on T_1 such that $S = T_1/\langle\overline{g}\rangle$.

3) If t_1 is a nowhere vanishing 2-form on T_1, then $t_1^{\overline{g}} = \xi_r t_1$ for a primitive r^{th} root ξ_r of 1.

4) The natural $\langle\overline{g}\rangle$ action on T_1 induces a faithful action on the space $H^2(\widetilde{T}_1, \mathbb{Q})/\operatorname{Pic}(\widetilde{T}_1) \otimes_{\mathbb{Z}} \mathbb{Q}$, where $\widetilde{T}_1$ is a minimal resolution of singularities on T_1.

Now part (3) of Proposition 2.1 follows from part (1) of the proposition and part (4) of the above claim (see the proof of Lemma 2.3 or its "added in proof" in [Z1]).

Next we prove part (4) of Proposition 2.1. Since $K_T \sim 0$, the first part of (4) follows from the canonical divisor formula for ramified coverings and the purity of branch locus.

Now suppose that $T \to S$ is unramified over $S^o := S - \operatorname{Sing} S$. Then the $\mathbb{Z}/r\mathbb{Z}$ cover $T_1 \to S$ is also unramified over S^o. So the pull back of K_S by this covering map, is equal to $K_{T_1}(\sim 0)$. Thus $rK_S \sim 0$ and S is a log Enriques surface of index $I(S)$ equal to a factor of r because $q(S) \leq q(T) = 0$. We also have

$$\mathbb{C} \cong H^0(S, \mathcal{O}(K_S)^{\otimes I(S)}) = H^0(T_1, \mathcal{O}(K_{T_1})^{\otimes I(S)})^{\langle g \rangle}.$$

So $r = I(S)$ by part (3) of the above claim, and T_1 is the canonical cover of S. Since $I(S) \leq 21$ (cf. [B], see also [Z1]), Proposition 2.1 (4) follows. □

Again we shall use the notations and the assumptions made at the beginning of §2. By the ramification formula, we have :

$$0 \sim K_T = \pi^* K_S + R$$

where R is the ramification divisor. Let $B = \sum_{i=1}^n B_i$ be the decomposition of the branch locus $B = \pi(R)$ into irreducible components.

Write $\pi^{-1} B_i = \sum_{j=1}^{n_i} R_{ij}$ where R_{ij} is irreducible. Then $(R)_{\text{red}} = \sum_{i,j} R_{ij}$. Let $\mathbb{Z}/m_i\mathbb{Z}$ be the maximal subgroup of $\mathbb{Z}/\sigma\mathbb{Z}$ stabilizing one of R_{ij}'s and hence all of them. Then $m_i n_i = \sigma$.

Let $\pi_* R_{ij} = \gamma_i B_i$ where γ_i is the degree of the finite morphism $R_{ij} \to B_i$. Then $\gamma_i | m_i$ and $b_i := m_i/\gamma_i$ (≥ 2) is the ramification index of π over B_i.

Now we can write

$$R = \sum_{i,j} (b_i - 1) R_{ij}.$$

So $-\sigma K_S = -\pi_* \pi^* K_S \sim \pi_* R = \sum_{i,j} (b_i - 1)\gamma_i B_i = \sum_i n_i (b_i - 1)\gamma_i B_i = \sigma \sum_i (1 - 1/b_i) B_i$. Thus we have proved the following result.

Lemma 2.2. *We have*

$$-K_S \sim_{\mathbb{Q}} \sum_i (1 - 1/b_i) B_i \geq 1/2 B \geq 0.$$

We now prove the following lemma for later use.

Lemma 2.3. *Let X be a normal projective surface with only log terminal singularities. Suppose that $\kappa(X) = -\infty$ and the fundamental group of $X^o := X - \operatorname{Sing} X$ is finite. Then X and the universal covering of X^o are all rational surfaces.*

Proof. Extend the universal covering of X^o to a Galois covering $\tau : U \to X$ unramified over X^o. Note that since X has only log terminal singularities, so does U (cf. the proof of [Ka, Proposition 1.7]). Thus, both X and U are $\mathbb{Q}$-factorial. Note also that $K_U = \tau^* K_X$ and hence $\kappa(U) = \kappa(X) = -\infty$.

Let $g : \widetilde{U} \to U$ be a minimal resolution of singularities of U. Since U has only log terminal singularities, we can write (cf. *e.g.* [Ko, Chapter 3])

$$g^* K_U = K_{\widetilde{U}} + D^*$$

where D^* is an effective $\mathbb{Q}$-divisor with support contained in $g^{-1}(\operatorname{Sing} U)$. Thus, we have $\kappa(\widetilde{U}) \leq \kappa(U) = -\infty$. Hence $\widetilde{U}$ is a ruled surface. Note that since $U - \operatorname{Sing} U$ is simply connected, so is $\widetilde{U}$. In particular, $q(\widetilde{U}) = 0$. Therefore, $\widetilde{U}$ is a rational surface. Since X is dominated by U and since $\dim U \leq 2$, X is also a rational surface by Lüroth's Theorem. □

For the next three theorems, for a variety X we shall let $X^o = X - \operatorname{Sing} X$.

Theorem 2.4. *With the notations and assumptions at the beginning of §2, the following are true:*

(1) *There exists a composition $S \to \overline{S}$ of contractions of extremal rays such that the $\mathbb{Z}/\sigma\mathbb{Z}$ Galois covering $T \to S$ induces a $\mathbb{Z}/\sigma\mathbb{Z}$ Galois covering $\overline{T} \to \overline{S}$, where $\overline{T}$ is a $K3$ surface with, at worst, rational double point singularities, so that the following diagram commutes:*

$$\begin{array}{ccc} T & \longrightarrow & \overline{T} \\ \downarrow & & \downarrow \\ S & \longrightarrow & \overline{S}. \end{array}$$

(2) *There exists a natural surjective homomorphism between the fundamental groups:*

$$\pi_1(\overline{S}^o) \longrightarrow \pi_1(S^o).$$

(3) *One of the following cases occurs.*

3-1) *The covering $\overline{T} \to \overline{S}$ is unramified over $\overline{S}^o$. Hence $\overline{S}$ is a log Enriques surface whose index $I(\overline{S})$ is a factor of σ and whose canonical cover is a $K3$ surface with, at worst, rational double point singularities (see Remark 1.2 in the introduction). We also have $\sigma \leq 8I(\overline{S}) \leq 8 \times 21$.*

3-2) *$-K_{\overline{S}}$ is $\mathbb{Q}$-linearly equivalent to a nonzero effective $\mathbb{Q}$-divisor. There exists a contraction $\varphi : \overline{S} \to \mathbb{P}^1$ of an extremal ray. Hence $\overline{S}$ has Picard number two and every fiber of φ is irreducible. Moreover, a general fiber of φ is $\mathbb{P}^1$ and $\overline{S}$ is a rational surface with only log terminal singularities.*

3-3) *$\overline{S}$ is a del Pezzo surface of Picard number one and with only log terminal singularities.*

Proof. Since S has only log terminal singularities, we can apply the log minimal model programme [KMM] for S. So we have a composition $f : S \to \overline{S}$ of contractions of extremal rays where $\overline{S}$ has only log terminal singularities (cf. [ibid. Proposition 5-1-6]) such that one of the three cases mentioned below occurs.

To prove Theorem 2.4 (1), we let $\overline{T}$ be the normalization of $\overline{S}$ in the function field of T. Then the Stein factorization guarantees the existence of the commutative diagram in Theorem 2.4 (1). Since T is a $K3$ surface with, at worst, rational double point singularities, the same is true for $\overline{T}$ because there is a birational morphism $T \to \overline{T}$. This proves Theorem 2.4 (1).

Theorem 2.4 (2) follows from the following :

$$S^o - (f - \text{ exceptional divisor}) = \overline{S}^o - \text{ (several smooth points of } \overline{S}).$$

Case (1). $K_{\overline{S}}$ is nef. We shall show that this fits Case (3-1) of Theorem 2.4. Since $-K_{\overline{S}} \sim_{\mathbb{Q}} f_* \sum_i (1 - 1/b_i) B_i \geq 0$ (cf. Lemma 2.2), $f_* B = 0$ and $K_{\overline{S}}$ is $\mathbb{Q}$-linearly equivalent to zero. Since the covering $T \to S$ is unramified outside a union of B and a finite set, the covering $\overline{T} \to \overline{S}$ is unramified outside a finite set and hence it is unramified over $\overline{S}$ − Sing $\overline{S}$ by the purity of branch locus. The first part of (3-1) is proved. The rest of (3-1) follows from Proposition 2.1.

Case (2). $K_{\overline{S}}$ is not nef and there exists a contraction $\varphi : \overline{S} \to B$ of an extremal ray where B is a smooth curve. We shall show that this fits Case (3-2) of Theorem 2.4. The first assertion of (3-2) follows from Lemma 2.3 because $K_{\overline{S}}$ is not nef and hence non-zero.

By [KMM, Lemma 3-2-5], $\rho(\overline{S}) = \rho(B) + 1 = 1$. Hence every fiber of φ is irreducible. Note that $g(B) \leq q(\overline{S}) = 0$. So $B \cong \mathbb{P}^1$. By [KMM, Remark 5-1-1], a general fiber of φ is $\mathbb{P}^1$. Thus $\overline{S}$ is rational.

All assertions in (3-2) are thus proved.

Case (3). $K_{\overline{S}}$ is not nef and the Picard number $\rho(\overline{S}) = 1$. Note that $-K_{\overline{S}} \sim_{\mathbb{Q}} f_* \sum_i (1 - 1/b_i) B_i$ and the latter one is zero or ample. Since $K_{\overline{S}}$ is not nef, $-K_{\overline{S}}$ is ample. This fits Case (3-3) of Theorem 2.4. □

We shall consider Cases (3-2) and (3-3) of Theorem 2.4 separately.

Theorem 2.5. *Suppose Case* (3-3) *in Theorem* 2.4 *occurs. Then the following statements are true.*

(1) *The fundamental groups $\pi_1(\overline{S}^o)$ and $\pi_1(S^o)$ are both finite groups.*

(2) *S and the universal covering of S^o are both rational surfaces.*

Proof. Since $\overline{S}$ is a del Pezzo surface with only log terminal singularities, $\pi_1(\overline{S}^o)$ is finite by [GZ1, 2] (see also [FKL] for a differential geometric proof). Now (1) follows from Theorem 2.4 (2).

In view of Lemma 2.3, we only have to prove that $\kappa(S) = -\infty$. By Lemma 2.2, we have $-K_S \sim_{\mathbb{Q}} \sum_i (1 - 1/b_i) B_i \geq 0$. If $\kappa(S) \geq 0$, then $K_S \sim_{\mathbb{Q}} 0$ and hence $K_{\overline{S}} = f_* K_S \sim_{\mathbb{Q}} 0$. This contradicts the fact that $-K_{\overline{S}}$ is ample. So $\kappa(S) = -\infty$. □

Theorem 2.6. *Assume that Case* (3-2) *in Theorem* 2.4 *occurs. Then the following two statements are true.*

(1) *Suppose that the fundamental group $\pi_1(\overline{S}^o)$ is finite. Then $\pi_1(S^o)$ is also finite, and both S and the universal covering of S^o are rational surfaces.*

(2) *Suppose $\pi_1(\overline{S}^o)$ is infinite. Then we have the following commutative diagram which satisfies* (2-1)–(2-6) *below.*

$$\begin{array}{ccccc} T & \longrightarrow & \overline{T} & \longleftarrow & \widehat{T} \\ \downarrow & & \downarrow & & \downarrow \\ S & \longrightarrow & \overline{S} & \longleftarrow & \widehat{S} \\ \downarrow & & \downarrow & & \downarrow \\ \mathbb{P}^1 & = & \mathbb{P}^1 & \longleftarrow & C. \end{array}$$

2-1) $\widehat{T}$ *is an abelian surface.*

2-2) $\widehat{S}$ *is a minimal (smooth) ruled surface with* $q(\widehat{S}) = g(C) = 1$ *or* 2. $\widehat{S} \to C$ *is a ruling induced from the* $\mathbb{P}^1$*-fibration* $\varphi : \overline{S} \to \mathbb{P}^1$.

2-3) $\widehat{S} \to \overline{S}$ *is a Galois covering unramified over* $\overline{S}^o$.

2-4) $\widehat{T} \to \overline{T}$ *is a Galois covering unramified over* $\overline{T}^o$ *such that* $G(\widehat{T}/\overline{T})$ *is a subgroup of* $\mathrm{Gal}(\widehat{S}/\overline{S})$.

2-5) $\widehat{T} \to \widehat{S}$ *is a Galois covering such that* $\mathrm{Gal}(\widehat{T}/\widehat{S})$ *is a subgroup of* $\mathrm{Gal}(\overline{T}/\overline{S}) = \mathbb{Z}/\sigma\mathbb{Z}$.

2-6) $\pi_1(\widehat{S}^o) = \pi_1(C)$ *is an infinite group.* $\pi_1(\overline{S}^o)/\pi_1(\widehat{S}^o) = \mathrm{Gal}(\widehat{S}/\overline{S})$.

Proof. (1) Assume that $\pi_1(\overline{S}^o)$ is finite. The first assertion follows from Theorem 2.4 (2). In view of Lemma 2.3, we only have to prove that $\kappa(S) = -\infty$. This can be done in a way similar to the proof of Theorem 2.5.

(2) Assume that $\pi_1(\overline{S}^o)$ is infinite. Let E be a general fiber of the $\mathbb{P}^1$-fibration $\varphi : \overline{S} \to \mathbb{P}^1$. Let $E_1, \ldots, E_r$ be all of its fibers with multiplicity $m_i \geq 2$. If $r \leq 2$, then [No, Lemma 1.5] implies the following exact sequence :

$$(1) = \pi_1(E) \longrightarrow \pi_1(\overline{S}^o - E_1 - E_2) \longrightarrow \pi_1(\mathbb{C}^*) = \mathbb{Z}.$$

Hence $\pi_1(\overline{S}^o - E_1 - E_2)$ is abelian. Thus, $\pi_1(\overline{S}^o)$, which is a surjective image of $\pi_1(\overline{S}^o - E_1 - E_2)$, is also abelian. So, $\pi_1(\overline{S}^o) = H_1(\overline{S}^o, \mathbb{Z})$, which is finite because $\overline{S}$ is rational (cf. *e.g.* the proof of [GZ2, Lemma 1.4]). This contradicts our assumption. Thus, $r \geq 3$. By the solution of Fenchel's conjecture due to [C, F], there exists a Galois covering $C \to \mathbb{P}^1$, ramified exactly over the $\varphi(E_i)$'s, such that every point lying over $\varphi(E_i)$ has ramification index m_i. Let $\widehat{S}$ be the normalization of $\overline{S} \times_{\mathbb{P}^1} C$. Then the natural Galois covering $\widehat{S} \to \overline{S}$ is unramified over $\overline{S}^o$. So (2-3) follows.

We also have

$$\pi_1(\overline{S}^o)/\pi_1(\widehat{S}^o) = \mathrm{Gal}(\widehat{S}/\overline{S}).$$

Moreover, if we let $\psi : \widehat{S} \to C$ be the $\mathbb{P}^1$-fibration induced from $\varphi : \overline{S} \to \mathbb{P}^1$, then every fiber of ψ is reduced.

Let F be a general fiber of ψ. Applying [No, Lemma 1.5], we have the following exact sequence :

$$(1) = \pi_1(F) \longrightarrow \pi_1(\widehat{S}^o) \longrightarrow \pi_1(C) \longrightarrow (1).$$

So $\pi_1(\widehat{S}^o) = \pi_1(C)$. Since $\pi_1(\overline{S}^o)$ is infinite, $\pi_1(\widehat{S}^o) = \pi_1(C)$ is infinite. Thus (2-6) is proved. We also have $g(C) \geq 1$.

Let $T^* = \overline{T} \times_{\overline{S}} \widehat{S}$ and let $\widehat{T}$ be the normalization of an irreducible component of T^*. We shall prove (2-1), (2-2), (2-4) and (2-5). For simplicity, we may assume that $\widehat{T}$ itself is an irreducible component of T^*.

Note that $\mathrm{Gal}(\widehat{S}/\overline{S})$ and $\mathrm{Gal}(\overline{T}/\overline{S})$ act faithfully on T^* such that

$$T^*/\,\mathrm{Gal}(\widehat{S}/\overline{S}) = \overline{T}, \qquad T^*/\,\mathrm{Gal}(\overline{T}/\overline{S}) = \widehat{S}.$$

Let $G_{\widehat{T}\to\overline{T}}$ and $G_{\widehat{T}\to\widehat{S}}$ be the stabilizer subgroups of $\widehat{T}$ with respect to the actions $\mathrm{Gal}(\widehat{S}/\overline{S})$ and $\mathrm{Gal}(\overline{T}/\overline{S})$ respectively. Then these two subgroups are also stabilizers of any irreducible component of T^* with respect to the corresponding actions. Moreover, we have :

$$\widehat{T}/G_{\widehat{T}\to\overline{T}} = \overline{T}, \qquad \widehat{T}/G_{\widehat{T}\to\widehat{S}} = \widehat{S},$$
$$G_{\widehat{T}\to\overline{T}} = \mathrm{Gal}(\widehat{T}/\overline{T}), \qquad G_{\widehat{T}\to\widehat{S}} = \mathrm{Gal}(\widehat{T}/\widehat{S}).$$

Thus (2-4) and (2-5) are proved by noting that $\widehat{S} \to \overline{S}$ is unramified outside the finite set $\mathrm{Sing}\,\overline{S}$ and by using the purity of branch locus.

Let $\widetilde{S}$ be a desingularization of $\widehat{S}$. Since $\overline{S}$ has only log terminal singularities, the same is true for $\widehat{S}$ (cf. (2-3) and the proof of [Ka, Proposition 1.7]), and hence $\widehat{S}$ has only rational singularities (cf. [KMM, Theorem 1-3-6]). Thus $q(\widetilde{S}) = q(\widehat{S})$.

In view of (2-4), the pull back of $K_{\overline{T}} (\sim 0)$, by the covering $\widehat{T} \to \overline{T}$, is equal to $K_{\widehat{T}}$. Hence $K_{\widehat{T}} \sim 0$ and $\widehat{T}$ is an abelian surface because $q(\widehat{T}) \geq q(\widehat{S}) = q(\widetilde{S}) = g(C) \geq 1$. This proves (2-1). We also have $g(C) = 1$ or 2 because now $q(\widehat{T}) = 2$.

We now prove (2-2). Since every fiber of the $\mathbb{P}^1$-fibration $\psi : \widehat{S} \to C$ is reduced by the above arguments, we only have to show that every fiber of ψ is irreducible, in order to finish the proof of (2-2). Suppose to the contrary that $F_1 = \sum_{i=1}^m F_{1i}$ is a reducible fiber of ψ. Note that the pull back of F_{1i}, by the covering $\widehat{T} \to \widehat{S}$, is contained in fibers of the (elliptic) fibration on $\widehat{T}$ induced by ψ, and hence it is a union of (smooth) fibers. This implies that $F_{1i}.F_{1j} = 0$ for all i, j. This is absurd because F_1 is connected. (2-2) is proved. □

References

[A] V. A. Alexeev, Theorems about good divisors on log Fano varieties (Case of index $r > d - 2$), Lecture Notes in Mathematics 1479, Springer-Verlag, 1989, 1–9.

[B] R. Blache, The structure of l.c. surfaces of Kodaira dimension zero, I, J. Algebraic Geom. 4 (1995), 137–179.

[C] T. C. Chau, A note concerning Fox's paper on Fenchel's conjecture, Proc. Amer. Math. Soc. 88 (1983), 584–586.

[CM] A. Conte and J. P. Murre, Algebraic varieties of dimension three whose hyperplane sections are Enriques surfaces, Ann. Scuola Norm. Sup. Pisa Cl. Sci. (4) 12 (1985), 43–80.

[F] R. H. Fox, On Fenchel's conjecture about F-groups, Math. Tidsskr. B. (1952), 61–65.

[FKL] A. Fujiki, R. Kobayashi and S. Lu, On the fundamental group of certain open normal surfaces, Saitama Math. J. 11 (1993), 15–20.

[GZ1,2] R. V. Gurjar and D. -Q. Zhang, π_1 of smooth points of a log del Pezzo surface is finite, I, II, J. Math. Sci. Univ. Tokyo 1 (1994), 137–180; 2 (1995), 165–196.

[Ka] Y. Kawamata, The cone of curves of algebraic varieties, Ann. of Math. 119 (1984), 603–633.

[KMM] Y. Kawamata, K. Matsuda and K. Matsuki, Introduction to the minimal model problem, Adv. Stud. Pure Math 10 (1987), 283–360.

[Ko] J. Kollár, Flip and abundance for algebraic threefolds, Astérisque 211 (1992).

[Ni] V. V. Nikulin, Finite automorphism groups of Kähler $K3$ surfaces, Trans. Moscow Math. Soc. 38 (1980), 71–135.

[No] M. V. Nori, Zariski's conjecture and related problems, Ann. Sci. Éc. Norm. Sup. 16 (1983), 305–344.

[OZ1,2] K. Oguiso and D. -Q. Zhang, On the most algebraic $K3$ surfaces and the most extremal log Enriques surfaces I, Amer. J. Math. 118 (1996); II, preprint 1995.

[P] Yuri G. Prokhorov, Remarks on algebraic threefolds whose hyperplane sections are Enriques surfaces, preprint 1994.

[R] M. Reid, Projective morphisms according to Kawamata, preprint, University of Warwick 1983.

[S] T. Sano, On classification of non-Gorenstein $\mathbb{Q}$-Fano 3-folds of index 1, J. Math. Soc. Japan 47 (1995), 369–380.

[Z1,2] D. -Q. Zhang, Log Enriques surfaces, I, II, J. Math. Kyoto Univ. 31 (1991), 419–466; 33 (1993), 357–397.

[Z3] D. -Q. Zhang, The fundamental group of the smooth part of a log Fano variety, Osaka J. Math. 32 (1995), 637–644.

On certain Laplace inequalities on manifolds

Detang Zhou

Abstract. We give an estimate for the size of the negative sets of solutions to the differential inequality $\Delta u \geq \lambda u$ which arises naturally when we study the length of the mean curvature of an isometric immersion f of M into another Riemannian manifold N with curvature bounded below.

1991 Mathematics Subject Classification: 58C40, 58G25.

1. Introduction

The aim of this paper is to analyze a kind of differential inequality and show that the corresponding result in [CIK] is incorrect without additional assumptions. For the sake of completeness we will give some details which may be contained in the references. Let M be a complete connected smooth Riemannian manifold of dimension $m > 2$. Let Δ denote the Laplace–Beltrami operator of M. We adopt the convention that Δ is a negative semidefinite operator. For $x_0 \in M$ and $r \in R_+$, we denote by $B^M(x_0, r)$ the open metric ball of radius r centered at x_0 and we denote its closure by $\bar{B}^M(x_0, r)$. We shall deal with negative solutions of the inequality

$$\Delta u \geq \lambda u. \tag{1.1}$$

We obtain

Theorem 1.1. *Let $U \subset M$ be a domain containing $B^M(x_0, r)$ for some r and x_0. Suppose that*

$$\mathrm{Ricci}\,|_x \geq -\delta \tag{1.2}$$

where $\delta \geq 0$ is a constant. Then for given $\lambda < 0$ satisfying

$$\lambda < \begin{cases} 0, & \text{when } \delta = 0, \\ -\frac{(m-1)^2\delta}{4}, & \text{when } \delta > 0, \end{cases} \tag{1.3}$$

* Supported partially by NNSFC and Natural Science Foundation of Shandong

there exists a constant $R > 0$ such that if u is a C^2 nonpositive function defined on U and satisfying (1.1), *then either $r < R$ or else $u \equiv 0$ on U and the constant R can be estimated in terms of m, δ, λ.*

Let $f : M \to N$ be a C^∞ isometric immersion of M into another Riemannian manifold N of dimension n. For $y \in N$, we denote by $i_N(y)$ the radius of injectivity of N at y and $i(N) = \inf_{y \in N} i_N(y)$. When we write $B^N(o, \rho)$, we assume that ρ is a number satisfying the following condition:

(1.4) The inverse of the exponential map at o is bijective on $B^N(o, \rho)$ and if the supremum κ of the sectional curvature K_N of N over $B^N(o, \rho)$ is positive, $\rho \leq \pi/(2\sqrt{\kappa})$.

Let $H : M \to R_+ \cup \{0\}$ represent the length of the mean curvature vector associated with f. Define $\mu : R \times (0, +\infty) \to R$ by

$$\mu(x, r) := \begin{cases} \sqrt{x}\cot(r\sqrt{x}), & \text{for } x > 0, \\ 1/r, & \text{for } x = 0, \\ \sqrt{-x}\coth(r\sqrt{-x}), & \text{for } x < 0. \end{cases} \tag{1.5}$$

We would like to present here some special cases of applications of Theorem 1.1, the proofs of which are straightforward by Theorem 4.1.

Theorem 1.2. *Let M be a complete Riemannian manifold with nonnegative Ricci curvature and let κ be the supremum of the sectional curvatures on N. Then if $\sup_{x \in M} H(x) < \mu(\kappa, \rho)$ then $f(M)$ cannot be contained in $B^N(o, \rho)$ for any $o \in M$ and any ρ satisfying condition* (1.4).

Theorem 1.3. *Let M be a complete Riemannian manifold with Ricci curvature bounded below by a constant $-\delta$ ($\delta > 0$) and let κ be the supremum of the sectional curvatures on N. Then if $\sup_{x \in M} H(x) < \mu(\kappa, \rho)$ then $f(M)$ cannot be contained in $B^N(o, \rho)$ for any $o \in M$ and any ρ satisfying conditions* (1.4) *and* (4.2).

It should be remarked that the conclusion of Theorem 1.1 was also proved in [CIK] without condition (1.3). In fact we can see that their result is not correct. By putting $u = -e^{-\rho}$ in $H^m(-1)$ and $\lambda = 2 - m$, we have a counterexample to Theorem 1 in [CIK]. This paper is organized as follows: in Section 2 we give some geometric preliminaries needed in this paper, in Section 3 we analyze a second order ordinary differential equation and prove Theorem 1.1. In Section 4 we apply Theorem 1.1 to study some geometric problems.

2. Geometric preparations

Let M be a complete connected Riemannian manifold of dimension $m \geq 2$. For $p \in M$ and $0 \neq v \in T_pM$, we define the Ricci curvature of M with the normalization

$$\mathrm{Ric}_M(v) := \mathrm{trace}\{T_pM \ni w \mapsto R^M(v, w)v\}/(m-1)|v|^2. \tag{2.1}$$

The following two lemmas are well known (see [BC]).

Lemma 2.1. *Assume that*

$$\mathrm{Ric}_M|_x \geq -\delta \tag{2.2}$$

where $\delta \geq 0$ *is a constant and* $\rho(x) = d(x, x_0)$. *If* x *is not on the cut locus of the point* x_0 *then*

$$\Delta\rho(x) \leq (m-1)\sqrt{\delta}\coth(\rho\sqrt{\delta}). \tag{2.3}$$

Define $V(t) = \mathrm{vol}(B^M(x_0, t))$; then we have

Lemma 2.2. *Suppose that* (2.2) *holds. Then for all* s, $r \geq s > 0$

$$\frac{V'(s)}{V(s)} \leq \frac{\sinh^{m-1}(s\sqrt{\delta})}{\int_0^s \sinh^{m-1}(t\sqrt{\delta})\,dt}.$$

The boundary of $B^M(x_0, s)$ is a smooth manifold outside its intersection with the cut locus of x_0. The singular points of $\partial B^M(x_o, s)$ have measure 0 with respect to the measure induced in $\partial B^M(x_0, s)$ which we denote by $d\alpha_s$. We let $\vec{n}_s$ denote the outward pointing unit normal vector field defined on the regular points of $\partial B^M(x_0, s)$. The mean curvature function of $\partial B^M(x_0, s)$ with respect to $\vec{n}_s$ will be denoted by η_s. The following formula is well known :

$$\frac{d}{ds}[d\alpha(s)] = (m-1)\eta_s d\alpha(s). \tag{2.4}$$

One can find the proof of the following lemma in [ES].

Lemma 2.3. *If* (2.2) *holds for some nonnegative constant* δ *on* U *then for all* $s > 0$,

$$\eta_s(x) \leq \mu(\sqrt{\delta}, s)$$

where μ *is the function defined in* (1.5).

Writing $p(s) = \int_0^s \sinh^{m-1}(t\sqrt{\delta})\,dt$ when $\delta > 0$ and $p(s) = s^m$ when $\delta = 0$, a direct calculation shows that

$$\frac{p''(s)}{p'(s)} = (m-1)\mu(\sqrt{\delta}, s). \tag{2.5}$$

3. Differential inequality

Following [CIK] we keep the notation of §2 and define $\bar{u} : (0, r) \to (-\infty, 0)$ by

$$\bar{u}(s) := \frac{1}{p(s)} \int_{B^M(p,s)} u \, d\mu, \tag{3.1}$$

where u is a nonpositive solution of (1.1) and $d\mu$ is the volume element with respect to the Riemannian metric of M.

Lemma 3.1. *The function $\bar{u}$ defined above is twice differentiable and satisfies*

$$\bar{u}''(s) \geq \lambda \bar{u}(s) - g(s)\bar{u}'(s), \tag{3.2}$$

where $g(s) = \frac{2p'(s)}{p(s)} - \frac{p''(s)}{p'(s)}$.

Proof. It is easy to see that $\bar{u}$ is differentiable and

$$\bar{u}'(s) = -\frac{p'(s)}{[p(s)]^2} \int_{B^M(p,s)} u \, d\mu + \frac{1}{p(s)} \int_{\partial B^M(p,s)} u \, d\alpha_s$$

then

$$\begin{aligned}
\bar{u}''(s) &= \left\{ 2\frac{[p'(s)]^2}{[p(s)]^3} - \frac{p''(s)}{[p(s)]^2} \right\} \int_{B^M(p,s)} u \, d\mu - \frac{2p'(s)}{[p(s)]^2} \int_{\partial B^M(p,s)} u \, d\alpha_s \\
&\quad + \frac{1}{p(s)} \int_{\partial B^M(p,s)} [\langle \nabla u, n_s \rangle + (m-1)u\eta_s] \, d\alpha_s \\
&= \frac{1}{p(s)} \int_{\partial B^M(p,s)} \langle \nabla u, n_s \rangle \, d\alpha_s + \left[\frac{2p'(s)}{p(s)} - \frac{p''(s)}{p'(s)} \right] \frac{p'(s)}{[p(s)]^2} \int_{B^M(p,s)} u \, d\mu \\
&\quad - 2\frac{p'(s)}{[p(s)]^2} \int_{\partial B^M(p,s)} u \, d\alpha_s + \frac{1}{p(s)} \int_{\partial B^M(p,s)} (m-1)u\eta_s \, d\alpha_s.
\end{aligned}$$

By the divergence theorem in the first term of $\bar{u}''$

$$\begin{aligned}
\frac{1}{p(s)} \int_{\partial B^M(p,s)} \langle \nabla u, n_s \rangle \, d\alpha_s &= \frac{1}{p(s)} \int_{B^M(p,s)} \Delta u \, d\mu \\
&\geq \frac{1}{p(s)} \int_{B^M(p,s)} \lambda u \, d\mu = \lambda \bar{u}(s),
\end{aligned}$$

then the last two terms add to

$$-2\frac{p'(s)}{[p(s)]^2}\int_{\partial B^M(p,s)} u\,d\alpha_s + \frac{1}{p(s)}\int_{\partial B^M(p,s)} (m-1)u\eta_s\,d\alpha_s$$
$$\geq -2\frac{p'(s)}{[p(s)]^2}\int_{\partial B^M(p,s)} u\,d\alpha_s + \frac{p''(s)}{p'(s)}\frac{1}{p(s)}\int_{\partial B^M(p,s)} u\,d\alpha_s$$
$$= -\left[\frac{2p'(s)}{p(s)} - \frac{p''(s)}{p'(s)}\right]\frac{1}{p(s)}\int_{\partial B^M(p,s)} u\,d\alpha_s.$$

Therefore, by applying the previous formula for $\bar{u}'$, we obtain the result. □

Lemma 3.2. *If the Ricci curvatures of M are bounded below by $-\delta$ for some $\delta \geq 0$ and λ satisfies* (1.3), *then there exists an a priori constant τ depending only m, δ, and λ such that if $\bar{u}(x)$ is a solution of* (3.2) *which is strictly negative on some interval $(0, T)$ then $T < \tau$.*

Proof.

$$\bar{u}''(s) \geq \lambda\bar{u}(s) - g(s)\bar{u}'(s).$$

Set $G(t) = e^{\int g(s)\,ds} = \frac{p^2(s)}{p'(s)}$. Then

$$(G(s)\bar{u}'(s))' = G(s)\bar{u}''(s) + G(s)g(s)\bar{u}'(s) \geq \lambda G(s)\bar{u}(s).$$

Defining $v(s) = -\frac{G(s)\bar{u}'(s)}{\bar{u}(s)}$, we have

$$-v'(s) + \frac{v^2}{G(s)} \leq \lambda G(s),$$
$$v'(s) \geq \frac{v^2(s)}{G(s)} - \lambda G(s)$$
$$= \frac{v^2(s)}{p^2(s)}p' - \lambda\frac{p^2(s)}{p'(s)}.$$

Write $w(s) = \frac{v(s)}{p(s)}$, then $v' = p'w + pw'$,

$$p'w + w'p \geq w^2p' - w'\frac{p^2}{p'},$$
$$w' \geq \frac{p'}{p}[w^2 - w - \lambda(\frac{p}{p'})^2]$$
$$\geq \frac{p'}{p}\left[(w-\frac{1}{2})^2 - \frac{1}{4} - \lambda(\frac{p}{p'})^2\right].$$

(i) If $\delta = 0$, $\frac{p}{p'} = s$ then for any $\lambda < 0$, we can find $s_0 > 0$ such that $-\lambda(\frac{p}{p'}) > \frac{1}{4}+1$ for $s > s_0$. Thus

$$\frac{w'}{w^2+1} \geq \frac{p'}{p},$$

$$\log \frac{p(s)}{p(s_0)} \leq \tan^{-1} w(s)\,\Big|_{s_0}^{s} \leq \pi\ .$$

Then we have proved the conclusion.

(ii) If $\delta < 0$, a direct calculation shows that

$$\lim_{s\to+\infty} \frac{p(s)}{p'(s)} = \lim_{s\to+\infty} \frac{\sinh^{m-1}(s\sqrt{\delta})}{(m-1)\sinh^{m-2}(s\sqrt{\delta})\cosh(s\sqrt{\delta})\sqrt{\delta}} = \frac{1}{(m-1)\sqrt{\delta}},$$

and then

$$\lim_{s\to+\infty} \left[\frac{p(s)}{p'(s)}\right]^2 = \frac{1}{(m-1)^2\delta}.$$

From the condition of the lemma we can find $s_0 > 0$, such that $-\lambda\left(\frac{p(s)}{p'(s)}\right)^2 > \frac{1}{4} + \epsilon_0$. Using a proof similar to that in (i) we can show our conclusion. □

The proof of Theorem 1.1 is straightforward from Lemma 3.1 and Lemma 3.2.

4. Applications

Let M be a connected Riemannian manifold of dimension m and let N be another manifold of dimension $n > m$. We assume that M is complete and has Ricci curvature bounded below. Let

$$f : M \to N \tag{4.1}$$

be a C^∞ isometric immersion. The mean curvature vector will be denoted by H and we set $|H|$ to be its length. For each real number k, we define a function F_k : $[0, +\infty) \to R$ after Markvorsen [M2]:

$$F_k(t) =: \begin{cases} [1-\cos(t\sqrt{k})]/k, & \text{if } k > 0, \\ t^2/2, & \text{if } k = 0, \\ [1-\cosh(t\sqrt{-k})]/k, & \text{if } k < 0. \end{cases}$$

Now let o be a fixed point of N. For each $x \in M$ we let r be the function defined on M by setting

$$r(x) := d_N(o, f(x)).$$

We also define $\varphi : M \to R_+ \cup \{0\}$ by $\varphi = F_k \circ r$.

Assume that κ is the supremum of the sectional curvature K_N over all points in $B^N(o, \rho)$ where $\rho > 0$ satisfies (1.2).

Proposition 4.1. [M2] *Under the assumptions above, for each $x \in M$ such that $f(x) \in B^N(o, \rho)$*

$$\Delta\varphi \geq mF'_\kappa \circ r(x)[\mu(\kappa, r(x)) - \langle g_f, H\rangle],$$

where the Laplacian is taken in M. Moreover equality holds if and only if $K_N \equiv K$ in all directions of the unique minimizing segment joining o and $f(x)$ in N.

Theorem 4.1. *Let U be a domain of M containing $B^M(x_0, r)$ for some x_0 and r satisfying* (1.4). *Suppose that the Ricci curvature of M is bounded below by $-\delta$ ($\delta \geq 0$) and o is a point of N so that the sectional curvature of N over the points in $B^N(o, \rho)$ is not greater than κ, where ρ is assumed to satisfy condition* (1.4) *and suppose $\delta > 0$ satisfies the additional condition:*

$$\begin{cases} \sin^2(\rho\sqrt{\kappa}) < \frac{4m\kappa}{(m-1)^2\delta}, & \text{for } \kappa > 0, \\ \rho^2 < \frac{4m}{(m-1)^2\delta}, & \text{for } \kappa = 0, \\ \sinh^2(\rho\sqrt{-\kappa}) < \frac{-4m\kappa}{(m-1)^2\delta}, & \text{for } \kappa < 0. \end{cases} \tag{4.2}$$

Then given any $r \in R_+$ there exists a constant R depending only on m, δ, and κ such that if

$$\sup_{x\in U} H(x) < \mu(\kappa, \rho), \tag{4.3}$$

then $f(U)$ cannot be contained in $B^N(o, \rho)$ for any $o \in N$ unless $r < R$.

For the proof of Theorem 4.1 see [CIK], the only difference lies in using Theorem 1.1 in this paper instead of the Theorem 1.1 in [CIK]. Using a proof similar to that of a corresponding result in [ES] one can easily get:

Theorem 4.2. *Let $f : M \to N$ be an immersion and suppose that the numbers ρ and κ satisfy the conditions in Theorem* 4.1. *Suppose that the scalar curvature of M is bounded below by $-C$. Assume that*

$$\sup_{x\in U} H(x) = \mu(\kappa, \rho)$$

and that $f(M) \subset \bar{B}^N(o, \rho)$ for some $o \in N$. Then $f(M)$ is a minimally immersed submanifold of the manifold $\partial B^N(o, \rho)$.

In the situation of Theorem 1.1, [BK], [JK] and [JX] showed that if condition (4.3) holds then M is not contained in $B^N(o, r)$.

References

[BK] Baikoussis, C. H., and Koufogiorgos, T. H., Harmonic map into a cone, Arch. Math. 40 (1983), 372–376.

[BC] Bishop, R. L. and Crittenden, R. J., Geometry of Manifolds, Academic Press, New York 1964.

[CE] Cheeger, J., and Ebin, D. G., Comparison Theorems in Riemann Geometry, North-Holland, Amsterdam 1975.

[CIK] Coghlan, L., Itokawa, Y., and Kosecki, R., On the solutions to certain Laplace inequalities with applications to geometry of submanifolds, J. Geom. Anal. 2 (1992), 195–211.

[CX] Chen, Q., and Xin, Y. L., A generalized maximum principle and its applications in geometry, Amer. J. Math. 114 (1992), 355–366.

[ES] Eschenburg, J.-H., and O'Sullivan, J. J., Jacobi tensors and Ricci curvature, Math. Ann. 252 (1980), 1–26.

[JK] Jorge, L., and Koutroufiotis, D. An estimate for the curvature of bounded submanifolds, Amer. J. Math. 103 (1981), 711–725.

[JX] Jorge, L., and Xavier, F., An inequality between the exterior diameter and the mean curvature of bounded immersions, Math. Z. 178 (1981), 77–82.

[M1] Markvorsen, S., A sufficient condition for a compact immersion to be spherical, Math. Z. 183 (1983), 407–411.

[M2] —, On the mean exit time from a minimal submanifolds, J. Differential Geom. 29 (1989), 1–8.

List of Participants

Aleksandrov, Aleksandr Grigorjevich, Russian Academy of Sciences, Russia

Aslaksen, Helmer, National University of Singapore, Singapore

Aravinda, C. S., Tata Institute of Fundamental Research, India

Au Wai Man, University of Hong Kong, Hong Kong

Berrick, A. J., National University of Singapore, Singapore

Chan Onn, National University of Singapore, Singapore

Chan Yiu Man, National University of Singapore, Singapore

Chang Jeongwook, Seoul National University, South Korea

Chang Sun-Yung, Alice, University of California at Los Angeles, USA

Chen Hsiao Yun, Louis, National University of Singapore, Singapore

Chen Roul Chwian, Roger, National University of Singapore, Singapore

Chen Zehua, National University of Singapore, Singapore

Choe Jaigyoung, POSTECH, South Korea

Choi Gundon, Seoul National University, South Korea

Choi Hyeong In, Seoul National University, South Korea

Choi Kwok Pui, National University of Singapore, Singapore

Chong Chi Tat, National University of Singapore, Singapore

Chou Kai Seng, Chinese University of Hong Kong, Hong Kong

Chua Kok Seng, German Singapore Institute, Singapore

Chua Seng Kee, National University of Singapore, Singapore

Chua Seng Kiat, National University of Singapore, Singapore

Datta, Basudeb, Indian Institute of Science, India

Dong Fang, National University of Singapore, Singapore

El Kacimi Alaoui, Aziz, Université de Lille, France

Farrell, Thomas, State University of New York at Binghamton, USA

Fukaya, Kenji, University of Kyoto, Japan

Gershkovich, Vladimir, The University of Melbourne, Australia

Gilkey, Peter Belden, University of Oregon, USA

Goh Geok Yen, National University of Singapore, Singapore

Goh Say Song, National University of Singapore, Singapore

Han Dongsoong, JeonJu University, South Korea

Hoàng Lê Minh, University of Hochiminh City, Vietnam

Hong Jia Xing, Fudan University, China

Hu Kaiyuan, National University of Singapore, Singapore

Hu Xinmin, Zhongshan University, China

Jayawardene, Ananda K, National University of Singapore, Singapore

Jeon Myungjin, Semyung University, South Korea

Jiang Mei-yue, Beijing University, China

Ki U-Hang, Kyungpook National University, South Korea

Kim Dong-Soo, Chonnam National University, South Korea

Kim Taesoon, Mokwon University, South Korea

Kim Young Ho, Kyungpook National University, South Korea

Kitano, Teruaki, Tokyo Institute of Technology, Japan

Koh Liang Khoon, National University of Singapore, Singapore

Kwong Hong Hin, National University of Singapore, Singapore

Lam Kee Yuen, University of British Columbia, Canada

Lang Mong Lung, National University of Singapore, Singapore

Lee Isaac Chaujun, National Sun Yat-Sen University, Taiwan

Lee Seng Hwang, National University of Singapore, Singapore

Lee Soo Teck, National University of Singapore, Singapore

Leong Yu Kiang, National University of Singapore, Singapore

Leung Ho Hon, Denny, National University of Singapore, Singapore

Leung Ka Hin, National University of Singapore, Singapore

Leung Man Chun, National University of Singapore, Singapore

Leung Pui Fai, Fred, National University of Singapore, Singapore

Li Xiaoying, Academia Sinica, China

Lim Lek Heng, National University of Singapore, Singapore

Lin Chang-Shou, National Chung Cheng University, Taiwan

Lin En-Bing, University of Toledo, USA

Lin Kai-Ching, University of Alabama, USA

Ling San, National University of Singapore, Singapore

Liu Shiping, National University of Singapore, Singapore

Liu Zhuangwei, National University of Singapore, Singapore

Lo Kwai Wing, The University of Hong Kong, Hong Kong

Loo, Bonaventure, National University of Singapore, Singapore

Lou Jiann Hua, National University of Singapore, Singapore

Lum Kai Hing, National University of Singapore, Singapore

Ma Chuanyu, Nanjing University, China

Ma Kwan Ho, Lawrence, National University of Singapore, Singapore

Mok Ngaiming, University of Hong Kong, Hong Kong

Mokhov, Oleg, Steklov Mathematical Institute, Russia

Nair, Sunil, State University of New York at Stony Brook, USA

Ng Kok Onn, National University of Singapore, Singapore

Ng Sze Kui, Hong Kong Baptist College, Hong Kong

Ng Tze Beng, National University of Singapore, Singapore

Ohba, Kiyoshi, Ochanomizu University, Japan

Pan Xingbin, Zhejiang University, China

Pang Yu Hin, Peter, National University of Singapore, Singapore

Peng Tsu Ann, National University of Singapore, Singapore

Qu Ruibin, National University of Singapore, Singapore

Raina, Ashok, Tata Institute of Fundamental Research, India

Rodríguez Sánchez, Gerardo, Universidad de Salamanca, Spain

Ross, Marty, University of Melbourne, Australia

Seo Sangwon, Seoul National University, South Korea

Shao Qi Man, National University of Singapore, Singapore

Shen Yi Bing, Hangzhou University, China

Sheng Qin, National University of Singapore, Singapore

Shi Yuguang, Academia Sinica, China

Simon, Leon, Stanford University, USA

Soret, Marc, Université de Tours, France

Sun Yeneng, National University of Singapore, Singapore

Tan Choon Ee, Roger, National University of Singapore, Singapore

Tan Eng Chye, National University of Singapore, Singapore

Tan Geok Choo, National University of Singapore, Singapore

Tan Hwee Huat, National University of Singapore, Singapore

Tan Kok Choon, National University of Singapore, Singapore

Tan Ser Peow, National University of Singapore, Singapore

Tang Wai Shing, National University of Singapore, Singapore

Tay Kian Boon, National University of Singapore, Singapore

Tay Tiong Seng, National University of Singapore, Singapore

Tkachev, Vladimir, Volgograd State University, Russia

To Wing Keung, National University of Singapore, Singapore

Ueno, Kenji, Kyoto University, Japan

Visus, Isabel, Universidad de Salamanca, Spain

Wan Yau Heng, Tom, The Chinese University of Hong Kong, Hong Kong

Wang Daoyi, National University of Singapore, Singapore

Wang Hongyu, National University of Singapore, Singapore

Wang Wenxiang, California State University, USA

Weng Lin, National University of Singapore, Singapore

Wong Yan Loi, National University of Singapore, Singapore

Wu Jyh Yang, National Chung Cheng University, Taiwan

Xu Kai, National University of Singapore, Singapore

Xu Xingwang, National University of Singapore, Singapore

Yan Min, Hong Kong University of Science and Technology, Hong Kong

Yang, Paul, University of Southern California, USA

Yang Yue, National University of Singapore, Singapore

Yap Von Bing, National University of Singapore, Singapore

Zhang De Qi, National University of Singapore, Singapore

Zhou Detang, Shandong University, China

Zhou Qing, East China Normal University, China

Zhu Chengbo, National University of Singapore, Singapore

Zul Kepli Mohd Desa, Universiti Kebangsaan Malaysia, Malaysia

List of Contributors

A. G. Aleksandrov, Institute of Control Sciences, Russian Academy of Sciences, Profsojunaja ul. 65 GSP-7, Moscow 117806, Russia, email: alexandr@icsan.msk.su

Helmer Aslaksen, Department of Mathematics, National University of Singapore, Singapore 119260, email: mathelmr@nus.sg

Sun-Yung A. Chang, Department of Mathematics, UCLA, 405 Hilgard Ave., Los Angeles, CA 90095-1555, U.S.A., email: chang@math.ucla.edu

Hyeong In Choi, Department of Mathematics, Seoul National University, Seoul 151–742, Korea, email: hichoi@alliant.snu.ac.kr

Kai-Seng Chou, Department of Mathematics, The Chinese University of Hong Kong, Shatin, Hong Kong, email: kschou@cuhk.hk

Basudeb Datta, Department of Mathematics, Indian Institute of Science, Bangalore 560012, India, email: dattab@math.iisc.ernet.in

F. T. Farrell, Department of Mathematical Sciences, SUNY Binghamton, Binghamton, NY 13902, USA

Kenji Fukaya, Department of Mathematics, Faculty of Science, Kyoto University, Kitashirakawa, Kyoto, Japan, email: fukaya@kusm.kyoto-u.ac.jp

D. Geng, Department of Mathematics, The Chinese University of Hong Kong, Shatin, Hong Kong, email: dgeng@ims.cuhk.hk

V. Gershkovich, Mathematics Department, Melbourne University, Parkville, Victoria 3052, Australia, email: vg@mame.mu.OZ.AU

Peter Gilkey, Department of Mathematics, University of Oregon, Eugene, OR 97403, USA, email: gilkey@bright.uoregon.edu

Hoàng Lê Minh, Department of Mathematics, University of Hochiminh City, Hochiminh City, Vietnam, email: hanoi!hlminh@mathdep.edu.vn

Hong Jiaxing, Institute of Mathematics, Fudan University, Shanghai, China

Hsueh-Ling Huynh, AT&T Bell Laboratories, Murray Hill, NJ 07974, USA, email: huynh@research.att.com

L. E. Jones, Department of Mathematics, SUNY Stony Brook, Stony Brook, NY 11794, USA

Kee Yuen Lam, Department of Mathematics, UBC Vancouver, B.C., Canada, V6T 1Z2, email: lam@math.ubc.ca

Isaac Chaujun Lee, Department of Applied Mathematics, National Sun Yat-sen University, Kaohsiung, Taiwan 80424, email: isaac@math.nsysu.edu.tw

Man Chun Leung, Department of Mathematics, National University of Singapore, Singapore 119260, email: matlmc@nus.sg

En-Bing Lin, Department of Mathematics, University of Toledo, Toledo OH 43606, U.S.A., email: elin@math.utoledo.edu

Kai-Ching Lin, Mathematics Department, University of Alabama, Tuscaloosa, AL 35487, USA, email: klin@ua1vm.ua.edu

Ma Chuanyu, Department of Mathematics, Nanjing University, Nanjing, 210093 Jiangsu, People's Republic of China

J. Muñoz Masqué, CSIC, C/ Serrano 144, 28006-Madrid, Spain, email: jaime@iec.csic.es

Ngaiming Mok, Department of Mathematics, The University of Hong Kong, Pokfulam Road, Hong Kong, email: nmok@hkucc.hku.hk

Oleg Mokhov, Department of Geometry and Topology, Steklov Mathematical Institute, ul. Vavilova, 42 Moscow, GSP-1 117966 Russia, email: mokhov@class.mian.su

A. K. Raina, Theoretical Physics Group, Tata Institute of Fundamental Research, Homi Bhabha Road, Bombay, India 400 005, email: raina@theory.tifr.res.in

G. Rodríguez Sánchez, Dpto. de Estadística y Matemática Aplicadas, Universidad de Salamanca, Avda. Fernández Ballesteros s/n, 37700-Béjar, Spain, email: gerardo@gugu.usal.es

Weimin Sheng, Mathematics Department, Hangzhou University, Hangzhou 310028, People's Republic of China

Leon Simon, Department of Mathematics, Stanford University, Stanford, CA 94305, USA, email: lms@gauss.stanford.edu

S. Tanabé, Laboratoire de Mathematique, Institut Elie Cartan, Université de Henri Poincaré, Nancy I B.P.239, 54506 Vandoeuvre les Nancy, France, email: tanabe@iecn.u-nancy.fr

Vladimir G. Tkachev, Department of Mathematical Analysis, Volgograd State University, ul. 2-ya Prodolnaya 30, Volgograd, Russia 400062, email: tkachev@math.vgu.tsaritsyn.su

Kenji Ueno, Department of Mathematics, Faculty of Science, Kyoto University, Kyoto, 606–01, Japan, email: ueno@kusm.kyoto-u.ac.jp

Tom Y. H. Wan, Mathematics Department, The Chinese University of Hong Kong, Shatin, N.T., Hong Kong, email: tomwan@cuhk.hk

Paul C. Yang, Department of Mathematics, USC, DRB 140, 36th Place, Los Angeles, CA 90089-1113

D.-Q. Zhang, Mathematics Department, National University of Singapore, Singapore 119260, email: matzdq@nus.sg

Detang Zhou, Mathematics Department, Shandong University, Jinan, Shandong 250100, People's Republic of China

Xiaolin Zhou, Department of Mathematics, University of Toledo, Toledo, OH 43606, USA

Bei Fragen zur Produktsicherheit wenden Sie sich bitte an:
If you have any questions regarding product safety,
please contact:

Walter de Gruyter GmbH
Genthiner Straße 13
10785 Berlin
productsafety@degruyterbrill.com